Machine Learning in Cardiovascular Medicine

Machine Learning in Cardiovascular Medicine

Edited by

Subhi J. Al'Aref

Assistant Professor of Medicine, University of Arkansas for Medical Sciences, Little Rock, Arkansas, United States

Gurpreet Singh

Senior Manager, Data Science Capabilities, Global Vx Tech, Glaxosmithkline, Philadelphia, United States

Lohendran Baskaran

Visiting Assistant Professor of Research in Radiology, Weill Cornell Medicine, New York, New York, United States

and

Consultant, Department of Cardiovascular Medicine, National Heart Centre, Singapore

Dimitris Metaxas

Rutgers School of Arts and Sciences, The State University of New Jersey, Piscataway, NJ, United States

ELSEVIER

ACADEMIC PRESS

An imprint of Elsevier

Academic Press is an imprint of Elsevier
125 London Wall, London EC2Y 5AS, United Kingdom
525 B Street, Suite 1650, San Diego, CA 92101, United States
50 Hampshire Street, 5th Floor, Cambridge, MA 02139, United States
The Boulevard, Langford Lane, Kidlington, Oxford OX5 1GB, United Kingdom

Notices

Library of Congress Cataloging-in-Publication Data
A catalog record for this book is available from the Library of Congress

British Library Cataloguing-in-Publication Data
A catalogue record for this book is available from the British Library

ISBN: 978-0-12-820273-9

For information on all Academic Press publications visit our website at https://www.elsevier.com/books-and-journals

Publisher: Stacy Masucci
Acquisitions Editor: Katie Chan
Editorial Project Manager: Susan Ikeda
Production Project Manager: Selvaraj Raviraj
Cover Designer: Christian J.Bilbow

Typeset by TNQ Technologies

Subhi J. Al'Aref, MD*: First and foremost, I would like to dedicate this textbook to my wife* Nadia *who is my soulmate, love of my life, and my better half. I would like to thank my parents,* Jamal *and* Nisreen, *and my parents-in-law, Dr. Samir Darwish and Ms. Sahar Darwish, for being my backbone in life and providing limitless love, support, and care. I would also like to dedicate this textbook to my family, especially* Omar, Jumana, Samik, Zeid, Reem, Rami & Hussein, Mustafa, Yasmine, Hassan, Sarah, and Sareen. *Words can never express how much I love you.*

Gurpreet Singh, PhD: *I would like to dedicate this book to my wife* Simran *for being a constant source of love, my mentor* Prof. Lakshminarayana Samavedham *for his encouragement, and my parents and parents-in-law,* S. Tarlochan Singh, Sdn. Narinder Kaur, S. Visheshver Singh Narula, *and* Late Sdn. Harmit Kaur Bajaj, *for their support. Also, to my friends and siblings, you all have been a source of inspiration that has always instilled confidence in myself and has kept me motivated. Thank you.*

Lohendran Baskaran, MBBS: *To my family, thank you for your support, and for letting me chart my own journey. To friends, colleagues and collaborators, thank you for your support, past, present and future.*

Dimitris Metaxas, PhD*: I would like to dedicate this book to my wife Geanne for her continuous love and support and my two daughters Ada and Christina. To my late parents Nickolas and Ada for teaching me how to lead, excel, inspire, and help others. To my colleagues and collaborators, especially Dr. Leon Axel for the close to 30 years of excellence in collaborative cardiac research and results. To my students for their hard work, innovation, perseverance, and support resulting in many awards and publications in prestigious conferences and journals.*

Contents

Contributors

Aaron D. Aguirre
Cardiology Division, Massachusetts General Hospital, Harvard Medical School, Boston, MA, United States; Wellman Center for Photomedicine, Massachusetts General Hospital, Boston, MA, United States; Center for Systems Biology, Massachusetts General Hospital, Boston, MA, United States

Mouaz H. Al-Mallah
Houston Methodist DeBakey Heart and Vascular Center, Houston, TX, United States

Jamal Al Ani
Department of Medicine and Medical Education, Weill Cornell Medicine-Qatar, Doha, Qatar

Subhi J. Al'Aref
Division of Cardiology, Department of Medicine, University of Arkansas for Medical Sciences, Little Rock, AR, United States

Ahmed M. Altibi
Harvard T.H. Chan School of Public Health, Harvard University, Boston, MA, United States

Leon Axel
Department of Radiology, New York University School of Medicine, New York, NY, United States

Andrea Baggiano
Centro Cardiologico Monzino, IRCCS, Milan, Italy

Lohendran Baskaran
Dalio Institute of Cardiovascular Imaging, Weill Cornell Medicine/NewYork-Presbyterian, New York, NY, United States; Department of Cardiovascular Medicine, National Heart Centre, Singapore

Jan-Walter Benjamins
University of Groningen, University Medical Center Groningen, Department of Cardiology, Groningen, The Netherlands

Laura J. Brattain
Human Health and Performance Systems, MIT Lincoln Laboratory, Lexington, MA, United States; Center for Ultrasound Research & Translation, Department of Radiology, Massachusetts General Hospital, Harvard Medical School, Boston, MA, United States

Qi Chang
Department of Computer Science, Rutgers The State University of New Jersey, Piscataway, NJ, United States

Gloria Cicala
Department of Medicine and Surgery, Azienda Ospedaliero-Universitaria di Parma, Università degli Studi, Parma, Italy

Kristin M. Corey
Duke Institute for Health Innovation, Durham, NC, United States

Jessica De Freitas
Hasso Plattner Institute for Digital Health at Mount Sinai, Icahn School of Medicine at Mount Sinai, New York, NY, United States; Department of Genetics and Genomic Sciences, Icahn School of Medicine at Mount Sinai, New York, NY, United States

Damini Dey
Biomedical Imaging Research Institute and Artificial Intelligence in Medicine, Cedars-Sinai Medical Center, Los Angeles, CA, United States

Abdallah Elshafeey
Department of Medicine and Medical Education, Weill Cornell Medicine-Qatar, Doha, Qatar

Mohamed B. Elshazly
Department of Medicine and Medical Education, Weill Cornell Medicine-Qatar, Doha, Qatar

Laura Fusini
Centro Cardiologico Monzino, IRCCS, Milan, Italy

Benjamin S. Glicksberg
Hasso Plattner Institute for Digital Health at Mount Sinai, Icahn School of Medicine at Mount Sinai, New York, NY, United States; Department of Genetics and Genomic Sciences, Icahn School of Medicine at Mount Sinai, New York, NY, United States

Andrea I. Guaricci
Institute of Cardiovascular Disease, Department of Emergency and Organ Transplantation, University Hospital Policlinico of Bari, Bari, Italy

Marco Guglielmo
Centro Cardiologico Monzino, IRCCS, Milan, Italy

Donghee Han
Department of Imaging and the Smidt Heart Institute, Cedars-Sinai Medical Center, Los Angeles, CA, United States

Kipp W. Johnson
Department of Genetics and Genomic Sciences, Icahn School of Medicine at Mount Sinai, New York, NY, United States

Luis Eduardo Juarez-Orozco
University of Groningen, University Medical Center Groningen, Department of Cardiology, Groningen, The Netherlands; Department of Cardiology, Division Heart and Lungs, University Medical Centre Utrecht, University of Utrecht, Utrecht, The Netherlands

Mohammad Kachuee
Computer Science Department, University of California, Los Angeles, CA, United States

Aman Kansal
Duke Institute for Health Innovation, Durham, NC, United States

Sehj Kashyap
Duke Institute for Health Innovation, Durham, NC, United States

Felix Y.J. Keng
Department of Cardiovascular Medicine, National Heart Centre, Singapore

Shaden Khalaf
Houston Methodist DeBakey Heart and Vascular Center, Houston, TX, United States

Pegah Khosravi
Computational Oncology, Department of Epidemiology and Biostatistics, Memorial Sloan Kettering Cancer Center, New York, NY, United States

Attila Kovács
Heart and Vascular Center, Semmelweis University, Budapest, Hungary

Viksit Kumar
Center for Ultrasound Research & Translation, Department of Radiology, Massachusetts General Hospital, Harvard Medical School, Boston, MA, United States

Benjamin C. Lee
Dalio Institute of Cardiovascular Imaging, Weill Cornell Medicine/NewYork-Presbyterian, New York, NY, United States

Andrew Lin
Biomedical Imaging Research Institute and Artificial Intelligence in Medicine, Cedars-Sinai Medical Center, Los Angeles, CA, United States

Pál Maurovich-Horvat
Department of Radiology, Medical Imaging Center, Semmelweis University, Budapest, Hungary

Dimitris N. Metaxas
Department of Computer Science, Rutgers The State University of New Jersey, Piscataway, NJ, United States

Omar Mhaimeed
Department of Medicine and Medical Education, Weill Cornell Medicine-Qatar, Doha, Qatar

Riccardo Miotto
Hasso Plattner Institute for Digital Health at Mount Sinai, Icahn School of Medicine at Mount Sinai, New York, NY, United States; Department of Genetics and Genomic Sciences, Icahn School of Medicine at Mount Sinai, New York, NY, United States

Giuseppe Muscogiuri
Centro Cardiologico Monzino, IRCCS, Milan, Italy

Aziz Nazha
Cleveland Clinic Taussig Cancer Center, Leukemia Department, Cleveland, OH, United States

Gianluca Pontone
Centro Cardiologico Monzino, IRCCS, Milan, Italy

Mark Rabbat
Medicine and Radiology, Division of Cardiology, Loyola University Chicago, Chicago, IL, United States

Mark G. Rabbat
Loyola University of Chicago, Chicago, IL, United States; Edward Hines Jr. VA Hospital, Hines, IL, United States

Nathan Radakovich
Cleveland Clinic Lerner College of Medicine, Cleveland, OH, United States

Mina Rezaei
Massachusetts General Hospital, Harvard Medical School, Boston, MA, United States

Francesca Ricci
Department of Biomedicine and Prevention Division of Diagnostic Imaging, University of Rome "Tor Vergata," Rome, Italy

Anthony E. Samir
Center for Ultrasound Research & Translation, Department of Radiology, Massachusetts General Hospital, Harvard Medical School, Boston, MA, United States

Majid Sarrafzadeh
Computer Science Department, University of California, Los Angeles, CA, United States

Alpana Senapati
Houston Methodist DeBakey Heart and Vascular Center, Houston, TX, United States

Mark Sendak
Duke Institute for Health Innovation, Durham, NC, United States

Piotr J. Slomka
Biomedical Imaging Research Institute and Artificial Intelligence in Medicine, Cedars-Sinai Medical Center, Los Angeles, CA, United States

Emily Tat
Department of Internal Medicine, Columbia University Medical Center, New York, NY, United States

Brian A. Telfer
Human Health and Performance Systems, MIT Lincoln Laboratory, Lexington, MA, United States

Márton Tokodi
Heart and Vascular Center, Semmelweis University, Budapest, Hungary

Pim van der Harst
University of Groningen, University Medical Center Groningen, Department of Cardiology, Groningen, The Netherlands; University of Groningen, University Medical Centre Groningen, Department of Genetics, Groningen, The Netherlands; Department of Cardiology, Division Heart and Lungs, University Medical Centre Utrecht, University of Utrecht, Utrecht, The Netherlands

Jake Vasilakes
Institute for Health Informatics and College of Pharmacy, University of Minnesota, Minneapolis, MN, United States

Ming Wai Yeung
University of Groningen, University Medical Center Groningen, Department of Cardiology, Groningen, The Netherlands

Rui Zhang
Institute for Health Informatics and College of Pharmacy, University of Minnesota, Minneapolis, MN, United States

Sicheng Zhou
Institute for Health Informatics and College of Pharmacy, University of Minnesota, Minneapolis, MN, United States

Preface

The field of cardiovascular medicine has undergone tremendous evolution over the last number of decades. From reliance on auscultation and clinical examination to now, a deluge of diagnostic information that has aided more informed treatment decision making and improved clinical outcomes. To date, the clinician has managed to integrate the ever-growing information into practice but in the last number of years, data are outstripping the capacity of a physician to process it. In addition, there is an increasing awareness of the richness of information that is embedded within imaging data, which extends beyond what the human eye can process and extract. Artificial intelligence and machine learning have rapidly transformed every aspect of our lives and are now being explored and implemented in a number of aspects of cardiovascular medicine.

Machine Learning in Cardiovascular Medicine is one of the first textbooks to highlight the ever-expanding applications of machine and deep learning within cardiovascular medicine. In this book, the editors have compiled an excellent overview of the current applications of machine learning for clinical data analysis, as well as highlighted the recent advances in deep learning for cardiovascular image analysis. From traditional to state-of-the-art algorithms, the first few chapters set the technical foundation for understanding the inner workings of these algorithms as well as appreciating available metrics for the evaluation of their performance for a specific clinical task. The book then delves into current applications of machine learning: from its role in natural language processing to various imaging modalities such as CT, MRI, and Echocardiography. The editors have provided a tour de force go to reference for all those in the field. Importantly, the editors seek to also expose the many limitations in the space and also the ethical challenges that we will all be facing as we seek to implement and refine these tools in our clinical practice. To do so, they have assembled an incredible group of authors with expertise in every aspect of artificial intelligence and cardiovascular medicine. As Andrew Ng famously stated: "*As leaders, it is incumbent on all of us to make sure we are building a world in which every individual has an opportunity to thrive. Understanding what AI can do and how it fits into your strategy is the beginning, not the end, of that process*," this book is a perfect fit to educate the clinical and research communities regarding current, and potential, future applications of artificial intelligence within cardiovascular medicine.

Jonathon A. Leipsic, M.D., F.R.C.P.C., F.S.C.C.T.

Acknowledgments

This textbook is inspired by the recent and ever-expanding applications of machine learning for cardiovascular applications. We would like to acknowledge the dedication, support, and contribution of every author to the content herein presented, as well as the editorial and technical support from the publisher (especially Ms. *Susan Ikeda* and Ms. *Swapna Praveen*). We would also like to express our sincere gratitude and appreciation for the help of Ms. *Zahra Fatima* in organizing the content of the textbook. We would like to thank our respective heads and directors for facilitating this endeavor whilst providing invaluable support: Leslee J. Shaw, Terrance Chua SJ, Felix Keng YJ, Tan SY and Paul Mounsey.

Nomenclature

P(A)	Probability of Event A
P(A|B)	Probability of Event A given B
P(A,B)	Joint probability of A and B
w_x	Weights tensor for parameter of machine learning model
f(x)	Function of x
(G)	Generator network for generalized adversarial network (GAN)
(D)	Discriminator network for generalized adversarial network (GAN)
°C	degree Celsius
mg/mL	Concentration
mL	Milliliter, unit of volume
x_t	Input layer of neural network
h_t	Hidden layers of a neural network
y_t	Output layer of a neural network
Msec	Millisecond, unit of time
cm^2	Centimeter square, unit of area
Hz	Hertz, unit of frequency
R_2	R-Squared coefficient
R_p	Pearson correlation coefficient

Introduction

Artificial Intelligence (AI) has become a topic for discussion within various medical domains, and especially within the cardiovascular community. The availability of large multidimensional datasets within medicine offers a framework for significant advances with the appropriate harnessing of AI. The integration of AI can occur throughout the continuum of medicine; from basic laboratory discovery to clinical application and healthcare delivery. Yet, integrating AI within medicine has been met with both excitement and skepticism. By understanding how AI works and developing an appreciation for both its limitations and strengths, clinicians can harness its power to streamline workflow and improve patient care. AI provides an opportunity to improve upon research methodologies, beyond what are currently available using traditional statistical approaches. Further, computers scientists and data analysts can provide solutions, but often lack easy access to clinical insight that may help focus their efforts. This textbook provides the platform to bring these two disciplines together.

Machine Learning in Cardiovascular Medicine addresses the ever-expanding applications of AI, and specifically machine learning (ML), in healthcare and within cardiovascular medicine. AI was initially born out of the need to solve nonlinear and combinatorially complex tasks. Over time, AI has transformed innumerable aspects of human life; the development of augmented reality, autonomous driving, robotics, as well as sophisticated predictive modeling are just a few examples of how AI has sparked the current innovative climate within the industry sector. While the industrial applications of ML are nearly ubiquitous, its introduction into the medical field has been much more gradual. The landscape, however, is rapidly changing as a result of the availability of computational power as well as the creation of large repositories of datasets. The computer science and medical communities are increasingly streamlining efforts at harnessing the potential of ML in order to solve complex tasks and optimize day-to-day workflow in the healthcare sector. This book is focused on emphasizing ML for biomedical applications and aims at providing a comprehensive summary of the past and present of AI, basics of ML as well as clinical applications of ML within cardiovascular medicine for predictive analytics and precision medicine.

Subhi J. Al'Aref, M.D., F.A.C.C.

CHAPTER

Technological advances within digital medicine

1

Ahmed M. Altibi[1], Subhi J. Al'Aref[2], Pegah Khosravi[3]

[1]*Harvard T.H. Chan School of Public Health, Harvard University, Boston, MA, United States;* [2]*Division of Cardiology, Department of Medicine, University of Arkansas for Medical Sciences, Little Rock, AR, United States;* [3]*Computational Oncology, Department of Epidemiology and Biostatistics, Memorial Sloan Kettering Cancer Center, New York, NY, United States*

Chapter outline

Machine Learning in Cardiovascular Medicine. https://doi.org/10.1016/B978-0-12-820273-9.00001-4

Introduction

Technological advances within digital medicine have been gradually changing the landscape of healthcare and biomedical research. Since William Withering utilized digitalis 250 years ago, from *Foxglove* plant, to treat "dropsy" (congestive heart failure) (Krikler, 1985), cardiovascular medicine had been leading the medical field in evidence generation and state-of-the-art technological innovations (Steinhubl and Topol, 2015). From the first "Hufnagel" artificial heart valve in 1951 (HUFNAGEL, 1951), to the first artificial heart transplantation in 1982 (World Health Organization, 2010), and the first "robotic" minimally invasive valve surgery in 2000 (Bush, Nifong, Alwair, & Chitwood, 2013)—technological advances within digital medicine have been continuously changing the face of medical care. Sensors and wearables are getting smaller and deep-learning algorithms are becoming more powerful, and as a result, artificial intelligence (AI) applications are becoming more sophisticated (Coravos et al., 2019; Fogel and Kvedar, 2018). In a healthcare epoch awash with digitized data, applications of AI have now expanded into provinces that were not previously thought to be attainable to humans (Yu, Beam, & Kohane, 2018).

In this chapter, we provide an overview of the key advances in digital technologies impacting healthcare in cardiovascular medicine, including: (i) wearable healthcare technology, (ii) mobile health and telehealth, and (iii) robotic technology. A plethora of wearable technologies currently exists and can be broadly classified into diagnostic versus therapeutic devices. Diagnostic wearables rely on digitized data acquisition for monitoring (e.g., smartwatches and fitness trackers). On the other hand, therapeutic devices have instrumental components enabling them to intervene to prevent, mitigate, or treat certain medical conditions (e.g., wearable cardioverter defibrillators).

Wearable sensing devices

Fitness trackers and smartwatches

Introduction

With the recent technological revolution involving nanocomputers, AI, big data, wireless networks, and cellular networks, the past decade has witnessed an explosion in the utility of wearable optic sensors technologies worldwide (Rawassizadeh, Price, &Petre, 2015; Strain, Wijndaele, &Brage, 2019). These technologies rely on implementing a plethora of sensors that are capable of monitoring an array of physiologic parameters, such as vital signs (heart rate, blood pressure, body temperature, and oxygen), cardiac electrical activity (normal and abnormal rhythms), and physical activity metrics (movement type, steps count, energy expenditure, and energy intensity). Wearable sensing devices come in a multitude of forms, with *Fitness Trackers* and *Smartwatches* being the most popular and commercially available technologies globally. More than 100 million devices have been sold in 2016 worldwide, with an estimated global market value of $18 billion for the same year (Henriksen et al., 2018). These devices work via a complex integrative system of sensors, algorithms, and synchronized mobile apps to collect and analyze the metrics of interest before deploying them in a more readable form to the user. A wide array of manufacturers exists in the market, but most of these devices are marketed via the top five brands: Apple, Samsung, Fitbit, Garmin, and Xiaomi. While smartwatches and activity trackers have greatly enhanced our ability to track and potentially increase our physical activity levels, only little evidence exists to support the fact that introducing such technologies would substantially impact health outcomes related to inactivity, such as obesity (Finkelstein et al., 2016; Jakicic et al., 2016).

Devices and sensors

Fitness trackers and Smartwatches are equipped with a multitude of sensors (accelerometer, gyroscope, magnetometer, altimeter, barometers, and GPS) that function to record fitness-related metrics such as movement acceleration, steps count, energy expenditure, and physical location for the individual (Balli, Sağbaş, & Peker, 2019). The accelerometer, also known as piezoelectric pedometer, is a triaxial sensor capable of measuring acceleration in three directions and provides measures of high accuracy for steps, distances, and energy expenditure (Cruz, Brooks, & Marques, 2017). This contrasts with the traditional uniaxial pedometer that measures acceleration in one axial plane of motion. Such difference in technicalities and the resultant accuracy gap have made accelerometers to be the current standard for motion-related data for wearable devices.

In addition to accelerometers, some wearables utilize additional sensors to increase the accuracy of the device, such as gyroscopes, magnetometers, barometers, and altimeters. Gyroscope sensor, or angular rate sensor, functions via sensing the angular velocity (change in rotational angle per unit of time), and hence, provides information on both orientation and rotation (Passaro, Cuccovillo, Vaiani, De Carlo, & Campanella, 2017). A gyroscope, when combined with the accelerometer, functions to provide three-dimensional representation of the workout environment (Wagenaar et al., 2011). Magnetometers function via measuring magnetism. Direction, strength, and relative change of a magnetic field are utilized in wearables to function as a digital compass. The three-axis accelerometer, three-axis gyroscope, and magnetometer are often combined into one contained system called the inertial measurement unit (IMU) (Iosa, Picerno, Paolucci, & Morone, 2016). IMU sensing system measures the linear acceleration, rotational velocities, orientation, and 3D positioning of the object via data integration from the aforementioned set of sensors (Muro-de-la-Herran, García-Zapirain, & Méndez-Zorrilla, 2014). IMUs are nowadays implemented in most wearable devices as inexpensive, noninvasive, and compatible systems to measure motion.

Finally, photoplethysmogram (PPG) is a noninvasive optical sensing technology that utilizes an infrared light to detect any minor volumetric changes in blood of the peripheral circulation (Ghamari, 2018; Tamura, Maeda, Sekine, & Yoshida, 2014). This measurement provides valuable information about the cardiovascular system enabling measurement of pulse rate (PR), blood pressure, respiration, oxygen saturation, and other physiologic parameters. While PPG technology is limited by the fact that PPG signals are prone to Motion Artifacts (MA) (Peng, Zhang, Gou, Liu, & Wang, 2014), the low cost and noninvasiveness of these sensors allow for wide adoption in wearable devices (Zhang et al., 2019).

Ultimately, the quality of a wearable device depends on the accuracy of its sensors and the algorithms utilized to operate data collected from these sensors. This explains the increased focus in revolutionizing sensory technologies in healthcare wearable devices (Aroganam, Manivannan, & Harrison, 2019).

Clinical applications

The ability of optical sensors to detect abnormal heart rhythms has been a topic of interest in healthcare. The *Apple Heart Study* was designed to evaluate the Apple Watch irregular rhythm notification's ability to accurately identify atrial fibrillation (AF) – the most common clinically significant arrhythmia (Fig. 1.1) (Perez et al., 2019). The trial, which enrolled about half a million Apple smartwatch users with no prior AF diagnosis, concluded that the irregular rhythm notification on Apple Watch had a positive predictive value of 84% for detecting AF in patients who had received an irregular pulse notification by the smartwatch. Data from the Apple Heart Study supports the ability of these algorithms to correctly identify AF, paving the way towards an increased applicability of this technology in the early detection of abnormal heart rhythms (Perez et al., 2019).

In addition to diagnostic and therapeutic applications, wearable devices can be implemented to collect aggregated data as part of large-scale public health research studies (Henriksen et al., 2018). Wearable technologies such as smartwatches and fitness trackers can be utilized as an innovative data collection method for physical activity and health metrics. The Apple Heart Study, for instance, represents a ground breaking study on how clinical trials can be conducted. Perez et al. (2019) The adopted pragmatic site-less study design in the Apple Heart Study, where the entire study from recruitment, enrollment, study intervention, and data collection were administered remotely, could lead to a paradigm shift on how clinical trials will be conducted in the future (Turakhia et al., 2019). Such pragmatic site-less study designs theoretically allow for randomization of a large number of participants, within the shortest time frame, and at a low cost. Nevertheless, experts dispute that utilizing wearable technology in clinical trials with such pragmatic site-less styles might introduce uncertainness, especially in the absence of blinding, and might represent a factual threat to the core concept of "randomization" – the hallmark of the traditional highly-explanatory real-world clinical trials.

Wearable cardioverter defibrillator (WCD)

Introduction

Sudden cardiac death (SCD) is responsible for half of all cardiac-related deaths (Adabag, Luepker, Roger, & Gersh, 2010; Stecker et al., 2014; Zheng, Croft, Giles,

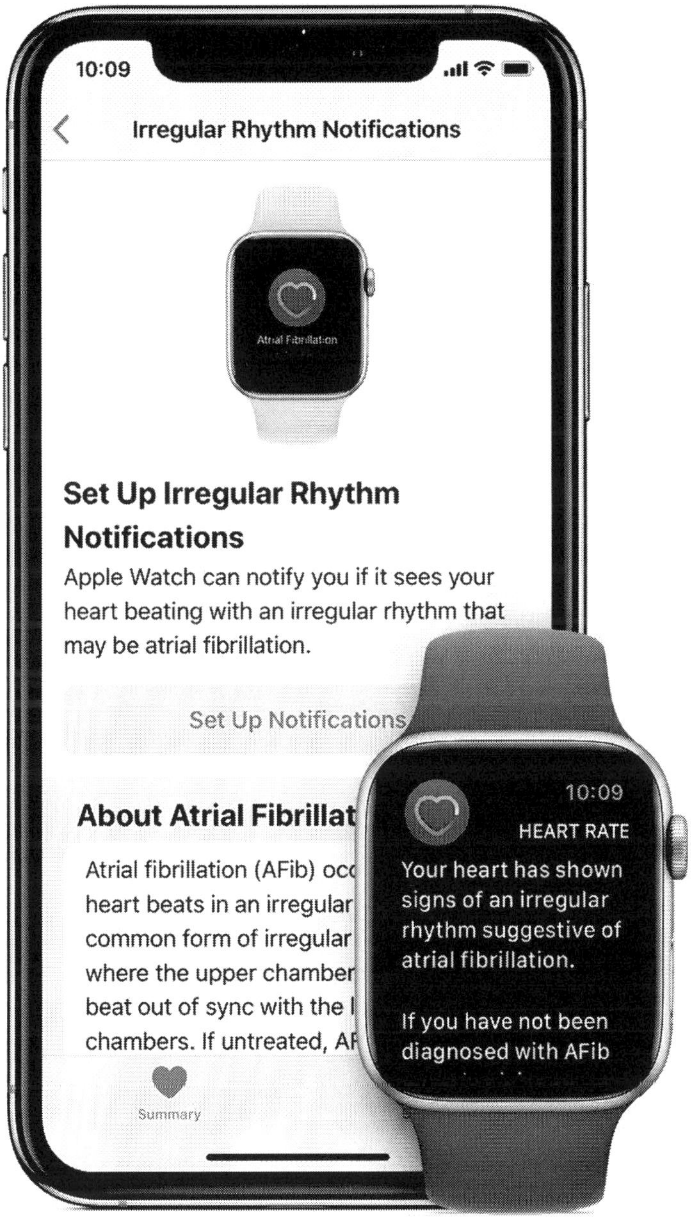

FIGURE 1.1 Apple Watch.

Using PPG sensors, Apple Watch monitors for abnormal heart rhythms. When a rhythm signaling atrial fibrillation is identified via the irregular rhythm notification, patients receive a notification on their Apple Watch.

Courtesy of Apple Inc.

& Mensah, 2001). The majority of SCD is caused by life-threatening arrhythmias: ventricular fibrillation (VF) and ventricular tachycardia (VT) (Demirovic and Myerburg, 1994; John et al., 2012; McElwee, Velasco, &Doppalapudi, 2016). While cardiopulmonary resuscitation (CPR) provides circulatory support for arrested patients, the only effective way to terminate these fatal rhythms is via electrical defibrillation. Thus, implantable cardioverter defibrillator (ICD) devices were developed in the 1980s to be used in patients with the greatest risk for SCD (Matchett et al., 2009; Mirowski et al., 1980). ICD devices, surgically implanted underneath the skin, constantly monitor the cardiac rhythm while delivering an electronic shock to terminate life-threatening arrhythmias of VT or VF when detected (DiMarco, 2003). ICDs have demonstrated efficacy in the primary and secondary prevention of SCD (Al-Khatib et al., 2018). Like ICDs, wearable cardioverter defibrillator (WCD) technology was developed in 2001 as an external device that is capable of preventing SCD via detecting and terminating VT/VF rhythms (Sharma, Bordchar, & Ellenbogen, 2017). The technology was initially developed as a bridge for ICDs when immediate ICD therapy was deemed unfeasible or contraindicated. The LifeVest WCD (Fig. 1.2), solely manufactured by Zoll (Zoll Inc., Chelmsford, MA), had been the first and only WCD technology to be approved by the United States Food and Drug Administration (FDA) in 2001 (Al-Khatib et al., 2018). Over the past decade, the technology has been widely utilized, with tens of thousands of high-risk patients using LifeVest on a single day to confer protection from SCD.

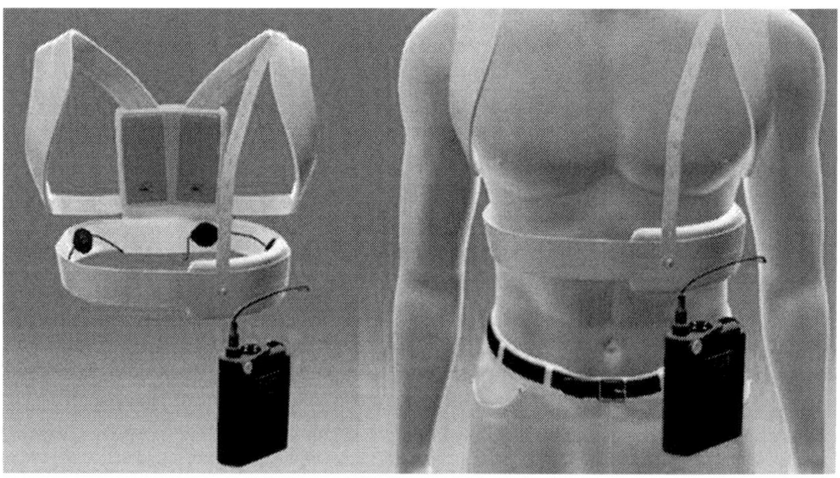

FIGURE 1.2 LifeVest wearable cardioverter defibrillator.

The device consists of a garment (worn underneath clothing) and a monitor. The three defibrillating electrodes are located on the back of the garment, while four sensing electrodes are located on the elastic belt around the chest. Both systems are connected to the monitor unit (carried around the waist).

Courtesy of Zoll Inc.

How does the WCD function?

The WCD device consists of two main components—a garment and a monitor. The garment is worn under clothing and functions to detect arrhythmias and to deliver life-saving shock when an abnormal heart rhythm is detected—via four sensing and three defibrillating electrodes (Delgado, Toquero, Mitroi, Castro, & Fernandez, 2013). The monitor, carried around the waist or from a shoulder strap, continuously monitors the heart rhythm, records arrhythmias, and determines when a treating shock is required to restore normal heart rhythm. WCD detection of arrhythmias occurs via complex algorithms that utilize predefined ECG criteria, based on rate and morphology (Dillon, Szymkiewicz, & Kaib, 2010). These algorithms have been reported to achieve a detection rate with sensitivity and specificity of >95%. When a life-threatening arrhythmia is detected by the WCD, the LifeVest system sounds an alarm to caution the patient on an "impending" shock. The alarm functions as a test for an individual's level of consciousness; if the individual is conscious, he/she ought to press on two response buttons to cease an impending "inappropriate" shock. Should there be an absence of a response, the device charges and delivers a shock after giving another warning for potential bystanders. If the shock fails to terminate the life-threatening arrhythmia, the LifeVest can deliver additional treatment shocks as needed (up to five). The WCD is programmed to shock at certain energy ranges with the maximum output being 150 J.

What are the indications for WCD?

Thus far, WCDs are used as bridging/temporizing interventions rather than destination measures substituting for ICDs. Clinical indications for WCDs were initially coined by American Heart Association (AHA) science advisory in 2016. Later, further endorsement was publicized by other scientific bodies (e.g., Heart Rhythm Society) (Piccini et al., 2016). Among others, the three primary indications for WCD are (Al-Khatib et al., 2018; Sharma et al., 2017):

- Bridging for interrupted ICD or "pending" ICD implant.
- Following myocardial infarction (MI) or coronary revascularization—with severely reduced ejection fraction (EF) (<35%).
- Following new diagnosis of nonischemic cardiomyopathy—with severely reduced EF (<35%).

ICD "interruption" refers to the state where an implanted ICD system is extracted for a compelling reason such as device infection or endocarditis—to achieve proper source control (Klein et al., 2010). ICD reinsertion is typically deferred until the infection is completely eradicated, leaving the patient exposed to SCD risk from ventricular arrhythmias in this window period. In such a scenario, WCD is clinically indicated to confer protection from VF/VT until ICD reimplantation is achieved (Ellenbogen et al., 2017). Similarly, patients with an indication for initial ICD placement might have a comorbid condition (e.g., active infection) that prohibits immediate ICD implantation. WCD is indicated in such cases as a bridge toward ICD placement (Mugnai, Lupo, Zerbo, Saccà, & Zoppo, 2019). Further, patients with recent MI with severe LV dysfunction (EF <35%) are at a high risk for SCD. However, since EF recovery (EF > 35%) occurs in more than half of the patients, current guidelines recommend a "wait time" of 40 days (or 90 days if the coronary arteries were revascularized) following an acute MI before considering ICD insertion (Al-Khatib et al., 2018; Tracy et al., 2013). The use of WCD within this 40-day window had been recommended for primary prevention against SCD.

Does the WCD technology save lives?

WCD therapy has been proven to be highly effective in terminating VT/VF rhythms when worn properly. Studies have shown the first-shock termination rate to be comparable with that for ICDs, ranging between 95% and 100% (Chung et al., 2010; Reek et al., 2003). SCD among patients wearing the device is a rare event with an estimated incidence of 0.7 per 100 individuals over 3 months (Masri et al., 2019). As a matter of fact, mortality among patients prescribed WCDs occurs for reasons other than an intrinsic "failure" of the device. These include noncompliance with wearing the device, improper positioning of the device, and improper abortion of the shock by a bystander (Feldman et al., 2004; Tanawuttiwat et al., 2014).

WCD has been shown to be highly efficient in terminating abnormal rhythms. However, many experts have raised concerns that the device is increasingly presumed to be the "de facto" standard of care in certain subsets of patients, despite the inconclusive evidence. The evidence to support an overall mortality benefit and cost-effectiveness is still limited. This becomes important knowing that leasing a LifeVest costs ≥ $3000 per month (Masri et al., 2019; Healy and Carrillo, 2015). This is particularly true for post-MI patients with EF ≤ 35%; this was explored by the investigators of the VEST trial (published in 2018) (Olgin et al., 2018). Thus far, the VEST trial (Vest Prevention of Early Sudden Death Trial) is the largest randomized, controlled trial examining the efficacy of WCD in preventing SCD as compared to guidelines-directed medical therapy. The VEST investigators concluded that WCD did not lead to a significant reduction in arrhythmic death compared to medical therapy in patients with recent MI and an EF ≤ 35% (Olgin et al., 2018). Clearly, further "indication-specific" clinical trials are ought to be conducted to precisely determine the efficacy and cost-effectiveness of WCDs.

Mobile health in cardiovascular medicine

Introduction

Cardiovascular disease is the leading cause of death globally among all ethnic and racial groups and is responsible for one in every four deaths in the United States (A nationwide framework for surveillance of cardiovascular and chronic lung diseases, 2011; Heron, 2016). The Center for Disease Control (CDC) estimates that heart diseases cost the United States more than $500 billion each year (RTI International, 2017). Two major recognized barriers for promoting cardiac health are the global deficit of health professionals and the ever-growing cost of healthcare services (Darzi and Evans, 2016). In fact, cardiovascular diseases were a leading driver of medical services spending growth between 2000 and 2013. Using AI and machine learning (ML), mobile health (mHealth) technologies might be one innovative solution to help address these current barriers and facilitate bridging the science of big data with the perpetuated needs within cardiovascular care (Kelli, Witbrodt, & Shah, 2017; Free et al., 2013).

mHealth incorporates all medical services and public health activities enabled via mobile devices. It is estimated that there are more than 300,000 mobile health applications available in the market—the majority are not FDA approved. Strikingly, more than 600 of these commercially available smartphone applications are specifically designed to improve adherence to therapy (Bates, Landman, & Levine, 2018). Applications of mHealth in cardiovascular medicine are numerous and can be used to promote preventative, diagnostic, and therapeutic aspects of medicine (Cowie et al., 2016). Via induction of behavioral modification, mHealth application might be effective to foster disease prevention via risk factors control (e.g., weight loss, smoking cessation, and physical activity) (Park, Beatty, Stafford, & Whooley, 2016; Piette et al., 2015). Similarly, mHealth can also promote therapeutic aspects of medicine via enhancing adherence to medications—a major reason for treatment failure in chronically ill individuals (Armitage, Kassavou, & Sutton, 2020; Bermon et al., 2019; Farmer et al., 2019).

In the era of miniaturization, great strides have also been made to revolutionize current diagnostic and imaging modalities by integrating AI systems in the vicinity of mobile health technologies. Smartphone-based electrocardiogram (EKG) and "pocket" echocardiography are two pioneering examples for this integration in diagnostic cardiology (Seetharam, Kagiyama, & Sengupta, 2019).

Smartphone-based electrocardiogram (EKG)

Over the past decade, mHealth technologies have witnessed a tremendous leap in the applicability of mHealth in the monitoring of chronic diseases (Nguyen and Silva, 2016). EKG is one example where mHealth tools can be utilized to promote early detection and treatment of abnormal heart rhythms (Whitehead and Seaton, 2016; Hickey et al., 2016). Willem Einthoven invented the first "string galvanometer" to measure cardiac electrical current in 1903, and he coined the term *electrocardiogram* to describe his device. This first room-sized immobile equipment required transmission of electrocardiogram signals for a distance via a protected telephone cable (Barold, 2003; De Luna, 2019; Zetterström, 2009). This has evolved over the past century into the current EKG machine consisting of a small cart with a laptop-sized device, reading screen, and a printer (Walker and Muhlestein, 2018).

The use of smartphones for cardiac rhythm monitoring is another leap for EKG development and utilization. Currently, there are two types of smartphone technologies used to detect and monitor heart rhythms: (i) PPG, and (ii) Single-lead platform technology (White and Flaker, 2017; Garabelli, Stavrakis, & Po, 2017). The

PPG technique utilizes the phone's LED light and camera for rhythm registration from the fingertip of an individual—as if the smartphone was turned into a PPG sensor without additional hardware (Sun and Thakor, 2016). The user has to place his/her finger over the camera lens to obtain the PPG signal. A smartphone software then acquires the measurements from the PPG signal and utilizes preprogrammed algorithms to analyze the heart rhythm (Koenig et al., 2016; Krivoshei et al., 2017). One novel mobile application that utilizes PPG technology is *Cardiio Rhythm* (Cardiio, Cambridge, MA, USA). Several validation studies indicated the *Cardiio Rhythm* to be capable of accurately detecting heart rates and distinguishing abnormal rhythms (e.g., AF) from normal sinus rhythm (Chan et al., 2016; Poh and Poh, 2017).

Over time, this technology has progressed to the more promising "single-lead" smartphone-based EKG—first FDA-approved in 2012. Single-lead technology combines an external EKG sensor with a smartphone application to generate a one lead rhythm strip (Frisch, Weiss, Dikdan, Keith, & Sarkar, 2019). One pioneering device in this technology is the *Kardia Mobile (*AliveCor, San Francisco, California, USA) (William et al., 2018). The *Kardia* device consists of two metal electrodes capable of generating a bipolar lead one electric signal when touched by the patient's right and left digits (Fig. 1.3). The EKG electric signal is then converted into an ultrasound frequency modulation (FM) sound signal (18–24 kHz) that in turn gets demodulated by the smartphone software into a digital EKG tracing (Garabelli et al., 2017; White and Flaker, 2017). The tracing can be reviewed in real-time on the smartphone but is also instantly stored on a secure cloud database for future review.

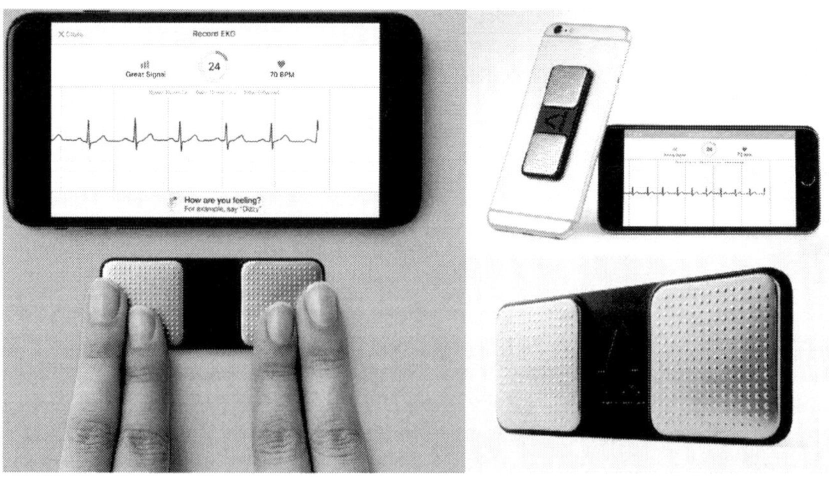

FIGURE 1.3 Kardia single-lead mobile EKG.

A single-lead mobile EKG utilizes two metal electrodes to generate a bipolar lead electric signal when pressure is applied by the patient's digits.

AliveCor Kardia is currently the most clinically validated mobile-based EKG technology. Studies have shown AliveCor to be highly sensitive (98%) and specific (97%) in capturing abnormal heart rhythms, such as AF (Halcox et al., 2017; Himmelreich et al., 2019). Clinical trials have also demonstrated it to be highly reliable when used in routine practice to remotely screen high-risk patients for AF (Goldenthal et al., 2019; Wegner et al., 2019). Noteworthy, comparing the two types of smartphone-based EKG technologies, PPG and single-lead, has shown them to have a comparable sensitivity and specificity in diagnosing abnormal rhythms (>95%), albeit with a higher positive predictive value with single-lead devices (Chan et al., 2016; Proesmans et al., 2019).

Interestingly, the world's first six-lead smartphone-enabled EKG was pioneered by *Kardia* in 2019. This functions via incorporating an innovative third electrode to allow visibility into certain arrhythmias that were previously undetected with single-lead EKG (Fig. 1.4). (Albert, 2019; "FDA Grants First Ever Clearance For Six-Lead Personal ECG Device | AliveCor," n.d.) This six-lead EKG product was FDA-cleared in 2019 and is expected to become the next substitute for the current single-lead technology.

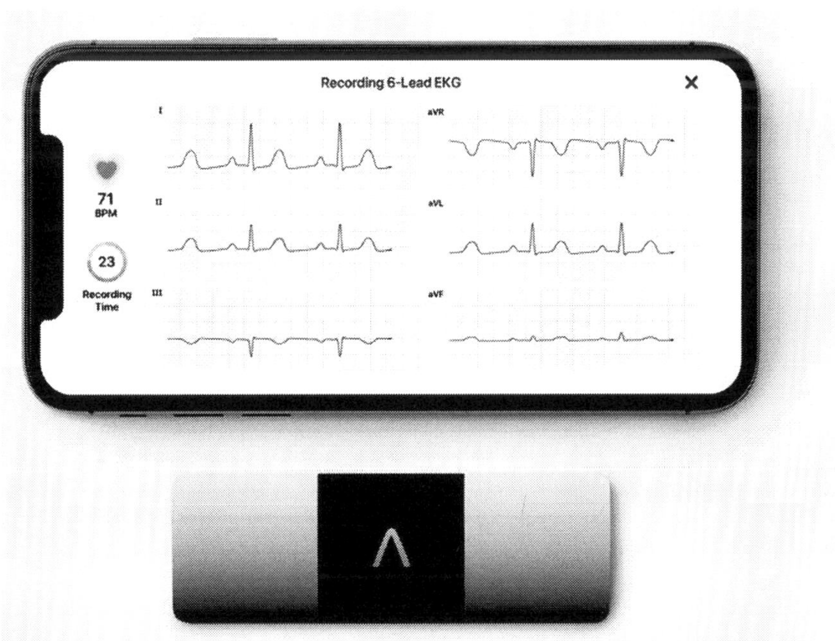

FIGURE 1.4 Kardia six-lead mobile EKG.

The world's first six-lead smartphone-enabled EKG was FDA approved in 2019. The device incorporates two electrodes on the top for fingers, and one on the bottom to contact the skin of the left leg. This Kardia EKG delivers leads I, II, III, aVL, aVR, and aVF.

Pocket-echocardiography

Echocardiography is the primary imaging modality to assess cardiac function and to diagnose a multitude of cardiac conditions. The fundamental concept of *"portable"* echocardiography goes back to the 1970s when the *"ultrasonic"* stethoscope first appeared (Ligtvoet, Rusterborgh, Kappen, & Bom; Roelandt, Wladimiroff, & Baars, 1978). Since then, technological advancement has led to the emergence of several *"miniaturized"* echocardiographic devices—ranging from the laptop-based devices to hand-carried systems and the most recent portable *"pocket"* echocardiography (Bhavnani, Narula, & Sengupta, 2016; Seraphim, Paschou, Grapsa, & Nihoyannopoulos, 2016; Tofield & Badano, 2011). Evidently, a substantial variation in functionality and capability exists among different types of portable echocardiography machines—especially with regard to color-flow and spectral Doppler (which display the blood flow measurements graphically over time, indicating flow velocities recorded over time) that is not available in all portable devices (Mirabel et al., 2015).

"Pocket" echocardiography machines are smartphone-sized devices that provide diagnostic quality 2D with or without color Doppler imaging in real-time (Chamsi-Pasha, Sengupta, & Zoghbi, 2017). The device, which is battery operated, consists of (i) broad-bandwidth ultrasound probe (1.7–5.0 MHz) that performs the image acquisition, reconstruction, and processing, and (ii) smartphone that functions to display and archive the echocardiographic images (Fig. 1.5) (Seraphim et al., 2016). The simplicity of use, easy transportability (e.g., small size and low weight), and the relatively cheap cost of the device have encouraged a spectrum of healthcare providers, other than cardiologists, to adopt this technology for prompt decision-making (Seraphim et al., 2016; Tofield and Badano, 2011).

The introduction of pocket echocardiography devices in clinical practice has raised questions over the diagnostic capacity of these devices when compared to the standard transthoracic echocardiography (sTTE) (Seraphim et al., 2016). Studies comparing the pocket-sized echocardiography devices with sTTE have demonstrated an excellent correlation between the two devices in assessing for basic echocardiographic parameters—such as left ventricular (LV) function, wall motion abnormalities, valvular heart diseases, pericardial effusion, and cardiac tamponade (Andersen et al., 2011; Culp, Mock, Chiles, & Culp, 2010; Filipiak-Strzecka, John, Kasprzak, Michalski, & Lipiec, 2013; Fukuda et al., 2009; Kitada et al., 2013; Kono et al., 2011; Liebo et al., 2011; Prinz & Voigt, 2011). For instance, studies demonstrated the pocket echocardiography to be adequate for interpretation of LV ejection fraction and LV end-diastolic dimensions in >95% of cases, with an accuracy level exceeding 90%, compared to the sTTE—when operated by an appropriately trained physician (Andersen et al., 2011; Liebo et al., 2011). Nonetheless, the limited image resolution and lack of spectral Doppler in many pocket devices make assessment of valvular heart diseases particularly more challenging (Kono et al., 2011; Prinz and Voigt, 2011).

In view of the increased evidence to support the valuable diagnostic utility of pocket echocardiography, the European Society of Cardiology (ESC) released

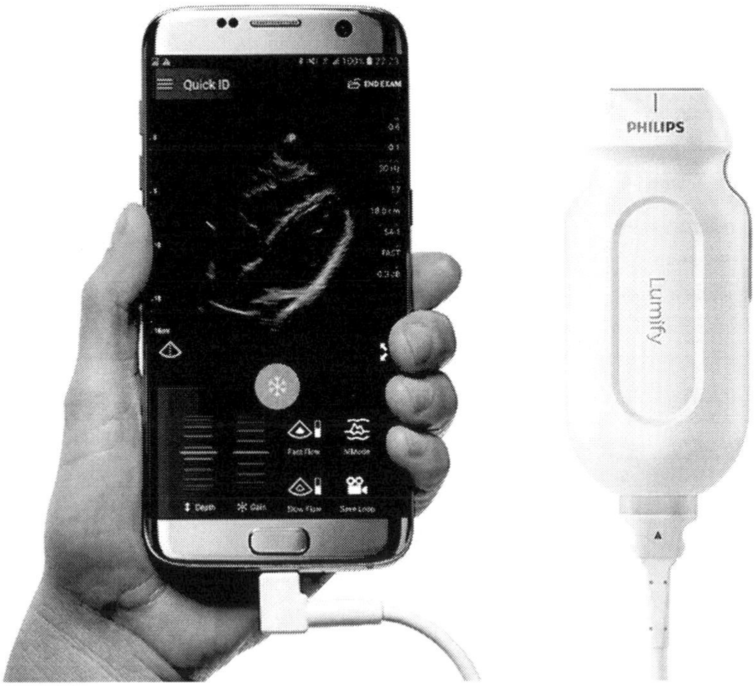

FIGURE 1.5 Philips Lumify pocket ultrasound.

A pocket echocardiography device consisting of a broad-bandwidth ultrasound probe that performs image acquisition, it is attached via a USB-probe to a smartphone that functions to display and archive echocardiographic images.

special recommendations for healthcare providers on the use of pocket-size imaging devices (Cardim et al., 2019; Tofield and Badano, 2011). The ESC emphasized that these devices do not provide complete diagnostic echocardiographic examination and cannot be used as a substitute for the sTTE. Pocket echocardiography should rather be portrayed as an extension for the standard cardiac physical examination with an aim to maximize diagnostic power for cardiac abnormalities (Cardim et al., 2019; Tofield and Badano, 2011). Nevertheless, with the projected evolution of sensor/transistor technologies, further miniaturization of ultrasound devices, and revolutionization of their functionality (e.g., adding spectral Doppler), more widespread use is certainly a matter of time.

Robotic technology in cardiology

Robotic technology has been used in medicine for several decades, primarily in surgical (Kim, 2014; Diodato and Damiano, 2003) specialties and radiation therapy (Ward and Koyfman, 2016; Avkshtol et al., 2016). The use of robotics in cardiovascular medicine was introduced in the 2000s when the first-in-man robotically assisted AF ablation was performed (Bolotin, Kypson, Nifong, & Chitwood, 2004; Gerosa, Bianco, Buja, & Di Marco, 2004). Robotic technology was developed to improve patient care via achieving superior precision and accuracy in various interventions; nonetheless, mitigation of occupational hazards associated with invasive cardiology also created a gigantic motivating force for these innovations (Smilowitz, Balter, & Weisz, 2013). Radiation risks for interventional cardiologist are significant; with radiation dose exposure approaching 50% of the total collective dose (Mohammadi, Danaee, &Alizadeh, 2017; Vano, Gonzalez, Canis, & Hernandez-Lezana, 2003). On average, radiation exposure from left ventriculography and coronary angiography is equivalent to 300 chest X-rays, coronary stent to 1000, radiofrequency ablation to 1500, and peripheral arterial intervention to 2500 chest X-rays (Coles et al., 2006; Kocinaj et al., 2006; Neofotistou et al., 2003). Further, interventionalists are prone for orthopedic and ergonomic injuries due to long standing hours and heavy-leaded aprons worn to shield them from radiation during procedures (Goldstein, Balter, Cowley, Hodgson, & Klein, 2004; Smilowitz et al., 2013).

The past few years have witnessed advances in robotic technology within cardiovascular medicine; leading to a wider utilization in both investigative and therapeutic domains (Mahmud, Pourdjabbar et al., 2017). The four primary uses of robotic technology in cardiovascular medicine are: (i) percutaneous coronary intervention (PCI), (ii) peripheral vascular intervention (PVI), (iii) catheter-based ablation for arrhythmias, and (iv) minimally invasive cardiac surgeries (Bonatti, Vetrovec, Riga, Wazni, & Stadler, 2014).

Robotics in PCI

When Andreas Gruntzig performed the first-in-human coronary angioplasty in 1977 in Zurich, Switzerland, the landscape of cardiovascular medicine was transformed forever (King and Schlumpf, 1993). Since its inception, coronary interventions have evolved tremendously. However, the fundamental technique of introducing a catheter and deploying an intracoronary balloon or stent remained largely unchanged (Mahmud, Naghi et al., 2017). With that, the inherent occupational hazards (e.g., ionized radiation exposure and orthopedic injury) associated with PCIs remained a major safety concern for interventionalists (Andreassi et al., 2016). Robotic PCI technology was innovated to both enhance the performance of the intervention and mitigate these risks. The first prototype for a robotic PCI system was declared in 2006 by Beyar et al. (2006), this was later refined to become the CorPath 200 (Cornidus Inc., MA, USA)—the only robotic system currently available for coronary interventions. The system consists of (i) remote radiation-shielded interventional cockpit, and (ii) robotic arm and cassette located at the bedside trail (Fig. 1.6). To control the robotic arm, the operating interventionalist uses a console joystick to advance, retract, and rotate the catheters, enabling him/her to perform the procedure remotely (Wegermann, Swaminathan, & Rao, 2019). PRECISE

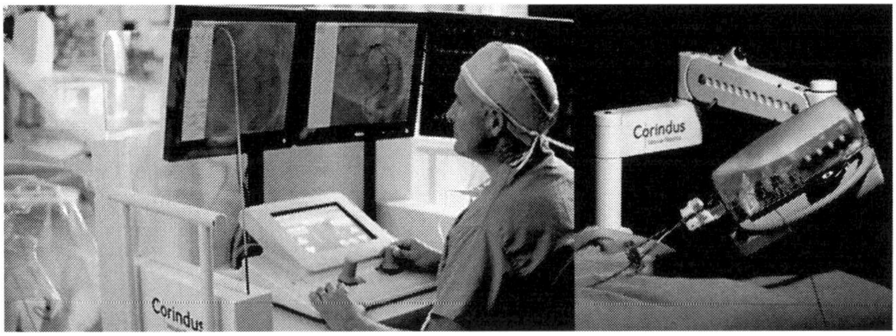

FIGURE 1.6 The CorPath 200 Robotic PCI System.

(*Left*) The operator sits in a remote radiation-shielded interventional cockpit to perform the procedure. A console joystick is used to advance, retract, and rotate the catheter. (*Right*) Articulating robotic arm translates operator's commands into precise movements and manipulations of guidewires and balloon/stent catheters.

(Percutaneous Robotically Enhanced Coronary Intervention) was a single-arm multicenter, nonrandomized trial to assess the performance and safety of CorePath 200 system (Weisz et al., 2013). The trial, which enrolled patients with "simple" coronary lesions, demonstrated >98% technical success, >97% clinical success, and >95% reduction in radiation exposure for operators (Weisz et al., 2013). It is based on the PRECISE trial that CorPath 200 system was FDA approved in 2012 as the first robotic PCI system. With further results from CORI-PCI trial, current evidence demonstrates safety and feasibility of robotic technology in both simple and complex coronary lesions (Mahmud, Naghi et al., 2017).

However, technical hurdles remain to deem certain complex coronary lesions ineligible for robotic interventions (e.g., bifurcated coronary lesions, severely calcified lesions, and chronic total occlusions), hampering global implementation of robotics in catheterization laboratories (Pourdjabbar, Ang, Reeves, Patel, & Mahmud, 2017). Future iterations of current robotic PCI system will require addressing these limitations. Finally, a newly emerging concept in the discipline of robotic PCI is "tele-stenting," where an interventional cardiologist can perform the procedure from geographically distanced locations via remotely placed operator cockpit. Through remote robotic assistance, tele-stenting might be the future enabling technology for hospitals with lower volume and low expertise in complex PCIs (Madder et al., 2019).

Robotics in PVI

Similar to PCI, peripheral vascular interventions (PVIs) have also been associated with increased risk for hazardous radiation exposure and orthopedic injuries. In fact, PVIs had been associated with a significantly greater (1.4 folds) radiation risk for operator as compared to PCI. Currently, two robotic systems are FDA-approved for PVIs: (i) CorPath 200 system (Corindus Vascular Robotics, MA,

USA), and (ii) Magellan System (Auris Surgical Robotics Inc., CA, USA). The RAPID (Robotic-Assisted Peripheral Intervention for Peripheral Artery Disease) trial was a prospective single-arm trial to assess the feasibility and safety of CorPath robotic technology in femoropopliteal PVIs (Mahmud et al., 2016). In this trial, the device technical success and clinical procedural success were both 100%. Further, no adverse events were associated with the use of the robotic system. In 2016, data from the RAPID trial led to the FDA approval of the system for femoropopliteal vascular interventions (Mahmud et al., 2016). However, the use of robotics in infrapopliteal lesions has not been well-established (Pourdjabbar et al., 2017), although successful robotic infrapopliteal PVIs were described in individual cases. Magellan is the second FDA-approved robotically assisted PVI system (Fig. 1.7). The system was utilized successfully to revascularize iliac and femoropopliteal lesions, with a reported clinical success rate of 95% and no adverse events (Bismuth, Duran, Stankovic, Gersak, & Lumsden, 2013). The demonstrated feasibility, safety, and effectiveness of this robotic system led to clearance by the FDA in 2012. Unlike the CorPath 200 system, the Magellan system has not been designed or approved for coronary interventions, rather its utility had been exclusive for PVIs (Mahmud et al., 2017b). In addition to peripheral arterial disease, utility of the Magellan system expanded to include a multitude of endovascular procedures, including: (i) endovascular aneurysm repair (EVAR), (ii) fenestrated EVAR, (iii) renal artery cannulation, and (iv) uterine artery embolization (Cochennec et al., 2015; Riga, Bicknell, Rolls, Cheshire, & Hamady, 2013; Rolls et al., 2014).

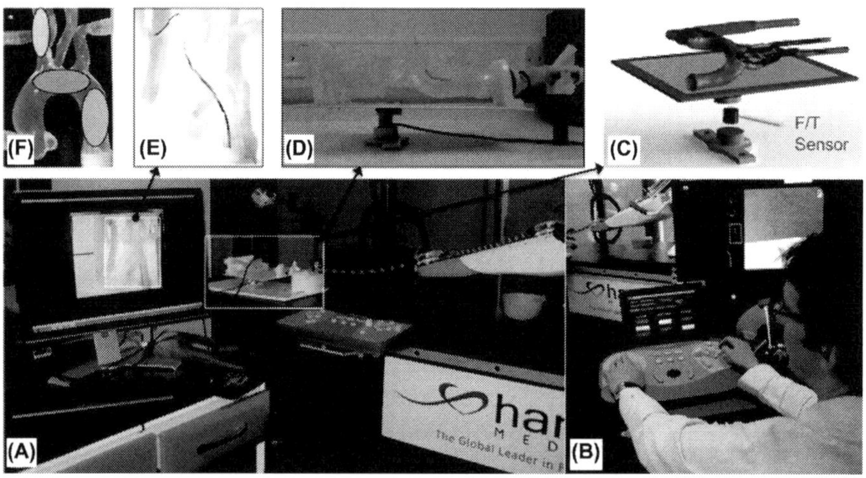

FIGURE 1.7 Magellan Robotic System.

Magellan Robotic system is used in the treatment of peripheral vascular disease and a multitude of endovascular procedures. The figure depicts experimental setup for the robotic catheter (A), console joystick used to control catheter (B), view of the force measurement platform (C and D), fluoroscopic imaging for guidance (E), and phases of the procedure (F).

Robotics in electrophysiology

Catheter-based ablation has become one of the preferred treatment modalities for sustained arrhythmias (i.e., atrial fibrillation) (Andrade, Macle, Nattel, Verma, & Cairns, 2017; January et al., 2014). Conventional (manual) catheter ablation procedures are technically challenging for nonexpert operators and had been associated with long procedural time, suboptimal success rate, and significant radiation exposure. Therefore, robotic magnetic navigation systems (MNS) were innovated over the past decade to tackle these challenges via enhancing catheter steerability, improving catheter stability, and catheter-tissue contact (Nölker et al., 2012; Steven et al., 2010). Conceptually, the robotic-guided MNS operates in a manner similar to the conventional ablation system—except that the ablation catheter is positioned within a robotic arm that can be manipulated remotely by the operator. During the procedure, a venous access (e.g., femoral vein) is used to advance an ablation catheter toward the right atrium to puncture the interatrial septum. Once a transseptal puncture is performed, the operating electrophysiologist leaves the patient's bedside to an adjacent control room, where a robotic-assisted remote-control system is used to accomplish the ablation procedure (Wutzler, Wolber, Haverkamp, & Boldt, 2015). Two MNS systems have been developed and currently utilized to facilitate robotic navigations: (i) Niobe II (third generation MNS) and (ii) Niobe ES (fourth generation MNS) (Da Costa et al., 2017). In the "Man and Machine" trial, the success rate of robotic-guided AF ablation was noninferior to that with manual AF ablation (Rillig et al., 2017). Despite the significantly longer procedural time, robotic ablation was associated with 50% reduction in major procedural complications compared with manual ablation (Virk and Kumar, 2019). Further, robotic ablation resulted in greater reduction of fluoroscopic exposure time—an increasingly recognized safety parameter in AF ablation (Zhang et al., 2014). In a nutshell, robotic magnetic navigation systems appear to be the technology of the future, demonstrating favorable outcomes for both patients and operators (Da Costa et al., 2017).

Future perspectives

Contemporary robotic-assisted cardiovascular interventions have enhanced procedural accuracy/precision, improved quality of clinical outcomes, and minimized occupational hazards (i.e., radiation exposure). Future iterations for robotic-assisted devices should take into consideration the technical challenges associated with more complex scenarios (e.g., robotic PCI for bifurcated or severely calcified coronary lesions). Further studies assessing cost-effectiveness of robotics for coronary and peripheral interventions are also warranted.

Conclusions

Innovations in digital cardiovascular technology continue to evolve with recent and ongoing advances in robotic technology, wearable technology, mobile health, telehealth, and others. Digitization of cardiovascular care holds promise to revolutionize diagnostic and imaging modalities, improve precision of therapeutic interventions, and mitigate occupational hazards on healthcare workers. With the powerful AI methods and ML algorithms, digital health technologies have the potential to transform the practice of cardiology. To outdistance the rapidly burgeoning digital medicine, development of ethical and regulatory frameworks ensuring appropriate implementation becomes a necessity (Chapter 18).

References

A nationwide framework for surveillance of cardiovascular and chronic lung diseases, (2011).

Adabag, A. S., Luepker, R. V., Roger, V. L., & Gersh, B. J. (2010). Sudden cardiac death: Epidemiology and risk factors. *Nature Reviews Cardiology, 7*(4), 216–225.

Al-Khatib, S. M., Stevenson, W. G., Ackerman, M. J., Bryant, W. J., Callans, D. J., Curtis, A. B., et al. (2018). 2017 AHA/ACC/HRS guideline for management of patients with ventricular arrhythmias and the prevention of sudden cardiac death. *Circulation, 138*(13).

Albert, D. (2019). A six-lead heart monitor on your smartphone: An interview with David Albert. *Future Cardiology, 16*(1), 9–11.

Andersen, G. N., Haugen, B. O., Graven, T., Salvesen, Ô., Mjôlstad, O. C., & Dalen, H. (2011). Feasibility and reliability of point-of-care pocket-sized echocardiography. *European Journal of Echocardiography, 14*(12), 1195–1202.

Andrade, J. G., Macle, L., Nattel, S., Verma, A., & Cairns, J. (2017). Contemporary atrial fibrillation management: A comparison of the current AHA/ACC/HRS, CCS, and ESC guidelines. *Canadian Journal of Cardiology, 33*(8), 965–976.

Andreassi, M. G., Piccaluga, E., Guagliumi, G., Del Greco, M., Gaita, F., & Picano, E. (2016). Occupational health risks in cardiac catheterization laboratory workers. *Circulation: Cardiovascular Interventions, 9*(4).

Armitage, L. C., Kassavou, A., & Sutton, S. (2020). Do mobile device apps designed to support medication adherence demonstrate efficacy? A systematic review of randomised controlled trials, with meta-analysis. *BMJ Open, 10*(1).

Aroganam, G., Manivannan, N., & Harrison, D. (2019). Review on wearable technology sensors used in consumer sport applications. *Sensors (Basel), 19*(9).

Avkshtol, V., Dong, Y., Hayes, S. B., Hallman, M. A., Price, R. A., Sobczak, M. L., et al. (2016). A comparison of robotic arm versus gantry linear accelerator stereotactic body radiation therapy for prostate cancer. *Research and Reports in Urology, 8*, 145–158.

Balli, S., Sağbaş, E. A., & Peker, M. (2019). Human activity recognition from smart watch sensor data using a hybrid of principal component analysis and random forest algorithm. *Measurement and Control, 52*(1–2).

Barold, S. S. (2003). Willem Einthoven and the birth of clinical electrocardiography a hundred years ago. *Cardiac Electrophysiology Review, 7*(1), 99–104.

Bates, D. W., Landman, A., & Levine, D. M. (2018). Health apps and health policy what is needed? *The Journal of the American Medical Association, 320*(19), 1975–1976.

Bermon, A., Uribe-Rodríguez, A. F., Pérez-Rivero, P. F., Prieto-Merino, D., Cáceres, R., Di, I., et al. (2019). Evaluation of the efficacy and safety of text messages targeting adherence to cardiovascular medications in secondary prevention: The txt2heart Colombia randomised controlled trial protocol. *BMJ Open, 9*(12).

Beyar, R., Gruberg, L., Deleanu, D., Roguin, A., Almagor, Y., Cohen, S., et al. (2006). Remote-control percutaneous coronary interventions: Concept, validation, and first-in-humans pilot clinical trial. *Journal of the American College of Cardiology, 8*(13).

Bhavnani, S. P., Narula, J., & Sengupta, P. P. (2016). Mobile technology and the digitization of healthcare. *European Heart Journal, 37*(18).

Bismuth, J., Duran, C., Stankovic, M., Gersak, B., & Lumsden, A. B. (2013). A first-in-man study of the role of flexible robotics in overcoming navigation challenges in the iliofemoral arteries. *Journal of Vascular Surgery, 57*(2 Suppl.).

Bolotin, G., Kypson, A. P., Nifong, L. W., & Chitwood, W. R. (2004). Robotically-assisted left atrial fibrillation ablation and mitral valve repair through a right mini-thoracotomy. *The Annals of Thoracic Surgery, 78*(4), e63–e64.

Bonatti, J., Vetrovec, G., Riga, C., Wazni, O., & Stadler, P. (2014). Robotic technology in cardiovascular medicine. *Nature Reviews Cardiology, 11*, 266–275.

Bush, B., Nifong, L. W., Alwair, H., & Chitwood, W. R. (2013). Robotic mitral valve surgery-current status and future directions. *Annals of Cardiothoracic Surgery, 2*(6), 814–817.

Cardim, N., Dalen, H., Voigt, J. U., Ionescu, A., Price, S., Neskovic, A. N., et al. (2019). The use of handheld ultrasound devices: A position statement of the European association of cardiovascular imaging (2018 update). *European Heart Journal - Cardiovascular Imaging, 20*(3), 245–252.

Chamsi-Pasha, M. A., Sengupta, P. P., & Zoghbi, W. A. (2017). Handheld echocardiography: Current state and future perspectives. *Circulation, 136*(22), 2178–2188.

Chan, P. H., Wong, C. K., Poh, Y. C., Pun, L., Leung, W. W. C., Wong, Y. F., et al. (2016). Diagnostic performance of a smartphone-based photoplethysmographic application for atrial fibrillation screening in a primary care setting. *Journal of the American Heart Association, 5*(7).

Chung, M. K., Szymkiewicz, S. J., Shao, M., Zishiri, E., Niebauer, M. J., Lindsay, B. D., et al. (2010). Aggregate national experience with the wearable cardioverter-defibrillator: Event rates, compliance, and survival. *Journal of the American College of Cardiology, 56*(3), 194–203.

Cochennec, F., Kobeiter, H., Gohel, M., Marzelle, J., Desgranges, P., Allaire, E., et al. (2015). Feasibility and safety of renal and visceral target vessel cannulation using robotically steerable catheters during complex endovascular aortic procedures. *Journal of Endovascular Therapy, 22*(2), 187–193.

Coles, D. R., Smail, M. A., Negus, I. S., Wilde, P., Oberhoff, M., Karsch, K. R., et al. (2006). Comparison of radiation doses from multislice computed tomography coronary angiography and conventional diagnostic angiography. *Journal of the American College of Cardiology, e247*(9), 1840–1845.

Coravos, A., Goldsack, J. C., Karlin, D. R., Nebeker, C., Perakslis, E., Zimmerman, N., et al. (2019). Digital medicine: A primer on measurement. *Digital Biomarkers, 3*, 31–71.

Cowie, M. R., Bax, J., Bruining, N., Cleland, J. G. F., Koehler, F., Malik, M., et al. (2016). E-Health: A position statement of the European Society of Cardiology. *European Heart Journal, 37*(1), 63–66.

Cruz, J., Brooks, D., & Marques, A. (2017). Accuracy of piezoelectric pedometer and accelerometer step counts. *The Journal of Sports Medicine and Physical Fitness, 57*(4), 426–433.

Culp, B. C., Mock, J. D., Chiles, C. D., & Culp, W. C. (2010). The pocket echocardiograph: Validation and feasibility. *Echocardiography, 27*(7), 759–764.

Da Costa, A., Guichard, J. B., Maillard, N., Romeyer-Bouchard, C., Gerbay, A., & Isaaz, K. (2017). Substantial superiority of Niobe ES over Niobe II system in remote-controlled magnetic pulmonary vein isolation. *International Journal of Cardiology, 230*, P319–P323.

Darzi, A., & Evans, T. (2016). The global shortage of health workers-an opportunity to transform care. *Lancet, 388*(10060), 2576–2577.

De Luna, A. B. (2019). Willem einthoven and the ECG. *European Heart Journal, 40*(41), 3381–3383.

Delgado, H., Toquero, J., Mitroi, C., Castro, V., & Fernandez, I. (2013). Principles of external defibrillators. In *Cardiac defibrillation.*

Demirovic, J., & Myerburg, R. J. (1994). Epidemiology of sudden coronary death: An overview. *Progress in Cardiovascular Diseases, 37*(1), 39–48.

Dillon, K. A., Szymkiewicz, S. J., & Kaib, T. E. (2010). Evaluation of the effectiveness of a wearable cardioverter defibrillator detection algorithm. *Journal of Electrocardiology, 43*(1), 63–67.

DiMarco, J. P. (2003). Implantable cardioverter-defibrillators. *New England Journal of Medicine, 349*(19), 1836–1847.

Diodato, M. D., & Damiano, R. J. (2003). Robotic cardiac surgery: Overview. *Surgical Clinics of North America, 83*(6), 1351–1367.

Ellenbogen, K. A., Koneru, J. N., Sharma, P. S., Deshpande, S., Wan, C., & Szymkiewicz, S. J. (2017). Benefit of the wearable cardioverter-defibrillator in protecting patients after implantable-cardioverter defibrillator explant: Results from the national registry. *JACC: Clinical Electrophysiology, 3*(3), 243–250.

Farmer, A., Allen, J., Bartlett, K., Bower, P., Chi, Y., French, D., et al. (2019). Supporting people with type 2 diabetes in effective use of their medicine through mobile health technology integrated with clinical care (SuMMiT-D feasibility): A randomised feasibility trial protocol. *BMJ Open, 9*(12).

FDA grants first ever clearance for six-lead personal ECG device | AliveCor, n.d. Retrieved from https://www.alivecor.com/press/press_release/fda-grants-first-ever-clearance-for-six-lead-personal-ecg-device. (accessed 17 3 2020).

Feldman, A. M., Klein, H., Tchou, P., Murali, S., Hall, W. J., Mancini, D., et al. (2004). Use of a wearable defibrillator in terminating tachyarrhythmias in patients at high risk for sudden death: Results of WEARIT/BIROAD. *Pacing and Clinical Electrophysiology.*

Filipiak-Strzecka, D., John, B., Kasprzak, J., Michalski, B., & Lipiec, P. (2013). Pocket-size echocardiograph-a valuable tool for nonexperts or just a portable device for echocardiographers. *Advances in Medical Sciences, 27*(1), 4–9.

Finkelstein, E. A., Haaland, B. A., Bilger, M., Sahasranaman, A., Sloan, R. A., Nang, E. E. K., et al. (2016). Effectiveness of activity trackers with and without incentives to increase physical activity (TRIPPA): A randomised controlled trial. *The Lancet Diabetes & Endocrinology, 4*(12), 983–995.

Fogel, A. L., & Kvedar, J. C. (2018). Artificial intelligence powers digital medicine. *npj Digital Medicine, 1*(5).

Free, C., Phillips, G., Watson, L., Galli, L., Felix, L., Edwards, P., et al. (2013). The effectiveness of mobile-health technologies to improve health care service delivery processes: A systematic review and meta-analysis. *PLoS Medicine, 10*(1).

Frisch, D. R., Weiss, M., Dikdan, S. J., Keith, S. W., & Sarkar, K. (2019). Improved accuracy and confidence with multiple-lead recordings from a single-lead mobile electrocardiographic device. *Pacing and Clinical Electrophysiology, 42*(9).

Fukuda, S., Shimada, K., Kawasaki, T., Fujimoto, H., Maeda, K., Inanami, H., et al. (2009). Pocket-sized transthoracic echocardiography device for the measurement of cardiac chamber size and function. *Circulation Journal, 73*(6), 1092–1096.

Garabelli, P., Stavrakis, S., & Po, S. (2017). Smartphone-based arrhythmia monitoring. *Current Opinion in Cardiology, 32*(1), 53–57.

Gerosa, G., Bianco, R., Buja, G., & Di Marco, F. (2004). Totally endoscopic robotic-guided pulmonary veins ablation: An alternative method for the treatment of atrial fibrillation. *European Journal of Cardio-Thoracic Surgery, 26*(2), 450–452.

Ghamari, M. (2018). A review on wearable photoplethysmography sensors and their potential future applications in health care. *International Journal of Biosensors & Bioelectronics, 4*(4), 195–202.

Goldenthal, I. L., Sciacca, R. R., Riga, T., Bakken, S., Baumeister, M., Biviano, A. B., et al. (2019). Recurrent atrial fibrillation/flutter detection after ablation or cardioversion using the AliveCor KardiaMobile device: iHEART results. *Journal of Cardiovascular Electrophysiology, 30*(11), 2220–2228.

Goldstein, J. A., Balter, S., Cowley, M., Hodgson, J., & Klein, L. W. (2004). Occupational hazards of interventional cardiologists: Prevalence of orthopedic health problems in contemporary practice. *Catheterization and Cardiovascular Interventions, 63*(4), 407–411.

Halcox, J. P. J., Wareham, K., Cardew, A., Gilmore, M., Barry, J. P., Phillips, C., et al. (2017). Assessment of remote heart rhythm sampling using the AliveCor heart monitor to screen for atrial fibrillation the REHEARSE-AF study. *Circulation, 136*(19), 1784–1794.

Healy, C. A., & Carrillo, R. G. (2015). Wearable cardioverter-defibrillator for prevention of sudden cardiac death after infected implantable cardioverter-defibrillator removal: A cost-effectiveness evaluation. *Heart Rhythm, 12*(7), 1565–1573.

Henriksen, A., Mikalsen, M. H., Woldaregay, A. Z., Muzny, M., Hartvigsen, G., Hopstock, L. A., et al. (2018). Using fitness trackers and smartwatches to measure physical activity in research: Analysis of consumer wrist-worn wearables. *Journal of Medical Internet Research.*

Heron, M. (2016). National vital statistics reports deaths: Leading causes for 2013 National Vital Statistics Reports: From the Centers for Disease Control and Prevention, National Center for Health Statistics. *National Vital Statistics System, 65*(2), 1–95.

Hickey, K. T., Hauser, N. R., Valente, L. E., Riga, T. C., Frulla, A. P., Creber, R. M., et al. (2016). A single-center randomized, controlled trial investigating the efficacy of a mHealth ECG technology intervention to improve the detection of atrial fibrillation: The iHEART study protocol. *BMC Cardiovascular Disorders, 16*(52).

Himmelreich, J. C. L., Karregat, E. P. M., Lucassen, W. A. M., van Weert, H. C. P. M., de Groot, J. R., Louis Handoko, M., et al. (2019). Diagnostic accuracy of a smartphone-operated, single-lead electrocardiography device for detection of rhythm and conduction abnormalities in primary care. *The Annals of Family Medicine, 17*(5), 403–411.

Hufnagel, C. A. (1951). Aortic plastic valvular prosthesis. *Bull Georgetown University Medical Center, 4*(5), 128–130.

Iosa, M., Picerno, P., Paolucci, S., & Morone, G. (2016). Wearable inertial sensors for human movement analysis. *Expert Review of Medical Devices, 13*(7), 641–659.

Jakicic, J. M., Davis, K. K., Rogers, R. J., King, W. C., Marcus, M. D., Helsel, D., et al. (2016). Effect of wearable technology combined with a lifestyle intervention on long-term weight loss: The IDEA randomized clinical trial. *The Journal of the American Medical Association, 316*(11), 1161–1171.

January, C. T., Wann, L. S., Alpert, J. S., Calkins, H., Cigarroa, J. E., Cleveland, J. C., et al. (2014). 2014 AHA/ACC/HRS guideline for the management of patients with atrial fibrillation. *Journal of the American College of Cardiology, 74*(1), 104–132.

John, R. M., Tedrow, U. B., Koplan, B. A., Albert, C. M., Epstein, L. M., Sweeney, M. O., et al. (2012). Ventricular arrhythmias and sudden cardiac death. *Lancet, 380*(9852), 1520–1529.

Kelli, H. M., Witbrodt, B., & Shah, A. (2017). The future of mobile health applications and devices IN cardiovascular health. *European Medical Journal Innovations.*

Kim, K. C. (2014). *Robotics in general surgery.*

King, S. B., & Schlumpf, M. (1993). Ten-year completed follow-up of percutaneous transluminal coronary angioplasty: The early Zurich experience. *Journal of the American College of Cardiology, 22*(2), 353–360.

Kitada, R., Fukuda, S., Watanabe, H., Oe, H., Abe, Y., Yoshiyama, M., et al. (2013). Diagnostic accuracy and cost-effectiveness of a pocket-sized transthoracic echocardiographic imaging device. *Clinical Cardiology, 36*(10), 603–610.

Klein, H. U., Meltendorf, U., Reek, S., Smid, J., Kuss, S., Cygankiewicz, I., et al. (2010). Bridging a temporary high risk of sudden arrhythmic death. Experience with the wearable cardioverter defibrillator (WCD). *Pacing and Clinical Electrophysiology, 33*(3), 353–367.

Kocinaj, D., Cioppa, A., Ambrosini, G., Tesorio, T., Salemme, L., Sorropago, G., et al. (2006). Radiation dose exposure during cardiac and peripheral arteries catheterisation. *International Journal of Cardiology, 113*(2), 283–284.

Koenig, N., Seeck, A., Eckstein, J., Mainka, A., Huebner, T., Voss, A., et al. (2016). Validation of a new heart rate measurement algorithm for fingertip recording of video signals with smartphones. *Telemedicine Journal and e-Health, 22*(8), 631–636.

Kono, Y., Fukuda, S., Shimada, K., Oe, H., Maeda, K., Kawasaki, T., et al. (2011). Pocket-sized echo for evaluation of mitral and tricuspid regurgitation. *JACC: Cardiovascular Imaging, 4*(8), 921.

Krikler, D. M. (1985). Withering and the foxglove: The making of a myth. *British Heart Journal, 54*(3), 256–257.

Krivoshei, L., Weber, S., Burkard, T., Maseli, A., Brasier, N., Kühne, M., et al. (2017). Smart detection of atrial fibrillation. *Europace, 19*(5), 753–757.

Liebo, M. J., Israel, R. L., Lillie, E. O., Smith, M. R., Rubenson, D. S., & Topol, E. J. (2011). Is pocket mobile echocardiography the next-generation stethoscope? A cross-sectional comparison of rapidly acquired images with standard transthoracic echocardiography. *Annals of Internal Medicine, 155*(1), 33–38.

Ligtvoet, C., Rusterborgh, H., Kappen, L., & Bom, N. (1978). Real time ultrasonic imaging with a hand-held scanner Part I-Technical description. *Ultrasound in Medicine and Biology, 4*(2), 91–92.

Madder, R. D., VanOosterhout, S., Mulder, A., Bush, J., Martin, S., Rash, A., et al. (2019). Feasibility of robotic telestenting over long geographic distances: A pre-clinical ex vivo and in vivo study. *EuroIntervention, 15*(6), E510–e512.

Mahmud, E., Naghi, J., Ang, L., Harrison, J., Behnamfar, O., Pourdjabbar, A., et al. (2017a). Demonstration of the safety and feasibility of robotically assisted percutaneous coronary intervention in complex coronary lesions: Results of the CORA-PCI study (complex robotically assisted percutaneous coronary intervention). *JACC: Cardiovascular Interventions, 10*(13), 1320–1327.

Mahmud, E., Pourdjabbar, A., Ang, L., Behnamfar, O., Patel, M. P., & Reeves, R. R. (2017b). Robotic technology in interventional cardiology: Current status and future perspectives. *Catheterization and Cardiovascular Interventions, 90*(6), 956–962.

Mahmud, E., Schmid, F., Kalmar, P., Deutschmann, H., Hafner, F., Rief, P., et al. (2016). Feasibility and safety of robotic peripheral vascular interventions: Results of the RAPID trial. *JACC: Cardiovascular Interventions.*

Masri, A., Altibi, A. M., Erqou, S., Zmaili, M. A., Saleh, A., Al-Adham, R., et al. (2019). Wearable cardioverter-defibrillator therapy for the prevention of sudden cardiac death: A systematic review and meta-analysis. *JACC: Clinical Electrophysiology, 5*(2), 152–161.

Matchett, M., Sears, S. F., Hazelton, G., Kirian, K., Wilson, E., & Nekkanti, R. (2009). The implantable cardioverter defibrillator: Its history, current psychological impact and future. *Expert Review of Medical Devices, 6*(1), 43–50.

McElwee, S. K., Velasco, A., & Doppalapudi, H. (2016). Mechanisms of sudden cardiac death. *Journal of Nuclear Cardiology, 23*(6), 1368–1379.

Mirabel, M., Celermajer, D., Beraud, A. S., Jouven, X., Marijon, E., & Hagège, A. A. (2015). Pocket-sized focused cardiac ultrasound: Strengths and limitations. *Archives of Cardiovascular Diseases, 108*(3), 197–205.

Mirowski, M., Reid, P. R., Mower, M. M., Watkins, L., Gott, V. L., Schauble, J. F., et al. (1980). Termination of malignant ventricular arrhythmias with an implanted automatic defibrillator in human beings. *New England Journal of Medicine, 303*(6), 322–324.

Mohammadi, M., Danaee, L., & Alizadeh, E. (2017). Reduction of radiation risk to interventional cardiologists and patients during angiography and coronary angioplasty. *The Journal of Tehran University Heart Center, 12*(3), 101–106.

Mugnai, G., Lupo, A., Zerbo, F., Saccá, S., & Zoppo, F. (2019). Single-center experience with wearable cardioverter-defibrillator as a bridge before definitive ICD implantation. *Indian Pacing and Electrophysiology Journal, 20*(2), 60–63.

Muro-de-la-Herran, A., García-Zapirain, B., & Méndez-Zorrilla, A. (2014). Gait analysis methods: An overview of wearable and non-wearable systems, highlighting clinical applications. *Sensors (Switzerland), 25*(2), 3362–3394.

Neofotistou, V., Vano, E., Padovani, R., Kotre, J., Dowling, A., Toivonen, M., et al. (2003). Preliminary reference levels in interventional cardiology. *European Radiology, 13*(10), 2259–2263.

Nguyen, H. H., & Silva, J. N. A. (2016). Use of smartphone technology in cardiology. *Trends in Cardiovascular Medicine, 26*(4), 376–386.

Nölker, G., Gutleben, K. J., Muntean, B., Vogt, J., Horstkotte, D., Dabiri Abkenari, L., et al. (2012). Novel robotic catheter manipulation system integrated with remote magnetic navigation for fully remote ablation of atrial tachyarrhythmias: A two-centre evaluation. *Europace.*

Olgin, J. E., Pletcher, M. J., Vittinghoff, E., Wranicz, J., Malik, R., Morin, D. P., et al. (2018). Wearable cardioverter-defibrillator after myocardial infarction. *New England Journal of Medicine, 379*(13), 1205–1215.

Park, L. G., Beatty, A., Stafford, Z., & Whooley, M. A. (2016). Mobile phone interventions for the secondary prevention of cardiovascular disease. *Progress in Cardiovascular Diseases, 58*(6), 639–650.

Passaro, V. M. N., Cuccovillo, A., Vaiani, L., De Carlo, M., & Campanella, C. E. (2017). Gyroscope technology and applications: A review in the industrial perspective. *Sensors (Switzerland), 17*(10), 2284.

Peng, F., Zhang, Z., Gou, X., Liu, H., & Wang, W. (2014). Motion artifact removal from photoplethysmographic signals by combining temporally constrained independent component analysis and adaptive filter. *BioMedical Engineering Online, 13*(50).

Perez, M. V., Mahaffey, K. W., Hedlin, H., Rumsfeld, J. S., Garcia, A., Ferris, T., et al. (2019). Large-scale assessment of a smartwatch to identify atrial fibrillation. *New England Journal of Medicine, 381*(20), 1909–1917.

Piccini, J. P., Allen, L. A., Kudenchuk, P. J., Page, R. L., Patel, M. R., & Turakhia, M. P. (2016). Wearable cardioverter-defibrillator therapy for the prevention of sudden cardiac death: A science advisory from the American Heart Association. *Circulation, 133*(17), 1715–1727.

Piette, J. D., List, J., Rana, G. K., Townsend, W., Striplin, D., & Heisler, M. (2015). Mobile health devices as tools for worldwide cardiovascular risk reduction and disease management. *Circulation, 132*(21), 2012–2027.

Poh, M. Z., & Poh, Y. C. (2017). Validation of a standalone smartphone application for measuring heart rate using imaging photoplethysmography. *Telemedicine Journal and e-Health, 23*(8), 678–683.

Pourdjabbar, A., Ang, L., Reeves, R. R., Patel, M. P., & Mahmud, E. (2017). The development of robotic technology in cardiac and vascular interventions. *Rambam Maimonides Medical Journal, 8*(3), E0030.

Prinz, C., & Voigt, J. U. (2011). Diagnostic accuracy of a hand-held ultrasound scanner in routine patients referred for echocardiography. *Journal of the American Society of Echocardiography, 24*(2), 111–116.

Proesmans, T., Mortelmans, C., Van Haelst, R., Verbrugge, F., Vandervoort, P., & Vaes, B. (2019). Mobile phone-based use of the photoplethysmography technique to detect atrial fibrillation in primary care: Diagnostic accuracy study of the FibriCheck App. *Journal of Medical Internet Research, 7*(3), E12284.

Rawassizadeh, R., Price, B. A., & Petre, M. (2015). Wearables: Has the age of smartwatches finally arrived? *Communications of the ACM, 58*(1), 45–47.

Reek, S., Geller, J. C., Meltendorf, U., Wollbrueck, A., Szymkiewicz, S. J., & Klein, H. U. (2003). Clinical efficacy of a wearable defibrillator in acutely terminating episodes of ventricular fibrillation using biphasic shocks. *Pacing and Clinical Electrophysiology, 26*(10), 2016–2022.

Riga, C. V., Bicknell, C. D., Rolls, A., Cheshire, N. J., & Hamady, M. S. (2013). Robot-assisted fenestrated endovascular aneurysm repair (FEVAR) using the magellan system. *Journal of Vascular and Interventional Radiology, 24*(2), 191–196.

Rillig, A., Schmidt, B., Di Biase, L., Lin, T., Scholz, L., Heeger, C. H., et al. (2017). Manual versus robotic catheter ablation for the treatment of atrial fibrillation: The man and machine trial. *JACC: Clinical Electrophysiology, 3*(8), 875–883.

Roelandt, J., Wladimiroff, J. W., & Baars, A. M. (1978). Ultrasonic real time imaging with a hand-held-scanner part II-initial clinical experience. *Ultrasound in Medicine and Biology, 4*(2), 93–97.

Rolls, A. E., Riga, C. V., Bicknell, C. D., Regan, L., Cheshire, N. J., & Hamady, M. S. (2014). Robot-assisted uterine artery embolization: A first-in-woman safety evaluation of the magellan system. *Journal of Vascular and Interventional Radiology, 25*(12), 1841–1848.

RTI International. (2017). Cardiovascular disease: A costly burden for America – projections through 2035. *American Heart Association*, 5–14.

Seetharam, K., Kagiyama, N., & Sengupta, P. P. (2019). Application of mobile health, telemedicine and artificial intelligence to echocardiography. *Echo Research and Practice, 6*(2), 41–52.

Seraphim, A., Paschou, S. A., Grapsa, J., & Nihoyannopoulos, P. (2016). Pocket-sized echocardiography devices: One stop shop service? *Journal of Cardiovascular Ultrasound, 24*(1), 1–6.

Sharma, P. S., Bordachar, P., & Ellenbogen, K. A. (2017). Indications and use of the wearable cardiac defibrillator. *European Heart Journal, 38*(4), 258–267.

Smilowitz, N. R., Balter, S., & Weisz, G. (2013). Occupational hazards of interventional cardiology. *Cardiovascular Revascularization Medicine, 14*(4), 223–228.

Stecker, E. C., Reinier, K., Marijon, E., Narayanan, K., Teodorescu, C., Uy-Evanado, A., et al. (2014). Public health burden of sudden cardiac death in the United States. *Circulation: Arrhythmia and Electrophysiology, 7*(2), 212–217.

Steinhubl, S. R., & Topol, E. J. (2015). Moving from digitalization to digitization in cardiovascular care why is it important, and what could it mean for patients and providers? *Journal of the American College of Cardiology, 66*(13), 1489–1496.

Steven, D., Servatius, H., Rostock, T., Hoffmann, B., Drewitz, I., Müllerleile, K., et al. (2010). Reduced fluoroscopy during atrial fibrillation ablation: Benefits of robotic guided navigation. *Journal of Cardiovascular Electrophysiology, 21*(1), 6–12.

Strain, T., Wijndaele, K., & Brage, S. (2019). Physical activity surveillance through smartphone apps and wearable trackers: Examining the UK potential for nationally representative sampling. *JMIR mHealth uHealth, 7*(1), E11898.

Sun, Y., & Thakor, N. (2016). Photoplethysmography revisited: From contact to noncontact, from point to imaging. *IEEE Transactions on Biomedical Engineering, 63*(3), 463–477.

Tamura, T., Maeda, Y., Sekine, M., & Yoshida, M. (2014). Wearable photoplethysmographic sensors—past and present. *Electron, 3*(2), 282–302.

Tanawuttiwat, T., Garisto, J. D., Salow, A., Glad, J. M., Szymkiewicz, S., Saltzman, H. E., et al. (2014). Protection from outpatient sudden cardiac death following ICD removal using a wearable cardioverter defibrillator. *Pacing and Clinical Electrophysiology, 37*(5), 562–568.

Tofield, A., & Badano, L. P. (2011). The use of pocket size imaging devices: A position statement by the European Association of Echocardiography. *European Heart Journal, 32*(4), 385–386.

Tracy, C. M., Epstein, A. E., Darbar, D., DiMarco, J. P., Dunbar, S. B., Estes, N. A. M., et al. (2013). 2012 ACCF/AHA/HRS focused update incorporated into the ACCF/AHA/HRS 2008 guidelines for device-based therapy of cardiac rhythm abnormalities. *Journal of the American College of Cardiology, 61*(3), E6–75.

Turakhia, M. P., Desai, M., Hedlin, H., Rajmane, A., Talati, N., Ferris, T., et al. (2019). Rationale and design of a large-scale, app-based study to identify cardiac arrhythmias using a smartwatch: The Apple Heart Study. *American Heart Journal, 207*, 66–75.

Vano, E., Gonzalez, L., Canis, M., & Hernandez-Lezana, A. (2003). Training in radiological protection for interventionalists. Initial Spanish experience. *British Journal of Radiology, 76*(904), 217–219.

Virk, S. A., & Kumar, S. (2019). Remote magnetic versus manual catheter navigation for atrial fibrillation ablation: A meta-analysis. *Circulation: Arrhythmia and Electrophysiology.*

Wagenaar, R. C., Sapir, I., Zhang, Y., Markovic, S., Vaina, L. M., & Little, T. D. C. (2011). Continuous monitoring of functional activities using wearable, wireless gyroscope and accelerometer technology. In *Proceedings of the annual international conference of the IEEE engineering in medicine and biology society*. EMBS, 4844–4847.

Walker, A., & Muhlestein, J. (2018). Smartphone electrocardiogram monitoring: Current perspectives. *Advanced Health Care Technologies, 2018*(4), 15–24.

Ward, M. C., & Koyfman, S. A. (2016). Transoral robotic surgery: The radiation oncologist's perspective. *Oral Oncology, 60*, 96–102.

Wegermann, Z. K., Swaminathan, R. V., & Rao, S. V. (2019). Cath lab robotics: Paradigm change in interventional cardiology? *Current Cardiology Reports, 21*(10), 119.

Wegner, F. K., Kochhäuser, S., Ellermann, C., Lange, P. S., Frommeyer, G., Leitz, P., et al. (2019). Prospective blinded evaluation of the smartphone-based AliveCor Kardia ECG monitor for atrial fibrillation detection: The PEAK-AF study. *European Journal of Internal Medicine, 73*, 72–75.

Weisz, G., Metzger, D. C., Caputo, R. P., Delgado, J. A., Marshall, J. J., Vetrovec, G. W., et al. (2013). Safety and feasibility of robotic percutaneous coronary intervention: PRECISE (percutaneous robotically-enhanced coronary intervention) study. *Journal of the American College of Cardiology, 61*(15), 1596–1600.

White, R. D., & Flaker, G. (2017). Smartphone-based arrhythmia detection: Should we encourage patients to use the ECG in their pocket? *Journal of Atrial Fibrillation, 9*(6), 1605.

Whitehead, L., & Seaton, P. (2016). The effectiveness of self-management mobile phone and tablet apps in long-term condition management: A systematic review. *Journal of Medical Internet Research, 18*(5), E97.

William, A. D., Kanbour, M., Callahan, T., Bhargava, M., Varma, N., Rickard, J., et al. (2018). Assessing the accuracy of an automated atrial fibrillation detection algorithm using smartphone technology: The iREAD Study. *Heart Rhythm*.

World Health Organization. (2010). *Medical devices: Managing the mismatch* (pp. 1–143). World Heal. Organ.

Wutzler, A., Wolber, T., Haverkamp, W., & Boldt, L. H. (2015). Robotic ablation of atrial fibrillation. *Journal of Visualized Experiments, 99*, E52560.

Yu, K. H., Beam, A. L., & Kohane, I. S. (2018). Artificial intelligence in healthcare. *Nature Biomedical Engineering, 2*(10), 719–731.

Zetterström, R. (2009). Nobel Prize to Willem Einthoven in 1924 for the discovery of the mechanisms underlying the electrocardiogram (ECG). *Acta Paediatrica, 98*(8), 1380–1382.

Zhang, W., Jia, N., Su, J., Lin, J., Peng, F., & Niu, W. (2014). The comparison between robotic and manual ablations in the treatment of atrial fibrillation: A systematic review and meta-analysis. *PloS One, 9*(5), E96331.

Zhang, Y., Song, S., Vullings, R., Biswas, D., Simões-Capela, N., Van Helleputte, N., et al. (2019). Motion artifact reduction for wrist-worn photoplethysmograph sensors based on different wavelengths. *Sensors (Switzerland), 104*(18), 2158–2163.

Zheng, Z. J., Croft, J. B., Giles, W. H., & Mensah, G. A. (2001). Sudden cardiac death in the United States, 1989 to 1998. *Circulation*.

CHAPTER

An overview of artificial intelligence: basics and state-of-the-art algorithms

2

Qi Chang[1], Leon Axel[2], Dimitris N. Metaxas[1]

[1]*Department of Computer Science, Rutgers The State University of New Jersey, Piscataway, NJ, United States;* [2]*Department of Radiology, New York University School of Medicine, New York, NY, United States*

Chapter outline

Machine Learning in Cardiovascular Medicine. https://doi.org/10.1016/B978-0-12-820273-9.00002-6

Introduction

In the early days of artificial intelligence (AI), the field rapidly tackled and solved problems that were intellectually difficult for humans but were relatively straightforward for computers— problems that could be described by a list of formal, mathematical rules (Goodfellow, Bengio, and Courville 2016). The real challenge to AI has been shown to involve tasks that are easy for individuals to perform but hard to formally describe—problems that we solve intuitively, that feel automatic, like recognizing spoken words or faces on pictures. As such, the main purpose of this chapter is to present several solutions for these intuitive tasks. Within AI, the approach has been to allow computers to learn from statistics and experiences and thereby understand the world in terms of a hierarchy of concepts, with each concept defined through its relationship to simpler concepts. By gathering knowledge from experience, this approach avoids the need for human operators to formally specify all the knowledge that the computer needs. The hierarchy of concepts enables the computer to learn complicated concepts by building them out of simpler ones.

The first part of Chapter 2 presents basic concepts within machine learning and AI. Specifically, this chapter will cover basic knowledge on the scope and approach of machine learning, with a focus on specific terms that are often misconstrued within the community. In the second part of this chapter, we will provide an overview of the state-of-the-art machine learning algorithms commonly used in academic research and the industry domain. By discussing various algorithms and approaches, we aim to provide the reader with a platform to navigate through different algorithms and select an approach that is best suited for a specific research or medical task.

Basic concepts

What are AI, machine learning, and deep learning? And what is the difference between them?

John McCarthy, widely recognized as one of the godfathers of AI, defined AI as "the science and engineering of making intelligent machines." As the name suggests, AI can be loosely interpreted to refer to the incorporation of human intelligence into machines. In machine learning, machines can learn on their own without being explicitly programmed to do so. It also allows for self-modifications based on data in real-time scenarios. In other words, machine learning is the study of computer algorithms that allow computer programs to automatically improve through experience.

Deep learning is one of many approaches to machine learning. Other approaches include decision tree learning, inductive logic programming, clustering, reinforcement learning, and Bayesian networks, among others. Deep learning was loosely inspired by the structure and function of the brain, namely the interconnection of many neurons. Artificial Neural Networks (ANNs) are algorithms that mimic the biological structure of the brain. In ANNs, there are "neurons" which have discrete layers and connections to other "neurons." Each layer picks out a specific feature to learn, such as curves/edges in image recognition. It's this layering that gives deep learning its name, depth is created by using multiple layers as opposed to a single layer.

Analogies between human and machine learning (ML) naturally bring to mind the term "AI" and the obvious question: "*What's the difference between AI and machine learning?*" There's no clear consensus on this matter, but most (not all) agree that ML is one form of AI, and that AI is a far broader subject encompassing such areas as robotics, language processing, and computer vision systems. To increase the ambiguity even further, machine learning is being applied with increasing frequency to many of those aforementioned fields. As such, we can conclude by stating that the discipline of ML refers to a specific body of knowledge and an associated set of techniques. It's fairly clear what is, and what isn't, machine learning, whereas the same can't always be said of artificial intelligence. Paraphrasing Tom Mitchell's often-cited definition, a computer program is said to learn if its performance of a certain task, as measured by a computable score, improves with experience (Mitchell, 1997).

Types of machine learning

There are some variations in the available definitions of machine learning algorithms. However, such algorithms are commonly divided into categories according to their purpose, with the main categories being:

- ***Supervised Learning***
  Supervised learning algorithms try to model relationships and dependencies between the target prediction output and the input features, such that one can predict the output values for new data based on those relationships learned from the previous datasets. Here, human experts act as teachers where one feeds the computer with training data containing the input/predictors while providing the correct answers (output); using which the computer should be able to learn underlying patterns. The process of learning from data, and subsequently using the acquired knowledge to inform future decisions, is extremely powerful. Indeed, machine learning is rapidly becoming the engine that powers the modern data-driven economy. Widely used supervised machine learning techniques include Classification and Regression algorithms.

  Classification: It is a Supervised Learning task where the output is having defined labels (discrete value). It can be either binary or multiclass classification. In binary classification, the model predicts either 0 or 1; yes or no, but in case of multiclass classification, a model predicts more than one class. As an example: Gmail classifies mail in more than one class like social, promotions, updates, and forums. It can either classify one class or multi classes per input or, as in image segmentation, classify classes over each pixel.

  Regression: It is a Supervised Learning task where the output has a continuous value. The goal here is to predict a value as close as possible to an actual output value, with subsequent model evaluation by calculating an error value. As expected, the smaller the error the greater is the accuracy of a regression model.

- ***Unsupervised Learning***
  In this paradigm, a machine learning model is trained with unlabeled data and without any explicit target output. The computer might be able to discover new knowledge regarding patterns within datasets that humans might have previously neglected. However, these models are particularly useful in cases where the human expert doesn't know what to look for within the dataset. Such algorithms try to use techniques on the input data to mine for rules, detect patterns, summarize and group the data points, which ultimately helps derive meaningful insights while better describing the data to the end-users. Commonly used unsupervised learning algorithms include K-Means, Self-Organizing Maps (SOM), Autoencoder, and Generative adversarial network (GAN) (Chang et al., 2020; Goodfellow et al., 2014), among many others.
- ***Semi-supervised Learning***
  In the previous two paradigms, either there were no labels for all observations or labels are present for all the observations. Semi-supervised learning falls in between these two methods. In many practical situations, the cost of performing expert labeling is significant. Hence, in the absence of output labels in the majority of observations, semi-supervised algorithms are best suited for model creation. These methods exploit the idea that even though the group memberships of the unlabeled data are unknown, such data carries important information about group parameters.
- ***Reinforcement Learning***
  Reinforcement learning aims at using observations gathered from the interaction with the computationally simulated environment to learn to perform actions that would maximize the reward or minimize the risk. Reinforcement learning algorithms (referred to as the agent) continuously learn from the environment in an iterative fashion (Fig. 2.1). In the process, an agent learns from its experiences

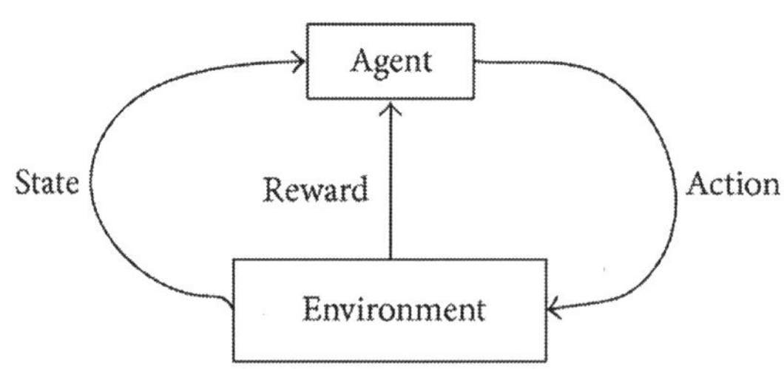

FIGURE 2.1 Reinforcement learning training process.

In order to produce intelligent programs (also called agents), reinforcement learning goes through the following steps: 1. Input state is observed by the agent. 2. Decision-making function is used to make the agent perform an action. 3. After the action is performed, the agent receives reward or reinforcement from the environment. 4. The state-action pair information about the reward is stored.

until it explores the full range of possible states. Reinforcement Learning allows machines and software agents to automatically determine the ideal behavior within a specific context, to maximize its performance. Simple reward feedback is required for the agent to learn its behavior; this is known as the reinforcement signal. Some applications of the reinforcement learning algorithms are computer played board games (Chess, Go), robotic hands, and self-driving cars.

Machine learning in practice

In the following sections, we will mainly focus on supervised learning tasks, which constitute a majority of the current applications of machine learning, and are the most frequently used methods in the medical domain. A generalized outline for a step-by-step approach to machine learning has been depicted in Fig. 2.2. To have a better understanding of the training process, the following technical terms are explained.

Generalization in machine learning: overfitting versus underfitting

Generalization refers to how well the concepts learned by a machine learning model apply to specific examples not seen by the model when it was learning. The goal of a good machine learning model is to generalize well from the training data to any data from the problem domain. This allows for accurate predictions when applying the model to never seen scenarios/data/patients. The cause of poor performance in

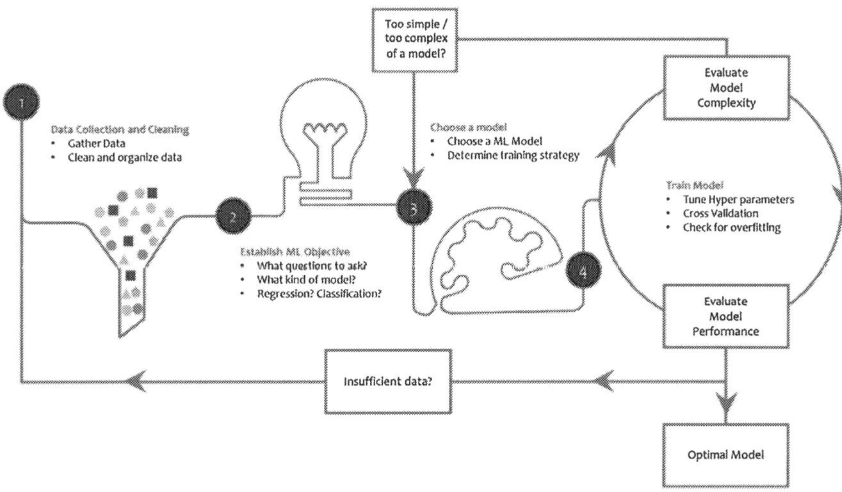

FIGURE 2.2 A general outline for the creation of machine learning–based models within a cardiology workflow (Al'Aref et al., 2018).

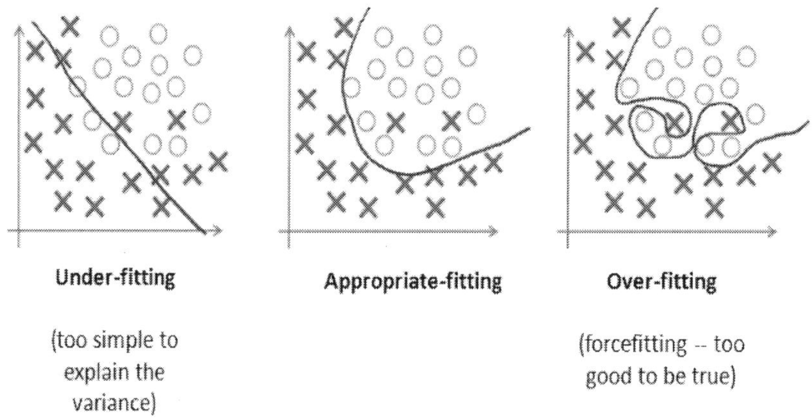

FIGURE 2.3 Underfitting versus Overfitting.

When the model is underfitting, the training error is high, while the training error is close to the test error. On the other hand, an overfitting model fits the training data too well, where the training error is lower than the test error.

machine learning is usually a result of either overfitting or underfitting of the model on the data. *Overfitting* refers to a phenomenon when the model learns the training data. The model learns both, the information and random noise, in the training data to an extent that it negatively impacts the performance of the model on new data. This means that the noise or random fluctuations in the training data is picked up and learned as concepts by the model. This negatively impacts the model's ability to generalize as noise being "random" is not replicated in the subsequent data. An overfitted model thus will not perform well on unseen new data.

Underfitting refers to the phenomenon when the model can neither model the training data nor generalize to new data. An underfitted machine learning model is not a suitable model as a result of poor performance on training data. Underfitting is often not discussed as it is easy to detect given the availability of good performance metrics. The remedy is to try alternate machine learning algorithms. Nevertheless, it does provide a good contrast to the problem of overfitting.

Both overfitting and underfitting can lead to poor model performance. But by far the most common problem in applied machine learning is overfitting. Overfitting is a significant issue since evaluation of model performance on training data is different from model performance on unseen data, which in principle is the reason for model creation to begin with (Fig. 2.3).

Dataset split: training versus validation versus testing

Training dataset is the sample of data used to fit/train the model. The *validation dataset* is used to evaluate a given model. By using this data to fine-tune the model hyperparameters, one could identify the best-fit model. Hence the model occasionally sees this data, but never does it "Learn" from it. The test dataset provides the gold standard which is used to evaluate the trained model. It's only used once a model is completely trained (using the train and validation sets). The test set is generally what is used to evaluate competing models. This is done to perform an independent validation of the model's performance.

In addition to this, several sampling can be used to prevent model overfitting. Cross-validation is one of the most commonly used resampling techniques wherein the entire dataset is split into two parts: train and test. Thereafter, the test set is kept aside, while a randomly chosen X% of the train dataset is used as the actual train set while the remaining (100-X)% is used as the validation set (X is a fixed number (typically 80%). The model is then iteratively trained and validated on these different sets. The use of cross-validation avoids overfitting, with K-fold cross-validation being the most popular method, where k-refers to the number of iterations.

Basic evaluation metrics

Evaluation metrics measure the performance of a particular model, which helps with optimal model selection in the testing phase. Different metrics may reflect the different goals when designing the task at hand. For instance, there are different metrics for the tasks of classification, regression, ranking, clustering, topic modeling, etc. Some metrics, such as precision-recall, are useful for multiple tasks. Classification, regression, and ranking are examples of supervised learning, which constitute a majority of machine learning applications. The choice of evaluation metrics heavily impacts the ability of the model to learn from the data. Hence, depending upon the characteristics of the dataset and the problem statement, the choice of an evaluation metric is made (Table 2.1).

- ***Model Accuracy***
  Model accuracy in terms of classification models can be defined as the ratio of correctly classified samples to the total number of samples:

$$\text{Accuracy} = \frac{\text{Number of correct predictions}}{\text{Total number of predictions}}$$

Table 2.1 Confusion matrix.

		Predicted	
Actual		Positive	Negative
	Positive	TP	FN
	Negative	FP	TN

True Positive (TP)—*A true positive is an outcome where the model correctly predicts the positive class.* ***True Negative (TN)***— *A true negative is an outcome where the model correctly predicts the negative class.* ***False Positive (FP)***— *A false positive is an outcome where the model incorrectly predicts the positive class.* ***False Negative (FN)***— *A false negative is an outcome where the model incorrectly predicts the negative class.*

Or for binary classification models, the accuracy can be defined as:

$$\text{Accuracy} = \frac{\text{TP} + \text{TN}}{\text{TP} + \text{TN} + \text{FP} + \text{FN}}$$

The use of the *accuracy* metric is justified when the number of examples for each class is somewhat balanced. However, when the dataset is imbalanced, accuracy should not be relied upon as it can be misleading. In those cases, metrics such as precision, recall, or F1 should be used.

- ***Precision and Recall***
  In a classification task, the precision for a class is the number of true positives divided by the total number of elements "labeled by the model" as belonging to the positive class:

$$\text{Precision} = \frac{\text{TP}}{(\text{TP} + \text{FP})}$$

  Recall is defined as the number of true positives divided by the total number of elements that actually belong to the positive class:

$$\text{Recall} = \frac{\text{TP}}{(\text{TP} + \text{FN})}$$

  In simple terms, *high precision* means that an algorithm returned substantially more relevant results than irrelevant ones, while *high recall* means that an algorithm returned most of the relevant results.
- ***F1 Score***
  The *F1 score* is the harmonic mean of the precision and recall, where an F1 score reaches its best value at 1 and worst at 0.

$$\text{F1} = \frac{2 \times (\text{Precision} \times \text{Recall})}{\text{Precision} + \text{Recall}}$$

  In segmentation tasks, F1 score is also referred to as the *Dice Coefficient*.
- ***ROC Curve and AUC (Area under the ROC Curve)***
  A receiver operating characteristic curve, or ROC curve, is a graphical plot that illustrates the diagnostic ability of a binary classifier system as its discrimination threshold is varied. The ROC curve is created by plotting the true positive rate (TPR) against the false positive rate (FPR) at various threshold settings. The ROC curve is a two-dimensional depiction of classifier performance. To compare classifiers, we may want to reduce ROC performance to a single scalar value representing an expected performance. A common method is to calculate the area under the ROC curve, abbreviated as AUC (Fig. 2.4). An excellent model has AUC close to 1 which means it has good measure of separability. A poor model has AUC close to 0 which means it has a poor measure of separability. In fact, an AUC of zero means that a specific model is reciprocating the result: it is predicting 0s as 1s and 1s as 0s. Finally, when AUC is equal to 0.5, it means the model has no class separation capacity at all.

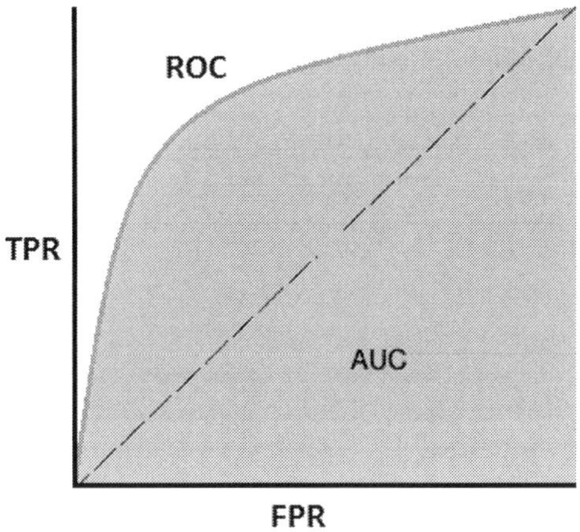

FIGURE 2.4 The relation between ROC (Receiver Operating Characteristics) and AUC (Area Under Curve).

ROC is a probability curve and AUC represents degree or measure of separability. It tells how much a model is capable of distinguishing between classes. Abbreviations: *FPR*, false positive rate; *TPR*, true positive rate.

State-of-the-art algorithms

Bayesian machine learning

Bayesian machine learning allows for the influence of prior knowledge and experience regarding the format and structure of a constructed model, independent of what the data tells us. This is especially useful when a particular dataset is too small to confidently train a particular model on probability theory. Bayes' theorem describes the probability of an event, based on prior knowledge of conditions that might be related to the event.

$$\mathrm{P(A|B)} = \frac{\mathrm{P(B|A)P(A)}}{\mathrm{P(B)}}$$

Here, P(A) and P(B) are independent probabilities of event A and B, respectively. P(A|B) means probability of observing A, given that event B has happened. In Bayesian machine learning, we use the Bayes' theorem to infer model parameters (theta) from data (D):

$$P(\theta|D) = \frac{P(D|\theta)P(\theta)}{P(D)}$$

$P(D)$ is a term that generally cannot be computed, but since it is just a normalizing constant, it doesn't matter that much. When comparing models, one is mainly interested in expressions containing theta, because $P(D)$ stays the same for each model. $P(\theta)$ is an estimation of model parameters based on expert opinion or previous experience. Most often, expert opinion is inconclusive and in the setting of ample data this expert opinion has minimal influence on model parameters. Inference should converge to probable theta as long as its not zero. $P(D|\theta)$ is called the likelihood of data given model parameters. The formula for likelihood is model-specific. Scientists often use likelihood for evaluation of models: a model that gives higher likelihood to real data is better. Finally, $P(\theta|D)$ is a probability distribution over model parameters obtained from prior beliefs and data.

Decision tree

Decision Trees are a type of Supervised Machine Learning where the data is continuously split according to a certain parameter. The tree can be explained by two entities, namely decision nodes and leaves. The leaves are the decisions or the final outcomes. And the decision nodes are where the data is split. A single decision tree is often not as performant as linear regression, logistic regression, etc. However, by introducing random forests and boosting, it can result in dramatic improvements in prediction accuracy at the expense of some loss in interpretation. Among decision support tools, decision trees have several advantages. Decision trees (Safavian & Landgrebe, 1991):

- Are simple to understand and interpret. People can intuitively understand decision tree models after a brief explanation.
- Have value even with relatively smaller datasets. Important insights can be generated based on experts describing a situation (its alternatives, probabilities, and costs) and their preference for outcomes.
- Help determine worst, best, and expected values for different scenarios.
- Use a white box model, if a given result is provided by a model.
- Can be combined with other decision techniques.

Decision tree versus random forest

If there was a way to generate a very large number of trees, averaging out their solutions, then one would likely get an answer that is going to be very close to the true answer. Enter the random forest—a collection of decision trees with a single, aggregated result. Random forests are commonly reported as the most accurate learning algorithm. Random forests reduce the variance seen in decision trees by:

- Using different samples for training,
- Specifying random feature subsets,
- Building and combining small (shallow) trees.

Random forest versus boosting

Like random forests, gradient boosting is a set of decision trees. The two main differences are:

- How trees are built: random forests build each tree independently while gradient boosting builds one tree at a time. This additive model (ensemble) works in a forward, stagewise manner, introducing a weak learner to improve the shortcomings of existing weak learners.
- Combining results: random forests combine results at the end of the process (by averaging or "majority rules"), while gradient boosting combines results along the way.

Regression

Regression is based on a hypothesis that can be linear, quadratic, polynomial, nonlinear, etc. The hypothesis is a function based on some hidden parameters and input values. In the training phase, the hidden parameters are optimized while the input values are presented in the training. The process that does the optimization is the gradient descent algorithm. If one is using neural networks, then there is an additional need for a back-propagation algorithm to compute gradient at each layer. Once the hypothesis parameters are trained (when they gave least error during the training), then the same hypothesis with the trained parameters is used with new input values to predict outcomes that will again be real values.

Linear regression versus logistic regression

- In case of Linear Regression, the outcome is continuous while in the case of Logistic Regression the outcome is discrete (not continuous).
- Linear Regression is a regression algorithm for Machine Learning while Logistic Regression is a classification algorithm for machine learning.

Support vector machine

Support Vector Machine, abbreviated as SVM, can be used for both regression and classification tasks. The objective of the support vector machine algorithm is to find a hyperplane in an N-dimensional space that distinctly classifies the data points. To separate the two classes of data points, there are many possible hyperplanes that could be chosen. The objective is to find a plane that has the maximum margin, i.e., the maximum distance between data points of both classes. Maximizing the margin distance provides some reinforcement, so that future data points can be classified with more confidence. The loss function that helps maximize the margin is the hinge loss:

$$l(y) = \max\left(0,\ 1 + \max_{y \neq t} w_y x - \ w_t x\right)$$

where t the target label, $w_t$and w_y the model parameters.

Deep learning and neural networks

Why neural networks work?

Since the inception of deep learning, few ideas have generated as much impact as convolutional neural networks (CNNs or ConvNets). Inspired by the experiments of David Hubel and Torsten Wiesel on the visual cortex (Hubel and Wiesel, 1962), these networks were introduced in the 1990s, where the first models were developed to recognize handwritten digits and showed promising performance (Bengio, LeCun, Nohl, & Burges, 1995). However, the great potential of these networks only captured the attention of the wider research community in 2012, during the "ImageNet Large Scale Visual Recognition Challenge" event. The CNN model, called "AlexNet" (Krizhevsky, Sutskever, & Hinton, 2012), achieved first place with an error rate of 16.4%, which was significantly better than the second place model with an error rate of 26.2%. In just 3 years, researchers progressed from the 8-layer "AlexNet" to the 152-layer "ResNet" (He, Zhang, Ren, & Sun, 2016), dropping the error rate of the ImageNet challenge to less than 4%. The deep CNN has become a major research tool and the backbone within the computer vision community.

The Neural Network layers are inspired by the basic structure of the brain. The computational unit of the brain is a neuron. Approximately 86 billion neurons can be found in the human nervous system, and they are connected with approximately 10^{14}–10^{15} synapses (Brenner and Sejnowski, 2011). Fig. 2.5 shows an illustration of a biological neuron (*left*) and a common mathematical model (*right*). Each neuron receives input signals from its dendrites and produces output signals along its (single) axon. The axon eventually branches out and connects via synapses to dendrites of other neurons. In the computational model of a neuron, the signals that travel along the axons (e.g., x_0) interact multiplicatively (e.g., w_0x_0) with the dendrites of the other neuron based on the synaptic strength at that synapse (e.g., w_0). The idea is that the synaptic strengths (the weights w) are learnable and control the strength of influence (and its direction: excitory (positive weight) or inhibitory

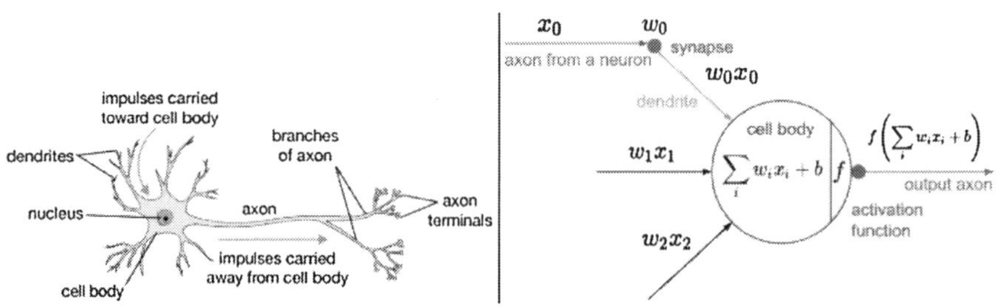

FIGURE 2.5 The basic structure of a neuron versus the structure of a neural network.

Image reproduced with permission from https://thenounproject.com/term/neuron/214105/.

(negative weight)) of one neuron on another. In the basic model, the dendrites carry the signal to the cell body where they all get summed. If the final sum is above a certain threshold, the neuron can fire, sending a spike along its axon. In computational models, we assume that the precise timing of the spikes does not matter and that only the frequency of the firing communicates information. Based on this rate code interpretation, we model the firing rate of the neuron with an activation function f, which represents the frequency of the spikes along the axon. Historically, a common choice of the activation function is the sigmoid function σ, since it takes a real-valued input (the signal strength after the sum) and squashes it to range between 0 and 1 (see Fig. 2.5).

Typical deep learning pipeline

The training of a deep learning task is similar to that of ML tasks, like classification, segmentation, or clustering. All pipeline steps are built upon a general framework, for instance, Tensorflow (Abadi et al.), Pytorch (Paszke et al., 2019), Theano (Team et al., 2016), Caffe (Jia et al., 2014). These frameworks provide a platform for running machine learning experiments: making it easy to try numerous ideas and techniques and manage various trials/experiments.

In the data preprocessing stage, the input data are collected and cleaned, while at this stage the ground truth is also defined and the quality of the data is validated. During the training stage, the prepared images and labels (if exist) are sent into a predefined convolutional neural network. Based on the output of the network, it calculates the loss function while the backpropagation will send the loss back to achieve optimal weights for each of the network layers. One *Epoch* is when an entire dataset is passed forward and backward through the neural network only once. *Iterations* are the number of batches needed to complete one epoch. During the testing stage, the platform should verify the test dataset, and decide which model is the best.

Neural network versus convolutional neural network

Convolutional Neural Networks are very similar to ordinary Neural Networks: they are made up of neurons that have learnable weights and biases. Each neuron receives some input, performs a dot product, and optionally follows it with a nonlinearity. The whole network still expresses a single differentiable score function: from the raw image pixels on one end to class scores at the other end. In addition, CNNs still have a loss function (e.g., SVM/Softmax) on the last (fully connected) layer, and all the tips/tricks that have been developed for learning regular Neural Networks still apply. Convolutional Neural Network architectures make the explicit assumption that the inputs are images, which allows us to encode certain properties into the architecture. Such a step allows efficient implementation of the forward function, as a result of vastly reduced parameters in the network.

Convolutional neural network layers

- ***Convolutional Layer***
  The convolutional (conv) layer is the core building block of a Convolutional Network (ConvNet) that does most of the computational heavy lifting. The conv layer's parameters consist of a set of learnable filters. Every filter is spatially small (in terms of width and height) but extends through the full depth of the input volume. For example, a typical filter on a first layer of a ConvNet might have a size of $5 \times 5 \times 3$ (i.e., 5-pixel width and height, and 3 because images have a depth of 3, which are the color channels). During the forward pass, one slides (more precisely, convolves) each filter across the width and height of the input volume and computes dot products between the entries of the filter and the input at any position. As one slides the filter over the width and height of the input volume, a 2-dimensional activation map is produced that gives the responses of that filter at every spatial position. Intuitively, the network will learn filters that activate when they see some type of visual feature such as an edge of some orientation or a blotch of some color on the first layer, or eventually entire honeycomb or wheellike patterns on higher layers of the network. Now, there will be an entire set of filters in each conv layer (e.g., 12 filters), and each of them will produce a separate two-dimensional activation map. Finally, these activation maps are stacked along the depth dimension thereby producing the output volume.
- ***Pooling Layer***
  It is common to periodically insert a pooling layer in-between successive conv layers in a ConvNet architecture. Its function is to progressively reduce the spatial size of the representation to reduce the number of parameters and computation in the network, and hence to also control for overfitting. The Pooling Layer operates independently on every depth slice of the input and resizes it spatially, using the MAX operation. The most common form is a pooling layer with filters of size 2×2 applied with a stride of 2 downsamples every depth slice in the input, and by 2 along both width and height, discarding 75% of the activations. Every MAX operation would, in this scenario, be taking a max over 4 numbers (little 2×2 region in some depth slice).
- ***Normalization Layer***
  Many types of normalization layers have been proposed for use in ConvNet architectures, sometimes with the intention of implementing inhibition schemes observed in the biological brain.
- ***Fully connected layer***
  Neurons in a fully connected layer have full connections to all activations in the previous layer, as seen in regular neural networks. Their activations can hence be computed with a matrix multiplication followed by a bias offset.
- ***Activation layer***
  Every activation function (or nonlinearity) takes a single number and performs a certain fixed mathematical operation on it. There are several activation functions one may encounter in practice:

Sigmoid Activation Layer: It takes a real-valued number and "squashes" it into a range between 0 and 1. In particular, large negative numbers become 0 and large positive numbers become 1. The mathematical form is:
$\sigma(x) = 1/(1 + e^{-x})$.
Tanh Activation Layer: It squashes a real-valued number to the range [-1, 1]. Like the sigmoid neuron, its activations saturate, but unlike the sigmoid neuron its output is zero-centered. Therefore, in practice the tanh nonlinearity is always preferred to the sigmoid nonlinearity. The mathematical form is:
$\tanh(x) = 2\sigma(2x) - 1$.
ReLU Activation Layer: The Rectified Linear Unit has become very popular in the last few years. It computes the function $f(x) = \max(0, x)$. In other words, the activation is simply thresholded at zero.
Leaky ReLU Activation Layer: Instead of the function being zero when x $<$ 0, a leaky ReLU will instead have a small negative slope (of 0.01, or so). That is, the function computes $f(x) = 1(x < 0)(ax) + 1(\geq 0)x \geq 0) \geq 0)(x)$, where α is a small constant.

Convolutional neural network architectures

The machine learning community has persisted in its effort at innovating a variety of neural network layers to achieve higher performance. Different settings and combination of layers significantly affect accuracy and performance. In this section, we discuss how these are commonly stacked together to form entire ConvNets.

- ***LeNet***
  The first successful applications of Convolutional Networks were developed by Yann LeCun in the 1990s. Of these, the best known is the LeNet architecture that was used to read zip codes, digits, etc.
- ***AlexNet***
  The first work that popularized Convolutional Networks in Computer Vision was the AlexNet, developed by Alex Krizhevsky, Ilya Sutskever, and Geoff Hinton. The AlexNet was submitted to the ImageNet ILSVRC challenge in 2012 and significantly outperformed the second runner-up (top five error of 16% compared to the runner-up with 26% error). AlexNet had a very similar architecture to LeNet, but was deeper, bigger, and featured Convolutional Layers stacked on top of each other (previously it was common to only have a single Conv layer always immediately followed by a POOL layer).
- ***ZF Net***
  The ILSVRC 2013 winner was a Convolutional Network from Matthew Zeiler and Rob Fergus. It became known as the ZFNet (short for Zeiler and Fergus Net). It was an improvement on AlexNet by tweaking the architecture hyperparameters, in particular by expanding the size of the middle convolutional layers and making the stride and filter size on the first layer smaller.

- ***GoogLeNet***
  The ImageNet Large Scale Visual Recognition Challenge (ILSVRC) 2014 winner was a Convolutional Network from Szegedy et al. from Google. Its main contribution was the development of an Inception Module that dramatically reduced the number of parameters in the network (4M, compared to AlexNet with 60M). Additionally, it used Average Pooling instead of Fully Connected layers at the top of the ConvNet, eliminating a large amount of parameters that do not seem to matter much. There are also several follow-up versions to the GoogLeNet, most recently Inception-v4.
- ***VGGNet***
  The runner-up in ILSVRC 2014 was the network from Karen Simonyan and Andrew Zisserman that became known as the VGGNet. Its main contribution was in showing that the depth of the network is a critical component of good performance. Their best network contained 16 CONV/FC layers and, appealingly, featured an extremely homogeneous architecture that only performed 3×3 convolutions and 2×2 pooling from the beginning to the end. Their pretrained model is available for plug and play use in Caffe. A downside of the VGGNet is that it is more expensive to evaluate and uses a lot more memory and parameters (140M). Most of these parameters are in the first fully connected layer, and it was since found that these FC layers can be removed with no performance downgrade, significantly reducing the number of necessary parameters.
- ***ResNet***
  Residual Network developed by (He et al., 2016) was the winner of ILSVRC 2015. It features special skip connections and a heavy use of batch normalization. The architecture is also missing fully connected layers at the end of the network. The reader is also referred to Kaiming's presentation (video, slides), and some recent experiments that reproduce these networks in Torch. ResNets are currently by far state-of-the-art Convolutional Neural Network models and are the default choice for using ConvNet in practice (as of May 10, 2016).

Attention model and interpretability of neural networks

One of the major hurdles of applying deep learning models to medical tasks is lack of interpretability. The machine learning model—Attention Model (AM)—first introduced for Machine Translation (Bahdanau, Cho, & Bengio, 2014), has now become a predominant concept in neural network literature. Attention has become enormously popular within the AI community as an essential component of neural architectures for a remarkably large number of applications in Natural Language Processing (NLP), statistical learning, speech and computer vision. A significant number of computer vision and medical research applications highlight the application of AM for the interpretation of the output of a neural networks model (Beecy et al., 2018; Oktay et al., 2018; Wang et al., 2019) (Fig. 2.6).

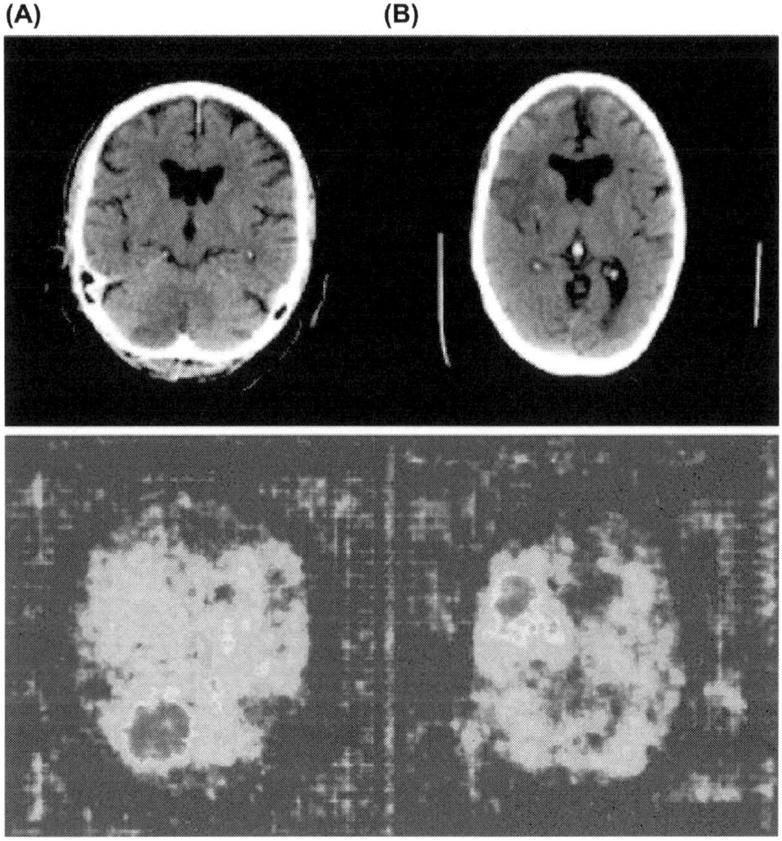

FIGURE 2.6 The stroke attention map shows the heatmap with the higher confidential score (red).

Conclusions

The application of traditional and advanced machine learning methods to solve cardiovascular tasks has witnessed significant strides over the past few years. This chapter covers the basic knowledge on the scope and approach of machine learning algorithms, specifically the utilization of convolutional neural networks to medical image tasks. Deep learning promises to revolutionize cardiovascular image analysis, and current efforts are focused at providing tools for interpretation of neural network output, as well as optimization for clinical applications.

References

Abadi, M., Barham, P., Chen, J., Chen, Z., Davis, A., Dean, J., et al. (2016). *Tensorflow: A system for large-scale machine learning. In12th symposium on operating systems design and implementation* (pp. 265–283).

Al'Aref, S. J., Anchouche, K., Singh, G., Slomka, P. J., Kolli, K. K., Kumar, A., et al. (2018). Clinical applications of machine learning in cardiovascular disease and its relevance to cardiac imaging. *European Heart Journal, 40*(24). https://doi.org/10.1093/eurheartj/ehy404.

Bahdanau, D., Cho, K., & Bengio, Y. (2014). *Neural machine translation by jointly learning to align and translate*. arXiv preprint arXiv:1812.07860.

Beecy, A. N., Chang, Q., Anchouche, K., Baskaran, L., Elmore, K., Kolli, K. K., et al. (2018). A novel deep learning approach for automated diagnosis of acute ischemic infarction on computed tomography. *Journal of the American College of Cardiology: Cardiovascular Imaging, 11*.

Bengio, Y., LeCun, Y., Nohl, C., & Burges, C. (1995). LeRec: A NN/HMM hybrid for on-line handwriting recognition. *Neural Computation*.

Brenner, S., & Sejnowski, T. J. (2011). Understanding the human brain. In *American association for the advancement of science*.

Chang, Q., Qu, H., Zhang, Y., Sabuncu, M., Chen, C., Zhang, T., & Metaxas, D. N. (2020). Synthetic learning: Learn from distributed asynchronized discriminator GAN without sharing medical image data. *The IEEE/CVF Conference on Computer Vision and Pattern Recognition (CVPR)*, 13856–13866. In press.

Goodfellow, I., Bengio, Y., & Courville, A. (2016). *Deep learning*. MIT Press.

Goodfellow, I., Pouget-Abadie, J., Mirza, M., Xu, B., Warde-Farley, D., Bengio, Y., et al. (2014). Generative Adversarial Nets. *Advances in neural information processing systems*, 2672–2680. In press.

He, K., Zhang, X., Ren, S., & Sun, J. (2016). Deep residual learning for image recognition. In *The IEEE conference on computer vision and pattern recognition*.

Hubel, D. H., & Wiesel, T. N. (1962). 'Receptive fields, binocular interaction and functional architecture in the cat's visual cortex. *The Journal of Physiology, 160*(1), 106–154.

Jia, Y., Shelhamer, E., Donahue, J., Karayev, S., Long, J., Girshick, R., et al. (2014). Caffe: Convolutional architecture for fast feature embedding. In *Proceedings of the 22nd ACM international conference on Multimedia*.

Krizhevsky, A., Sutskever, I., & Hinton, G. E. (2012). Imagenet classification with deep convolutional neural networks. *Advances in Neural Information Processing Systems*.

Mitchell, T. (1997). *Machine learning*. McGraw Hill.

Oktay, O., Schlemper, J., Folgoc, L. L., Lee, M., Heinrich, M., Misawa, K., et al. (2018). *Attention u-net: Learning where to look for the pancreas*. arXiv preprint arXiv: 1804.03999.

Paszke, A., Gross, S., Massa, F., Lerer, A., Bradbury, J., Chanan, G., et al. (2019). PyTorch: An imperative style, high-performance deep learning library. *Advances in Neural Information Processing Systems*, 8024–8035.

Safavian, S. R., & Landgrebe, D. (1991). A survey of decision tree classifier methodology. *IEEE transactions on systems, man, and cybernetics, 21*(3), 660–674.

Team, T. T., Al-Rfou, R., Alain, G., Almahairi, A., Angermueller, C., Bahdanau, D., et al. (2016). *Theano: A Python framework for fast computation of mathematical expressions*.

Wang, L., Wu, Z., Karanam, S., Peng, K.-C., Singh, R. V., Liu, B., et al. (2019). *Sharpen focus: Learning with attention separability and consistency*.

CHAPTER

Machine learning for predictive analytics

3

Sehj Kashyap, Kristin M. Corey, Aman Kansal, Mark Sendak
Duke Institute for Health Innovation, Durham, NC, United States

Chapter outline

Introduction

Predictive models have long been used in medicine for clinical tasks such as risk-stratification, patient prognostication, and treatment planning. The Framingham risk score, ASCVD risk score, and NEWS score are examples of predictive models that use logistic regression and scoring-based methods, and have already been integrated into routine clinical care (Royal College of Physicians, 2017; Prytherch et al., 2010; D'Agostino et al., 2008; Goff et al., 2014). Recently, advances in the field of machine learning (ML) have unlocked new opportunities for significantly improving traditional predictive modeling by enabling pattern learning from large and heterogeneous health data (Topol, 2019; Rajkomar, Dean, &Kohane, 2019; He et al., 2019) with minimalistic requirements for manual tuning.

Machine Learning in Cardiovascular Medicine. https://doi.org/10.1016/B978-0-12-820273-9.00003-8

What is unique about machine learning methods for predictive analytics as compared to traditional methods like statistical modeling and rules-based systems? It is important to acknowledge that there is no well-defined line separating machine learning methods from statistical and rule-based methods (Beam & Kohane, 2018). Beam et al. suggest that algorithms exist on a continuum between fully human-guided (expert rule-based systems) and fully machine-guided data analysis. An algorithm is more grounded in machine learning if its structure and parameters are less predetermined by humans and more learned directly from the data. The key promise of machine learning is to learn the parameters and structure of complex functions, accurately mapping inputs to outputs, with minimal to no human supervision (Rajkomar et al., 2019; Faes et al., 2019; Shah, Milstein, & Bagley, 2019). Whereas in traditional predictive modeling, a statistician would manually choose each variable to include in the model, including but not limited to what interactions and bias terms to include, a machine learning-based approach enables learning the structure and terms of a complex function through a minimally guided learning process. This allows learning from a broad array of data types with high dimensionality, which is increasingly the type of healthcare data that is available. The caveat is that while ML methods enable the learning of complex functions, there may still be cases where the complexity of the function confers little marginal increase in effectiveness over simpler models (Christodoulou et al., 2019). Thus, the performance of machine learning models should be contextualized with the performance of alternative statistical methods as well as clinical heuristics currently used in practice. There are increasing calls for strong and neutral comparisons against existing models (Topol, 2019; Rajkomar et al., 2019; Liu et al., 2019).

Developing a predictive model

In this section, we highlight four key steps in developing ML predictive models: (1) defining the opportunity, (2) creating the dataset, (3) developing the model, and (4) evaluating model performance. For more detailed outlines for developing machine learning models, we refer readers to the following reviews (Topol, 2019; Rajkomar et al., 2019; Chen et al., 2019) and best practices (Wiens et al., 2019). The highlighted steps do not include the processes involved to enable the integration of a model into a clinical workflow nor do they describe the process of evaluating the impact of a model on outcomes. These steps are still rare, and we consider an in-depth discussion to be out of our intended scope. Instead, readers can find guidelines that have been proposed elsewhere (Wiens et al., 2019; Berwick, 2003; Greenhalgh et al., 2017).

Defining the opportunity

The first critical step in developing a machine learning model is to clarify whether there is an opportunity for the intended clinical benefit. As pointed out by several

Table 3.1 A summary of key questions that should be asked at the outset of model development to clarify the opportunity of a model-integrated workflow.

Defining the opportunity
• What action(s) will be triggered or changed as a result of the model prediction? Can these actions be taken?
• How much lead time is needed between the prediction time and the window of time over which an outcome is predicted to occur?
• How will workflows be affected or need to change to utilize the model predictions?
• Will the benefits realized exceed the costs of taking action?

researchers (Chen et al., 2019; Emanuel & Wachter, 2019), a performant model alone is not sufficient for creating a clinical impact. Model developers and clinical experts should arrive at a working theory of how and why the model's predictions would bring about the clinical benefit.

Often, a predictive model triggers an action or a series of actions to realize these benefit(s). Identifying what action would be triggered or if those actions could be taken is the first important step. The second step is to determine the minimum lead time required between when the prediction is made and when the outcome occurs to take the identified steps. The range for clinically useful lead times can vary between close to zero, to months or years. For example, predicting an incident episode of atrial fibrillation right before onset could still be useful as an intervention can be rapidly deployed. Third, it is important to consider how workflows may be affected by clinical and operational constraints; what can be done to make taking the desired action as easy as possible in the real-world setting? Finally, will the benefits be realized in the true positive cases identified by the model exceed the costs of taking action on the false-positive cases? How does the model performance compare with and without deployment of the model? If a model predicts cardiac decompensation and triggers intensive care unit (ICU) admission, is the escalation of care for that particular patient the most beneficial use of an ICU bed? These are some of the considerations that need to be thought through before beginning model development work and several authors have proposed doing utility assessments before and during model development (Shah et al., 2019). Table 3.1 summarizes these key questions.

Creating the dataset

After the opportunity has been defined, a dataset is prepared to train a model. Health data can come in many forms or modalities, for example, clinical registry data, medical device data, imaging data, and lab results data (Weber, Mandl, & Kohane, 2014; Topol, 2019). Ultimately, all the health data are converted into mathematical representations in the form of vectors and matrices. For example, an X-ray could be represented as a fixed-length vector where each element of the vector is a value of a pixel in the image. Each pixel is referred to as a feature or a variable. Multiple X-rays would then be represented as a matrix, where one dimension is the number of samples and the other dimension is the number of pixels of each image. Health data can be described by three dimensions: the number of samples, each sample's features, and the number of measurements of each sample over time. Big data can refer to largeness in any one or multiple of these dimensions (Shah, 2014). It is important to know in which dimension the data are large because it influences what machine learning methods are used and the opportunities and limitations of the data.

Besides mathematically representing health data, it is also useful to organize the data onto a patient timeline. This has been proposed by multiple authors and most recently used in the paper by Rajkomar et al. for scalable deep learning in health (Rajkomar, 2018b). By taking a timeline view of patient health, a model developer can identify when and what data become available at a patient-level so that in prediction modeling only data up till the prediction time-point is used for model training. Different data stream update at different times and sometimes come from different sources. Electronic health records (EHRs), adopted en masse since the passing of the HITECH Act (Adler-Milstein & Jha, 2017), have become a major source for health data and thus training predictive models in healthcare. As of 2017, 96% of hospitals and 86% of physician offices in the United States had access to electronic health records (Hecht, 2019). Traditionally, EHRs have only stored those modalities of data that healthcare providers have access to in the course of a clinical encounter: labs and vitals data, patient demographic data, imaging, procedure data, diagnosis, and billing data. Curating data from EHRs to develop machine learning methods and implementing an analytical application requires significant investment and effort (Sendak, Balu, & Schulman, 2017). Electronic health records, like other data sources, can be limited and for certain analyses may need to be linked to other data sources, for example, insurance claims databases, surveillance databases, registries, patient's health-related internet queries, personal fitness device data, or even similar data collected at separate healthcare centers. Fig. 3.1 shows an example of patient data represented in a patient timeline.

Once the data have been gathered, organized, and represented mathematically, data quality checks must be done. These checks can reveal unusual entries in the data that need to be corrected or discarded before model training (Chen et al., 2019). The data may need to be transformed in other ways; for example, individual diagnosis codes of the patient may need to be grouped into umbrella terms like hypertension. Lab tests or vitals like weight entered with different units may need to be standardized. These transformations may be done to improve the interpretability of the model or its robustness (Nestor et al., 2018, 2019).

An important but subtle point is the validity of the labels. Labels are the clinical endpoints that a model is built to predict. If each patient record is converted into a mathematically represented input, the label is the output that the model must learn to map to from the input. If a model is built to detect hyperkalemia from electrocardiograph (ECG) recordings, the input is the ECG recordings and the label is whether that particular ECG reading was taken from a patient with hyperkalemia or not. In clinical medicine, nearly all endpoints to be predicted may have an element of subjective interpretation. For example, cardiac decompensation may be hard to

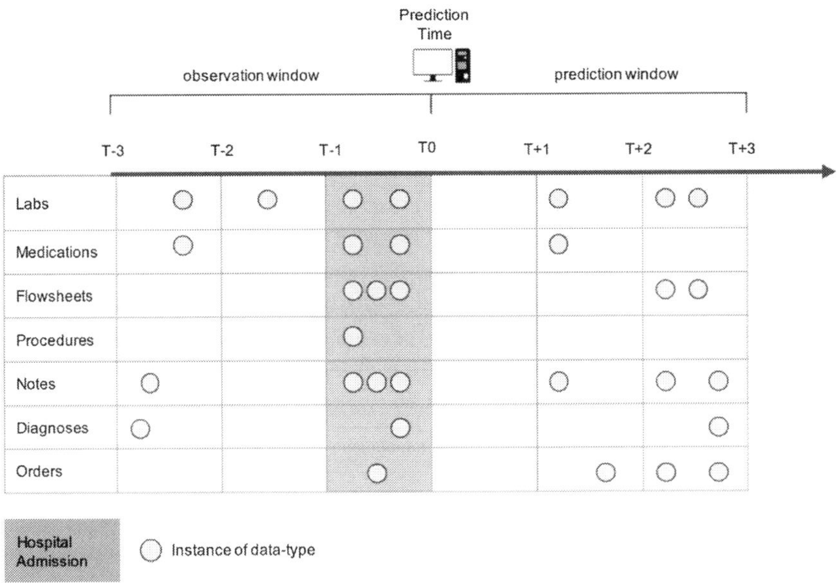

FIGURE 3.1 Patient timeline view of health data.

Patient health data can be organized into a timeline. Different health data streams may update at different times on the timeline (as indicated by the gray circles). When developing a predictive model, it is important to identify what is the observation window, the prediction time, and the prediction window. Only data within the observation window (before the prediction) should be used to train a model. Accidently including data from beyond the prediction time can cause information leakage and inappropriately overoptimistic model performance on prediction tasks.

The figure adapted from Rajkomar, A. et al. (2018b). Ensuring fairness in machine learning to advance health equity, Annals of Internal Medicine, *169(12), pp. 866–872. doi: 10.7326/M18-1990; Avati et al., (2018).* Predicting inpatient discharge prioritization with electronic health records. *Retrieved from http://arxiv.org/abs/1812.00371 (Accessed: 1 March 2020).*

define (Tehrani et al., 2018; Vallabhajosyula et al., 2019; Berg et al., 2019; Shah et al., 2016). In these cases, consensus labels may need to be created for the data (Krause et al., 2018), or alternative endpoints may also be explored Navar et al., 2019). The former method is costly and can be time-consuming, while the latter can still have limitations (Da Graca et al., 2013; Navar, 2019; Banda et al., 2017; Curtis et al., 2018).

Developing a model

Developing a model entails selecting a model type and learning the model parameters. Many model types exist and often many models are developed and evaluated before selecting a model to implement. One of the first steps is usually understanding whether the model will be predicting a real number (a regression problem) or probability of the patient data belonging to a certain class (a classification problem). Several different models are available to handle these problems. Further, the selection of a model for implementation depends on various considerations (Chen et al., 2019), including interpretability, data type, and inference time. Each of these models might not only have parameters that need to be learned by analyzing the data, they also have manually tunable parameters that are configured at the discretion of the model developer. Generally, models with a lot of trainable parameters require more data to train effectively than models with fewer parameters. Neural networks, lasso models, and ensemble models like boosted trees and random forests are models that are commonly used for high-dimensional data. Additionally, sometimes it may be desirable to maintain the sequence structure of time-series or text data, which could be done with recurrent neural networks (RNNs) while other times, this may not be necessary.

In learning the model parameters, the key goal is to learn, from the data, a mathematical function that maps inputs to outputs and can generalize to unseen inputs in the future. To enable this, the data are divided into two main parts: a development set and a validation set. The validation set is intended to act as unseen data; therefore, it should not be explored in any way until the final model has been selected and trained. The development set on the other hand is where model exploration can occur; different models can be tried, trained, and compared to select the best performing models. To ensure that the development data are not simply memorized, the development set is further subdivided into a training and tuning set. The model is trained on the training set and simultaneously, as the model parameters are being learned, the tuning set is used to check if the performance is generalizable to the patient population being studied. Model generalizability is directly linked to the clinical usability of these models. A model that does not generalize well is seldom of any use because, as powerful as these models are at recognizing patterns, the model would have only "memorized" the presented datapoints but would not have "learned" the actual problem at hand. Some of the most popular sampling techniques for model evaluation include K-Fold, Leave-one-out, and others.

Different parameters and models can be tested for performance using the development set, but the validation set can only be tested once. The way in which development and validation splits are made can also be important. Ken et al. and others have found that splits on patients and time can improve the robustness of fitted models to nonstationarity of health data (Nestor et al., 2018; Nestor et al., 2019; Jung and Shah, 2015). Fig. 3.2 summarizes the workflow of data preparation and model development.

Evaluating model performance

Once a model has been learned, its performance is evaluated. The most common metric on which a model is evaluated is its ability to discriminate between patients who have an outcome and patients who do not. Discrimination performance is

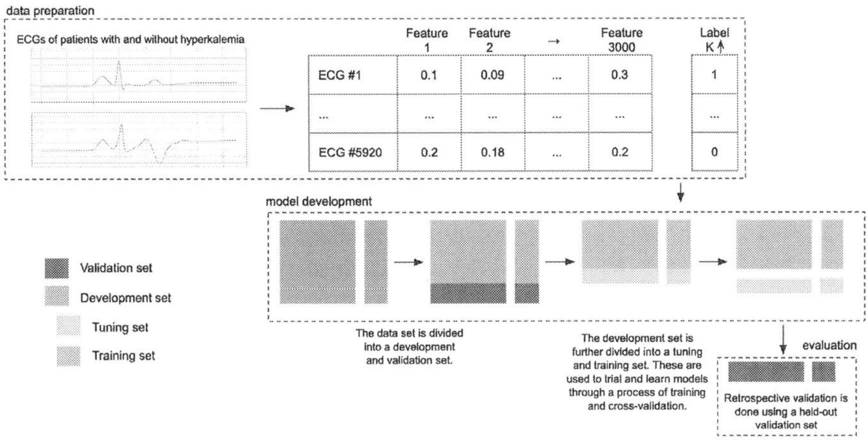

FIGURE 3.2 Data representation and model development.

This figure shows the data preparation and model development phase of predictive model development. Health data are first represented mathematically, each ECG is represented as a fixed-length vector of length 3000. As there are 5920 ECGs in this example, the input is a matrix of 5920 × 3000. Each ECG has a label indicating whether it belongs to a patient with hyperkalemia or not. This data is then split into the development and validation set. The development set is used to train a model by being further split into a training and tuning set. The validation set is used for evaluating the final model that is performant in the development set; it gives an estimate of the model's performance on unseen data. The process of validation, as well as the test set, may be repeated with different data splits to get confidence intervals for the performance metrics.

ECG example images from Galloway, C. D. et al. (2019). Development and validation of a deep-learning model to screen for hyperkalemia from the electrocardiogram, JAMA Cardiology, *4(5), pp. 428–436. doi: 10.1001/jamacardio.2019.0640.*

measured by metrics such as the area under the receiver operating curve (AUROC), which summarizes discriminative performance across thresholds, and area under the precision-recall curve, sensitivity, specificity, and F1 score. No one metric is considered sufficient for evaluating a model; for example, the AUROC has limitations both in interpretation and its equal weighting of false positives and false negatives. Furthermore, the results of a prediction model often trigger binary choices, and the selection of a threshold and performance of a model at a clinical threshold is important to examine. In addition, most of the clinical datasets have a class imbalance, that is, the number of examples of one class is far more in number than the other class or classes. In this instance, it is important to implement techniques such as undersampling or oversampling before training a model. Further, depending upon these attributes of a dataset, the choice of which performance parameter to look at while evaluating a model performance becomes very important as well. For example, a hypothetical dataset with two classes may have the number of examples for each class in the ratio of 99:1. A model trained and evaluated based on "classification accuracy" could be misleading as most of the time the goal to develop a model that is good at predicting the outcome for the minority class. For such datasets, the metrics should be carefully chosen, for example, the choice of Mathew's correlation coefficient might have been a better evaluation metric in the above-stated problem.

Another set of metrics tests the calibration of the trained model: how accurate are the estimated probability outputs by the model? If cases are prioritized by predicted probabilities, it is important to ensure that the probabilities correlate strongly to outcome incidence rates. Beyond the above metrics, other evaluations stress-test the reliability of the model: robustness checks, fragility checks, and differential performance (Saria & Subbaswamy, 2019). These checks are part of the predeployment testing that can be done to ensure that the models are safely implemented into clinical workflows.

Model evaluations are iterative and may depend on the context in which they are deployed. Furthermore, ongoing monitoring after integrating the model into clinical workflow is important for ensuring continued performance and safety. Changes in the data generating process may negatively affect the model performance and corrective steps like model retraining may need to be taken periodically (Davis et al., 2019; Davis et al., 2017). Finally, quantitative and qualitative evaluations should be conducted after a model is integrated into the clinical workflow to determine its impact on improving patient outcomes and clinical care.

Machine learning applications in cardiovascular medicine

The applications of machine learning in cardiology span across a diverse set of specialties and tasks (Fig. 3.3). Although cardiology has made significant strides in assessing individual patient risk and establishing evidence-directed guidelines, there is still tremendous opportunity for machine learning to improve both patient outcomes and quality of life. In this section, we highlight recent innovations integrating machine learning across four subspecialties within cardiology: preventive care, heart failure, interventional cardiology, and electrophysiology.

Preventive care

Cardiovascular disease (CVD) is the leading cause of morbidity and mortality. Multiple CVD risk prediction tools have been constructed (Hippisley-cox et al., 2008; Weng et al., 2017; D'Agostino et al., 2008; Goff et al., 2014; Ridker et al., 2007); yet, there are still a large number of cases that go undetected by these methods. One hypothesis is that accurately predicting the risk of CVD entails accounting for complex interactions between social determinants of health, environmental risk factors, family history, previous medical history, biomarkers, and access to healthcare among many more variables; which existing modeling approaches may

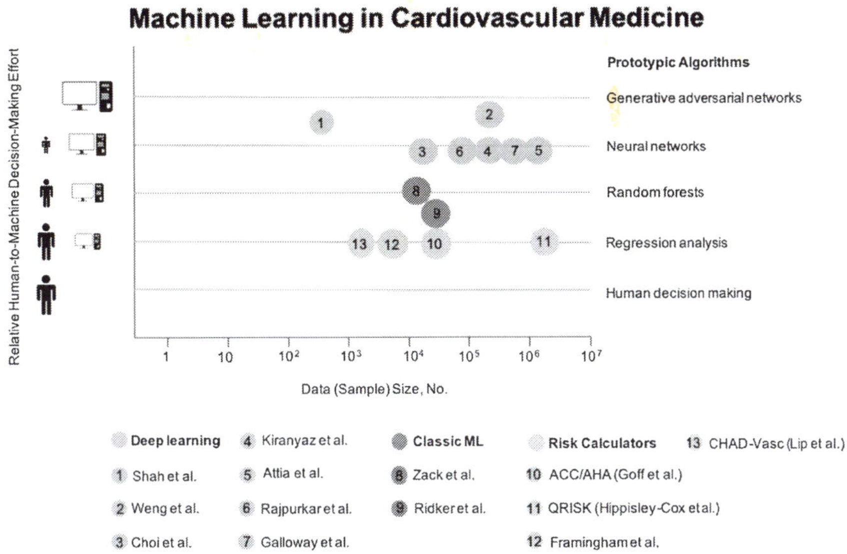

FIGURE 3.3 Examples of predictive modeling in cardiovascular medicine.

Predictive models in cardiovascular medicine have been developed using hundreds to millions of data points and techniques spanning traditional statistical modeling to deep learning approaches. Traditional risk calculators entail significant prespecification of the model structure. Machine learning approaches enable empirical learning of the structure and parameters of functions through a minimally guided learning process. Deep neural networks and other deep learning models have significantly more trainable parameters than other ML approaches like random forests and ensemble methods. Generative adversarial networks: Neural networks that learn everything from data with no human input. These networks learn complex statistical patterns in the data that can be used to generate new synthetic data points that look like real data. Convolutional neural networks: Neural networks that learn directly from raw images, which allows for the discovery of complex and nuanced relationships not known beforehand. Random forests: A collection of "if-then-else" statements learned from data; allows for nonlinear variable interactions to be automatically discovered. Regression analysis: A statistical model that assumes a specified relationship between the dependent and independent variables. Popular choices include linear and logistic regression models. Human decision making: Decisions that are made without formal analyses of data, such as wisdom, experience, intuition, and heuristics. Abbreviations: *ML*, machine learning.

This figure is adapted from and the following definitions of prototypical algorithms are laid out from Rajkomar, A., Dean, J., & Kohane, I. (2019) Machine learning in medicine, New England Journal of Medicine, *pp. 1347–1358. doi: 10.1056/NEJMra1814259.*

not model adequately. Recent studies have explored ML-based models to improve cardiovascular risk prediction.

Weng et al. used electronic health record data of 370,000 patients receiving care in the United Kingdom to build machine learning models for predicting incident

CVD (Weng et al., 2017). They compared four machine learning methods (logistic regression, random forest, gradient boosted decision trees, and neural network) to the established algorithm (ACC guideline) for predicting incident CVD over a 10-year time period. Their results showed that the machine learning algorithms performed superiorly to the established ACC/AHA risk prediction algorithm on model discrimination metrics. In some cases, features that were not in existing algorithms were weighted highly by the ML algorithm. Additionally, missingness in the EHR was informative for predicting the outcome. This suggests that the machine learning algorithms were able to detect novel patterns in the data.

Additional studies have supplemented the clinical data with biological data. Halim et al. investigated which of 53 previously reported protein biomarkers were most predictive of CVD (Halim & et al., 2015). They trained multiple L1-penalized logistic regression models, with varying numbers of clinical variables, to uncover the most important biomarkers for predicting death or myocardial infarction. By using machine learning methods, they were able to identify a smaller panel of relevant proteins that can be investigated for the development of adjunctive therapy for cardiovascular care.

Another application of machine learning in the detection of CAD has been in aiding selection of patients for noninvasive diagnostic imaging. Al'Aref et al. explored whether a machine learning model can improve the estimation of a patient's pretest probability for obstructive coronary artery disease (CAD) and therefore increase the diagnostic yield of noninvasive imaging modalities like coronary computed tomography angiography (Al'Aref et al., 2020).

Heart failure

In recent years, machine learning has been leveraged to better diagnose and prognose heart failure (HF). Choi et al. explored whether the use of deep learning to model temporal relations among events in EHRs could improve the prediction of incident diagnosis of HF (Choi et al., 2017). EHR captures detailed hospital encounter data and includes predictors such as diagnoses, medications, procedures, and provider notes that are all associated with timestamps. The investigators compared RNNs to other machine learning approaches on data from primary care patients at Sutter Palo Alto Medical Foundation. They found that an RNN outperformed alternative machine learning approaches that did not account for the temporal nature of patient historical data. Furthermore, they found that model performance was improved when using longer observation windows (time before the prediction of 18 vs. 12 months) which is in line with the hypothesis that signs of incident HF manifest over a period of time.

Beyond diagnosis, Mortazavi et al. and Frizzel et al. studied whether machine learning models could be used to accurately prognosticate HF-related sequelae in HF patients, specifically HF-related and all-cause readmissions (Frizzel et al., 2017; Mortazavi et al., 2016). Although their results differed in the observed performances of different models, both groups found it a challenge to accurately predict readmissions. Still, other efforts have been used to detect and classify subtypes of HF. This is an important task given HF with reduced ejection fraction (HFrEF) has established, effective goal-directed therapy whereas HF with preserved ejection fraction (HFpEF) does not. Borlaug et al. and Shah et al. investigated whether machine learning could help identify phenotypically distinct HFpEF categories (Borlaug & Redfield, 2011; Shah et al., 2015). They prospectively studied 397 patients with HFpEF from the outpatient clinic of the Northwestern University HFpEF program and integrated multiple sources of data across the clinical, laboratory, and imaging inputs. They employed unsupervised learning techniques, specifically hierarchical clustering and penalized clustering, to define mutually exclusive groups. They were able to classify participants into three distinct groups that differed not only in input characteristics but also in outcome risk. This study has several strengths including (1) large, well-phenotyped HFpEF cohort; (2) first to demonstrate unbiased categorization of a cardiovascular disorder; (3) successful validation of phenomapping on an independent sample of patients with HFpEF at Northwestern. Opportunities for future research include validation of this phenomapping technique with different datasets and further investigation into outcomes for these distinct classification groups. Additionally, this study provides a roadmap to phenomap other disease processes in cardiology.

Interventional cardiology

Machine learning in interventional cardiology (IC) has been limited thus far and has been largely focused on assessing the risk of complications and outcomes following procedures (Zack et al., 2019). Al'Aref et al. used a boosted ensemble of decision trees to identify novel factors that are associated with in-hospital mortality following percutaneous coronary intervention (PCI) (Al'Aref et al., 2019). Zack et al. (2019) investigated how to risk-stratify patients post PCI specifically for death or HF rehospitalization. The dataset they used was the Mayo Clinic PCI registry of 11,709 distinct patients with 14,349 PCI procedures spanning from January 2004 to December 2013. The results were presented in the form of pairwise comparison of AUROC and calculation of net classification indices for random forest versus logistic regression. This study was notable due to (1) size: large dataset with over 14,000 procedures; (2) interpretability: authors ranked predictors on importance using out-of-bag importance estimation (e.g., for 180-day mortality, age, creatinine, COPD history, and HF history were most important); (3) accounting for nonstationarity: authors divided the dataset into two sets of 5 years for both training and validation to help take into account changes in feature importance over time.

An area of IC that has attracted increasing attention is the diagnosis and management of cardiogenic shock (CS). CS is a state of cellular and tissue hypoxia due to cardiac dysfunction. The causes of CS are diverse and include myocardial infarction, postoperative low cardiac output syndrome, and decompensated chronic HF (Thiele et al., 2015). Despite numerous medical advancements, the mortality rate of patients with CS is exceedingly high at 40%–50% (Babaev et al., 2005; Goldberg et al.,

2016, 2009, 1999). These high mortality rates and the critical status of CS patients translates to a large cost burden for healthcare systems. The mean hospitalization cost for a CS admission involving advanced mechanical support devices is $146,942 (Stretch et al., 2014). These outcomes have not changed considerably in the last 30 years despite the development of various mechanical circulatory support devices, including intraaortic balloon pumps, left ventricle to aorta axial pumps, and extracorporeal membrane oxygenation devices (Baran et al., 2019). These devices can bridge patients with CS to either recovery or long-term solutions such as durable left ventricular assist devices or heart transplantation surgeries. Efforts have been made to better diagnose and manage CS including the development of a robust classification system (Baran et al., 2019) and assessment of a standardized team-based approach to improve CS outcomes (Tehrani et al., 2019).

Future research applying machine learning to CS is attractive for multiple reasons. First, mortality is high and outcomes have not appreciably changed. CS entails a state of rapid decline and anticipating CS is difficult and not well studied. Second, current clinical assessment often integrates data across multiple datasets including EHR, continuous hemodynamics, angiography, and imaging that can often be difficult even for the most seasoned clinicians. Finally, with improved diagnosis and management, it will be important to design workflows that maximize efficiency and enhance communication among interdisciplinary teams.

Electrophysiology

Machine learning has been instrumental in advancing the interpretation of ECGs. Precise classification of ECGs is crucial in detecting life-threatening arrhythmias and electrolyte imbalances. Recent work has focused on developing algorithms for feature extraction and ECG classification using wavelet transform methods (Qibin & Liqing, 2005; Mahmoodabadi et al., 2005). These algorithms have then been validated on a popular public ECG repository—the Massachusetts Institute for Technology-Beth Israel Hospital (MIT-BIH) arrhythmia database. For example, Zhao et al. first used wavelet transform and autoregressive modeling for feature extraction and then a support vector machine with Gaussian kernel to classify normal heartbeat and five different types of arrhythmias. More recent studies have incorporated the use of convolutional neural networks. For example, Kiranyaz et al. applied a 1-D CNN to the same MIT-BIH database and achieved improved performance in discerning ventricular ectopic beats and supraventricular ectopic beats (Kiranyaz, Ince, & Gabbouj, 2016).

In another example, Attia et al. sought to develop an algorithm for identifying patients with atrial fibrillation from normal sinus rhythm ECG strips. Atrial fibrillation is common, underdiagnosed, and associated with an increased risk of stroke and mortality (Attia et al., 2019). A major challenge in diagnosis is the paroxysmal nature of the rhythm, often undetected on a single ECG. There has been literature to suggest that certain features of even normal ECGs are predictive of underlying atrial fibrillation. Such an analysis used a CNN to identify the electrocardiographic signature of atrial fibrillation in a normal sinus rhythm and was able to accurately predict patients with atrial fibrillation. Another recent study by the Stanford Machine Learning Group trained a 34-layer CNN to detect 12 different arrhythmias, normal sinus, and noise (Rajpurkar et al., 2017). In contrast to the MIT-BIH database, this study collected and annotated a dataset of about 30,000 unique patients from a pool of nearly 300,000 patients. In this dataset, abnormal rhythms were sampled more frequently than normal rhythms to improve the balance between classes. They tested their model against board-certified cardiologists on a test set of 336 examples. Their model outperformed the average cardiologist on most rhythms, including atrial fibrillation and different types of atrioventricular block.

In addition to arrhythmia detection, recent work has also been focused on developing deep learning models to screen for life-threatening electrolyte abnormalities from ECG. A study by Galloway et al. used a deep CNN to detect hyperkalemia from ECGs of patients with chronic kidney disease (CKD). Hyperkalemia, especially in this patient population, can be life-threatening and screening is severely underperformed (Galloway et al., 2019). They trained their model on 450,000 patients from the Mayo Clinic in three states between 1994 and 2017 and validated on over 60,000 patients with stage 3 or greater CKD. Their model showed high discrimination with an AUC between 0.853 and 0.901, which suggests that analysis of electrophysiological signals might allow noninvasive screening hyperkalemia and other electrolyte imbalances.

Challenges and opportunities

Despite the potential impact machine learning may have on the field of cardiovascular medicine, there remain important challenges and opportunities to build and implement these types of models into everyday practice (Sendak et al., 2020). Researchers and clinicians should be aware of these challenges and opportunities as they greatly impact model development and model outputs, in turn determining the setting in which a model can be effectively and safely used. The main challenges and opportunities for machine learning in cardiovascular medicine are outlined briefly here.

Data quality

A model is only as good as the data from which it is derived. Data quality can be defined in terms of breadth, completeness, conformance, and plausibility and can drive a large portion of the quality of the model itself (Kahn et al., 2016). Model development often requires large amounts of data to train (Krittanawong et al., 2017). Fortunately, in healthcare large amounts of data are often captured digitally, making this less of a limitation. However, these large health data resources are commonly limited to EHR systems, registry data, and claims data. Obtaining these resources is difficult, often requiring internal and external partnerships as well as an immense amount of labor for data cleaning and feature engineering (Sendak et al., 2017). Patient bedside monitors are also able to capture data digitally, but additional technology infrastructure is usually required to store, aggregate, and integrate data with EHR systems. Beyond the different difficulties in obtaining these types of data, each one influences model performance in a different way. Training data that have features representing incorrect values as well as missing highly predictive features all together has the potential to hurt model performance. For example, 30-day HF readmission prediction models developed by Frizzell et al. demonstrated that despite trying multiple different types of model algorithms, model performance was limited by not including data on social determinants of health (Frizzell et al., 2017). Hospital readmissions are notoriously difficult to predict due to the complexity of social determinants of health factors not captured well by healthcare systems as well as an overall poor understanding of the exact mechanisms by which patients return to the hospital. This lack of data negatively influenced the readmission models developed despite having 5 years' worth of data on over 56,000 patients.

Apart from having the type of data needed for a prediction task, data quality also extends to the type of features engineered for a specific task. In developing HF prediction models, Choi et al. used EHR features such as orders, medications, and Common Procedural Terminology codes along with their corresponding timestamps to incorporate temporality into the development of the model (Choi et al., 2017). Using laboratory timestamps as a feature alone has been shown to improve survival prediction tasks compared to using the laboratory values (Agniel, Kohane, & Weber, 2018). Thus, understanding the context in which different types of data are captured is important for learning and predicting complex patient trajectories.

Additionally, inadequate outcome prevalence can compromise supervised model development just as much as inadequate input features. Unfortunately, it is challenging to ascertain the amount of data required to train a performant model before developing and evaluating machine learning models. Models trained on data that lack appropriate amounts of the outcome prevalence or labeled events may experience overfitting, causing underestimation of risk in low-risk populations and overestimation of risk in high-risk populations (Pavlou et al., 2015). For example, overfitting due to small amounts of data observations was identified as a limitation in Mortazavi et al. because the training dataset included 1653 patients, limiting the performance of their models to predict hospital readmissions in HF patients (Mortazavi et al., 2016). Li et al. and Rajpukar et al. described the difficulties in finding large enough ECG datasets with fine labeled arrhythmias for model training purposes (Rajpurkar et al., 2017; Li, Rajagopalan, & Clifford, 2014). Historically, most ECG datasets only labeled a handful of different types of arrhythmias. Because of this, most rhythm prediction models can only distinguish between a few rhythms. However, Rajpukar et al. were able to develop an extensive training database housing approximately 29,000 patients with 14 different types of outcome labeled arrhythmias, allowing for high-performing, enhanced model development in this area of research (Li et al., 2014). However, that is to say that there is no way of knowing exactly how much data are required to develop machine learning models for a particular task. Because of this, in-depth model performance analyses as well as data quality analyses should be performed during model development.

Generalizability

Generalizability has been a challenge in model dissemination and it is important for end-users to understand appropriate scenarios where models should and should not be used (Amarsingham et al., 2014; Sniderman, D'Agostino, & Pencina, 2015; Krumholz, 2014). Shah et al. described the development of an unsupervised phenotype classification task for patients with HF with preserved ejection fraction (Shah et al., 2015). However, one noted study limitation is the generalizability of this phenotype outside of the institution from which the training data were derived (Shah et al., 2015). This is a challenge for many institutionally developed models. Although other sources of data house regional or national patient data such as registry data and claims data, these data sources have challenges distinct from those associated with EHR systems. However, sharing EHR-based data across hospital systems is extremely difficult due to patient privacy regulations, making EHR data sources that span multiple health systems or states scarce. Notable examples of successful efforts to aggregate data across health systems are the patient-centered outcomes research network (Fleurence et al., 2014; Collins et al., 2014) and Cerner's Health Facts database and Observational Health Data Sciences and Informatics (Rhee et al., 2019a; Steyerberg et al., 2013). Few models are externally validated on patient cohorts outside of the data source institution (Steyerberg & et al., 2013). Because of this, the generalizability of many models is unknown, limiting their usefulness (Mortazavi et al., 2016; Zack et al., 2019). Galloway et al. demonstrated model validation on external cohorts consisting of patient populations from different locations. They designed a deep learning model to predict hyperkalemia from ECGs. This model was trained on a cohort of patients from Minnesota and then tested on cohorts from Florida and Arizona. These true external validation analyses test the generalizability of their model to different locations (Galloway et al., 2019). Although their model performed slightly worse in these external cohorts, the overall model performance was exceptional with AUCs ranging between 0.853 and 0.885 demonstrating that this model is generalizable to the two other clinical sites. Although challenging to perform due to the additional data requirements needed, this type of analysis is essential for testing generalizability and successful model deployment across health systems.

Apart from the challenge of demonstrating model generalizability to external sites, models are rarely validated over different time periods. The majority of model validation techniques perform validation of models using a held-out test set that is randomly selected from the source dataset. Very rarely are models tested on data outside the window time frame of the data source cohort due to difficulties in obtaining additional data. One way to estimate a model's performance over time is by holding-out a subset of data that belongs to the most recent time in the data period as a validation set. Fig. 3.4 demonstrates a type of temporal validation where

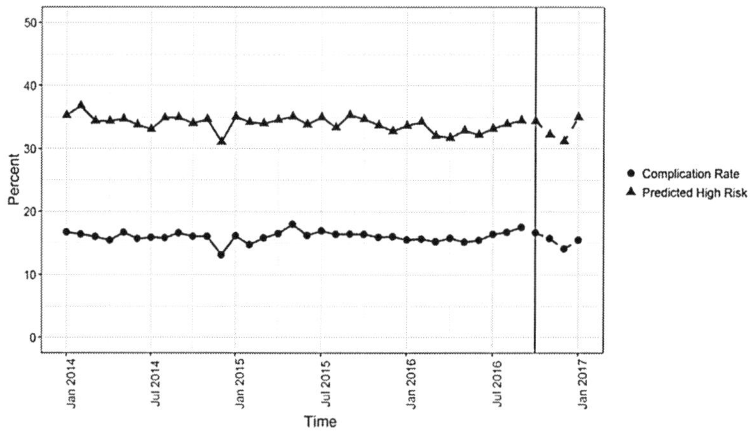

FIGURE 3.4 Assessing temporal stability of predictive model.

January 2014–September 2016 shows model predicted complication rates of training data and October 2016–January 2017 displays projected complication rates based on nonrandom, an out-of-sample test set from a different time period. This is an example of an analysis displaying temporal stability.

The figure is originally published in Corey, K. M. et al. (2018). Development and validation of machine learning models to identify high-risk surgical patients using automatically curated electronic health record data (Pythia): A retrospective, single-site study, PLoS Medicine, *15(11). doi: 10.1371/journal.pmed.1002701.*

prediction trends are displayed over time on a held-out validation set of the most recent data. These are then compared to trends of the actual outcomes within the same dataset (Corey et al., 2018). Similarly, Zack et al. developed machine learning models to predict mortality and hospitalization for congestive heart failure after PCI. As a part of the model's validation, they created two held out datasets, each one with a different consecutive 5-year window. They demonstrated that their models performed similarly on both temporal validation sets showing model stability. Although they demonstrated model stability over the course of those specific time blocks, they did not demonstrate temporal stability outside of their data source (2004–13) (Zack et al., 2019). To accomplish this task, an external validation analysis testing model performance on more recent data would be required. Generalizability of model performance to external timeframes and patient cohorts is a major limitation in a model's broad application within clinical practice. Efforts to mitigate these limitations through system partnerships for data pooling and continuous temporal model validation are important but are difficult barriers to overcome due to their labor intensity and resource requirement.

Furthermore, models have been shown to exhibit calibration drift over time. Davis et al. demonstrated that despite proper model calibration, over time calibration declined causing overpredicted risk. These drifts can be due to changing disease prevalence or phenotype as well as changes in systematic data capture (Davis et al., 2017). There have been several studies that have provided important insights into correcting model calibration drift (Davis et al., 2019; Gong et al., 2017). The need to continually assess the quality of input data as well as update model calibration is essential in maintaining "accuracy of, user confidence in, and safety of personalized risk predictions to support decision-making" (Davis et al., 2017).

Bias

Eloquently stated by Rajkomar et al. "All human activity is marred by unwanted and unconscious bias" (Rajkomar et al., 2018a). Machine learning algorithms learn from large amounts of historical data. Therefore any type of bias represented within the dataset is learned and translated into model performance (Rajkomar et al., 2018a). This has been described repeatedly among recidivism prediction models used for court sentencing in judicial systems as well as predicting investigations to Child Protective Services (Angwin, 2016; Chouldechova et al., 2018). Human bias represented in historical health data has also been studied and demonstrated that vulnerable patient populations experience higher rates of healthcare disparities, including lack of adequate care compared to others (Vayena, Blasimme, & Cohen, 2018). Algorithmic bias usually occurs in one of two ways: care biases which are then represented in the data or biases with the data collection itself. For example, a meta-analysis focusing on six different databases and cardiovascular health indicators found that there was a significant relationship between stigma and cardiovascular health indicators among socially stigmatized groups (Tomasev et al., 2019). If this data were used to train models to predict the risk of cardiovascular events among vulnerable populations, the model would propagate this data's biases.

Data collection biases come from an overrepresentation of a certain population within the data. Unbalanced training cohorts can lead to algorithmic biases when models are used on populations that are underrepresented in the training data. For example, Tomasev et al. developed a model to predict acute kidney injury using patients from the US Department of Veterans Affairs Healthcare system, where 6.38% of the population were female (Tomasev et al., 2019). Due to the imbalanced cohort, to be able to use this model on a female population, separate validation analyses would be required on female testing cohorts. To recognize and remove these types of biases from training data for model development, methods to identify data and model bias are essential before systems adoption and implementation. This type of data and model validation is still fairly new and a standardized way to mitigate model bias has yet to be identified. The Food and Drug Administration and the American Medical Association have drafted policy recommendations and quality criteria to identify and minimize bias in training data and model performance for models being systematically implemented into healthcare (FDA, 2019; AMA, 2018). Large machine learning groups such as the Fairness, Accountability, and Transparency in Machine Learning host international academic conferences to conduct academic teachings and collaborations to address bias issues in data. Academic researchers have also proposed key guidelines with regards to diminishing disparities in model performance, such as engaging with ethicists, social scientists, as well as regulation scholars during model development (Wiens et al., 2019). Rajkomar et al. described key recommendations to ensure fairness in machine learning including "incorporating marginalized populations and considering distributive justice within specific clinical and organizational contexts" during data and model development (Rajkomar et al., 2018a). To achieve fairness, they proposed three goals: equal outcomes, equal performance, and equal allocation. Equal outcomes ensure that all patients have equal benefit from model predictions, equal performance ensures that the model performs at the same accuracy for all patients, and equal allocation ensures that resources are given equally to all patients for similar model decisions (Rajkomar et al., 2018a). As new policy and standards arise with regards to bias regulation, monitoring, and safety, clinicians should be aware of algorithmic biases when developing or interpreting model predictions regarding the health of their patients and the proposed methods to mitigate those biases.

Interpretability

As machine learning in healthcare continues to grow, the role of model interpretability is evolving. Often called as the "black box," machine learning algorithms (classically deep learning) have been criticized for their lack of "explain-ability" in their decision-making process. Some see a lack of model interpretability as a safety concern (He et al., 2019; Rudin, 2019) whereas others propose that regulation of safety and efficacy can guide the appropriate use of models that lack interpretation (Verghese, Shah, & Harrington, 2018; Holm, 2019). There is also currently a lack of consensus around what exactly makes a model interpretation or explanation sufficient and best practices for interpretability research are emerging (Lipton, 2018). In 2016, the European Union passed a regulation that requires algorithms to provide explanations for automated decisions (Goodman & Flaxman, 2017). Due to the complexity of health-related data and prediction tasks, more complex modeling is used, making interpretability a difficult task (Johnson et al., 2018). Lipton et al. suggest that model interpretability is made up of two main aspects: transparency and post hoc explanations (Lipton, 2018). Transparency is broken up into three different levels, transparency at the level of the model, at the level of the individual components, and lastly at the level of the training algorithm. Post hoc interpretability refers to model result explanations. Although model explanations are not able to demonstrate exactly how a model works, they provide insight to end-users about how a model prediction or decision was reached. Machine learning methods to incorporate visual or text explanations are outlined in Lipton et al. (Lipton, 2018). Further research is needed to define and explore different types of model interpretability as well as regulatory policies to standardize interpretability for models used in healthcare settings.

Conclusions

In this chapter, we have covered methods for developing machine learning models for prediction, their applications in cardiovascular methods, and the main challenges and opportunities for their expanded use. As is evident from the examples shared, machine learning methods offer applications to an array of existing and new problems in cardiovascular medicine. These methods could significantly improve prediction and prognosis as well as offer new methods for diagnosis. However, the vast majority of models that have been described in the literature have not been implemented into clinical practice let alone become integrated into routine clinical practice. No one pathway exists for translating machine learning models into healthcare delivery (Sendak et al., 2020), but successful validation and deployment of machine learning models within the healthcare system requires immense collaboration between all stakeholders including healthcare providers and the machine learning community. As experts in healthcare, clinicians are well positioned to identify opportunities for machine learning to improve healthcare delivery and should be a part of model development as well as implementation. By working together, these stakeholders can leverage machine learning to bring efficiencies to healthcare systems across the country.

References

Adler-Milstein, J., & Jha, A. K. (2017). Hitech act drove large gains in hospital electronic health record adoption. *Health Affairs, 36*(8), 1416–1422. https://doi.org/10.1377/hlthaff.2016.1651

Agniel, D., Kohane, I. S., & Weber, G. M. (2018). Biases in electronic health record data due to processes within the healthcare system: Retrospective observational study. *BMJ (Online)*, 361. https://doi.org/10.1136/bmj.k1479

Al'Aref, S. J., et al. (2019). Determinants of in-hospital mortality after percutaneous coronary intervention: A machine learning approach. *Journal of the American Heart Association, 8*(5). https://doi.org/10.1161/JAHA.118.011160

Al'Aref, S. J., et al. (2020). Machine learning of clinical variables and coronary artery calcium scoring for the prediction of obstructive coronary artery disease on coronary computed tomography angiography: analysis from the CONFIRM registry. *European Heart Journal, 41*(3), 359–367. https://doi.org/10.1093/eurheartj/ehz565

AMA. (2018). *AMA passes first policy recommendations on augmented intelligence*. American Medical Association. Retrieved from https://www.ama-assn.org/press-center/press-releases/ama-passes-first-policy-recommendations-augmented-intelligence%0Ahttps://www.ama-assn.org/press-center/press-release/ama-passes-first-policy-recommendations-augmented-intelligence. (Accessed 1 March 2020).

Amarasingham, R., et al. (2014). Implementing electronic health care predictive analytics: Considerations and challenges. *Health Affairs, 33*(7), 1148–1154. https://doi.org/10.1377/hlthaff.2014.0352

Angwin, J., et al. (2016). Machine Bias. ProPublica. https://www.propublica.org/article/machine-bias-risk-assessments-in-criminal-sentencing. Accessed September 2020.

Attia, Z. I., et al. (2019). 'Screening for cardiac contractile dysfunction using an artificial intelligence–enabled electrocardiogram'. *Nature Medicine*, 70–74. https://doi.org/10.1038/s41591-018-0240-2

Avati, A., et al. (2018). *Predicting inpatient discharge prioritization with electronic health records*. Retrieved from http://arxiv.org/abs/1812.00371. (Accessed 1 March 2020).

Babaev, A., et al. (2005). Trends in management and outcomes of patients with acute myocardial infarction complicated by cardiogenic shock. *Journal of the American Medical Association, 294*(4), 448–454. https://doi.org/10.1001/jama.294.4.448

Banda, J. M., et al. (2017). Electronic phenotyping with APHRODITE and the Observational Health Sciences and Informatics (OHDSI) data network. *AMIA Joint Summits on Translational Science proceedings, 2017*, 48–57. Retrieved from http://www.ncbi.nlm.nih.gov/pubmed/28815104%0Ahttp://www.pubmedcentral.nih.gov/articlerender.fcgi?artid=PMC5543379.

Baran, D. A., et al. (2019). SCAI clinical expert consensus statement on the classification of cardiogenic shock: This document was endorsed by the American College of Cardiology (ACC), the American Heart Association (AHA), the Society of Critical Care Medicine (SCCM), and the Society of Thoracic Surgeons (STS) in April 2019. *Catheterization and Cardiovascular Interventions, 94*(1), 29–37. https://doi.org/10.1002/ccd.28329

Beam, A. L., & Kohane, I. S. (2018). Big data and machine learning in health care. *Journal of the American Medical Association*, 1317–1318. https://doi.org/10.1001/jama.2017.18391

Berg, D. D., et al. (2019). 'Epidemiology of shock in contemporary cardiac intensive care units. *Circulation: Cardiovascular quality and outcomes, 12*(3), e005618. https://doi.org/10.1161/CIRCOUTCOMES.119.005618

Berwick, D. M. (2003). Disseminating innovations in health care. *Journal of the American Medical Association*, 1969–1975. https://doi.org/10.1001/jama.289.15.1969

Borlaug, B. A., & Redfield, M. M. (2011). Diastolic and systolic heart failure are distinct phenotypes within the heart failure spectrum. *Circulation*. https://doi.org/10.1161/CIRCULATIONAHA.110.954388

Chen, P. H. C., Liu, Y., & Peng, L. (2019). How to develop machine learning models for healthcare. *Nature Materials*, 410–414. https://doi.org/10.1038/s41563-019-0345-0

Chouldechova, A., et al. (2018). A case study of algorithm-assisted decision making in child maltreatment hotline screening decisions. *Proceedings of Machine Learning Research, 81*, 1–15.

Christodoulou, E., et al. (2019). A systematic review shows no performance benefit of machine learning over logistic regression for clinical prediction models. *Journal of Clinical Epidemiology*, 12–22. https://doi.org/10.1016/j.jclinepi.2019.02.004

Choi, E., et al. (2017). Using recurrent neural network models for early detection of heart failure onset. *Journal of the American Medical Informatics Association, 24*(2), 361–370. https://doi.org/10.1093/jamia/ocw112

Collins, F. S., et al. (2014). PCORnet: Turning a dream into reality. *Journal of the American Medical Informatics Association, 21*(4), 576–577. https://doi.org/10.1136/amiajnl-2014-002864

Corey, K. M., et al. (2018). Development and validation of machine learning models to identify high-risk surgical patients using automatically curated electronic health record data (Pythia): A retrospective, single-site study. *PLoS Medicine, 15*(11). https://doi.org/10.1371/journal.pmed.1002701

Curtis, M. D., et al. (2018). Development and validation of a high-quality composite real-world mortality endpoint. *Health Services Research, 53*(6), 4460–4476. https://doi.org/10.1111/1475-6773.12872

D'Agostino, R. B., et al. (2008). General cardiovascular risk profile for use in primary care: The Framingham heart study. *Circulation, 117*(6), 743–753. https://doi.org/10.1161/CIRCULATIONAHA.107.699579

Da Graca, B., Filardo, G., & Nicewander, D. (2013). Consequences for healthcare quality and research of the exclusion of records from the death master file. *Circulation: Cardiovascular Quality and Outcomes*, 124–128. https://doi.org/10.1161/CIRCOUTCOMES.112.968826

Davis, S. E., et al. (2017). Calibration drift in regression and machine learning models for acute kidney injury. *Journal of the American Medical Informatics Association, 24*(6), 1052–1061. https://doi.org/10.1093/jamia/ocx030

Davis, S. E., et al. (2019). A nonparametric updating method to correct clinical prediction model drift. *Journal of the American Medical Informatics Association, 26*(12), 1448–1457. https://doi.org/10.1093/jamia/ocz127

Emanuel, E. J., & Wachter, R. M. (2019). Artificial intelligence in health care: Will the value match the hype? *Journal of the American Medical Association*, 2281–2282. https://doi.org/10.1001/jama.2019.4914

Faes, L., et al. (2019). Feasibility of automated deep learning design for medical image classification by healthcare professionals with limited coding experience. *The Lancet Digital Health*. https://doi.org/10.1016/S2589-7500(19)30108-6

FDA. (2019). *Developing a software precertification program: A working model*. Retrieved from https://www.fda.gov/downloads/MedicalDevices/DigitalHealth/DigitalHealthPreCertProgram/UCM629276.pdf. (Accessed 1 March 2020).

Fleurence, R. L., et al. (2014). Launching PCORnet, a national patient-centered clinical research network. *Journal of the American Medical Informatics Association, 21*(4), 578–582. https://doi.org/10.1136/amiajnl-2014-002747

Frizzell, J. D., et al. (2017). Prediction of 30-day all-cause readmissions in patients hospitalized for heart failure: Comparison of machine learning and other statistical approaches. *JAMA Cardiology, 2*(2), 204–209. https://doi.org/10.1001/jamacardio.2016.3956

Galloway, C. D., et al. (2019). Development and validation of a deep-learning model to screen for hyperkalemia from the electrocardiogram. *JAMA Cardiology, 4*(5), 428–436. https://doi.org/10.1001/jamacardio.2019.0640

Goff, D. C., et al. (2014). 2013 ACC/AHA guideline on the assessment of cardiovascular risk: A report of the American College of Cardiology/American Heart Association Task Force on Practice Guidelines'. *Circulation*. https://doi.org/10.1161/01.cir.0000437741.48606.98

Goldberg, R. J., et al. (1999). Temporal trends in cardiogenic shock complicating acute myocardial infarction. *New England Journal of Medicine, 340*(15), 1162–1168. https://doi.org/10.1056/NEJM199904153401504

Goldberg, R. J., et al. (2009). Thirty-year trends (1975 to 2005) in the magnitude of, management of, and hospital death rates associated with cardiogenic shock in patients with acute myocardial infarction a population-based perspective. *Circulation, 119*(9), 1211–1219. https://doi.org/10.1161/CIRCULATIONAHA.108.814947

Goldberg, R. J., et al. (2016). Decade-Long Trends (2001–2011) in the Incidence and hospital death rates associated with the in-hospital development of cardiogenic shock after acute myocardial infarction. *Circulation: Cardiovascular Quality and Outcomes, 9*(2), 117–125. https://doi.org/10.1161/CIRCOUTCOMES.115.002359

Gong, J. J., et al. (2017). Predicting clinical outcomes across changing electronic health record systems. In *Proceedings of the ACM SIGKDD international conference on knowledge discovery and data mining* (pp. 1497–1506). https://doi.org/10.1145/3097983.3098064

Goodman, B., & Flaxman, S. (2017). European union regulations on algorithmic decision making and a "right to explanation". *AI Magazine, 38*(3), 50–57. https://doi.org/10.1609/aimag.v38i3.2741

Greenhalgh, T., et al. (2017). Beyond adoption: A new framework for theorizing and evaluating nonadoption, abandonment, and challenges to the scale-up, spread, and sustainability of health and care technologies. *Journal of Medical Internet Research, 19*(11). https://doi.org/10.2196/jmir.8775

Halim, S. A., et al. (2015). Simultaneous consideration of multiple candidate protein biomarkers for long-term risk for cardiovascular events. *Circulation: Cardiovascular Genetics, 8*(1), 168–177. https://doi.org/10.1161/CIRCGENETICS.113.000490

He, J., et al. (2019). The practical implementation of artificial intelligence technologies in medicine. *Nature Medicine*, 30–36. https://doi.org/10.1038/s41591-018-0307-0

Hecht, J. (2019). The future of electronic health records. *Nature*, S114–S116. https://doi.org/10.1038/d41586-019-02876-y

Hippisley-Cox, J., et al. (2008). Predicting cardiovascular risk in England and Wales: Prospective derivation and validation of QRISK2. *BMJ, 336*(7659), 1475–1482. https://doi.org/10.1136/bmj.39609.449676.25

Holm, E. A. (2019). In defense of the black box. *Science*, 26–27. https://doi.org/10.1126/science.aax0162

Johnson, K. W., et al. (2018). Artificial intelligence in cardiology. *Journal of the American College of Cardiology*, 2668–2679. https://doi.org/10.1016/j.jacc.2018.03.521

Jung, K., & Shah, N. H. (2015). Implications of non-stationarity on predictive modeling using EHRs. *Journal of Biomedical Informatics, 58*, 168–174. https://doi.org/10.1016/j.jbi.2015.10.006

Kahn, M. G., et al. (2016). A harmonized data quality assessment terminology and framework for the secondary use of electronic health record data. *eGEMs (Generating Evidence & Methods to improve patient outcomes), 4*(1), 18. https://doi.org/10.13063/2327-9214.1244

Kiranyaz, S., Ince, T., & Gabbouj, M. (2016). Real-time patient-specific ECG classification by 1-d convolutional neural networks. *IEEE Transactions on Biomedical Engineering, 63*(3), 664–675. https://doi.org/10.1109/TBME.2015.2468589

Krause, J., et al. (2018). Grader variability and the importance of reference standards for evaluating machine learning models for diabetic retinopathy. *Ophthalmology, 125*(8), 1264–1272. https://doi.org/10.1016/j.ophtha.2018.01.034

Krittanawong, C., et al. (2017). Artificial intelligence in precision cardiovascular medicine. *Journal of the American College of Cardiology*, 2657–2664. https://doi.org/10.1016/j.jacc.2017.03.571

Krumholz, H. M. (2014). Big data and new knowledge in medicine: The thinking , training , and tools needed for a learning health system. *Health Affairs, 33*(7), 1163–1170. https://doi.org/10.1377/hlthaff.2014.0053

Li, Q., Rajagopalan, C., & Clifford, G. D. (2014). A machine learning approach to multi-level ECG signal quality classification. *Computer Methods and Programs in Biomedicine, 117*(3), 435–447. https://doi.org/10.1016/j.cmpb.2014.09.002

Lipton, Z. C. (2018). The mythos of model interpretability. *Communications of the ACM, 61*(10), 35–43. https://doi.org/10.1145/3233231

Liu, X., et al. (2019). Reporting guidelines for clinical trials evaluating artificial intelligence interventions are needed. *Nature Medicine*, 1467–1468. https://doi.org/10.1038/s41591-019-0603-3

Mahmoodabadi, S. Z., Ahmadian, A., & Abolhasani, M. D. (2005). ECG feature extraction using daubechies wavelets. In *Proceedings of the 5th IASTED international conference on visualization, imaging, and image processing, VIIP 2005* (pp. 343–348).

Mortazavi, B. J., et al. (2016). Analysis of machine learning techniques for heart failure readmissions. *Circulation: Cardiovascular Quality and Outcomes, 9*(6), 629–640. https://doi.org/10.1161/CIRCOUTCOMES.116.003039

Navar, A. M. (2019). Electronic health record data quality issues are not remedied by increasing granularity of diagnosis codes. *JAMA Cardiology*, 465. https://doi.org/10.1001/jamacardio.2019.0830

Navar, A. M., et al. (2019). Evaluation of mortality data from the social security administration death master file for clinical research. *JAMA Cardiology, 4*(4), 375–379. https://doi.org/10.1001/jamacardio.2019.0198

Nestor, B., et al. (2018). *Rethinking clinical prediction: Why machine learning must consider year of care and feature aggregation*. Retrieved from http://arxiv.org/abs/1811.12583. (Accessed 1 March 2020).

Nestor, B., et al. (2019). *Feature robustness in non-stationary health records: Caveats to deployable model performance in common clinical machine learning tasks*. Retrieved from http://arxiv.org/abs/1908.00690. (Accessed 1 March 2020).

Pavlou, M., et al. (2015). How to develop a more accurate risk prediction model when there are few events. *BMJ (Online)*, 351. https://doi.org/10.1136/bmj.h3868

Prytherch, D. R., et al. (2010). ViEWS-Towards a national early warning score for detecting adult inpatient deterioration. *Resuscitation, 81*(8), 932–937. https://doi.org/10.1016/j.resuscitation.2010.04.014

Qibin, Z., & Liqing, Z. (2005). ECG feature extraction and classification using wavelet transform and support vector machines. In *Proceedings of 2005 international conference on neural networks and brain proceedings, ICNNB'05* (pp. 1089–1092). https://doi.org/10.1109/icnnb.2005.1614807

Rajkomar, A., et al. (2018a). Scalable and accurate deep learning with electronic health records. *npj Digital Medicine, 1*(1). https://doi.org/10.1038/s41746-018-0029-1

Rajkomar, A., et al. (2018b). Ensuring fairness in machine learning to advance health equity. *Annals of Internal Medicine, 169*(12), 866–872. https://doi.org/10.7326/M18-1990

Rajkomar, A., Dean, J., & Kohane, I. (2019). Machine learning in medicine. *New England Journal of Medicine*, 1347–1358. https://doi.org/10.1056/NEJMra1814259

Rajpurkar, P., et al. (2017). *Cardiologist-level arrhythmia detection with convolutional neural networks*. Retrieved from http://arxiv.org/abs/1707.01836. (Accessed 1 March 2020).

Rhee, C., et al. (2019a). Epidemiology of hospital-onset versus community-onset sepsis in U.S. hospitals and association with mortality: A retrospective analysis using electronic clinical data. *Critical care medicine, 47*(9), 1169–1176. https://doi.org/10.1097/CCM.0000000000003817

Ridker, P. M., et al. (2007). Development and validation of improved algorithms for the assessment of global cardiovascular risk in women: The Reynolds Risk Score. *Journal of the American Medical Association, 297*(6), 611–619.

Royal College of Physicians. (2017). *National Early Warning Score (NEWS) 2 | RCP London, News 2*. Retrieved from https://www.rcplondon.ac.uk/projects/outputs/national-early-warning-score-news-2. (Accessed 1 March 2020).

Rudin, C. (2019). Stop explaining black box machine learning models for high stakes decisions and use interpretable models instead. *Nature Machine Intelligence, 1*(5), 206–215. https://doi.org/10.1038/s42256-019-0048-x

Saria, S., & Subbaswamy, A. (2019). *Tutorial: Safe and reliable machine learning*. Retrieved from http://arxiv.org/abs/1904.07204. (Accessed 1 March 2020).

Sendak, M. P., Balu, S., & Schulman, K. A. (2017). Barriers to achieving economies of scale in analysis of EHR data: A cautionary tale. *Applied Clinical Informatics, 8*(3), 826−831. https://doi.org/10.4338/ACI-2017-03-CR-0046

Sendak, M., et al. (2020). A path for translation of machine learning products into healthcare delivery. *EMJ Innovations*. https://doi.org/10.33590/emjinnov/19-00172

Shah, N. H. (2014). Realizing the promise of knowledge-driven healthcare. In P. R. O. Payne, & P. J. Embi (Eds.), *Translational informatics: Realizing the promise of knowledge-driven healthcare. Illustrated* (pp. 119−128). Springer.

Shah, S. J., et al. (2015). Phenomapping for novel classification of heart failure with preserved ejection fraction. *Circulation, 131*(3), 269−279. https://doi.org/10.1161/CIRCULATIONAHA.114.010637

Shah, R. U., et al. (2016). Post-hospital outcomes of patients with acute myocardial infarction with cardiogenic shock findings from the NCDR. *Journal of the American College of Cardiology, 67*(7), 739−747. https://doi.org/10.1016/j.jacc.2015.11.048

Shah, N. H., Milstein, A., & Bagley, S. C. (2019). Making machine learning models clinically useful. *Journal of the American Medical Association*, 1351−1352. https://doi.org/10.1001/jama.2019.10306

Sniderman, A. D., D'Agostino, R. B., & Pencina, M. J. (2015). The role of physicians in the era of predictive analytics. *Journal of the American Medical Association, 314*(1), 25−26. https://doi.org/10.1001/jama.2015.6177

Steyerberg, E. W., et al. (2013). Prognosis research strategy (PROGRESS) 3: Prognostic model research. *PLoS Medicine, 10*(2). https://doi.org/10.1371/journal.pmed.1001381

Stretch, R., et al. (2014). National trends in the utilization of short-term mechanical circulatory support: Incidence, outcomes, and cost analysis. *Journal of the American College of Cardiology, 64*(14), 1407−1415. https://doi.org/10.1016/j.jacc.2014.07.958

Tehrani, B., et al. (2018). Implementation of a cardiogenic shock team and clinical outcomes (INOVA-SHOCK Registry): Observational and Retrospective Study. *JMIR Research Protocols, 7*(6), e160. https://doi.org/10.2196/resprot.9761

Tehrani, B. N., et al. (2019). Standardized team-based care for cardiogenic shock. *Journal of the American College of Cardiology, 73*(13), 1659−1669. https://doi.org/10.1016/j.jacc.2018.12.084

Thiele, H., et al. (2015). Management of cardiogenic shock. *European Heart Journal*, 1223−1230. https://doi.org/10.1093/eurheartj/ehv051

Tomašev, N., et al. (2019). A clinically applicable approach to continuous prediction of future acute kidney injury. *Nature, 572*(7767), 116−119. https://doi.org/10.1038/s41586-019-1390-1

Topol, E. J. (2019). High-performance medicine: the convergence of human and artificial intelligence. *Nature Medicine*, 44−56. https://doi.org/10.1038/s41591-018-0300-7

Vallabhajosyula, S., et al. (2019). Cardiogenic shock in takotsubo cardiomyopathy versus acute myocardial infarction: an 8-year national perspective on clinical characteristics, management, and outcomes. *JACC: Heart Failure, 7*(6), 469−476. https://doi.org/10.1016/j.jchf.2018.12.007

Vayena, E., Blasimme, A., & Cohen, I. G. (2018). Machine learning in medicine: Addressing ethical challenges. *PLoS Medicine, 15*(11). https://doi.org/10.1371/journal.pmed.1002689

Verghese, A., Shah, N. H., & Harrington, R. A. (2018). What this computer needs is a physician humanism and artificial intelligence. *Journal of the American Medical Association*, 19−20. https://doi.org/10.1001/jama.2017.19198

Weber, G. M., Mandl, K. D., & Kohane, I. S. (2014). Finding the missing link for big biomedical data. *Journal of the American Medical Association*, 2479–2480. https://doi.org/10.1001/jama.2014.4228

Weng, S. F., et al. (2017). Can Machine-learning improve cardiovascular risk prediction using routine clinical data? *PLoS One, 12*(4). https://doi.org/10.1371/journal.pone.0174944

Wiens, J., et al. (2019). Do no harm: a roadmap for responsible machine learning for health care. *Nature Medicine*. https://doi.org/10.1038/s41591-019-0548-6

Zack, C. J., et al. (2019). Leveraging machine learning techniques to forecast patient prognosis after percutaneous coronary intervention. *JACC: Cardiovascular Interventions, 12*(14), 1304–1311. https://doi.org/10.1016/j.jcin.2019.02.035

CHAPTER

Deep learning for biomedical applications

4

Jessica De Freitas[1,2], Benjamin S. Glicksberg[1,2], Kipp W. Johnson[2], Riccardo Miotto[1,2]

[1]*Hasso Plattner Institute for Digital Health at Mount Sinai, Icahn School of Medicine at Mount Sinai, New York, NY, United States;* [2]*Department of Genetics and Genomic Sciences, Icahn School of Medicine at Mount Sinai, New York, NY, United States*

Chapter outline

Introduction

Healthcare is approaching a new era where clinical decisions and patient behaviors will be influenced by the analysis of large quantities of biomedical data. On one side, physicians are capturing the status of their treated patients, their thoughts, orders, and outcomes in electronic health records (EHRs), -omics biobanks, and images. On the other side, patients are becoming more aware of the importance of prevention and keeping continuous track of their health status using wearable devices and preventative visits. Advances in machine learning (ML) are then taking advantage of this data in an attempt to supplement clinical medicine.

However, there are many challenges in making full use of biomedical data, owing to their high-dimensionality, heterogeneity, temporal dependency, sparsity, and irregularity (Ching et al., 2018; Miotto et al., 2017). Sometimes, the same clinical phenotype is expressed in different ways across the data. As an example, a patient with "type 2 diabetes mellitus" can be identified in the EHRs by laboratory values of hemoglobin A1c > 6.5, presence of E11.9 ICD-10 code, "type 2 diabetes mellitus" mentioned in the free-text clinical notes, and so on. Consequently, it is

Machine Learning in Cardiovascular Medicine. https://doi.org/10.1016/B978-0-12-820273-9.00004-X

nontrivial to harmonize all these medical concepts to build a higher-level semantic structure and understand their correlations to inform biomedical applications.

A common approach in biomedical research is to have a domain expert specifying the phenotypes to use in an ad hoc manner. However, the supervised definition of the feature space scales poorly and misses the opportunity to discover novel patterns. Alternatively, representation learning methods allow automatic discovery of representations needed for prediction from the raw data (Bengio, Courville, & Vincent, 2013). Deep learning (DL) methods are representation-learning algorithms with multiple levels of abstractions, obtained by composing simple but nonlinear modules that each transform the representation at one level (starting with the raw input) into a representation at a higher, slightly more abstract level (LeCun, Bengio, & Hinton, 2015).

DL, a subfield of ML, has seen a dramatic resurgence in the past 6 years, largely driven by increases in computational power and the availability of massive new datasets. The field has witnessed striking advances in the ability of machines to manipulate data, including images (Russakovsky et al., 2015), language (Hirschberg and Manning, 2015), and speech (Hinton et al., 2012). Healthcare and medicine are starting to benefit from DL as well, mostly because of the large volume of data being generated (150 exabytes or 10^{18} bytes in the United States alone, growing 48% annually (Esteva et al., 2019)) as well as the increasing proliferation of medical devices and digital record systems.

This chapter discusses recent and forthcoming applications of DL in medicine highlighting the key aspects to significantly impact healthcare. We do not aim to provide a comprehensive background on technical details or general application of deep learning (see, e.g., Goodfellow, Bengio, & Courville, 2016). Instead, we focus on biomedical applications that originated from clinical imaging, EHRs, genomes, and digital health monitor devices. In the following, we briefly introduce DL, we review some of its applications with biomedical data and we discuss the opportunities, challenges, and applications related to these methods when used in the context of personalized medicine.

A brief overview on deep learning

ML is a general-purpose paradigm of artificial intelligence that can learn relationships from the data without the need to define them a priori (Bishop, 2006). ML is distinct from other types of computer programming in that it transforms the inputs of an algorithm into outputs using statistical, data-driven rules that are automatically derived from a large set of examples, rather than being explicitly specified by humans (Esteva et al., 2019). Historically, the typical ML workflow involves four steps: data harmonization, representation learning, model fitting, and evaluation (Jordan and Mitchell, 2015). For decades, constructing an ML system required careful engineering and domain expertise to transform the raw data into a suitable internal representation from which the learning subsystem, often a classifier, could detect patterns in the dataset.

In contrast, DL is a form of representation learning, that is, a machine is fed with raw data and develops its own representations needed for pattern recognition, which is composed of multiple layers of representations, usually derived with neural networks. These layers are typically arranged sequentially and composed of a large number of basic, nonlinear operations, such that the representation of one layer (beginning with the raw data input) is fed into the next layer and transformed into a more abstract representation (LeCun et al., 2015). As data flows through the layers of the system, the input space becomes iteratively warped until data points become distinguishable. In this manner, highly complex functions can be learned.

Different important advances established DL as the state-of-the-art solution for several tasks. These models can scale to large datasets, mostly because of their ability to run on specialized computing hardware, and continue to improve with more data, enabling them to outperform many classical ML approaches. Easy-to-use software packages have brought these techniques to a broad community of computational scientists. Additionally, recent improvements for regularizations, strategies that aim to reduce generalization error, have made networks more robust, even when the number of parameters is very large. The larger datasets now available are also sufficient for fitting the many parameters characterizing deep neural networks. The convergence of all these factors currently makes DL extremely adaptable and capable of addressing the differences of each domain to which it is applied. Moreover, DL systems can accept multiple data types as input, which is of particular relevance for heterogeneous healthcare data.

The most common models are trained using supervised learning, in which datasets are composed of input data points (e.g., a brain MRI scan, lab results) and corresponding output data labels (e.g., disease diagnosis, "abnormal" vs. "normal"). Unsupervised (or self-supervised) learning is usually used to draw inferences from datasets consisting of input data without labeled responses and to derive robust representations of the input space. Reinforcement learning, on the other hand, relies on computational agents that learn by trial and error or by expert demonstration and has recently been applied in medicine to learn wound suture for robotic-assisted surgery (Kassahun et al., 2016). Different types of neural networks can be used depending on the task (see Fig. 4.1). If the data have a natural and invariant adjacency structure, a convolutional neural network (CNN) can take advantage of that structure

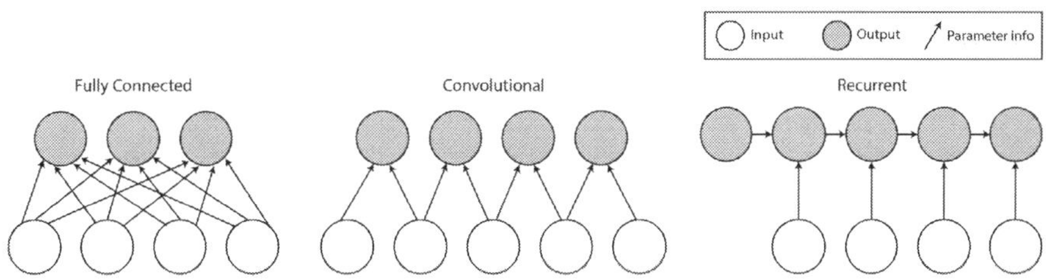

FIGURE 4.1

Schema of the most common neural networks shaping deep learning architectures.

by emphasizing local relationships, especially when convolutional layers are used in early layers of the neural network (LeCun et al., 1999). When the data have a strong time component, such as time series of lab values, recurrent neural networks (RNN), such as long short-term memory (LSTM) nets, can model the time sequences of the events (Hochreiter and Schmidhuber, 1997). Other neural network architectures such as autoencoders require no labels and are regularly used for unsupervised tasks (Vincent et al., 2010). Generative adversarial networks (GANs) are composed of two nets: a generator that generates new examples and a discriminator that classifies examples as either real (from the domain) or fake (generated) (Goodfellow et al., 2014). The two models are trained together in a zero-sum game (i.e., adversarial), until the discriminator model is fooled about half the time, meaning the generator model is generating new data with the same statistics as the training set. For example, a GAN trained on MRI scans can generate new MRI images that look at least superficially authentic to human observers, having many realistic characteristics. Although originally proposed for unsupervised learning, GANs have proven useful also for semisupervised (Salimans et al., 2016), supervised (Isola et al., 2017), and reinforcement learning (Ho and Ermon, 2016).

Deep learning for healthcare

The application of DL in medicine has been constantly increasing since 2012. Although initial publications only reported pilot studies and proof of concepts, more recent works are considering all the processes leading up to clinical deployment (Topol, 2019a,b). As an example, Fig. 4.2 lists all the different ML-based methods that have already been approved by the Food and Drug Administration (FDA) for clinical use updated to mid-2019.

The literature of this growing field is summarized in different excellent review papers and books (e.g., Chen et al., 2019; Ching et al., 2018; Esteva et al., 2019; Johnson et al., 2018; Miotto et al., 2017; Rajkomar, Dean, & Kohane, 2019; Sheikhalishahi et al., 2019; Topol, 2019a,b; Wu et al., 2019; Xiao, Choi, & Sun, 2018; Yu, Beam, & Kohane, 2018). The big advantage of using DL with biomedical data is the natural ability of processing different types of data, leading to effective and efficient heterogeneous learning frameworks that can leverage personalized medicine and next-generation healthcare (Shameer et al., 2017). Fig. 4.3 highlights these opportunities. The different data sources can be integrated into a common resource and jointly modeled end to end; alternatively, data sources can be first preprocessed separately and independently and then postprocessed together using neural networks toward joint representations leveraging all information.

In the following sections, we review a few relevant applications of DL in medicine, in particular related to medical imaging, EHRs, biomedical data, and health monitoring data. For each data type, we provide a comprehensive description of at least one study from the literature and briefly overview other relevant work. To the best of our knowledge, there are only a few applications that harness the ability

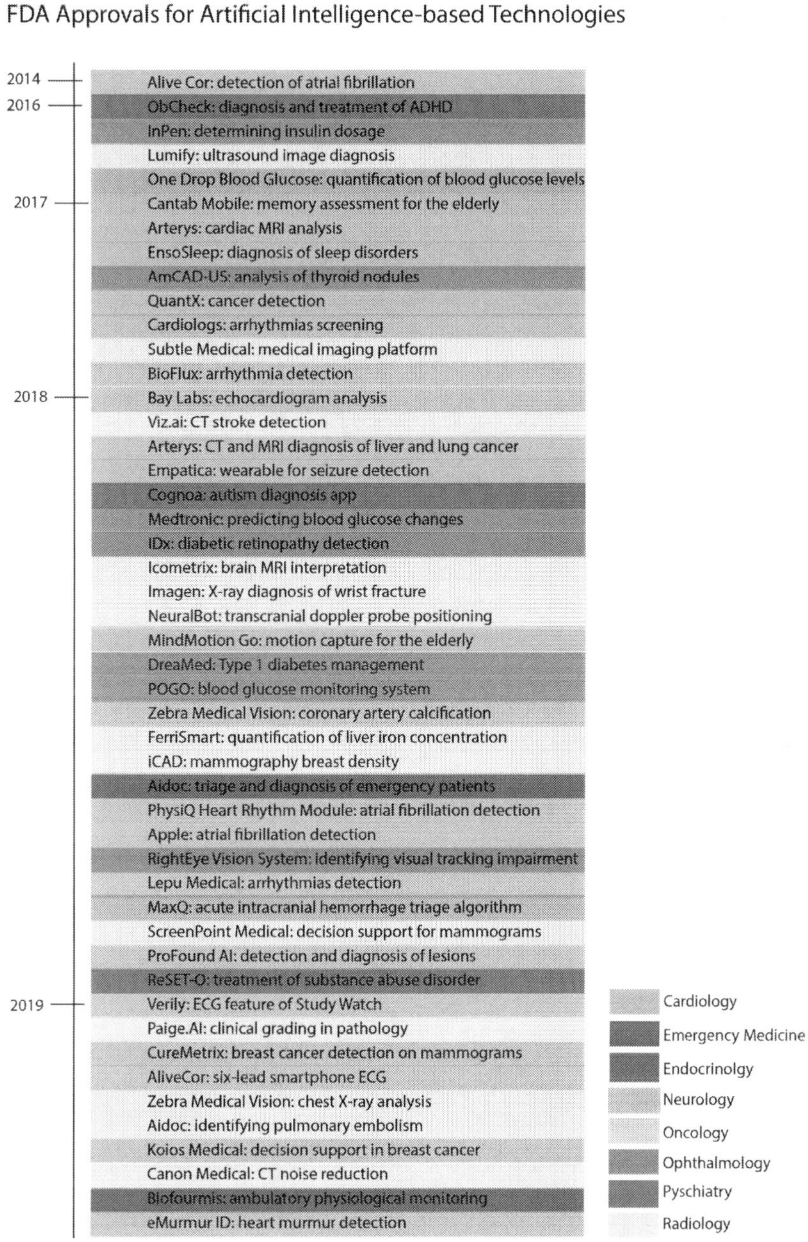

FIGURE 4.2

A timeline of FDA approvals for the clinical use of artificial intelligence-based technologies updated to mid-2019.

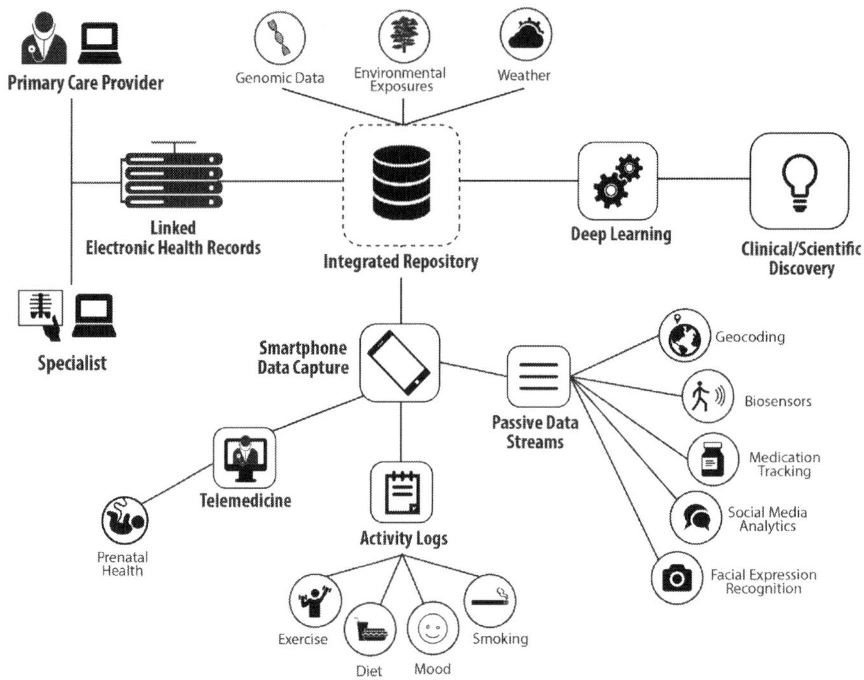

FIGURE 4.3

A framework that leverages different data sources for clinical and scientific discovery. Here, patient data stemming from hospital-linked electronic health records can be integrated together with other sources of data including genomics, environmental exposures, and elements captured from smartphones and other devices, such as activity logs and geocoding information. Deep learning algorithms can then be implemented to learn from such data to answer clinically relevant questions.

of DL to combine various data sources or a part of them in a joint representation for medical analysis and prediction. We address this point in the discussion section.

Clinical imaging

Some of the greatest success of DL has been in the field of computer vision, which focuses on deriving insight from images and videos, and deals with tasks such as object classification, detection, and segmentation. In the medical domain, computer vision can be used to determine, for example, if a patient's MRI scan contains an abnormal lesion. CNNs have grown to be central in this field, as they are good in modeling data with natural spatial invariance (i.e., images do not change if rotated or shifted). There are several examples of powerful applications of DL and CNNs in medical image analysis and computer-aided diagnosis (CAD) (Shen, Wu, & Suk, 2017).

CNNs were used to create an algorithm for automated detection of diabetic retinopathy and diabetic macular edema in retinal fundus photographs (Gulshan et al., 2016). Retinal photography with manual interpretation is a widely accepted screening tool for diabetic retinopathy, with a performance that usually exceeds that of in-person dilated eye examinations. Automated grading of diabetic retinopathy can increase efficiency, reproducibility, and coverage of screening programs, reduce barriers to access, and improve patient outcomes by providing early detection and treatment. DL predictions were compared with manual annotations by a team of ophthalmologists, who graded photos as referable diabetic retinopathy (moderate or severe), referable diabetic macular edema, or both. The automated algorithm obtained high sensitivity and specificity and promises to improve care and outcomes compared with the current ophthalmologic assessment. In a more recent study, Arcadu et al. showed that DL also obtained promising results in predicting the progression of diabetic retinopathy from color fundus photographs acquired at a single visit from an affected patient (Arcadu et al., 2019). The first medical device based on DL approved by the FDA also analyzes images of the retina to detect "greater than a mild level diabetic retinopathy" in adults who have diabetes (U.S Food and Drug Administration, 2018). DL was also used to extract new knowledge from retinal fundus images. In particular, Poplin et al. derived cardiovascular risk factors not previously thought to be present or quantifiable in retinal images, such as age, gender, smoking status, systolic blood pressure, and major adverse cardiac events (Poplin, Varadharajan et al., 2018). They also show that the trained DL models used anatomical features, such as the optic disc or blood vessels, to generate each prediction, reinforcing the significance of the result.

Skin cancer is primarily diagnosed visually, beginning with an initial screening and followed potentially by dermoscopic analysis, a biopsy, and histopathological examination. Automated classification of skin lesions using images promises to accelerate this process and to reduce subjectivity biases. Esteva et al. used a single CNN, trained end to end from images directly, using only pixels and disease labels as inputs to classify skin lesions (Esteva et al., 2017). They utilized a GoogleNet Inception v3 CNN architecture that was pretrained on about 1.28 million general images and trained in a dataset of about 130,000 skin lesion images via transfer learning. The images came from 18 different clinician-curated, open-access online repositories, as well as from clinical data from Stanford University Medical Center. They tested its performance against 21 board-certified dermatologists on biopsy-proven clinical images with two critical binary classification use cases: keratinocyte carcinomas versus benign seborrheic keratoses (i.e., identification of the most common cancers); and malignant melanomas versus benign nevi (i.e., identification of the most morbid skin cancer). The CNN obtained performances on par with all tested experts across both tasks, demonstrating the automatic capability of classifying skin cancer with a level of competence comparable to dermatologists. Embedding this type of deep neural networks on mobile devices promises to extend the reach of dermatologists outside of the clinic, potentially providing low-cost universal access to vital diagnostic care.

Zech and Badgeley et al. built deep neural networks to automatically detect pneumonia in chest radiographs (Zech et al., 2018). Their cross-sectional study involved over 100,000 images from three separate hospitals and obtained robust results by building a model that combined data from all sites. One of the major findings from this work, apart from the ability to detect pneumonia, was that models can learn other factors related to healthcare processes that are not relevant to the disease, which resulted in variable performance between sites. For example, CNNs were able to reliably identify what hospital system the image came from, raising concerns for generalizability. In a follow-up work, Badgeley et al. further investigated the contribution of confounding patient and healthcare variables for another computer-aided detection/diagnosis (CAD) task, specifically predicting hip fracture (Badgeley, Zech et al., 2019). In this study, the authors analyzed over 17,000 radiographs to not only classify fractures but to also determine if a DL model could correctly identify five patient traits (i.e., age, gender) and 14 hospital process variables (i.e., order date, scanner model, technician, department). If the models can do so, it would imply overfitting which should have no bearing on general predictive capabilities to detect the hip fracture. Overall, the model was able to detect fracture moderately well. Interestingly, however, the model could reliably detect all other 19 confounder variables individually. The CNN was able to perfectly predict the scanner model and robustly predict the scanner brand, meaning the models were then routinely using these patient and process variables as their main predictive source. When balancing a test set based on patient and hospital process variables the model performed no better than random. These results clearly show that DL for CAD requires careful calibration and control for such hospital process variables to be useful for all patients.

In addition to discovery, ML with images can be integrated as part of a "learning healthcare system" to facilitate hospital processes (Norgeot, Glicksberg, & Butte, 2019). As an example, Titano and Badgeley et al. applied automated deep neural network surveillance of cranial images to more rapidly detect acute neurological events (Titano et al., 2018). In the routine radiological workflow, images are typically presented to radiologists in the order in which they are generated for interpretation (with exceptions of course). One issue with this system is that certain events, such as hemorrhagic stroke, are time critical and necessitate immediate attention. However, based on the amount and complexity of cases in the queue, it could be hours until these scans are read. To address this limitation, the authors built a DL-based triage system where images were analyzed immediately after generation and flagged based on detected severity into an order based on clinical urgency. The authors performed a small double-blinded randomized control trial in a simulated environment with over 37,000 head CT images. Using their triage system, they found that the average time from preprocessing to being able to potentially raise an alarm was 150 times faster than humans. This work shows an interesting promise of how DL can effectively be embedded in healthcare operations.

DL has also been used to facilitate the analysis of cell images and tissue samples, where finding a trade-off between imaging quality, speed, noise reduction, and consistency in the analysis is difficult and often is a cause of missed opportunities and biased analysis. Lee et al. used CycleGAN (Zhu et al., 2017) to enhance the quality of high-throughput imaging data to achieve fast speed and high quality simultaneously (Lee et al., 2019). In particular, they developed an image model from low-throughput, high-resolution images to enhance features, such as microtubule networks, in high-throughput low-resolution images. Results showed promising performances and satisfactory robustness across different levels of noise. The tool aims to facilitate the understanding of microtubule responses within cells to targeted interventions, in the search for effective cancer therapeutics. Chen, Gadepalli, & MacDonald et al. (2019) proposed the augmented reality microscope (ARM), a cost-effective solution to improve the accuracy and efficiency of cancer diagnosis. The ARM overlays DL-based information onto the current view of the cell sample in real time, enabling seamless integration of machine learning into routine workflows.

Electronic health records

EHRs from millions of patients are routinely collected across diverse healthcare institutions. They consist of heterogeneous data elements, including demographic information, diagnosis, laboratory results, medication prescriptions, and free text clinical notes (Glicksberg, Johnson, & Dudley, 2018; Jensen, Jensen, & Brunak, 2012). EHRs provide snapshots of a patient's state of health and have created new opportunities to investigate the properties of health-related events across large populations. At the individual level, patient trajectories can foster personalized medicine (Glicksberg et al., 2016); across multiple individuals, EHRs can provide a vital resource to understand population health management and help make better decisions for healthcare operation policies.

A common approach of ML with EHRs is to have a domain expert designate the patterns to look for (i.e., the learning task and the targets) and to specify clinical variables in an ad hoc manner. Although appropriate in some situations, defining the feature space in a supervised manner does not scale or generalize well and does not allow for the discovery of novel patterns and features. DL can mitigate these limitations by reducing the need of manual feature engineering. In particular, deep neural networks are generally used to process aggregated EHRs, including both structured (e.g., diagnosis, medications, laboratory tests/results, procedures) and unstructured (e.g., free-text clinical notes) data, to obtain vector-based representations that are used for specific clinical inferences.

In one of the first applications of DL to EHRs, Miotto et al. presented "Deep Patient," an unsupervised deep feature learning framework to derive a general-purpose patient representation from EHR data that facilitates clinical predictive modeling (Miotto et al., 2016). Specifically, a deep neural network composed of a stack of denoising autoencoders was used to process EHRs in an unsupervised manner that captured stable structures and regular patterns in the data, which, grouped together, composed the deep patient representation. The Deep Patient framework is domain free (i.e., not related to any specific task since it learned over a large multidomain dataset), does not require any additional human effort, and can be easily applied to different predictive applications, both supervised and unsupervised.

Fig. 4.4 includes high-level conceptual architecture to derive deep patient representations. All EHRs are first extracted from the clinical data warehouse, preprocessed to identify and normalize clinically relevant phenotypes, and grouped in patient vectors. Each patient can be described by just a single vector or by a sequence of vectors computed in, for example, predefined temporal windows. The collection of vectors obtained from all the patients is used as input of the deep unsupervised neural network to discover a set of high-level general descriptors. Every patient in the data warehouse is then represented using these features and such deep representations can be efficiently and effectively applied to different clinical tasks (Miotto, Li, & Dudley, 2016).

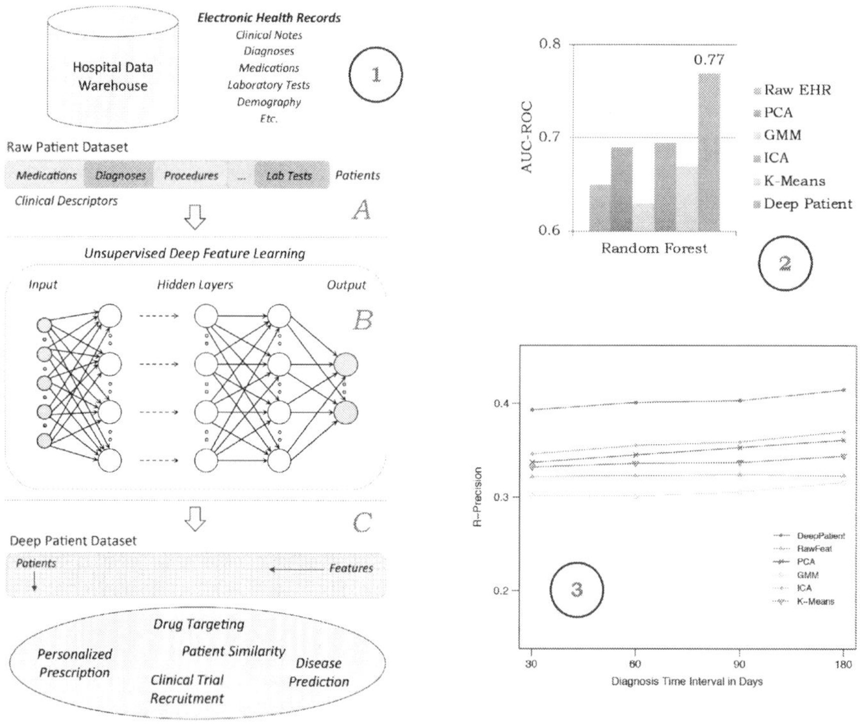

FIGURE 4.4

Conceptual framework used to derive the deep patient representations through unsupervised deep learning of a large EHR data warehouse (1). Experimental results on determining the probability of patients to develop various diseases within different time windows showed significant improvements over raw EHR data and alternative representation learning models in terms of AUC-ROC, that is, area under the ROC curve (2) and R-precision (3) (Miotto et al., 2016).

To prove the effectiveness of the idea, Deep Patient was used to capture hierarchical regularities and dependencies in the aggregated EHRs of about 700,000 patients from Mount Sinai Hospital. Patient data included some general demographic details (e.g., age, gender, and race), and common clinical descriptors available in a structured format such as diagnoses (ICD-9 codes), medications, procedures, and lab tests, as well as preprocessed clinical notes. The authors evaluated this representation as broadly predictive of health states by assessing the probability of patients to develop various diseases within different time windows. The evaluation included 76,214 test patients comprising 78 diseases from diverse clinical domains and temporal windows. Results significantly outperformed those achieved using representations based on raw EHR data and alternative feature learning strategies (see Fig. 4.4). Interestingly, prediction performance for severe diabetes, schizophrenia, and various cancers obtained the largest improvements compared to other algorithms. This is likely related to the chronic nature of these diseases, which lead to a larger and more consistent quantity of data.

These findings showed that the deep learned features describe patients in a way that is general and conducive to processing by automated methods in different domains and for different applications, such as therapy recommendation, disease subtyping, clinical trial recruitment, patient similarity, and readmission risks. This work represented a first step toward the next generation of clinical systems that can (1) scale to include many millions to billions of patient records and (2) use a single, distributed patient representation to effectively support clinicians in their daily activities—rather than multiple systems working with different patient representations derived for different tasks.

A large number of studies built upon this idea and addressed some of the limitations of Deep Patient (i.e., no explicit modeling of time, difficult interpretability of the predictions). Doctor AI uses patient history to predict diagnosis and medications for subsequent encounters using RNNs (Choi et al., 2015); DeepCare is an end-to-end deep dynamic network based on LSTMs to infer current illness states and predict future medical outcomes, which showed promising performance in the domains of mental health and type 2 diabetes for disease progression modeling, intervention recommendations, and future risk prediction (Pham et al., 2016, pp. 30–41); RETAIN is an interpretable predictive model based on attention and LSTMs that was used for heart failure prediction (Choi et al., 2016); Deepr uses CNNs to detect and combine clinical motifs in the longitudinal EHRs to predict readmission within 6 months (Nguyen et al., 2016); Lipton et al. used LSTMs to recognize patterns in multivariate time series of clinical measurements and classify different diagnoses (Lipton et al., 2015); and Rajkomar et al. used DL to predict in-hospital mortality, 30-day unplanned readmission, prolonged length of stay, and final discharge diagnosis (Rajkomar et al., 2018).

All these works focus on modeling patients to leverage clinical decision systems. In a separate but related research direction, DL has been used to derive EHR-based disease definitions (i.e., phenotypes) to facilitate cohort definitions. Similar to prospective study designs, high-quality EHR-based research requires rigorous selection criteria to designate case/control status particular to each disease. However, using a single clinical concept, such as a disease ICD code, often is not sufficient to produce reliable distinctions. Advanced EHR phenotyping of diseases is best understood as an "expert system," where researchers with advanced knowledge of a particular disease phenotype design a list of criteria that may be used to identify affected (i.e., cases) and sometimes assuredly nonaffected (i.e., controls) individuals by excluding those with ambiguous clinical status or insufficient data. This typically takes the form of intricate rule-based algorithms specifying the presence and/or absence of particular billing codes, predefined ranges for laboratory tests, prescription of characteristic medications and processing of clinical notes, among others (Kirby et al., 2016). However, these approaches are time-consuming, leading to only a relatively small amount of algorithms for diseases developed. Several ML-based phenotyping approaches have been proposed to speed-up automatic phenotype definition (e.g., Halpern et al., 2016; Pivovarov et al., 2015; Wang et al., 2015). In the specific context of DL, Glicksberg and Miotto et al. used neural networks to create unsupervised embeddings of the phenotype space within a large EHR system (Glicksberg & Miotto, 2018). In particular, they used the skip-gram algorithms of word2vec (Mikolov et al., 2013) to learn embeddings of the medical concepts reported in the structured EHRs of over a million patients. For each patient, they organized the clinical concepts into an irregularly sampled temporal sequence, where concepts adjacent to each other in the sequence should cluster together in the learned metric space. To this end, they first partitioned the patient data in consecutive time intervals composed of 15 days. Second, they removed duplicates from each time interval, and third, they random-shuffled the concepts in each interval (Choi, Chiu and Sontag, 2016). Each time interval represented as a sequence of unique medical concepts was then considered as a "sentence" to be given to the word2vec algorithm, which was trained using stochastic gradient descent and using as dynamic context the number of concepts in each sentence. In the end, every medical concept was represented as an embedded vector, with all the medical concepts mapped in the same metric space. Disease definitions can be created from these embeddings by exploring the neighborhood of a specific seed query. For example, Fig. 4.5 shows the phenotype that can be created for schizophrenia, starting from the ICD-9 code 295.9 "Unspecified schizophrenia." As it can be seen, most of the concepts are clinically related to the condition and create a "disease definition" that can be used to better identify a cohort of case/control patients. To this aim, Glicksberg and Miotto et al. used a weighted average of these embeddings to summarize the clinical history of patients. For each disease of interest, they then selected only a single meaningful seed clinical concept to use as a query, extended it using the closest items in the embedding space, and ranked patients based on their distance from this extended embedded query. From here, putative cohorts were systematically generated for the disease concept based on patient embeddings with the highest similarity and compared against the gold standards derived using PheKB (Kirby et al., 2016). Experiments showed good overlap with manually derived phenotypes for five diseases (i.e., ADHD, dementia, herpes zoster, sickle cell, and type 2 diabetes mellitus). Despite being a

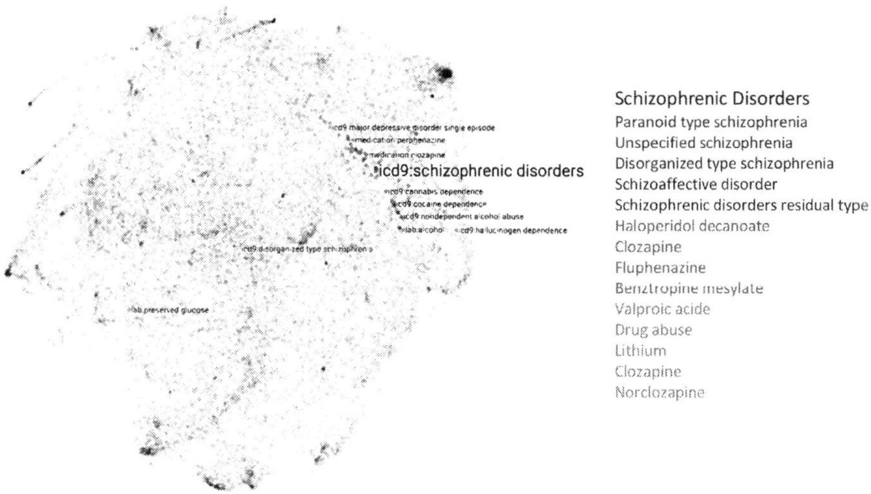

FIGURE 4.5

Disease phenotype for schizophrenic disorders derived using medical concepts embeddings (Glicksberg & Miotto, 2018). The list of concepts shows the most related ICD-9 codes (red), medications (blue), and procedure and lab tests (green).

proof of concept, this study portends the benefits of applying neural networks to derive vector-based representations of patients and medical concepts from the EHRs, which can then be used to efficiently identify meaningful insights for different clinical domains.

An important issue of EHRs is that a wealth of relevant clinical information is locked behind free-text clinical notes, requiring time-consuming chart review to obtain the complete clinical status of patients. Natural Language Processing (NLP), a branch of computer science that enables machines to understand and process human language for applications like machine translation (Wu et al., 2016), text generation (Kannan et al., 2016), and image captioning (Vinyals et al., 2015), has been widely used to parse clinical notes to extract relevant insights that can integrate structured EHRs and can guide clinical decisions (Sheikhalishahi et al., 2019; Wu et al., 2019). DL applied to clinical NLP has obtained promising results and is receiving growing acceptance in the medical community (Goldberg, 2015; Wu et al., 2019). Common applications include classification of clinical notes by disease codes (e.g., ICD-9/10 codes) (Baumel et al., 2017; Mullenbach et al., 2018; Shi et al., 2017) and prediction of disease onset (Liu, Zhang, & Razavian, 2018). Miotto et al. used multikernel CNNs to identify the presence of acute low back pain episodes in the clinical notes to inform an automated framework that can help front line primary care providers in the development of targeted strategies for low back pain in clinical practice (Miotto et al., 2020). This type of work promises to reduce well-known problems of structured EHRs, such as billing biases and incompleteness, improving the use of patient data for translational research and data-driven analysis.

Other promising applications of deep learning to EHRs include generation of synthetic clinical data using GANs (Beaulieu-Jones et al., 2019; Choi et al., 2017; Yahi et al., 2017); prediction of disease onsets from longitudinal lab tests (Razavian, Marcus, & Sontag, 2016); deidentification of patient records (Dernoncourt et al., 2016; Johnson et al., 2019); and subtyping of patients with Parkinson's disease (Baytas et al., 2017).

Biological data

DL in high-throughput biology is used to capture the internal structure of increasingly larger and high-dimensional datasets (e.g., DNA sequencing, RNA measurements). DL has been successfully applied in functional genomics, a field that tries to understand the relationship between genotype and phenotype by studying processes such as transcription, translation, and epigenetic regulation. The development of high throughput sequencing assays has provided massive amounts of -omics data, on the scale that DL requires for optimal performance. Databases like the Encyclopedia of DNA Elements (ENCODE) contain a wealth of knowledge on functional elements related to DNA binding, DNA accessibility, DNA methylation, 3D chromatin structure, transcription, and RNA binding (ENCODE Project Consortium, 2012). Using resources such as ENCODE, deep models enable the discovery of high-level features, improving performance over traditional models, increasing interpretability, and providing additional understanding about the structure of the biological data. CNNs have been particularly useful for input features like DNA sequences that require local dependencies to be taken into account. They have been used for various molecular phenotyping tasks such as DNA-binding prediction, predicting regulatory regions, DNA methylation levels, gene expression, variant calling, and predicting pathogenicity.

DNA-binding proteins modulate DNA's function in many ways such as controlling transcription, regulating cell growth and division, and modifying 3D structure. CNNs have been used to predict DNA-binding protein sites (Alipanahi et al., 2015; Zeng et al., 2016) and transcription factor binding sites (Quang, Chen, & Xie, 2015; Wang et al., 2018). Gene expression is also influenced by DNA methylation that changes chromatin accessibility. More methylated regions are associated with more open chromatin states. CNNs were used to identify DNA methylation patterns in single-cell data (Angermueller et al., 2017) and predict the effects of variants on DNA methylation (Zeng and Gifford, 2017). Regions of the genome beyond protein-encoding genes, such as enhancers and *cis*-regulatory regions, play important roles. DL has been used to identify promoters and more distal regulatory elements to make predictions about gene expression (Kelley et al., 2018) and to predict the effects of variants on gene expression (Zhou et al., 2018).

Although there are about 20,000 known genes, the number of proteins is much higher due to alternative splicing, rearrangements in the pattern of introns and exons when they are joined together to become mRNA. SpliceAI is a deep neural network that predicts splicing from pre-mRNA sequence (Jaganathan et al., 2019). The authors were able to show that mutations that lead to alterations in splicing were enriched in cases of patients with autism and intellectual disability.

One of the first applications of CNNs to genomics data, DeepSEA, is a model trained to predict the effects of noncoding variants solely on genomic sequence (Zhou and Troyanskaya, 2015). Noncoding regions of the genome are known to play important regulatory roles, though they do not form part of the gene that ultimately becomes a protein. Because most trait-associated single nucleotide polymorphisms (SNPs) are in noncoding regions, understanding their function remains an important challenge. DeepSEA has demonstrated its ability to predict effects on transcription factor (TF) binding, DNA accessibility, and histone marks of sequences. The training data came from ENCODE and Roadmap Epigenomics projects (which includes TF binding profiles for 160 different TFs, 125 DNase I-hypersensitive sites, and 104 histone-mark profiles). The pipeline integrates sequence information from a 1 kbp sequence context and learns sequence code at multiple spatial and hierarchical scales. DeepSEA's multitask joint learning architecture increases computational efficiency and allows predictive strength to be shared across the chromatin feature profiles. In a subsequent study, the authors use DeepSEA to investigate the role of noncoding mutations in autism spectrum disorder. They found that affected individuals had transcriptional and posttranscriptional regulation-disrupting de novo mutations of a significantly higher functional impact than their unaffected siblings. They were able to analyze specific tissues and pathways that were affected and prioritize variants for experimental validation (Zhou et al., 2019).

DL has also been used in more technical tasks such as calling genetic variants. Functional genomic assays rely on accurate base calling from next-generation sequencing data but this remains a challenge because of the error-prone process. Instrumentation, data processing tools, and even the genomic sequence itself introduce difficulties in establishing a ground truth. A majority of widely used base calling tools assume that read errors are independent, even though this is known to be an invalid assumption. DeepVariant is a CNN that has approximated the true but unknown interdependent relationship between read errors, thereby avoiding the independent read error assumption in variant calling (Poplin, Chang et al., 2018). The variant caller first aligns the reads to the reference genome and finds candidate SNPs with high sensitivity but low specificity. The DL model determines probabilities for each possible genotype at a locus using a pileup image of the reference and read data around each candidate variant. The authors show that DeepVariant performs better than state-of-the-art methods when evaluated on the held-out chromosomes of a ground truth set. The variant caller also demonstrated its generalizability by calling variants on a blinded sample, on reads aligned to different genome builds, and on different species data. This is invaluable for tasks for which there is insufficient data to train a DL model from scratch.

In cases where datasets are not sufficiently large enough to train a model on, transfer learning is an approach that incorporates the parameters from another model that was trained on a similar task. Transfer learning has been applied to predictive models of chromatin accessibility trained on DNA sequence data. Researchers trained the multitask model Basset on 149 cell types and then using these parameters for initialization, trained single-task models of chromatin accessibility in 15 other cell types. The performance of the models with transferred parameters performed better than models with randomly initialized parameters (Kelley, Snoek, & Rinn, 2016).

Differing data modalities can be used together as inputs in DL models to make use of complementary information. Features can be concatenated if raw data are not very different (early integration). Otherwise, if data modalities are very different, each can be first processed with separate layers, then the outputs can be concatenated, with subsequent layers integrating the features from each data modality (intermediate integration). Information on DNA sequences, gene expression, and chromatin accessibility were integrated to predict transcription factor binding in various cell types (Quang and Xie, 2019).

Unsupervised DL aims to characterize unlabeled data by extracting meaningful patterns from the data. These methods have been used to impute missing data (Arisdakessian et al., 2019; Qiu, Zheng, & Gevaert, 2018), extract gene expression signatures (Tan et al., 2017), and reduce dimensionality in single-cell RNA-seq data (Lin et al., 2017). An unsupervised method using a deep denoising autoencoder model was trained on SNP genotypes from yeast to predict gene expression values (Xie et al., 2017). Although the model was trained on yeast data, the authors speculate that a similar type of model trained on data from the GTEx project could be used to predict tissue-specific gene expression patterns in humans.

Digital health monitoring devices

Sensor-equipped smartphones, wearables, inexpensive embedded computing, and the related Internet-of-Things technologies are transforming health monitoring (Shameer et al., 2017). As the boundary between products for consumers and medical devices begins to soften, it is now possible for a single personal device to monitor a range of medical risk factors (Sandstrom et al., 2016). Potentially, these devices could give patients direct access to personal analytics that can contribute to their health, facilitate preventive care, and aid in the management of ongoing illness (Piwek et al., 2016).

ML is considered to be a key element in analyzing this type of data, leveraging other fields such as time series modeling and computer vision (Dunn, Runge, & Snyder, 2018; Randhawa, Shanthagiri, & Kumar, 2017). As an example, ML was used to generate a mobile Parkinson's disease score that objectively weighs features derived from smartphone activities (e.g., stride length from the gait activity) (Zhan et al., 2018). Although current Parkinson's disease measures are subjective, rater-dependent, and assessed in the clinic, a smartphone-derived severity score can provide an objective measure of motor symptoms inside and outside the clinic that could be valuable for clinical care and therapeutic development.

However, there are just a few preliminary studies applying DL on commercial wearable devices for health monitoring. This is likely due to hardware limitations and difficulties in collecting large quantities of data that can benefit from DL methodologies. Most of the work in the field is on human activity recognition. Deep stacked autoencoders, for example, were used to correlate smartphone sensor data and personal health (Li and Trocan, 2019). In particular, DL was used to classify accelerometer data in different user motion states such as sitting, standing, walking, and running. These classification values were then used to quantify health. For example, walking and running for a suitable amount of time are considered beneficial to human health and thus increased the health status score of an individual. Hammerla et al. evaluated DL to predict the freezing of gait in Parkinson's disease patients (Hammerla, Halloran, & Ploetz, 2016). Freezing is a common motor complication in Parkinson's disease, where affected individuals struggle to initiate movements such as walking. Results based on accelerometer data from above the ankle, above the knee, and on the trunk of 10 patients showed that RNNs obtained the best results, with a significantly large improvement over the other models, including CNNs.

DL, specifically computer vision, promises to leverage all digital health monitoring applications involving cameras, especially in passive environments. As an example, an emerging intelligent house device with potentially wide-ranging applications for healthcare outside of traditional medical settings is the smart mirror. Miotto et al. (2018) envision that smart mirrors comprising a combination of intelligent hardware and software could identify subtle, yet clinically relevant changes in physique and appearance (e.g., skin cancer detection, retinal fundus photos analysis). Similarly, a smart mirror could record and evaluate body position and motion to identify posture and movement issues, as well as offer feedback for corrective actions (e.g., during workouts, physical therapy, yoga sessions). Although smart mirrors promise to benefit healthcare, it is still uncertain when or how the prototypes will permeate the marketplace and become an integral component to monitor an individual's health. The current digital revolution that leverages DL to drive improvements in software and hardware is potentially taking the smart mirror to this envisioned wide adoption in health-related applications in the clinic and at home in the next few years, while maintaining their cost and security at a wide access level.

Discussion

In this chapter, we have discussed several important applications for DL in biomedicine across many clinical domains and data types. Further, we have provided some key examples of innovative studies that utilize DL effectively for biomedical research. It is clear that the trend will continue for many more studies utilizing such approaches. With the expected excitement and buzz around this field, it is important to consider how these types of work can have real-world impact. Here, we will discuss some high priority topics of DL in biomedicine that must be addressed going forward.

To effectively personalize medicine to the individual, it is important to characterize all the various states that make us unique. This involves information from our genetic makeup to our environmental exposures all the way to social graphs and networks (Topol, 2014). As such, the amount and variety of data required for such characterization will continue to expand. One of the biggest challenges will be how such information can be appropriately and effectively integrated. Furthermore, each type of data will have their own limitations, biases, and missingness that will have to be managed. One approach is to consider each data type separately and analyze them together in stages (Glicksberg et al., 2019). Another approach would be to learn from them jointly and concurrently. This strategy, while potentially more complex, could better enable identifying complex patterns between disparate aspects of the phenome as well as compensating for limitations of each individual data modality. For example, EHR-based disease subtyping is an important area that can be helped with ML of large-scale heterogeneous biomedical data. In fact, if clinical data are solely used to do this subtyping, the ensuing clusters may be biased by latent factors irrespective of a disease state such as cryptic relatedness that is traditionally hard to capture in EHR.

Although it is clear that DL represents a new frontier in biomedicine to enable more nuanced discoveries, one of the most important issues will be how these findings can be utilized. This will be critical for ML models that seek to augment the healthcare process. For these models to be integrated into practice, several barriers must be addressed. First and foremost, the results of these models must be robust and extensible to all populations. As we have seen, ML models of EHR data can learn factors that are unrelated to phenotype studied, such as healthcare process variables that are specific to the hospital setting. In such cases, this overfitting could result in suboptimal performance when applied to other populations (Norgeot et al., 2019). To overcome such issues, it will be imperative that models are assessed in as many sites as possible. It would of course be ideal if data from multiple sites could be investigated concurrently, but this is unlikely to happen for many reasons including infrastructural barriers and patient privacy. Another option to get around data access issues would be to build a model on one site and then fine-tune it through applying and refining it separately on other sites. At the very least, it would be necessary to gauge the performance of these models on other patient datasets, even if no model refinement is possible. One issue for replication is the discrepant ways in which data are stored and recorded in different health systems. This can be overcome through the utilization of common data model (CDM) frameworks such as the Observational Medical Outcomes Partnership CDM from the Observational Health Data Sciences and Informatics consortium or the Fast Healthcare Interoperability Resources CDM from Health Level Seven International (HL7), which standardize both the language and structure of the data.

Once a robust model is built and sufficiently replicated, the next challenge is to convince those involved in the healthcare process to utilize it. It is imperative that practitioners (e.g., nurses, clinicians), hospital administration, and the patients themselves all be involved in this decision and that all of these individuals are properly trained on how to best utilize the output from DL algorithms (Badgeley, Liu et al., 2019). It is also necessary that these changes be embedded in the system without disrupting it. One option for this is to run "silent trials" where prebuilt ML models are applied prospectively on new patient data to determine how they would perform after a certain interval of time, but without the actual intervention. The next step would be to run active prospective trials and pilot studies. The FDA has already constructed a regulatory framework for approval of ML software as a medical device (i.e., SaMD: Artificial Intelligence/Machine Learning-Based Software as a Medical Device). Going forward, getting models approved as SaMD will be the gold standard for full integration into the clinic. Although the steps outlined earlier to move ML into the clinic represent significant challenges, we envision that there will be an increasing amount of success stories in the foreseeable future. Of course, new and unexpected barriers will continue to emerge but as long as the field is diligent and rigorous in their assessment, they will be overcome.

Acknowledgments

The authors would like to thank the support from the Hasso Plattner Foundation. R.M. also acknowledges the support from the Alzheimer's Drug Discovery Foundation.

References

Alipanahi, B., et al. (2015). Predicting the sequence specificities of DNA- and RNA-binding proteins by deep learning. *Nature Biotechnology*, 831–838. https://doi.org/10.1038/nbt.3300

Angermueller, C., et al. (2017). DeepCpG: Accurate prediction of single-cell DNA methylation states using deep learning. *Genome Biology, 18*(1), 67.

Arcadu, F., et al. (2019). Deep learning algorithm predicts diabetic retinopathy progression in individual patients. *NPJ Digital Medicine, 2*, 92.

Arisdakessian, C., et al. (2019). DeepImpute: An accurate, fast, and scalable deep neural network method to impute single-cell RNA-seq data. *Genome Biology, 20*(1), 211.

Badgeley, M. A., Liu, M., et al. (2019). CANDI: an R package and Shiny app for annotating radiographs and evaluating computer-aided diagnosis. *Bioinformatics, 35*(9), 1610–1612.

Badgeley, M. A., Zech, J. R., et al. (2019). Deep learning predicts hip fracture using confounding patient and healthcare variables. *NPJ Digital Medicine, 2*, 31.

Baumel, T., et al. (2017). Multi-Label classification of patient notes a case study on ICD code assignment. *arXiv [cs.CL]*. Retrieved from http://arxiv.org/abs/1709.09587.

Baytas, I. M., et al. (2017). Patient subtyping via time-aware LSTM networks. In *Proceedings of the 23rd ACM SIGKDD international conference on knowledge discovery and data mining* (pp. 65–74). ACM.

Beaulieu-Jones, B. K., et al. (2019). Privacy-preserving generative deep neural networks support clinical data sharing. *Circulation. Cardiovascular Quality and Outcomes, 12*(7), e005122.

Bengio, Y., Courville, A., & Vincent, P. (2013). Representation learning: A review and new perspectives. *IEEE Transactions on Pattern Analysis and Machine Intelligence, 35*(8), 1798–1828.

Bishop, C. M. (2006). *Pattern recognition and machine learning*. Springer.

Chen, D., et al. (2019). Deep learning and alternative learning strategies for retrospective real-world clinical data. *NPJ Digital Medicine, 2*, 43.

Chen, P. C., Gadepalli, K., MacDonald, R., et al. (2019). An augmented reality microscope with real-time artificial intelligence integration for cancer diagnosis. *Nature Medicine, 25*, 1453–1457. https://www.nature.com/articles/s41591-019-0539-7.

Ching, T., et al. (2018). Opportunities and obstacles for deep learning in biology and medicine. *bioRxiv*. https://doi.org/10.1101/142760

Choi, Y., Chiu, C. Y.-I., & Sontag, D. (2016). Learning low-dimensional representations of medical concepts. *AMIA Joint Summits on Translational Science proceedings, 2016*, 41–50. AMIA Joint Summits on Translational Science.

Choi, E., et al. (2015). Doctor AI: Predicting clinical events via recurrent neural networks. *arXiv [cs.LG]*. Retrieved from http://arxiv.org/abs/1511.05942v11.

Choi, E., et al. (2017). Generating multi-label discrete patient records using generative adversarial networks. *arXiv [cs.LG]*. Retrieved from http://arxiv.org/abs/1703.06490.

Choi, E., et al. (2016). RETAIN: An interpretable predictive model for healthcare using reverse time attention mechanism. In D. D. Lee, et al. (Eds.), *Advances in neural information processing systems* (Vol. 29, pp. 3504–3512). Curran Associates, Inc.

Dernoncourt, F., et al. (2016). De-identification of patient notes with recurrent neural networks. *Journal of the American Medical Informatics Association*. https://doi.org/10.1093/jamia/ocw156

Dunn, J., Runge, R., & Snyder, M. (2018). Wearables and the medical revolution. *Personalized Medicine*. https://doi.org/10.2217/pme-2018-0044

ENCODE Project Consortium. (2012). An integrated encyclopedia of DNA elements in the human genome. *Nature, 489*(7414), 57–74.

Esteva, A., et al. (2017). Dermatologist-level classification of skin cancer with deep neural networks. *Nature*. https://doi.org/10.1038/nature21056

Esteva, A., et al. (2019). A guide to deep learning in healthcare. *Nature Medicine, 25*(1), 24–29.

U.S Food and Drug Administration. (2018). FDA permits marketing of artificial intelligence-based device to detect certain diabetes-related eye problems. *News Release*. Retrieved from https://www.fda.gov/news-events/press-announcements/fda-permits-marketing-artificial-intelligence-based-device-detect-certain-diabetes-related-eye.

Glicksberg, B. S., et al. (2016). Comparative analyses of population-scale phenomic data in electronic medical records reveal race-specific disease networks. *Bioinformatics, 32*(12), i101–i110.

Glicksberg, B. S., et al. (2019). Integrative analysis of loss-of-function variants in clinical and genomic data reveals novel genes associated with cardiovascular traits. *BMC Medical Genomics, 12*(Suppl. 6), 108.

Glicksberg, B. S., Johnson, K. W., & Dudley, J. T. (2018). The next generation of precision medicine: Observational studies, electronic health records, biobanks and continuous monitoring. *Human Molecular Genetics, 27*(R1), R56–R62.

Glicksberg, B. S., Miotto, R., et al. (2018). Automated disease cohort selection using word embeddings from Electronic Health Records. *Pacific Symposium on Biocomputing, 23*, 145–156.

Goldberg, Y. (2015). A primer on neural network models for natural Language Processing. *arXiv [cs.CL]*. Retrieved from http://arxiv.org/abs/1510.00726.

Goodfellow, I., Bengio, Y., & Courville, A. (2016). *Deep learning*. MIT Press.

Goodfellow, I., et al. (2014). Generative adversarial nets. In Z. Ghahramani, et al. (Eds.), *Advances in neural information processing systems* (Vol. 27, pp. 2672–2680). Curran Associates, Inc.

Gulshan, V., et al. (2016). Development and validation of a deep learning algorithm for detection of diabetic retinopathy in retinal fundus photographs. *Journal of the American Medical Association*. https://doi.org/10.1001/jama.2016.17216

Halpern, Y., et al. (2016). Electronic medical record phenotyping using the anchor and learn framework. *Journal of the American Medical Informatics Association, 23*(4), 731–740.

Hammerla, N. Y., Halloran, S., & Ploetz, T. (2016). Deep, convolutional, and recurrent models for human activity recognition using wearables. *arXiv [cs.LG]*. Retrieved from http://arxiv.org/abs/1604.08880.

Hinton, G., et al. (2012). Deep neural networks for acoustic modeling in speech recognition. *IEEE Signal Processing Magazine, 29*(6), 82–97.

Hirschberg, J., & Manning, C. D. (2015). Advances in natural language processing. *Science, 349*(6245), 261–266.

Hochreiter, S., & Schmidhuber, J. (1997). Long short-term memory. *Neural Computation, 9*(8), 1735–1780.

Ho, J., & Ermon, S. (2016). Generative adversarial imitation learning. In D. D. Lee, et al. (Eds.), *Advances in neural information processing systems* (Vol. 29, pp. 4565–4573). Curran Associates, Inc.

Isola, P., et al. (2017). Image-to-image translation with conditional adversarial networks. In *Proceedings of the IEEE conference on computer vision and pattern recognition* (pp. 1125–1134).

Jaganathan, K., et al. (2019). Predicting splicing from primary sequence with deep learning. *Cell, 176*(3), 535–548.e24.

Jensen, P. B., Jensen, L. J., & Brunak, S. (2012). Mining electronic health records: Towards better research applications and clinical care. *Nature Reviews Genetics, 13*(6), 395–405.

Johnson, K. W., et al. (2018). Artificial intelligence in cardiology. *Journal of the American College of Cardiology, 71*(23), 2668–2679.

Johnson, K. W., et al. (2019). Evaluation of patient re-identification using laboratory test orders and mitigation via latent space variables. *Pacific Symposium on Biocomputing, 24*, 415–426.

Jordan, M. I., & Mitchell, T. M. (2015). Machine learning: Trends, perspectives, and prospects. *Science, 349*(6245), 255–260.

Kannan, A., et al. (2016). Smart reply: Automated response suggestion for Email. In *Proceedings of the 22Nd ACM SIGKDD international conference on knowledge discovery and data mining* (pp. 955–964). New York, NY, USA: ACM (KDD '16).

Kassahun, Y., et al. (2016). Erratum to: Surgical robotics beyond enhanced dexterity instrumentation: A survey of machine learning techniques and their role in intelligent and autonomous surgical actions. *International Journal of Computer Assisted Radiology and Surgery*, 847. https://doi.org/10.1007/s11548-015-1340-9

Kelley, D. R., et al. (2018). Sequential regulatory activity prediction across chromosomes with convolutional neural networks. *Genome Research, 28*(5), 739–750.

Kelley, D. R., Snoek, J., & Rinn, J. L. (2016). Basset: Learning the regulatory code of the accessible genome with deep convolutional neural networks. *Genome Research, 26*(7), 990–999.

Kirby, J. C., et al. (2016). PheKB: A catalog and workflow for creating electronic phenotype algorithms for transportability. *Journal of the American Medical Informatics Association, 23*(6), 1046–1052.

LeCun, Y., Bengio, Y., & Hinton, G. (2015). Deep learning. *Nature, 521*(7553), 436–444.

LeCun, Y., et al. (1999). Object recognition with gradient-based learning. In D. A. Forsyth, et al. (Eds.), *Shape, contour and grouping in computer vision* (pp. 319–345). Berlin, Heidelberg: Springer Berlin Heidelberg.

Lee, H.-C., et al. (2019). Enhancing high-content imaging for studying microtubule networks at large-scale. *arXiv [eess.IV]*. Retrieved from http://arxiv.org/abs/1910.00662.

Lin, C., et al. (2017). Using neural networks for reducing the dimensions of single-cell RNA-Seq data. *Nucleic Acids Research, 45*(17), e156.

Lipton, Z. C., et al. (2015). Learning to diagnose with LSTM recurrent neural networks. *ICLR*, 1–18.

Li, H., & Trocan, M. (2019). Deep learning of smartphone sensor data for personal health assistance. *Microelectronics Journal, 88*, 164–172.

Liu, J., Zhang, Z., & Razavian, N. (2018). Deep EHR: Chronic disease prediction using medical notes. *arXiv [cs.LG]*. Retrieved from http://arxiv.org/abs/1808.04928.

Mikolov, T., et al. (2013). Distributed representations of words and phrases and their compositionality. In C. J. C. Burges, et al. (Eds.), *Advances in neural information processing systems* (Vol. 26, pp. 3111–3119). Curran Associates, Inc.

Miotto, R., et al. (2017). Deep learning for healthcare: Review, opportunities and challenges. *Briefings in Bioinformatics*. https://doi.org/10.1093/bib/bbx044

Miotto, R., et al. (2018). Reflecting health: Smart mirrors for personalized medicine. *NPJ Digital Medicine, 1*, 62.

Miotto, R., et al. (2016). Deep patient: An unsupervised representation to predict the future of patients from the electronic health records. *Scientific Reports, 6*, 26094.

Miotto, R., et al. (2020). Identifying acute low back pain episodes in primary care practice from clinical notes. *JMIR Medical Informatics, 8*(2), e16878.

Miotto, R., Li, L., & Dudley, J. T. (2016). Deep learning to predict patient future diseases from the electronic health records. In N. Ferro, et al. (Eds.), *Advances in information retrieval* (pp. 768–774). Cham: Springer International Publishing (Lecture Notes in Computer Science).

Mullenbach, J., et al. (2018). Explainable prediction of medical codes from clinical text. *arXiv [cs.CL]*. Retrieved from http://arxiv.org/abs/1802.05695.

Nguyen, P., et al. (2016). Deepr: A convolutional net for medical records. *IEEE Journal of Biomedical and Health Informatics*. https://doi.org/10.1109/JBHI.2016.2633963

Norgeot, B., et al. (2019). Assessment of a deep learning model based on electronic health record data to forecast clinical outcomes in patients with rheumatoid arthritis. *JAMA Network Open, 2*(3), e190606.

Norgeot, B., Glicksberg, B. S., & Butte, A. J. (2019). A call for deep-learning healthcare. *Nature Medicine, 25*(1), 14–15.

Pham, T., et al. (2016). *DeepCare: A deep dynamic memory model for predictive medicine*. Springer International Publishing.

Pivovarov, R., et al. (2015). Learning probabilistic phenotypes from heterogeneous EHR data. *Journal of Biomedical Informatics, 58*, 156–165.

Piwek, L., et al. (2016). The rise of consumer health wearables: Promises and barriers. *PLoS Medicine, 13*(2), e1001953.

Poplin, R., Chang, P.-C., et al. (2018). A universal SNP and small-indel variant caller using deep neural networks. *Nature Biotechnology, 36*(10), 983–987.

Poplin, R., Varadarajan, A. V., et al. (2018). Prediction of cardiovascular risk factors from retinal fundus photographs via deep learning. *Nature Biomedical Engineering, 2*(3), 158–164.

Qiu, Y. L., Zheng, H., & Gevaert, O. (2018). *A deep learning framework for imputing missing values in genomic data*. https://doi.org/10.1101/406066

Quang, D., Chen, Y., & Xie, X. (2015). DANN: A deep learning approach for annotating the pathogenicity of genetic variants. *Bioinformatics, 31*(5), 761–763.

Quang, D., & Xie, X. (2019). FactorNet: A deep learning framework for predicting cell type specific transcription factor binding from nucleotide-resolution sequential data. *Methods, 166*, 40–47.

Rajkomar, A., Dean, J., & Kohane, I. (2019). Machine learning in medicine. *New England Journal of Medicine, 380*(14), 1347–1358.

Rajkomar, A., et al. (2018). Scalable and accurate deep learning with electronic health records. *NPJ Digital Medicine, 1*(1), 18.

Randhawa, P., Shanthagiri, V., & Kumar, A. (2017). A review on applied machine learning in wearable technology and its applications. In *2017 international conference on intelligent sustainable systems* (pp. 347–354). ICISS.

Razavian, N., Marcus, J., & Sontag, D. (2016). Multi-task prediction of disease onsets from longitudinal lab tests. *arXiv [cs.LG]*. Retrieved from http://arxiv.org/abs/1608.00647.

Russakovsky, O., et al. (2015). ImageNet large scale visual recognition challenge. *International Journal of Computer Vision, 115*(3), 211–252.

Salimans, T., et al. (2016). Improved techniques for training GANs. In *Advances in neural information processing systems* (pp. 2234–2242).

Sandstrom, G. M., et al. (2016). Opportunities for smartphones in clinical care: The future of mobile mood monitoring. *Journal of Clinical Psychiatry*, e135–e137. https://doi.org/10.4088/jcp.15com10054

Shameer, K., et al. (2017). Translational bioinformatics in the era of real-time biomedical, health care and wellness data streams. *Briefings in Bioinformatics, 18*(1), 105–124.

Sheikhalishahi, S., et al. (2019). Natural Language Processing of clinical notes on chronic diseases: Systematic review. *JMIR Medical Informatics, 7*(2), e12239.

Shen, D., Wu, G., & Suk, H.-I. (2017). Deep learning in medical image analysis. *Annual Review of Biomedical Engineering, 19*, 221–248.

Shi, H., et al. (2017). Towards automated ICD coding using deep learning. *arXiv [cs.CL]*. Retrieved from http://arxiv.org/abs/1711.04075.

Tan, J., et al. (2017). ADAGE signature analysis: Differential expression analysis with data-defined gene sets. *BMC Bioinformatics, 18*(1), 512.

Titano, J. J., et al. (2018). Automated deep-neural-network surveillance of cranial images for acute neurologic events. *Nature Medicine, 24*(9), 1337–1341.

Topol, E. J. (2014). Individualized medicine from prewomb to tomb. *Cell, 157*(1), 241–253.

Topol, E. J. (2019a). *Deep medicine: How artificial intelligence can make healthcare human again*. Hachette UK.

Topol, E. J. (2019b). High-performance medicine: The convergence of human and artificial intelligence. *Nature Medicine, 25*(1), 44–56.

Vincent, P., et al. (2010). Stacked denoising autoencoders: Learning useful representations in a deep network with a local denoising criterion. *Journal of Machine Learning Research, 11*(Dec), 3371–3408.

Vinyals, O., et al. (2015). Show and tell: A neural image caption generator. In *Proceedings of the IEEE conference on computer vision and pattern recognition* (pp. 3156–3164).

Wang, Y., et al. (2015). Rubik: Knowledge guided tensor factorization and completion for health data analytics. *KDD: Proceedings/International Conference on Knowledge Discovery & Data Mining, 2015*, 1265–1274.

Wang, M., et al. (2018). DeFine: Deep convolutional neural networks accurately quantify intensities of transcription factor-DNA binding and facilitate evaluation of functional non-coding variants. *Nucleic Acids Research*, e69. https://doi.org/10.1093/nar/gky215

Wu, Y., et al. (2016). 'Google's neural machine translation system: Bridging the gap between human and machine translation'. *arXiv [cs.CL]*. Retrieved from http://arxiv.org/abs/1609.08144.

Wu, S., et al. (2019). Deep learning in clinical natural Language Processing: A methodical review. *Journal of the American Medical Informatics Association*. https://doi.org/10.1093/jamia/ocz200

Xiao, C., Choi, E., & Sun, J. (2018). Opportunities and challenges in developing deep learning models using electronic health records data: A systematic review. *Journal of the American Medical Informatics Association*. https://doi.org/10.1093/jamia/ocy068

Xie, R., et al. (2017). A deep auto-encoder model for gene expression prediction. *BMC Genomics, 18*(Suppl. 9), 845.

Yahi, A., et al. (2017). Generative adversarial networks for electronic health records: A framework for exploring and evaluating methods for predicting drug-induced laboratory test trajectories. *arXiv [cs.LG]*. Retrieved from http://arxiv.org/abs/1712.00164.

Yu, K.-H., Beam, A. L., & Kohane, I. S. (2018). Artificial intelligence in healthcare. *Nature Biomedical Engineering, 2*(10), 719–731.

Zech, J. R., et al. (2018). Variable generalization performance of a deep learning model to detect pneumonia in chest radiographs: A cross-sectional study. *PLoS Medicine, 15*(11), e1002683.

Zeng, H., et al. (2016). Convolutional neural network architectures for predicting DNA–protein binding. *Bioinformatics, 32*(12), i121–i127.

Zeng, H., & Gifford, D. K. (2017). Predicting the impact of non-coding variants on DNA methylation. *Nucleic Acids Research, 45*(11), e99.

Zhan, A., et al. (2018). Using smartphones and machine learning to quantify Parkinson disease severity: The mobile Parkinson disease score. *JAMA Neurology*. https://doi.org/10.1001/jamaneurol.2018.0809

Zhou, J., et al. (2018). Deep learning sequence-based ab initio prediction of variant effects on expression and disease risk. *Nature Genetics, 50*(8), 1171–1179.

Zhou, J., et al. (2019). Whole-genome deep-learning analysis identifies contribution of non-coding mutations to autism risk. *Nature Genetics, 51*(6), 973–980.

Zhou, J., & Troyanskaya, O. G. (2015). Predicting effects of noncoding variants with deep learning-based sequence model. *Nature Methods, 12*(10), 931–934.

Zhu, J.-Y., et al. (2017). Unpaired image-to-image translation using cycle-consistent adversarial networks. In *Proceedings of the IEEE international conference on computer vision* (pp. 2223–2232).

CHAPTER

5 Generative adversarial network for cardiovascular imaging

Mina Rezaei

Massachusetts General Hospital, Harvard Medical School, Boston, MA, United States

Chapter outline

Introduction

Cardiovascular disease (CVD) remains the leading cause of death in the United States, being responsible for 840,768 deaths (635,260 cardiac) in 2016 as reported by the American College of Cardiology (Benjamin, Muntner, & Bittencourt, 2019). The total annual cost of CVD in the United States was estimated at $351.2 billion in 2014–15, with an indirect cost of $213.8 billion, including 46% for inpatient care.

Cardiac image analysis plays an important role in cardiovascular disease diagnosis, treatment planning, and clinical monitoring. Cardiac imaging can provide useful information for the detection of different pathological conditions such as valvular disease, coronary artery disease, and myocardial disease. Further, imaging is often essential for the evaluation of therapy success. However, it is often difficult to interpret the raw data extracted from images due to the high variability in image

Machine Learning in Cardiovascular Medicine. https://doi.org/10.1016/B978-0-12-820273-9.00005-1

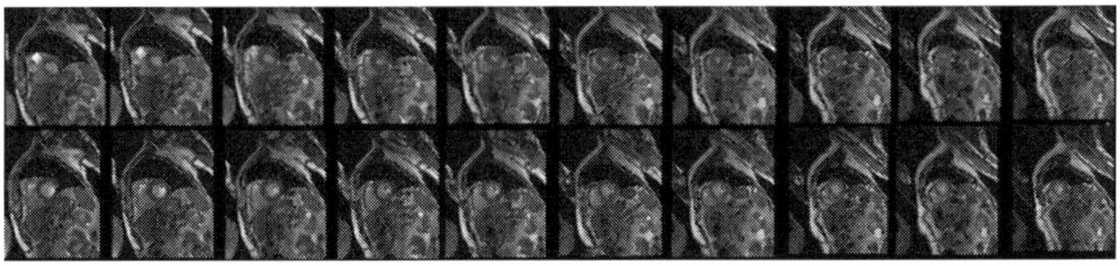

FIGURE 5.1 Short-axis MRI images from patient 35 in the ACDC dataset [20] shown as 2D.

The top row shows sequential systolic phase, with the last sub-figure in the top row showing the end of systole (myocardial contraction). The diastolic phase is shown in lower row.

quality and structure appearance among various individuals. Fig. 5.1 illustrates the diversity of myocardial tissue, vessels, and atrial/ventricular cavities in the systolic and diastolic phases on cardiac magnetic resonance imaging (MRI).

Machine learning has been applied to cardiac images and clinical data to improve prediction accuracy and efficiency, which could result in more precise and targeted therapy while streamlining the diagnostic process. For instance, the number of investigations that have utilized machine learning approaches for automating medical image segmentation, for example, cardiac MRI segmentation (Peng et al., 2016), has increased significantly in the last few years. One can roughly divide the current machine learning algorithms into two categories: (1) those based on *generative* probabilistic models and (2) those based on *discriminative* probabilistic models.

Generative approaches build models based on prior domain knowledge regarding the appearance and spatial distribution of the different tissue types. In the scenario of supervised learning, a generative model estimates the joint probability distribution of data $P(X,Y)$ between the observed image X and the corresponding segmentation mask or label Y. Examples of popular generative models are Naive Bayes, Hidden Markov Models, Latent Dirichlet Allocation, and Boltzmann Machines. Traditionally, generative probabilistic models have been popular where simple conditionally independent Gaussian models (Sabuncu et al., 2010) or Bayesian learning (Pohl et al., 2006) have been used for tissue appearance. Later, segmentation has been improved with the addition of the geometrical shape model (Folkesson, Samset, Kwong, & Westin, 2008; Yang, Staib, & Duncan, 2004), which encourages certain topological properties.

On the contrary, discriminative models directly learn the relationship between the local features of images to specific labels without any domain knowledge. Examples are logistic regression, scalar vector machine, neural networks, nearest neighbor, and conditional random fields. These local features can be raw input pixel values (Cocosco et al., 2008; Mahapatra, 2014a,b), histogram intensities (Queirós et al., 2014), texture features such as Gabor filter masks (Subbanna, Precup, Collins, & Arbel, 2013), or alignment-based features such as interimage gradient, region shape difference, and symmetry analysis (Menze et al., 2010). Traditional discriminative approaches such as support vector machines (Afshin et al., 2013), random forests (Rezaei & Tajeripour, 2013), and guided random walks (Eslami, Karamalis, Katouzian, & Navab, 2013) have also been used in cardiac image segmentation.

Another popular discriminative model is the neural network, in which the machine learns the hierarchical representation of features (LeCun, Bengio, & Hinton, 2015). This method has been successfully applied to the supervised classification task, where the classifier is trained with sufficient data to learn the healthy from nonhealthy labels (Bernard et al., 2018; Radford, Metz, & Chintala, 2015; Rezaei, Yang, & Meinal, 2017; Rezaei, Yang, & Meinal, 2018a; Rezaei, Yang, & Meinal, 2018b; Rezaei, Yang, & Meinal, 2018c). In the field of medical image segmentation, deep convolution neural networks (CNNs) have achieved promising results. Ronneberger, Fischer and Brox (2015) presented a fully convolutional neural network architecture, named UNet, for segmenting neuronal structures in electron microscopic stacks. The UNet has achieved outstanding performance in medical image segmentation. Later, 2D-UNet with shape embedded layer (Zotti et al., 2017), and 3D-UNet (Duan et al., 2019; Ye, Wang, Zhang, & Zhang, 2019) were used for cardiac image segmentation. Wang et al. (2018) improved the segmentation results by two-stage 3D-UNet wherein they applied for a multiatlas registration aside of 3D-UNet architecture for the segmentation of myocardial chambers and vessels. As an example, Rohé, Sermesant and Pennec (2017) used multiatlas registration besides 2D-UNet for whole heart segmentation. Lately, Goodfellow et al. introduced generative adversarial networks (GANs) (Goodfellow et al., 2014), which were composed of two models: a generative model and a discriminative model, as shown in Fig. 5.2. The discriminator model in GAN is a classifier that determines whether a given image looks like a real image from the dataset or like an artificially created image. The generator model takes random input values and transforms them into images through a deconvolution neural network.

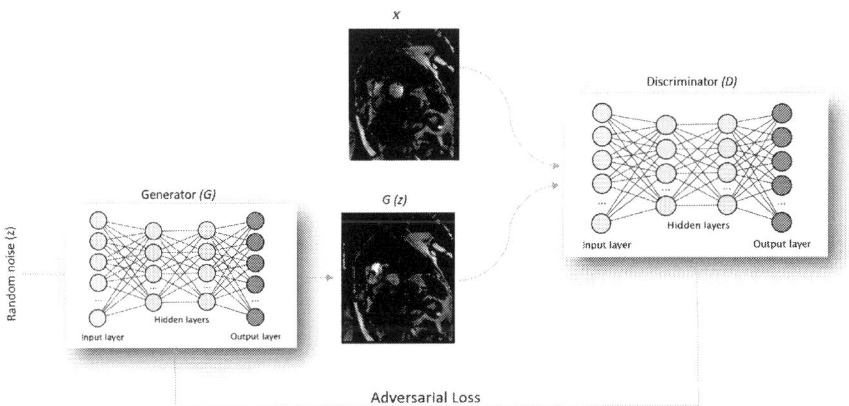

FIGURE 5.2 Schematic view of a conventional GAN for the synthesis of cardiac MR images.

The figure shows the input, output and the internal feature representation of the generator G and discriminator D. G transforms a sample z from $p(z)$ into a generated nodule xg.

Recently, Litjens et al. (2017) reviewed and summarized 300 papers, up until June 2017, focusing on deep learning and medical image analysis. Further, we add 300 recent high citation papers to the collection by considering the same topics of machine learning application for medical imaging, published from 2017 until October 2019 in MICCAI, ISBI, SPIE, IPMI, MIDL conferences and workshops as well as IEEE Transaction Medical Imaging and Medical Imaging journals. Figs. 5.3, 5.4, and 5.5 show the detailed statistics according to publication year, image modality, and application area. Medical image segmentation using machine learning and deep learning is the most popular research topic as shown in Fig. 5.5, in which classification and detection tasks are second and third, respectively.

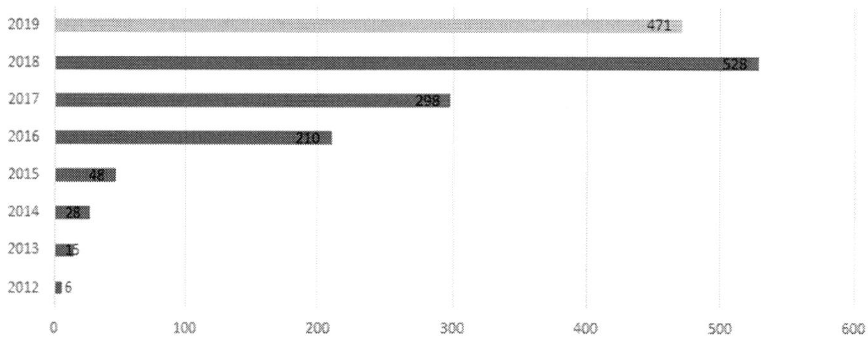

FIGURE 5.3 Number of medical image analysis papers using deep learning approaches published between 2012 and June of 2019.

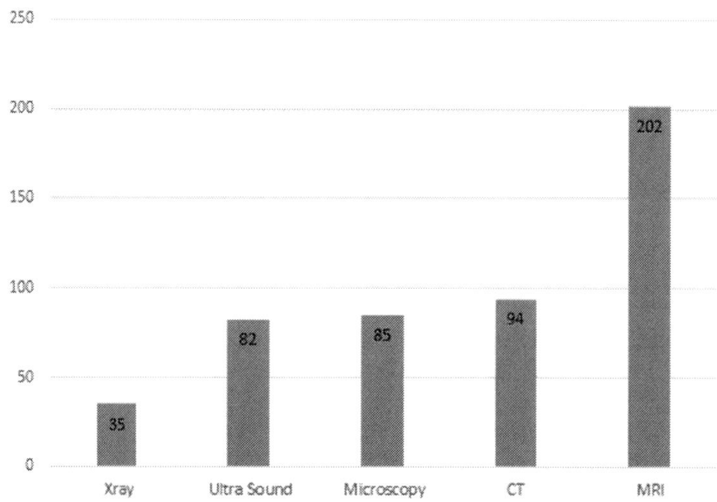

FIGURE 5.4 Categorization of deep learning papers according to imaging modality.

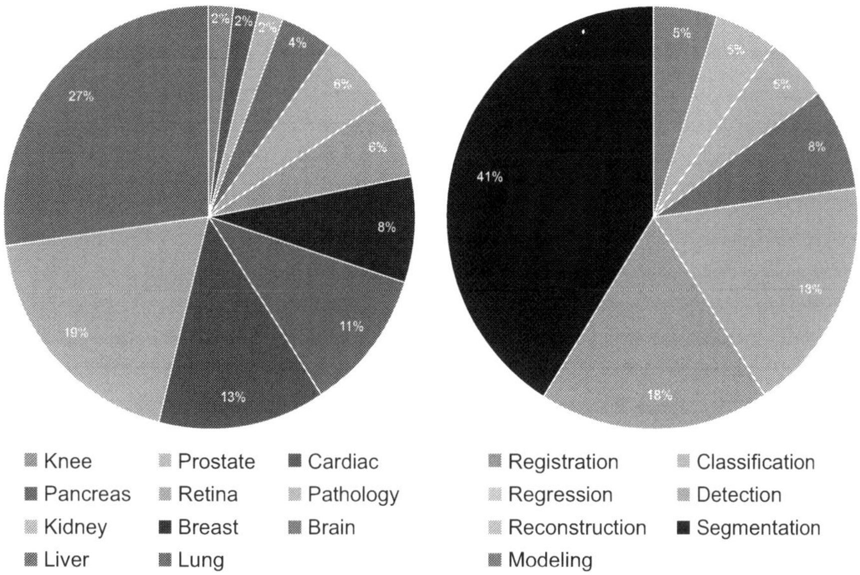

FIGURE 5.5 Left panel: distribution of publications stratified by organ system. Right panel: distribution of publications stratified by task.

The remainder of the chapter is structured as follows: We begin with a brief introduction of the principles of GANs, overview the challenges associated with conventional GANs, and discuss the motivation behind some GAN variants in Research background section. This is followed by a comprehensive review of medical image analysis tasks using GANs in Deep learning in cardiac medical imaging section. We categorize all the work according to machine learning canonical tasks such as reconstruction, image synthesis, segmentation, classification, detection, registration, and multitask learning and highlight recent work on cardiac imaging. Section GAN application in medical imaging is a summary section followed by a discussion of prospective applications and forthcoming challenges.

Research background

Generative adversarial networks

The GAN framework (Goodfellow et al., 2014) is composed of a generator (G) and a discriminator (D) network where both networks are deep neural networks. The G takes random noise z as input and tries to generate samples $\hat{x} = G(z)$ from the desired data distribution P_{data}. Meanwhile, D classifies whether the generated samples $\hat{x}$ are from the real data distribution P_{data} or fake samples of P_z. Thereby, $D(x, \hat{x})$ is the scalar output of the discriminator network that shows the probability of the input being from a real or fake sample. Fig. 5.2 shows an illustration of a conventional GAN's configuration where the training process for G and D is similar to mini-max game. G in this example is generating a 2D slice of cardiac MRI images. D's objective is to differentiate these two groups of images whereas the generator G is trained to confuse the discriminator in an alternative view, and one can perceive G as receiving a reward signal from D depending upon whether the generated data is accurate or not. The gradient information is back-propagated from D to G, so G adapts its parameters in to produce an output image that can fool D. The training objectives of D and G can be expressed mathematically as follows:

$$L_G = \min_G E_{z\sim P(z)} \tag{5.1}$$

$$L_D = \max_D E_{x\sim P_{data}(x)} \tag{5.2}$$

$$L_{adv}: \min_G \max_D V(G, D) = E_{x\sim P_{data}(x)}[\log D(x)] + E_{z\sim p(z)}[1 - \log D(G(z))] \tag{5.3}$$

The generator G implicitly defines a probability distribution P_z as the distribution of the samples (G) obtained by z. Therefore, we would like Eq. (5.3) to converge to a good estimator of P_{data}, if given enough capacity and training time. If the discriminator D is trained to optimality before the next generator G updates, then minimizing L_G is proven to be equivalent to minimizing the Jensen–Shannon divergence (JSD) between $P_{data}(x)$ and $P_g(z)$ as mentioned by Goodfellow et al. The desired outcome after training is that samples formed by $\hat{x}$ should approximate the real data distribution $P_{data}(x)$.

Challenges for GAN optimization

The training procedure by GAN is hard and different from neural networks because we simultaneously optimize the generator and the discriminator for opposing objectives. Therefore, many methods address this problem by proposing variant losses, training with different architecture, or even different optimization algorithm. Another problem by classical GAN is nonconvergence where the model parameters oscillate, destabilize, and never converge. There has been nontrivial progress on how and when one can prove that GANs are globally convergent such as (1) game theory, (2) simplifying assumption, and (3) using normal neural network techniques.

(1) *Game theory*: These techniques try to model GAN training process using notions from game theory. Current studies (Heusel et al., 2017; Oliehoek et al., 2018) show training procedures that provably converge to some kind of approximate Nash equilibrium, but do so using unreasonably large resource constraints. The next step is to try and reduce those resource constraints.

(2) *Simplifying assumptions:* Some studies have shown that by simplifying assumptions of the generator and discriminator, one can move away the nonconvergence problem. For example, the simplified LGQ GAN

(Feizi, Farnia, Ginart, & Tse, 2017) trained a linear generator with Gaussian data, and quadratic discriminator to be globally convergent under training of n-player differentiable games described by Balduzzi et al. (2018).

(3) *Using techniques from normal neural networks*: Other studies have shown the success of convergence by applying normal and nonconvex neural networks. Choromanska et al. (2015) argued that the nonconvexity of deep neural networks is not a problem because low-quality local minima of the loss function become exponentially rare as the network gets larger.

Mode collapse is one of the hardest problems to solve in GAN. Here, the generator collapses, which produces limited varieties of samples. For example, in ACDC dataset (Bernard et al., 2018), there are five categories of cardiac images from 30 patients including abnormal right ventricles, hypertrophic cardiomyopathy, dilated cardiomyopathy, myocardial infarction, and healthy subjects. If one aims to generate some images similar to these five categories, there could be a situation where samples are generated for one or two categories. This problem is called mode collapse, when only one or a few modes of data are generated.

The nonconvergence and mode collapse are often explained as an imbalance between the discriminator and the generator. The obvious solution is to balance their training to avoid overfitting. However, little progress has been made but not due to a lack of effort. Some researchers even believe that this is not a feasible or a desirable goal because a good discriminator gives good feedback. Some of the attention is, therefore, shifted for cost functions with nonvanishing gradients instead.

In a discriminative model, loss measures the accuracy of the prediction, which is then used to monitor the progress of the training. However, the adversarial loss in GAN measures how well the discriminator is performing compared to the generator. Often, the generator cost increases but the image quality is improving. The next step is to examine the generated images manually to verify the progress. This makes model comparison harder, which in turn leads to difficulties in picking the best model in a single run. It also complicates the tuning process. Therefore, the next sections aim to cover investigations that have aimed at solving this issue.

Loss-variant GANs

Another design decision in GANs that significantly impacts performance is the choice of loss function in Eq. (5.3), which is an essential step to match the generated distribution to real data distribution. The objective function of original GAN is to minimize the JSD, wherein the JSD is derived from the Kullback–Liebler divergence (KLD).

Wasserstein GAN (WGAN) (Gulrajani et al., 2017) does not use the JSD to measure divergence, instead, it uses the Earth-Mover (EM) distance (also known as the Wasserstein distance). EM distance calculates the minimal cost of transforming one probability distribution into another. Least squares GAN (LSGAN) (Mao et al., 2017) has a setup similar to WGAN. However, instead of learning a critic function, LSGAN learns a loss function, where the loss for real samples should be lower than the loss for fake samples. This allows LSGAN to place a high focus on fake samples that have a very high margin. In summary, LSGAN uses l2 loss instead of log loss for proportional penalization and applied weight decay regularization to bound loss function.

Relaxed Wasserstein GAN (RWGAN) (Guo, Hong, Lin, & Yang, 2017) is another variation of the WGAN framework that instead of symmetric clamping of weights (like in WGAN) or a gradient penalty (like the one proposed for improved WGAN), RWGAN utilizes an asymmetric clamping strategy. The Mean and covariance Feature Matching GAN (McGAN) (Mroueh, Sercu, & Goel, 2017) is part of the same family of GANs as WGAN. This family is dubbed as the integral probability metric (IPM) family. These GANs use a critic architecture instead of an explicit discriminator. The critic function for McGAN measures the mean or the covariance features of the generated data distribution and the target data distribution.

Cramer GAN (Bellemare et al., 2017) addresses training limitations associated with WGAN. It claims that sum invariance, scale sensitivity, and unbiased sample gradients are three properties that a probability divergence should satisfy. They demonstrate that this is an issue in practice, and propose a new distance: the Cramer distance. Thus, the Cramer distance has been utilized, instead of EM distance, in addition to an improved unbiased sample gradient that is used in WGAN.

The Fisher GAN (Mroueh & Sercu, 2017) is another iteration on IPM GAN's claim to surpass McGAN and WGAN. Indeed, the Fisher GAN measures the Mahalanobis distance, which is the distance between two points that have correlated variables, relative to a centroid that is believed to be the mean of the distribution of the multivariate data. The training stability, unconstrained capacity, and efficient computation are addressed by Fisher GAN. As the parameterized critic approaches infinite capacity, it becomes an actual estimate of the Chi-square distance.

Energy-based GAN (EBGAN) (Zhao, Mathieu, & LeCun, 2016) replaces an autoencoder instead of a discriminator and the objective function estimated reconstruction loss with the following steps: (1) train an autoencoder on the original data; (2) run generated images through the trained autoencoder; while (3) poorly generated images will have low reconstruction loss. These steps have been shown to produce robust results (Table 5.1).

Table 5.1 Examples of loss variant GANs.

GAN variant	Objective
GAN (Goodfellow et al., 2014)	JSD derived from KLD
WGAN (Gulrajani et al., 2017)	EM distance
LSGAN (Mao et al., 2017)	L2 loss objective
RWGAN (Guo et al., 2017)	Relaxed WGAN framework
McGAN (Mroueh, Sercu, & Goel, 2017)	Mean/covariance minimization objective
Cramer GAN (Bellemare et al., 2017)	Cramer distance
Fisher GAN (Mroueh & Sercu, 2017)	Chi-square objective
EBGAN (Zhao et al., 2016)	Autoencoder instead of discriminator

Architecture-variant GANs

Motivated by different challenges in GAN training procedure described in Generative adversarial networks section, there have been several types of architecture variants proposed in the literature (Kurach et al., 2018; Wang, She, & Ward, 2019). Architecture-variant GANs are mainly proposed for different applications, for example, image to image transfer (Zhu, Park, Isola, & Efros, 2017), image super-resolution (Ledig et al., 2017), image completion (Isola, Zhu, Zhou, & Efros, 2017), and text-to-image generation (Qiao, Zhang, Xu, & Tao, 2019). In this section, we provide a review of architecture variants that help improve the performance of GANs from three aspects previously discussed, namely improving image diversity, improving image quality, and improving training stability. We overview the recent GANs architecture used for medical applications:

DCGAN (Radford, Metz, & Chintala, 2015): To address the instability of the conventional GAN architecture, Radford et al. proposed the deep convolutional GAN (DCGAN), in which both the generator and discriminator follow a deep convolutional network architecture. These networks can extract hierarchical features of the image by learning down/up-sampling as a result of feature location. In this way, the extracted features can be used to generate new ones. Key components of the DCGAN that affect the stability of the network are batch normalization and leaky-ReLU. Although DCGAN is more stable than GAN, it is still prone to mode collapse.

Conditional GAN (Mirza & Osindero, 2014): With the existence of labeled training data, this architecture helps generate specific images from the generator. With this architecture, one has some conditional information Y that describes some aspect of the data. Most commonly, the class label is given as conditional information. For example, if we are dealing with the ACDC dataset, Y could describe the spectrum of cardiac diseases on the MRI images. Then, this attribute information is inserted in both the generator and the discriminator. Once the training is done, if we want to generate an image of a particular label, we can simply feed the generator with the corresponding label as Y. The conditional GAN setting has been used in medical imaging for different purposes such as abnormality segmentation (Xue et al., 2018).

Recurrent GAN (Rezaei, Yang, & Meinel, 2019a; Rezaei, Yang, & Meinel, 2019b): Consequently, 3D medical images are often represented as a sequence of 2D slices. A long short-term memory (LSTM) unit is an effective model to process sequential data. In this regard, we utilize an LSTM unit inside the generative and discriminative network to get intraslice features as well as previous slice (interslice features). The generator network provides a solution to sequencing medical images (2D slices) into 2D segmentation masks. The discriminator takes the sequence of 2D generator output and the ground truth provided by an expert and determines whether the generator output is real or fake. The training procedure is similar to a two-player mini-max game, where a generator and a discriminator network are alternately trained to respectively minimize and maximize an objective function.

Pix2Pix (Isola et al., 2017): A conditional GAN-based architecture is used to learn the joint distribution of the paired data. Here, the generator is an autoencoder with a skip connection in an up convolutional layer similar to U-Net architecture. It has a discriminator called PatchGAN discriminator, which classifies individual patches in the image as "real versus fake," instead of classifying the entire image as "real versus fake." The authors argue that this helps in ensuring sharp high-frequency detail in the generated images. Finally, in the loss objective function, in addition to the usual adversarial loss in GANs, it uses an L1 distance term between the generated image and target image.

CycleGAN (Zhu et al., 2017): This architecture works even when there is no paired training data. All that is required is a set of images in the source domain and a set of images in the target domain. No one-one correspondence is required between the images. The key idea is to use two generator-like architectures, where the first one translates a source domain image into the target domain, and the second one tries to reconstruct the source domain image from the target domain image. An L2 reconstruction loss is used between the source image and its reconstructed image, which is added to the usual adversarial loss function of the GANs.

SRGAN (Ledig et al., 2017): Superresolution GAN applies a deep network in combination with an adversary network to produce higher resolution images. SRGAN is more appealing to a human given the greater depth of detail compared to a similar design without GAN (SRResNet). During training, a high-resolution (HR) image is downsampled to a low-resolution (LR) image, while a GAN generator upsamples LR images to superresolution (SR) images. We use a discriminator to distinguish the HR images and backpropagate the GAN loss to train the discriminator and the generator.

BigGAN (Donahue & Simonyan, 2019): The proposed BigGAN applied three following modifications on conventional GANs:

Scalability: As the authors discovered that GANs benefit dramatically from scaling, they introduced two architectural changes to improve scalability (described in detail in the paper's Appendix B), while at the same time improved conditioning by applying orthogonal regularization to the generator.

Robustness: The orthogonal regularization applied to the generator makes the model amenable to the "truncation trick" so that fine control of the trade-offs between fidelity and variety is possible by truncating the latent space.

Stability: The authors discovered and characterized instabilities specific to large-scale GANs and devised solutions to minimize the instabilities, although these involve a relatively high trade-off on performance.

In addition to its performance boost at 128×128 resolution, BigGAN also outperformed the previous state of the art at 256×256 and 512×512 resolution on ImageNet. The model was also tested on the larger image dataset, JFT-300M, to demonstrate its transferability.

GAN evaluation

When it comes to evaluating GANs, there are several methods that can be divided into quantitative or qualitative measures (Table 5.2). In continuum, we briefly overview some of these measurements:

Inception score: The inception score (IS) measures the quality of the generated samples. To this end, Salimans et al. (2016) used a pretrained Inception-v3 network on ImageNet to generate samples and then compare the conditional label distribution with the marginal label distribution:

$$IS(G) = exp\left(E_{xP_{data}(x)} D_{KL}(p(y|x)\|p(y))\right) \quad 5.4$$

Higher scores are better, corresponding to a larger KL-divergence between the two distributions.

Fréchet inception distance (FID) (Heusel et al., 2017)*:* The FID is designed to improve on the IS by actually comparing the statistics of generated samples to real samples, instead of evaluating generated samples:

$$FID(x, g) = \|\mu_x - \mu_g\| + Tr(\Sigma_x + \Sigma_g - 2(\Sigma_x \Sigma_g)) \quad 5.5$$

Table 5.2 Examples of GAN evaluation metrics.

	Measure	Objective
Quantitative	Inception score (Salimans et al., 2016)	$IS(G) = exp\left(E_{x\,P_{data}(x)} D_{KL}(p(y\|x)\|p(y))\right)$
	Fréchet inception distance (Heusel et al., 2017)	$FID = \|\mu_x - \mu_g\| + (\Sigma_x + \Sigma_g - 2(\Sigma_x \Sigma_g))$
	The wasserstein critic (Gulrajani et al., 2017)	$W(P_r, P_g) = \inf_{\eta \in \Pi(P_r, P_g)} E_{(x,y)\eta} \|[x - y]\|$
	Reconstruction error (Xiang & Li, 2017)	
	Geometry score (Lucic et al., 2018)	Geometrical comparisons between real data and generated data
	Precision-recall (Khrulkov & Oseledets, 2018)	Quantify the degree of overfitting in GANs, often over dataset.
Qualitative	Preference judgment (Chen et al., 2016)	Ask from participant to rank models in terms of the fidelity of their generated images.
	Mode drop and collapse (Srivastava et al., 2017)	Over datasets with known modes (e.g., a GMM or a labeled dataset), modes are computed as by measuring the distances of generated data to mode centers.
	Network internal (Zhang et al., 2017)	Illustrating the internal representation and dynamics of models (e.g., space continuity) as well as visualizing learned features.

Indeed, a lower FID is better, corresponding to more similar real and generated samples as measured by the distance between their activation distributions. A common criticism is that the FID and IS scores measure *sample quality* and do not really capture sample diversity.

The Wasserstein critic (Gulrajani et al., 2017)*:* The critic (e.g., a neural network [NN]) is trained to produce high values at real samples and low values at generated samples.

$$W(P_r, P_g) = \inf_{\eta \in \Pi(P_r, P_g)} E_{(x,y)\eta} \| [x - y] \| \tag{5.6}$$

$\Pi(P_r,P_g)$ denote the set of all joint distributions $n(x, y)$ whose marginals are, respectively, Pr and Pg.

Reconstruction error (Xiang & Li, 2017)*:* Measures the reconstruction error (e.g., L2 norm) between a test image and its closest generated image.

Geometry score (Khrulkov & Oseledets, 2018)*:* Suggests computing geometric properties of the generated data manifold and comparing those properties to the real data.

Precision and recall (Lucic et al., 2018)*:* Attempts to measure both the "precision" and "recall" of GANs. These measures are used to quantify the degree of overfitting in GANs.

Preference judgment (Zhang et al., 2017)*:* Participants are asked to rank models in terms of fidelity of their generated images (e.g., pairs, triples).

Mode drop and collapse (Srivastava et al., 2017)*:* Over datasets with known modes (e.g., a GMM or a labeled dataset), modes are computed by measuring the distances of generated data to mode center.

Network internal (Chen et al., 2016): Explores and illustrates the internal representation and dynamics of models (e.g., space continuity) as well as visualizing learned features.

Deep learning in cardiac medical imaging

In this section, we provide an overview of the current state of the art in the application of deep learning algorithms to cardiovascular imaging. This section is organized by imaging modality, with most papers focusing on ultrasound, CT, and MRI. Applications of deep learning for cardiovascular imaging in detection, classification, segmentation, and synthetization (Table 5.3).

Table 5.3 Deep Learning application in heart diseases.

Deep learning model	Application	Image modality	Data/NS
Multiscale CNNs (Wu et al., 2019)	LV segmentation	Ultrasound	Private/51
CNN (Levin, Parker, Halpern, & Rao, 2019)	Classification of IVUS frames as normal versus TCFA	Ultrasound	Private/100

Table 5.3 Deep Learning application in heart diseases.—*cont'd*

Deep learning model	Application	Image modality	Data/NS
Recurrent GAN (Rezaei et al., 2018a,b,c)	Segmentation MYO, LV, RV, classification of 5 cardiac diseases	MRI	ACDC/150
Cascade GANs (Rezaei et al., 2018a,b,c)	Segmentation WH, GV	MRI	HVSMR/30
CNN (Abdi et al., 2017)	Classification of seven cardiac diseases	Ultrasound	Private/1000
FCN and deformable CNN (Gulshan et al., 2016)	LV segmentation	3D echocardiography	Private/15
Deep sparse (De Fauw et al., 2018)	3D aortic valve detection and segmentation	Ultrasound	Private/865
3D GAN (Wolterink et al., 2015)	Noise reduction	CT	Private/56
3D and 2D CNNs (Metz et al., 2009)	Coronary calcium detection	CT	Private/200
Autoencoder and CNN (Gandhi, Mosleh, Shen, & Chow, 2018)	Identification of patients with functionally significant stenosis	CT	Private/252
Multiscale CNN (Zhu et al., 2017)	Bounding box localization of LV	MRI	Private/33
CNN (Wolterink, Leiner, Viergever, & Išgum, 2017)	Segmentation of stents	X-ray	Private/63
Autoencoder (Lessmann et al., 2017)	Real-time denoising of angiography frames	X-ray	Private/20
Cascade unet (Petitjean et al., 2015)	Segmentation of five cardiac structures	MRI	Private/100
Cycle-GAN (Moradi et al., 2016)	Noise and artifact reduction in low-doses CT	CT	100
CNN (Oktay et al., 2016)	Registration of cardiac models from CT to intraoperative X-ray	X-ray	Private/821
CNN (Toth et al., 2018)	Classification of plaques in intravascular	OCT	Private/66
CNNs (Radford et al., 2015)	Prediction of obstructive disease from myocardial perfusion imaging	SPECT	Private/2276
CNNs (Betancur et al., 2018)	Plaque detection in intravascular	OCT	Private/49

CNN, *convolutional neural networks;* GAN, *generative adversarial networks;* LV, *left ventricle;* MYO, *myocardium vessel;* NS, *number of samples;* RV, *right ventricle;* SPECT, *single-photon emission tomography;* WH, *whole heart.*

GAN application in medical imaging

In this section, we provide an overview of the current state of the art in the application of deep learning algorithms to medical imaging. We cover details of GANs application for cardiovascular imaging. This section focuses on GAN application for canonical tasks such as image synthetization, reconstruction, classification, detection, segmentation, and registration.

Medical image synthesis

GANs have been widely proposed for image generation. We can group the GAN-based approaches into two categories: unsupervised generative framework and conditional GANs.

Unsupervised GAN maps random noise to synthetic, realistically looking images following the training data distribution while the conditional GAN framework has a condition on prior knowledge for both the generator and the discriminator rather than solely utilizing noise. Fig. 5.6 shows the difference between unsupervised GAN and conditional GAN framework. Unsupervised GAN has been widely used to synthesize certain types of medical images either from noise alone or from prior knowledge by setting the conditions of image data for handling imbalanced classes (Rezaei et al., 2020), missing medical image modalities (Abdi et al., 2017), or image augmentation (Sandfort, Yan, Pickhardt, & Summers, 2019).

In an unsupervised setting, different GAN frameworks have been implemented for 2D or 3D synthetic image generation. Thereafter, the generated images in addition to the training samples are fed to the deep learning models with the goal of providing balanced, augmented, and diverse data.

Hence, the diversity gained from synthesized images can introduce more variability into the training set and considerably improve classification results. In continuum, we describe the GAN framework used for generating minority class examples. For instance, Rezaei et al. proposed different 2D and 3D conditional and unsupervised DCGAN for MRI image synthesis to mitigate imbalanced class problem. The researchers then provided different experiments on balanced datasets with GAN generated samples, classical data augmentation, and imbalanced dataset. As expected, the diversity gained from synthesized images introduced more variability into the training set and considerably improved classification results. Figs. 5.7 and 5.8 show different settings by unsupervised and conditional configuration for image synthetization, respectively.

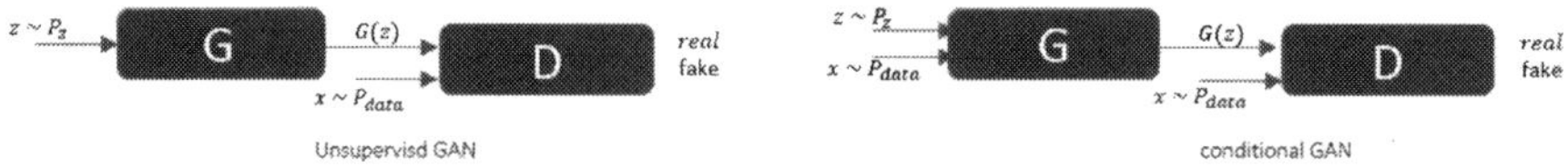

FIGURE 5.6 Comparision of unsupervised GAN and semisupervised or conditional GAN.

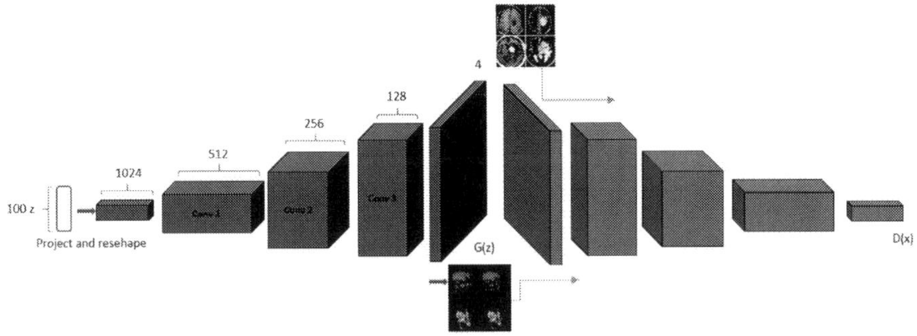

FIGURE 5.7 Unsupervised deep convolutional generative adversarial architecture for synthetic minority oversampling.

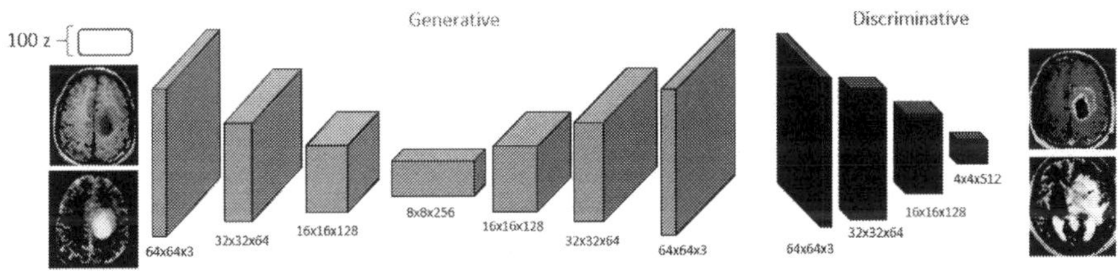

FIGURE 5.8 Deep conditional GAN architecture for synthetic minority oversampling regarding missing modality.

In a similar study, PGAN (Golany & Radinsky, 2019) has been proposed to generate personalized electrocardiogram (ECG) signals and improve ECG classification results by a convolutional generative adversarial network (DCGAN). Sandfort et al. (2019) showed that artificially generated cardiac CT could improve segmentation accuracy better than both imbalanced and balanced real datasets.

Cross modality synthesis (such as generating CT-like images based on MR images) (Moradi et al., 2016) is deemed to be useful for multiple reasons, one of which is to reduce the extra acquisition time and cost. Another reason is to generate new training samples with the appearance being constrained by the anatomical structures delineated in the available modality. Most of the methods reviewed in this section share many similarities to those in Deep learning in cardiac medical imaging section. Pix2Pix-based frameworks have been used in cases where different imaging modality data can be coregistered to ensure data fidelity. CycleGAN-based frameworks are used to handle more general cases where registration is challenging such as in cardiac applications. Wolterink, Leiner, Viergever and Išgum (2017) synthesized CT images from MR images and found that training using unpaired images was even better than using aligned images.

Similar to Zhu et al. (2017), Sandfort et al. (2019) successfully leveraged Cycle-GANs for unpaired image-to-image translation, specifically for synthesizing pairs of cardiac MRI images and a segmentation mask from pairs of cardiac CT slices with the ground-truth segmentation mask. The authors showed that the performance of a segmentation model can be improved by 16% when additionally trained on the synthetic data and that synthetic data alone are sufficient for training a model that performs merely 5% lower than a model trained on real data. Oktay et al. (2016) generated superresolution images from MRI and improved classification accuracy using newly generated samples.

PET. Positron emission tomography (PET) images are frequently used for diagnosis and staging in oncology, and the combined acquisition of PET and CT images is a standard procedure in clinical practice. Furthermore, PET/CT imaging is becoming an important evaluation tool for new drug therapies. However, PET studies are expensive and lead to radiation exposure, which carries an increased risk of future malignancy. Consequently, the imaging community has been working on synthesizing PET images directly from CT data. In this context, GANs have also shown outstanding performance. Initial promising results for synthesizing liver PET images from CT data with conditional GANs have been obtained by Mirza and Osindero (2014). The conditional GAN was able to synthesize very realistic looking PET images, however, at the cost of low response to underrepresented tumor regions. This led to poor tumor detection performance in a set of use-case experiments. In contrast, the authors found that a fully connected network (FCN) for PET image synthesis has been capable of synthesizing tumors but produces blurry images in general. By blending corresponding synthetic PET images from the conditional GAN and the FCN, they were able to achieve very high tumor detection performance. Similarly, Ben-Cohen et al. (2017) utilized a conditional GAN for synthesizing 200×200 pixel-sized PET images from pairs of CT images and binary label maps. Although CT images alone would be sufficient as input, the addition of the label map provides the tumor location, and as such the investigators were able to obtain globally more realistic, synthetic output. Because of the two-channel input to the generator, they referred to their network as the multichannel GAN.

MRI. PET from MRI has been used for monitoring disease of progression, understanding the physiopathology while evaluating treatment efficacy in multiple sclerosis (MS). In addition, the measurement of myelin content in PET images of the human brain has recently been shown to be clinically useful (Ben-Cohen et al., 2019). Unfortunately, PET imaging for MS is costly and invasive, as it requires the injection of a radioactive tracer. Bi et al. (2017) successfully utilized a cascade of two conditional GANs for synthesizing such PET images from a set of different MR modalities. Their approach operates directly on volumetric data, leveraging a 3D U-Net for the generator networks and discriminator networks with 3D convolutions. Interestingly, the authors noted that a single conditional GAN was insufficient for the task at hand as it produced blurry images. Splitting the synthesis task into smaller, more stable subproblems seemed to drastically improve the results.

Ultrasound. Tom and Sheet (2018) proposed a conditional GAN architecture for synthesizing 2D ultrasound (US) images of a fetus phantom, as produced by a freehand US probe, given 3D spatial pixel location within the anatomy. In contrast to the standard conditional GAN, the authors found it necessary to transform the pixel locations into feature maps and to concatenate them with the produced feature maps at each level of the generator to facilitate training. In their experiments, they demonstrate the capability of simulating US images at locations unseen to the network, quantify the generation of ultrasound images by comparing the location of clinically relevant anatomical landmarks in synthetic and real images, and verifying the realism of the generated images in a usability study. The quantitative results showed that anatomical landmarks were roughly synthesized at the right location with a mean error of 6.1 mm. In the usability study, a sonographer was able to correctly distinguish between real and generated samples, mostly as a result of the checkerboard artifact in the synthetic images. After blurring the images using a Gaussian kernel with $\sigma = 1.5$, the sonographer was not able to reliably tell the difference anymore. The interested reader is also referred to the NiftyNet framework (Gibson et al., 2018), in which this conditional GAN is contained. Tom and Sheet (2018) applied GANs for intravascular ultrasound (IVUS) simulation in a multistage setup. A first generator conditioned on physically simulated tissue maps produced speckle images, which in turn acted as the conditioning input to a second residual network-based generator. The second generator mapped the speckle images to low resolution, synthetic 64×64 pixels sized US images. A third generator transformed these low-resolution images into high-resolution samples at a resolution of 256×256 pixels. In a visual test, the synthetic images could not reliably be distinguished from real ones.

X-ray. Bozorgtabar et al. (2019) trained conditional GANs for chest X-ray synthesis from perturbed segmentation maps and combined it with Bayesian neural networks for active learning. More precisely, they proposed an uncertainty-based synthetic sample selection approach for training segmentation neural networks on samples that are informative to a specific model.

Microscopy. Han and Yin (2017) proposed a conditional GAN framework similar to Pix2Pix for transferring between phase contrast and differential interference contrast microscopy images, however with two discriminator networks rather than one. A U-net like generator was trained to synthesize the image of a certain modality from an image of the source modality and a cell mask. Table 5.3 summarizes the GAN application in medical image analysis and cardiac images.

Classification and detection

Classification is one of the most successful applications of deep neural networks and generative adversarial networks that have been applied to various medical imaging benchmarks. Hierarchical image features can be extracted from a deep neural network and trained with imagewise class labels. GANs have been used for classification problems as well, either using part of the generator and/or discriminator as a feature extractor or directly using the discriminator as a classifier. Examples of classification by conditional GAN on cardiac images are included (Lau et al., 2018; Madani et al., 2018; Rezaei et al., 2018a,b,c).

Computational detection of anomalies from images using supervised algorithms require a large amount of annotated training data—extremely difficult and time consuming to be provided by experts. Unsupervised algorithms, such as GANs, have the potential to be effective in such a setting. The main driving idea of anomaly detection is to train GAN on normal images, so that inference on diseased images would lead to incomplete reconstructions. Such incompleteness is typically associated with the regions of an anomaly. Schlegl et al. (2019) showed that such an idea could be effective in detecting anomalies in optical coherence tomography images of the retina. They proposed an unsupervised GAN-based architecture (AnoGAN) that during testing, given an input diseased image, generates the corresponding healthy image version. The difference between generated and input images is considered as an anomaly. Chen, Li, Zhang and Menze (2019), Watanabe, Togo, Ogawa and Haseyama (2019) and Zenati et al. (2018) modified the architecture of Schlegl and colleagues and applied it for anomaly detection in brain MRI images.

Segmentation

Conditional GANs (cGANs) have the advantage of being able to provide better representations for multimodal data generation because there is a control on the modes of the data being generated. This makes cGANs suitable for image semantic segmentation task, where we condition on an observed image and generate a corresponding output image.

Cascade-GANs (Rezaei et al., 2018a,b,c) presented an ensemble of conditional GAN on the HVSMR dataset for great vessel and myocardium segmentation with the idea of multiscale L1 loss function in the generative network.

Datasets, challenges, and current public benchmarks

In this section, we overview the most popular benchmark and cardiac image challenges from 2011 until 2020 (Fig. 5.9).

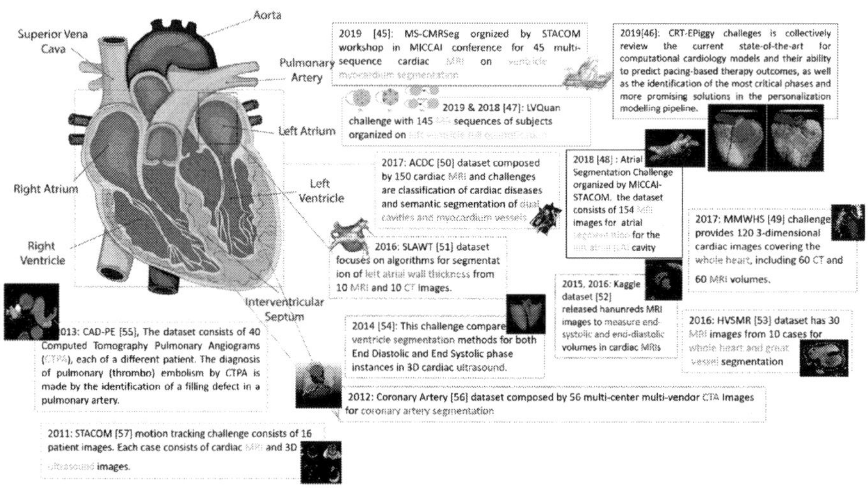

FIGURE 5.9 Different challenges for cardiac anatomy tasks.

MS-CMRSeg (Zhuang et al., 2019): The multisequence cardiac MRI segmentation (MS-CMRISeg) challenge aims at creating an open competition for various research groups to test and validate their methods, particularly for the multisequence ventricular and myocardial segmentation. This challenge provides 45 multisequence cardiac MRI images from patients with underlying cardiomyopathy. Each patient had been scanned using three cardiac MRI sequences, that is, late gadolinium enhancement (LGE), T2, and balanced steady-state free precession (bSSFP). The task of this challenge is to segment the ventricles and myocardium from LGE, combining it with the two other sequences (T2 and bSSFP) from the same patients, which can then be used to assist the LGE segmentation. This challenge is in conjunction with STACOM and MICCAI 2019 conferences.

CRT-EPiggy[1]: The aim of this challenge is to collectively review current state-of-the-art computational models and their ability to predict pacing-based therapy outcomes, as well as the identification of the most critical phases and more promising solutions in the personalization modeling pipeline. This challenge is in conjunction with the STACOM workshop as part of the MICCAI 2019 conference. The data are available on the webpage of challenge.[2]

LVQuan[3] (Tobon-Gomez et al., 2013): This challenge is an extension of Left Ventricle Full Quantification Challenge MICCAI 2018 (LVQuan18), the main difference is that this challenge (LVQuan19) will provide original data without preprocessing for training and testing phases, which is more clinically relevant than the data provided by LVQuan18. This challenge focuses on the analysis of the left ventricle in patients undergoing cardiac MRI, utilizing morphology and temporal dynamics to determine LV function. Such an analysis would be time consuming within a clinical routine and is prone to error and interobserver variability. In this challenge, the extraction of the LV cavity as well as myocardium, and subsequently the regression of regional wall thicknesses, LV dimensions, and the classification of the phase of the cardiac cycle are to be performed. These are parameters that are commonly used to assess LV function. As such, the major emphasis of this challenge is to train effective machine learning models that can estimate a set of clinically relevant LV indices (regional wall thicknesses, cavity dimensions, area of cavity and myocardium, cardiac phase) directly from MR images. No intermediate segmentation will be required throughout the whole process.

Atrial segmentation[4]: A total of 154 of 3D MRIs from patients with atrial fibrillation are to be used for this challenge. The participants are required to produce a computational framework capable of performing fully automatic segmentation of the left atrial (LA) cavity from 3D GE-MRIs without any manual assistance (data resolution is $0.625 \times 0.625 \times 0.625$ mm^3). Each 3D MRI patient data has been

[1] https://zenodo.org/record/3249511#.Xl9iam5FyUk.
[2] https://crt-epiggy19.surge.sh/.
[3] https://lvquan19.github.io/.
[4] http://atriaseg2018.cardiacatlas.org/.

acquired using a clinical whole-body MRI scanner, with each dataset containing the raw images as well as the corresponding expert-annotated ground truth labels for the LA cavity. The raw MRIs are in grayscale and the segmentation labels are binary (255 = positive, 0 = negative). The dimensions of the MRIs may vary between patients; however, all MRIs contain exactly 88 slices in the *Z*-axis. The dataset is split such that 100 scans are used for training, while 54 scans are used for testing and evaluation. Participants are to have access to all MRIs and their respective labels (LA cavity mask) in the training set, and only MRIs in the testing set.

MMWHS[5]: The organizers are providing 20 CT and 20 MRI volumes with corresponding manual segmentations of seven whole heart substructures. The volumes were acquired in clinics with different scanners, resulting in varying image quality, resolution, and voxel spacing.

ACDC (Bernard et al., 2018). The dataset covers several well-defined pathologies with enough cases to (1) properly train machine learning methods and (2) assess the variations of the main physiological parameters obtained from cine MRI (in particular diastolic volume and ejection fraction). The dataset is composed of 150 exams (all from different patients) divided into five evenly distributed subgroups (four pathological plus one healthy subject groups) as described later. Furthermore, each patient comes profile has the following additional information: weight, height, as well as the diastolic and systolic phase instants. The database is made available to participants through two datasets from the dedicated online evaluation website: (1) a training dataset of 100 patients along with the corresponding manual references based on the analysis of one clinical expert; (2) a testing dataset composed of 50 additional patients, without manual annotations but with detailed patient information. The raw input images are provided through the Nifti format.

SLAWT (Jia, Cadour, Cochet, & Sermesant, 2016): This challenge focused on algorithms for the segmentation of LA wall thickness from MRI and CT. It was part of the Statistical Modeling and Computational Modeling of the Heart (STACOM) 2016 workshop held at MICCAI 2016. Images included in the challenge consisted of MRI ($n = 10$) and CT ($n = 10$) datasets. The MRI images had been acquired at 1.4 mm isotropic resolution, while the CT images were acquired at 0.5 mm in-plane resolution with a slice thickness of 1 mm.

Kaggle[6]: The challenge of using MRI to measure cardiac volumes and derive ejection fraction is the fact that the process is manual and time consuming. For instance, a skilled cardiologist typically analyzes and interprets MRI scans to determine the ejection fraction. The process can take up valuable time, which a provider could otherwise be using for direct patient care to 20 min to complete the time the cardiologist could be spending with his or her patients. As a result, making this measurement process more efficient will enhance the physicians' ability

[5] http://www.sdspeople.fudan.edu.cn/zhuangxiahai/0/mmwhs/.

[6] https://www.kaggle.com/c/second-annual-data-science-bowl.

to streamline the clinical workflow for MRI scan interpretation while providing greater opportunity for direct patient contact as well as efficient clinical work.

The 2015 Data Science Bowl challenge aimed at creating a benchmark of machine learning algorithms for the automated measurement of end-systolic and end-diastolic volumes on cardiac MRI. The database includes more than 1000 MRI images in 2D cine frames each containing around 30 images across the cardiac cycle. This dataset was compiled by the National Institutes of Health and the Children's National Medical Center and is an order of magnitude larger than any cardiac MRI dataset previously released. The competition task had been to create an automated method capable of determining the left ventricular volume at two time points: after systole, when the heart is contracted and the ventricles are at their minimum volume, and after diastole, when the heart is at its largest volume.

HVSMR[7]: The 3D cardiovascular magnetic resonance images had been acquired through clinical practice at Boston Children's Hospital, Boston. Cases included a variety of congenital heart defects, while some subjects had undergone interventions. Imaging had been done in an axial view on a 1.5T scanner (Phillips Achieva) without contrast administration using a steady-state free precession (SSFP) pulse sequence. Subjects breathed freely during the scan, while ECG and respiratory-navigator gating were used to remove cardiac and respiratory motion (TR = 3.4 ms, TE = 1.7 ms, $\alpha = 60°$). Image dimension and image spacing varied across subjects, an average of $390 \times 390 \times 165$ and $0.9 \times 0.9 \times 0.85$ mm, respectively, in the full-volume training dataset.

CETUS[8]: The goal of this contest is to compare left-ventricle segmentation methods for both end-diastolic and end-systolic phase instances. The global dataset is composed of 45 sequences of 3D ultrasound volumes of one cardiac cycle from 45 patients. Ultrasound recordings had been acquired at three different hospitals with three different manufacturers of ultrasound machines. The database will be made available to participants in three stages through the dedicated MIDAS website after a personal registration. A training dataset of 15 patients along with the corresponding manual contours has been provided from March of 2014. The two remaining test datasets, each containing 15 patients, without manual annotations will then be provided to the participants. The evaluation procedure is done fully and automatically on the MIDAS website.

CAD-PE[9]: A pulmonary (thrombo)-embolism (PE) refers to the situation when a blood clot becomes lodged in a pulmonary artery. PE is usually diagnosed with computed tomography pulmonary angiography (CTPA). Delay of treatment or lack of treatment results in significant morbidity and mortality. In comparison to an expert panel, the average radiologist has a sensitivity between 77% and 94% for detecting a PE on CTPA. Computer-aided detection (CAD) algorithms have

---

[7] http://segchd.csail.mit.edu/index.html.

[8] https://www.creatis.insa-lyon.fr/Challenge/CETUS/index.html.

[9] http://www.cad-pe.org/.

been shown to increase radiologists' sensitivity. However, CAD has not been adopted into clinical practice due to the large amount of false positive detections that those systems produce. This challenge aims at fostering research in the field of CAD for PE detection and to provide an objective benchmark with which to compare the already existent algorithms.

Coronary artery (Kirişli et al., 2013): Forty-eight multicenter multivendor computed tomography angiography (CTA) datasets of symptomatic patients are used. Eighteen of the CTA images, with associated reference standard quantifications obtained from quantitative coronary angiography and consensus reading of CTA, are available for optimizing the algorithms (i.e., training set). The remaining 30 studies are used to validate the algorithms (i.e., testing sets); for this set, the reference standard is not shared with the participants.

Motion tracking challenge[10]: The objective of this challenge is to evaluate the accuracy and reproducibility of different motion analysis algorithms applied to a multimodal cardiac database. The database includes a dataset from a dynamic phantom 1 and 15 datasets from healthy volunteers. The data were acquired at the Division of Imaging Sciences and Biomedical Engineering, King's College London, United Kingdom, and the Department of Internal Medicine II—Cardiology, University of Ulm, Germany. The Cardiac Atlas Project hosts the challenge data. Each case consists of cardiac MRI and 3D ultrasound images. The MR acquisition includes (1) cine SSFP sequences in 2-chamber, 4-chamber, and short-axis views, (2) a whole-heart SSFP sequence gated at end-diastole and end-expiration; and (3) a 4D tagged MRI sequence. Sequences including three orthogonal tagged directions will be provided together with a fused grid-tagged volume sequence, in VTK and NIFTI formats. The ultrasound acquisition includes a 4D apical volume acquisition. All datasets can be provided in an anonymized DICOM format.

Challenges, limitations, and future perspectives

It is clear that deep learning and generative adversarial neural networks are very useful when one is tasked with producing accurate decisions based on complicated datasets. However, there are some major challenges and limitations of GANs that were discussed in Research background section. Emphasis should be placed on technical challenges such as mode collapse, none convergence, lack of interpretability and theoretical underpinnings of many GAN models and techniques, and the resulting difficulty in deciding exactly what it is that makes one model better than another, specifically in medical datasets. Other challenges are a result of difficulties in evaluation and assessment of performance.

The progress of the GANs model in medical imaging and the clinical dataset is not quite as settled. One issue is that medical images are often three-dimensional,

[10] http://stacom.cardiacatlas.org/motion-tracking-challenge/.

and three-dimensional convolutional neural networks are not as well developed as their 2D counterparts. A significant challenge relates to memory and compute consumption when using CNNs with higher-dimensional imaging data; challenges that researchers are trying various approaches to deal with (treating 3D as stacks of 2Ds, patch- or segment-based training and inference, downscaling, etc.). Other important challenges are related to data, trust, interpretability, workflow integration, and regulations.

References

Abdi, A. H., et al. (2017). Automatic quality assessment of echocardiograms using convolutional neural networks: Feasibility on the apical four-chamber view. *IEEE Transactions on Medical Imaging, 36*(6), 1221–1230.

Afshin, M., et al. (2013). Regional assessment of cardiac left ventricular myocardial function via MRI statistical features. *IEEE Transactions on Medical Imaging, 33*(2), 481–494.

Balduzzi, D., et al. (2018). The mechanics of n-player differentiable games. *arXiv*. preprint arXiv:1802.05642.

Bellemare, M. G., et al. (2017). The cramer distance as a solution to biased wasserstein gradients. *arXiv*. preprint arXiv:1705.10743.

Ben-Cohen, A., et al. (2017). *Virtual PET images from CT data using deep convolutional networks: Initial results*. s.l., s.n. (pp. 49–57)

Ben-Cohen, A., et al. (2019). Cross-modality synthesis from CT to PET using FCN and GAN networks for improved automated lesion detection. *Engineering Applications of Artificial Intelligence, 78*, 186–194.

Benjamin, E. J., Muntner, P., & Bittencourt, M. S. (2019). Heart disease and stroke statistics-2019 update: A report from the American heart association. *Circulation, 139*(10), e56–e528.

Bernard, O., et al. (2018). Deep learning techniques for automatic MRI cardiac multi-structures segmentation and diagnosis: Is the problem solved? *IEEE Transactions on Medical Imaging, 37*(11), 2514–2525.

Betancur, J., et al. (2018). Deep learning for prediction of obstructive disease from fast myocardial perfusion SPECT: A multicenter study. *Journal of the American College of Cardiology: Cardiovascular Imaging, 11*(11), 1654–1663.

Bi, L., et al. (2017). Synthesis of positron emission tomography (PET) images via multi-channel generative adversarial networks (GANs). In *Molecular imaging, reconstruction and analysis of moving body organs, and stroke imaging and treatment* (pp. 43–51). Springer.

Bozorgtabar, B., et al. (2019). Informative sample generation using class aware generative adversarial networks for classification of chest Xrays. *Computer Vision and Image Understanding, 184*, 57–65.

Chen, X., et al. (2016). *Infogan: Interpretable representation learning by information maximizing generative adversarial nets*. s.l., s.n. (pp. 2172–2180)

Chen, J., Li, H., Zhang, J., & Menze, B. (2019). Adversarial convolutional networks with weak domain-transfer for multi-sequence cardiac MR images segmentation. *arXiv*. preprint arXiv:1908.09298.

Choromanska, A., et al. (2015). *The loss surfaces of multilayer networks*. s.l., s.n. (pp. 192–204)

Cocosco, C. A., et al. (2008). Automatic image-driven segmentation of the ventricles in cardiac cine MRI. *Journal of Magnetic Resonance Imaging: An Official Journal of the International Society for Magnetic Resonance in Medicine, 28*(2), 366–374.

De Fauw, J., et al. (2018). Clinically applicable deep learning for diagnosis and referral in retinal disease. *Nature Medicine, 24*(9), 1342–1350.

Donahue, J., & Simonyan, K. (2019). *Large scale adversarial representation learning*. s.l., s.n. (pp. 10541–10551)

Duan, J., et al. (2019). Automatic 3D bi-ventricular segmentation of cardiac images by a shape-refined multi-task deep learning approach. *IEEE Transactions on Medical Imaging, 38*(9), 2151–2164.

Eslami, A., Karamalis, A., Katouzian, A., & Navab, N. (2013). Segmentation by retrieval with guided random walks: Application to left ventricle segmentation in MRI. *Medical Image Analysis, 17*(2), 236–253.

Feizi, S., Farnia, F., Ginart, T., & Tse, D. (2017). Understanding gans: The lqg setting. *arXiv*. preprint arXiv:1710.10793.

Folkesson, J., Samset, E., Kwong, R. Y., & Westin, C.-F. (2008). Unifying statistical classification and geodesic active regions for segmentation of cardiac MRI. *IEEE Transactions on Information Technology in Biomedicine, 12*(3), 328–334.

Gandhi, S., Mosleh, W., Shen, J., & Chow, C.-M. (2018). Automation, machine learning, and artificial intelligence in echocardiography: A brave new world. *Echocardiography, 35*(9), 1402–1418.

Gibson, E., et al. (2018). NiftyNet: A deep-learning platform for medical imaging. *Computer Methods and Programs in Biomedicine, 158*, 113–122.

Golany, T., & Radinsky, K. (2019). *PGANs: Personalized generative adversarial networks for ECG synthesis to improve patient-specific deep ECG classification*. s.l., s.n. (pp. 557–564)

Goodfellow, I., et al. (2014). *Generative adversarial nets*. s.l., s.n. (pp. 2672–2680)

Gulrajani, I., et al. (2017). *Improved training of wasserstein gans*. s.l., s.n. (pp. 5767–5777)

Gulshan, V., et al. (2016). Development and validation of a deep learning algorithm for detection of diabetic retinopathy in retinal fundus photographs. *The Journal of the American Medical Association, 316*(22), 2402–2410.

Guo, X., Hong, J., Lin, T., & Yang, N. (2017). Relaxed wasserstein with applications to gans. *arXiv*. preprint arXiv:1705.07164.

Han, L., & Yin, Z. (2017). *Transferring microscopy image modalities with conditional generative adversarial networks*. s.l., s.n. (pp. 99–107)

Heusel, M., et al. (2017). *Gans trained by a two time-scale update rule converge to a local nash equilibrium*. s.l., s.n. (pp. 6626–6637)

Isola, P., Zhu, J.-Y., Zhou, T., & Efros, A. A. (2017). *Image-to-image translation with conditional adversarial networks*. s.l., s.n. (pp. 1125–1134)

Jia, S., Cadour, L., Cochet, H., & Sermesant, M. (2016). *STACOM-SLAWT challenge: Left atrial wall segmentation and thickness measurement using region growing and marker-controlled geodesic active contour*. s.l., s.n. (pp. 211–219)

Khrulkov, V., & Oseledets, I. (2018). Geometry score: A method for comparing generative adversarial networks. *arXiv*. preprint arXiv:1802.02664.

Kirişli, H. A., et al. (2013). Standardized evaluation framework for evaluating coronary artery stenosis detection, stenosis quantification and lumen segmentation algorithms in computed tomography angiography. *Medical Image Analysis, 17*(8), 859–876.

Kurach, K., et al. (2018). *The gan landscape: Losses, architectures, regularization, and normalization.*

Lau, F., et al. (2018). Scargan: Chained generative adversarial networks to simulate pathological tissue on cardiovascular mr scans. In *Deep learning in medical image analysis and multimodal learning for clinical decision support* (pp. 343–350). Springer. s.l.

LeCun, Y., Bengio, Y., & Hinton, G. (2015). Deep learning. *Nature, 521*(7553), 436–444.

Ledig, C., et al. (2017). *Photo-realistic single image super-resolution using a generative adversarial network*. s.l., s.n. (pp. 4681–4690)

Lessmann, N., et al. (2017). Automatic calcium scoring in low-dose chest CT using deep neural networks with dilated convolutions. *IEEE Transactions on Medical Imaging, 37*(2), 615–625.

Levin, D. C., Parker, L., Halpern, E. J., & Rao, V. M. (2019). Coronary CT angiography: Reversal of earlier utilization trends. *Journal of the American College of Radiology, 16*(2), 147–155.

Litjens, G., et al. (2017). A survey on deep learning in medical image analysis. *Medical Image Analysis, 42*, 60–88.

Lucic, M., et al. (2018). *Are gans created equal? A large-scale study*. s.l., s.n. (pp. 700–709)

Madani, A., Moradi, M., Karargyris, A., & Syeda-Mahmood, T. (2018). *Chest x-ray generation and data augmentation for cardiovascular abnormality classification*. s.l., s.n. (p. 105741M)

Mahapatra, D. (2014a). Analyzing training information from random forests for improved image segmentation. *IEEE Transactions on Image Processing, 23*(4), 1504–1512.

Mahapatra, D. (2014b). Automatic cardiac segmentation using semantic information from random forests. *Journal of Digital Imaging, 27*(6), 794–804.

Mao, X., et al. (2017). *Least squares generative adversarial networks*. s.l., s.n. (pp. 2794–2802)

Menze, B. H., et al. (2010). *A generative model for brain tumor segmentation in multi-modal images*. s.l., s.n. (pp. 151–159)

Metz, C. T., et al. (2009). Coronary centerline extraction from CT coronary angiography images using a minimum cost path approach. *Medical Physics, 36*(12), 5568–5579.

Mirza, M., & Osindero, S. (2014). Conditional generative adversarial nets. *arXiv*. preprint arXiv:1411.1784.

Moradi, M., et al. (2016). *A cross-modality neural network transform for semi-automatic medical image annotation*. s.l., s.n. (pp. 300–307)

Mroueh, Y., & Sercu, T. (2017). *Fisher gan*. s.l., s.n. (pp. 2513–2523)

Mroueh, Y., Sercu, T., & Goel, V. (2017). Mcgan: Mean and covariance feature matching gan. *arXiv*. preprint arXiv:1702.08398.

Oktay, O., et al. (2016). *Multi-input cardiac image super-resolution using convolutional neural networks*. s.l., s.n. (pp. 246–254)

Oliehoek, F. A., et al. (2018). *Beyond local nash equilibria for adversarial networks*. s.l., s.n. (pp. 73–89)

Peng, P., et al. (2016). A review of heart chamber segmentation for structural and functional analysis using cardiac magnetic resonance imaging. *Magnetic Resonance Materials in Physics, Biology and Medicine, 29*(2), 155–195.

Petitjean, C., et al. (2015). Right ventricle segmentation from cardiac MRI: A collation study. *Medical Image Analysis, 19*(1), 187–202.

Pohl, K. M., et al. (2006). A Bayesian model for joint segmentation and registration. *NeuroImage, 31*(1), 228–239.

Qiao, T., Zhang, J., Xu, D., & Tao, D. (2019). *Mirrorgan: Learning text-to-image generation by redescription*. s.l., s.n. (pp. 1505–1514)

Queirós, S., et al. (2014). Fast automatic myocardial segmentation in 4D cine CMR datasets. *Medical Image Analysis, 18*(7), 1115–1131.

Radford, A., Metz, L., & Chintala, S. (2015). Unsupervised representation learning with deep convolutional generative adversarial networks. *arXiv*. preprint arXiv:1511.06434.

Rezaei, M., et al. (2020). *Generative synthetic adversarial network for internal bias correction and handling class imbalance problem in medical image diagnosis*. s.l., s.n. (p. 1095027)

Rezaei, M., & Tajeripour, F. (2013). 31st international EuroPACS meeting. *International Journal of Computer Assisted Radiology and Surgery, 8*(1), S397–S402.

Rezaei, M., Yang, H., & Meinel, C. (2017). *Deep neural network with l2-norm unit for brain lesions detection*. s.l., s.n. (pp. 798–807)

Rezaei, M., Yang, H., & Meinel, C. (2018a). *Generative adversarial framework for learning multiple clinical tasks*. s.l., s.n. (pp. 1–8)

Rezaei, M., Yang, H., & Meinel, C. (2018b). *Instance tumor segmentation using multitask convolutional neural network*. s.l., s.n. (pp. 1–8)

Rezaei, M., Yang, H., & Meinel, C. (2018c). Whole heart and great vessel segmentation with context-aware of generative adversarial networks. In *Bildverarbeitung für die Medizin 2018* (pp. 353–358). Springer. s.l.

Rezaei, M., Yang, H., & Meinel, C. (2019a). *Learning imbalanced semantic segmentation through cross-domain relations of multi-agent generative adversarial networks*. s.l., s.n. (p. 1095027)

Rezaei, M., Yang, H., & Meinel, C. (2019b). Recurrent generative adversarial network for learning imbalanced medical image semantic segmentation. In *Multimedia tools and applications* (pp. 1–20).

Rohé, M.-M., Sermesant, M., & Pennec, X. (2017). *Automatic multi-atlas segmentation of myocardium with svf-net*. s.l., s.n. (pp. 170–177)

Ronneberger, O., Fischer, P., & Brox, T. (2015). *U-net: Convolutional networks for biomedical image segmentation*. s.l., s.n. (pp. 234–241)

Sabuncu, M. R., et al. (2010). A generative model for image segmentation based on label fusion. *IEEE Transactions on Medical Imaging, 29*(10), 1714–1729.

Salimans, T., et al. (2016). *Improved techniques for training gans*. s.l., s.n. (pp. 2234–2242)

Sandfort, V., Yan, K., Pickhardt, P. J., & Summers, R. M. (2019). Data augmentation using generative adversarial networks (CycleGAN) to improve generalizability in CT segmentation tasks. *Scientific Reports, 9*(1), 1–9.

Schlegl, T., et al. (2019). f-AnoGAN: Fast unsupervised anomaly detection with generative adversarial networks. *Medical Image Analysis, 54*, 30–44.

Srivastava, A., et al. (2017). *Veegan: Reducing mode collapse in gans using implicit variational learning*. s.l., s.n. (pp. 3308–3318)

Subbanna, N. K., Precup, D., Collins, D. L., & Arbel, T. (2013). *Hierarchical probabilistic Gabor and MRF segmentation of brain tumours in MRI volumes*. s.l., s.n. (pp. 751–758)

Tobon-Gomez, C., et al. (2013). *Left atrial segmentation challenge: A unified benchmarking framework*. s.l., s.n. (pp. 1–13)

Tom, F., & Sheet, D. (2018). *Simulating patho-realistic ultrasound images using deep generative networks with adversarial learning*. s.l., s.n. (pp. 1174–1177)

Toth, D., et al. (2018). 3D/2D model-to-image registration by imitation learning for cardiac procedures. *International Journal of Computer Assisted Radiology and Surgery, 13*(8), 1141–1149.

Wang, C., et al. (2018). A two-stage 3D Unet framework for multi-class segmentation on full resolution image. *arXiv.* preprint arXiv:1804.04341.

Wang, Z., She, Q., & Ward, T. E. (2019). Generative adversarial networks: A survey and taxonomy. *arXiv.* preprint arXiv:1906.01529.

Watanabe, H., Togo, R., Ogawa, T., & Haseyama, M. (2019). *Bone metastatic tumor detection based on AnoGAN using CT images*. s.l., s.n. (pp. 235–236)

Wolterink, J. M., et al. (2015). Automatic coronary calcium scoring in non-contrast-enhanced ECG-triggered cardiac CT with ambiguity detection. *IEEE Transactions on Medical Imaging, 34*(9), 1867–1878.

Wolterink, J. M., Leiner, T., Viergever, M. A., & Išgum, I. (2017). Generative adversarial networks for noise reduction in low-dose CT. *IEEE Transactions on Medical Imaging, 36*(12), 2536–2545.

Wu, D., et al. (2019). Automated anatomical labeling of coronary arteries via bidirectional tree LSTMS. *International Journal of Computer Assisted Radiology and Surgery, 14*(2), 271–280.

Xiang, S., & Li, H. (2017). On the effects of batch and weight normalization in generative adversarial networks. *arXiv.* preprint arXiv:1704.03971.

Xue, Y., et al. (2018). Segan: Adversarial network with multi-scale l 1 loss for medical image segmentation. *Neuroinformatics, 16*(3–4), 383–392.

Yang, J., Staib, L. H., & Duncan, J. S. (2004). Neighbor-constrained segmentation with level set based 3-D deformable models. *IEEE Transactions on Medical Imaging, 23*(8), 940–948.

Ye, C., Wang, W., Zhang, S., & Wang, K. (2019). Multi-depth fusion network for whole-heart CT image segmentation. *IEEE Access, 7*, 23421–23429.

Zenati, H., et al. (2018). Efficient gan-based anomaly detection. *arXiv.* preprint arXiv: 1802.06222.

Zhang, H., et al. (2017). *Stackgan: Text to photo-realistic image synthesis with stacked generative adversarial networks*. s.l., s.n. (pp. 5907–5915)

Zhao, J., Mathieu, M., & LeCun, Y. (2016). Energy-based generative adversarial network. *arXiv.* preprint arXiv:1609.03126.

Zhuang, X., et al. (2019). Evaluation of algorithms for multi-modality whole heart segmentation: An open-access grand challenge. *Medical Image Analysis, 58*, 101537.

Zhu, J.-Y., Park, T., Isola, P., & Efros, A. A. (2017). *Unpaired image-to-image translation using cycle-consistent adversarial networks*. s.l., s.n. (pp. 2223–2232)

Zotti, C., et al. (2017). *GridNet with automatic shape prior registration for automatic MRI cardiac segmentation*. s.l., s.n. (pp. 73–81)

CHAPTER

Natural language processing

6

Jake Vasilakes, Sicheng Zhou, Rui Zhang

Institute for Health Informatics and College of Pharmacy, University of Minnesota, Minneapolis, MN, United States

Chapter outline

Machine Learning in Cardiovascular Medicine. https://doi.org/10.1016/B978-0-12-820273-9.00006-3

Introduction to biomedical natural language processing

It is widely assumed that a key component of intelligence is the ability to understand and use natural language. *Natural languages* are methods of communication that have developed naturally within human society without explicit planning. These include languages such as English, Chinese, and Pashto. *Natural language processing* (NLP) is a subfield of artificial intelligence whose goal is to program computers to understand and analyze natural language. The main challenge NLP faces in accomplishing this goal is *ambiguity*, which abounds in natural language. Words often have multiple meanings (e.g., "duck" in "We saw her duck."), as do sentences (e.g., "No one does what we do well."), and ungrammatical English is often still quite meaningful (e.g., "Me want cheeseburger" still gets the point across). Natural language thus stands in contrast to *formal language* such as computer programming languages and languages of logic. Formal languages are completely unambiguous. For example, a statement in C++ has exactly one meaning and all statements that are not grammatical C++ have no meaning at all.

To address the ambiguity in natural language, NLP borrows extensively from the fields of linguistics, computer science, and artificial intelligence. Adopting this interdisciplinary approach, NLP has been successful for widely used languages such as English, specifically the "typical" use of English as in news articles and social media posts. However, as the field matures and NLP techniques become more successful, researchers have started applying NLP to various new domains.

In general, the goal of NLP is to understand what a given collection of texts, called a *corpus*, is about. The biomedical domain, which includes journal articles and clinical notes from electronic health records (EHRs), is one of the most exciting domains to which NLP has been applied in recent years. NLP has found much success here, having been used to automatically extract knowledge from the biomedical literature, to improve clinical outcomes via clinical decision support, and to discover new drug interactions. However, the analysis of a biomedical corpus (collection of texts) brings forth an additional set of challenges over traditional NLP, in the form of increased ambiguity and patient data privacy.

Ambiguity is a major challenge in biomedical text. Biomedical text has its own *sublanguage*, that is, sets of words, phrases, and structures that differ significantly from typical language (Friedman, Kra, & Rzhetsky, 2002). It contains a number of words that do not occur elsewhere (e.g., "eIF2α Kinase GCN2") and clinical text uses a shorthand that abandons much of the structure of typical language (e.g., "68yo WM c CHF and DMII admitted for edema and DOE"). As such, many of the NLP systems developed for typical language do not generalize to biomedical text and thus perform poorly. These sublanguages can, furthermore, differ between subdomains and clinical sites making the task even more difficult.

Indeed, a major challenge in clinical NLP is generalizability: systems developed using data from one clinical site or specialty (e.g., surgery or cardiovascular medicine) often do not perform well when applied to data from another (Carrell et al., 2017).

Patient data privacy is another critical issue for NLP applied to the clinical text. The use of clinical text corpora requires prior approval from institutional review boards (IRBs) and often requires *de-identification*, the process of removing all protected health information (e.g., names, dates, and addresses) that could allow one to identify the patient discussed in the note. Obtaining IRB approval and data deidentification are lengthy processes that significantly affect the speed with which new clinical NLP systems can be developed. Even once these processes are complete, sharing data across institutions is difficult, which further hampers the development of NLP systems that perform well across clinical sites.

The above challenges, while specific to biomedical NLP, pose questions that are important for the field of NLP as a whole to answer: *How should we approach sentence structure and meaning for ungrammatical or semi-grammatical text? How can we efficiently learn the meanings of previously unseen words? How can we develop NLP systems that respect personal (patient) data privacy?* We are a long way from definitive answers to these questions and so these are crucial points to consider when deploying NLP systems in the real world.

In the following sections, we discuss (1) The typical tasks when working with biomedical NLP and (2) The methods commonly used to address them. We are careful to illustrate where biomedical NLP differs significantly from traditional NLP, why it does so, and any implications.

Tasks in biomedical NLP

Creating an NLP model that understands a given corpus requires the completion of a number of tasks, which can be broadly categorized into *high-level* and *low-level* tasks. High-level tasks in NLP are often an end in themselves, while low-level tasks are part of the pipeline in addressing a high-level task.

High-level tasks

High-level tasks include but are not limited to automatic language translation, summarization, and question answering. The most prevalent high-level task in biomedical NLP, however, is *information extraction* (IE). IE aims to discover information or knowledge that is, in a sense, hidden within the text. IE is therefore crucial for a variety of time- and life-saving clinical processes. For example, IE can help facilitate chart review by extracting information from clinical notes (Wu et al., 2013), it can identify potential drug interactions in the literature (Herrero-zazo, Segura-Bedmar, Martínez, & Declerck, 2013; Zhang et al., 2014; Liu, Tang, Chen, & Wang, 2016), and it can identify risk factors from clinical text to aid in clinical decision support (Demner-Fushman, Chapman, & McDonald, 2009). IE comprises a number of subtasks in practice—named-entity recognition, relationship extraction, and entity linking—which are described below. To help illustrate these subtasks, we employ the example of extracting adverse drug reactions from clinical reports.

Named-entity recognition

The goal of named-entity recognition (NER) is to find all mentions of certain types of concepts or entities in a given corpus. For example, an NER system might find mentions of symptoms and drugs within clinical notes. That is, it reads through the input text and outputs the entity type (symptom or drug or null) of each word. The main hurdle NER must face is *polysemy* or words with multiple meanings. In cases of polysemy, NER turns to one of two subtasks: *word-sense disambiguation* (WSD) and *acronym disambiguation*. The goal of WSD is to determine the intended meaning of a word as used in the text. WSD systems leverage context around the target word as well as syntactic information. For example, the word "cold" is polysemous, having different meanings according to the context and its role in the sentence, e.g., "I feel cold" (adjective) versus "I have a cold" (noun). *Acronym/abbreviation disambiguation* is related to WSD in that it aims to disambiguate acronyms by finding their correct expansion. For example, "MAP" expands to *mitogen-activated protein* but also *mean arterial pressure*.

Although polysemy is pervasive in text, many of the entities in the biomedical text are much less susceptible to misinterpretation. For example, the word "tamoxifen", especially in a clinical note, almost certainly refers to the breast cancer drug. In some cases, this means that rather than employ complex context- and syntax-aware methods, biomedical NER can proceed as a simple dictionary lookup. If the given word exists in a list of drugs, then it is annotated as such. For certain entity types, this method works relatively well in practice.

Entity linking

Biomedical entities often have a large variety of synonymous terms. For example, *vitamin C, L-ascorbic acid,* and *sodium ascorbate* all refer to the same entity. Giving a word or phrase a precise meaning is called *grounding* and it is often accomplished by the task of *entity linking*, which links entity mentions (such as those found by NER) to entries in terminology or database. Entity linking serves two purposes: (1) it is a means of normalizing all synonymous mentions into a single form (such as a specific database ID) and (2) it enables access to the other information stored about that entity in the database. For example, by linking the term *sodium ascorbate* to its PubChem entry 54,670,067, we gain access to its chemical structure, safety information, uses, etc. Entity linking also allows the knowledge generated by an IE system to interface with and inform other biomedical information systems.

Entity linking also faces issues with ambiguity. Even if the NER or WSD system correctly identifies a word's meaning, there may be multiple candidate database entries to link to. For example, should "increased mean blood pressure" be linked to the Unified Medical Language System (UMLS) entry `increased mean arterial pressure` (C0520853) or `mean arterial pressure increased` (C4087413)?

Relationship extraction

Relationship extraction (RE) aims to discover relationships between entities in the text. The entities are often obtained by running an NER system beforehand. Using the example of discovering adverse drug reactions, an RE system should discern whether (1) the given symptom is an adverse reaction of the drug, (2) the drug treats the symptom, or (3) the symptom and the drug are unrelated in the text. The output of an RE system is thus a set of triples with a <*subject, verb, object*> structure, such as <*tamoxifen, has_adverse_reaction, chest pain*>. RE systems use context, such as the words between the two entities and the sentence structure, to determine whether two entities are related. Both are crucial for an effective RE system as clues in the context can be negated by sentence structure. Take for example the pair of sentences "We reviewed the charts of patients who were administered tamoxifen. It has been reported to cause chest pain but not vomiting." First notice that the drug name "tamoxifen" does not occur in the second sentence, but is rather referred to by "It." Resolving the meaning of pronouns like "It" across sentences is a task called *anaphora resolution*. Second, the context "reported to cause" between the drug ("It") and the symptom "vomiting" indicates a potential adverse reaction, yet the "but not" indicates the opposite is true. The task of determining the scope of negation words such as this is called *negation detection* and *negation resolution*, respectively.

The set of triples output by an RE system can be considered hypotheses, which can inform real-world clinical outcomes such as those related to pharmacovigilance, precision medicine, and drug repurposing.

Low-level tasks

Low-level NLP tasks are those that are not an end in themselves but are rather an integral part of larger NLP systems and crucial to the systems that perform the high-level tasks given earlier. Some low-level tasks are *tokenization, sentence boundary detection*, *part-of-speech tagging, syntactic parsing,* and *language modeling*. Many modern and very capable systems exist to perform these tasks, even given the challenges of biomedical text, so this section will provide only a brief overview.

Tokenization

Tokenization is the process of splitting a text into its constituent word instances or *tokens*. Although in English most tokens are separated by whitespace or certain punctuation marks, biomedical text often contains exceptional tokens such as "q.i.d." or "L-ascorbic acid."

Sentence boundary detection

Even in grammatical language, determining where a sentence starts and ends can be difficult. The period "." which is used in many languages to denote the end of a sentence, occurs in other places, such as in acronyms (M.D.). This ambiguity is increased in clinical note shorthand, which can leave out punctuation altogether.

Systems for sentence segmentation thus employ a variety of contextual clues to determine whether a given character, such as a period, is indeed an end-of-sentence marker.

Part-of-speech tagging

The part of speech (POS) of a word indicates the role it plays in a sentence, such as verb, noun, and preposition. Knowing a word's POS can thus provide crucial clues about its meaning. Knowing, for example, that "Buffalo" is being used as a proper noun allows an NER system to narrow down the possible labels for it. POS tagging systems assign one of a number of tags, such as those from the Penn Treebank, to each word in the input.

Syntactic parsing

Although POS tagging gives the roles of individual words, syntactic parsing determines how words fit together syntactically. Syntactic parsing can be *shallow* or *deep*. Shallow syntactic parsing groups words into phrases, such a noun or verb phrases. Deep syntactic parsing goes a step further by determining how words and phrases fit together to form a grammatical sentence. In both cases, a syntactic parser starts with the POS tags and combines them according to a set of grammatical rules to form a *parse tree*.

A deep syntactic parse of the sentence "The patient reports experiencing chest pain." is given in Fig. 6.1. The structure of a sentence illuminated by a syntactic parse shows how phrases are related to each other and can thus provide crucial clues to an RE system. For example, the parse tree shows that there is indeed a connection via a verb phrase between "The patient" and "chest pain."

NLP methods

A number of methods exist for accomplishing the tasks described in the last section, which can be broadly grouped into *rule-based* methods and *statistical* or *machine-*

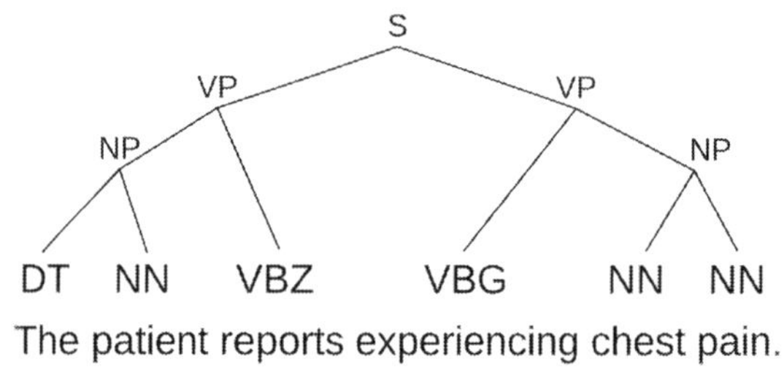

FIGURE 6.1

The full syntactic parse for the sentence "The patient reports experiencing chest pain." The parser recursively combines the POS tags according to predefined grammatical rules until it obtains a full parse. The abbreviations are defined as follows. *DT*, determiner; *NN*, singular noun; *NP*, noun phrase; *S*, sentence; *VBG*, gerund; *VBZ*, third-person singular verb; *VP*, verb phrase.

learning methods. This section describes some common methods in each of these categories.

Rule-based methods

As discussed in the introductory section, natural language often breaks the rules of grammaticality. It is therefore impossible to come up with a set of rules that would allow computers to understand every utterance in a given language. Nevertheless, rules can account for a number of linguistic phenomena. Although the general NLP community has generally moved away from the use of rule-based models for analyzing text in favor of more complex machine learning models, biomedical NLP continues to benefit from them. Rule-based systems such as MetaMap (National Library of Medicine, 2019) and NegEx (Chapman et al., 2013) are still widely used. Additionally, rule-based models are often straightforward to implement and use, and their decision-making processes are readily interpretable, both of which are vital attributes in the biomedical domain where NLP systems often carry the weight of affecting patients' health outcomes.

One of the most popular and powerful ways to implement rule-based NLP is by using *regular expressions* (regex). Regex is a small programming language designed to match and extract text. A given regex specifies a pattern to search for and is composed of literal characters and metacharacters. Literal characters allow for exact matches, for example, `cat`. Metacharacters have special meanings that either match multiple characters or which operate on another character. For example, the period "." matches any character except newline and the plus "+" means *match the previous character one or more times.* Thus the regex `cat.+` matches the literal string "cat" followed by any number of other characters such as "catamaran," "catch," etc.

There are a number of existing expert-curated biomedical terminologies and ontologies that lend even greater power to pattern matching systems. Resources such as the UMLS (Bodenreider, 2004) and the Medical Dictionary of Regulatory Activities (Maintenance and Support Services Organization, 2019) provide standard naming systems and taxonomies for a large number of biomedical concepts. By augmenting keywords obtained from these resources using regular expressions, an NLP system is often able to handle much of the linguistic variation in biomedical text, such as plural forms.

Statistical and machine learning methods

Despite the continued relevance of rule-based models in biomedical NLP, the rise of machine learning (ML), specifically deep learning, has allowed for unprecedented performance on a variety of biomedical NLP tasks. We here describe a few common ML models for NLP, but in general one can think of an ML model for NLP as taking in a collection of numbers representing the text and outputting a collection of probabilities or *predictions* corresponding to the specified hypotheses (e.g., "at risk of heart attack" vs "not at risk of heart attack"), which can then be used to make a decision.

Features

As mentioned earlier, the input to an ML model must be a numerical representation of the input, but what does it mean to represent a word or a sentence numerically? There are a number of specific implementations, but in general the text is transformed into a *vector*, that is, a list of numbers, that in some way represents its meaning.

Bag-of-words (BOW) is one of the simplest ways to represent text numerically. Each word in the *corpus* (collectively called the *vocabulary*) is assigned a place in a large vector. A "1" is marked down for each word that occurs in a given document. For example, say the vocabulary extracted from a *corpus* is the following:

[*the, patient, arrhythmia, doctor, reports, saw*]

Then the following sentences would be represented as

The patient reports arrhythmia [1, 1, 1, 0, 1, 0]
The doctor saw the patient [1, 0, 0, 1, 0, 1]

Note that BOW does not maintain the order of the words, nor does it indicate how many times a word occurs. Additionally, words such as "the" that provide little insight into a document's meaning are treated the same as more illuminating words, such as "arrhythmia." *Term frequency inverse document frequency (TF-IDF)* accounts for variation in word importance by weighting words according to a combination of their frequency in the document and their frequency in the entire corpus (Manning, Raghavan, & Schütze, 2008). "Important" words occur frequently in a few documents but infrequently across the entire corpus; these will receive a high score. Words that occur frequently throughout the corpus are "unimportant" and as such receive a low score.

Word embeddings or *word vectors* are one of the most important advances in NLP in recent years. As opposed to BOW and TF-IDF, which represent each word by a single number, word embeddings represent each word as a vector of numbers. Specifically, the numeric vector of a word is computed by a neural network model (discussed later) and is determined by the contexts in which that word occurs in a corpus. The assumption here is that words with similar contexts have similar meanings. In practice, this assumption is a powerful one. When estimated using a sufficiently large amount of text data, word embeddings capture phenomena such as synonymy and analogy surprisingly well. Fig. 6.2 provides a geometric interpretation of how word embeddings capture analogous meanings between words.

The invention of word embeddings has led to significant advances in NLP across tasks and domains. Indeed, it was a major motivation for the rise of neural network models in NLP. The original word embedding model, Google's *Word2Vec* (Mikolov, Sutskever, Chen, Corrado, & Dean, 2013), is still widely used and has been adapted to biomedical text (Pyysalo, Ginter, Moen, Salakoski, & Ananiadou, 2013; Zhang, Chen, Yang, Lin, & Lu, 2019). Other implementations are fastText (Bojanowski, Grave, Joulin, & Mikolov, 2017), which allows embeddings to be determined for out of vocabulary words, and GloVe (Pennington, Socher, & Manning, 2014), which

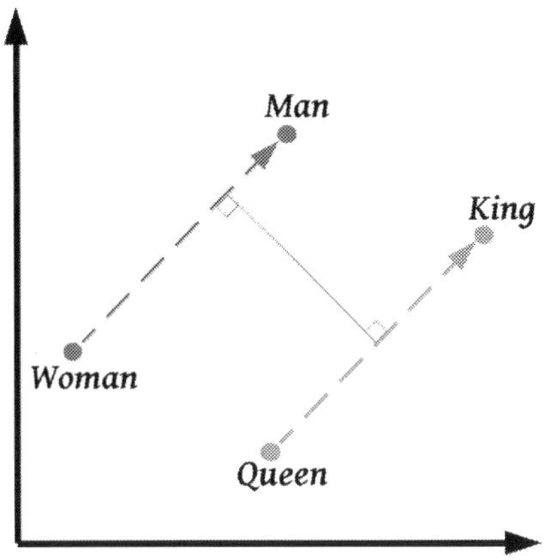

FIGURE 6.2

An illustration of how word embeddings capture similar meanings between concepts. The word embeddings for "Woman," "Man," "Queen," and "King" can be visualized as points in space. When estimated on sufficient data, the difference vectors between the embeddings of the word pairs are parallel and of the same magnitude, showing that the embeddings capture the distinction between woman and man.

estimates the embeddings from a word cooccurrence matrix, rather than directly from the text.

n-Gram language model

An *n*-gram language model is one that predicts the next word given the $n - 1$ previous words using a simple count-and-divide method. The predictive model is estimated for a given sequence of n words by dividing the number of times that sequence occurs in the corpus by the number of times the previous $n - 1$ words occur in the corpus. For example, say the 4-gram "She fed the dog" occurs two times in a *corpus* and the 3-gram "She fed the" occurs eight times. Then the probability that the next word is "dog" given "She fed the" is $\frac{2}{8} = 0.25$. Language models allow us to compute the overall probability of a series of words, which is useful for a variety of prediction tasks.

Hidden Markov model

Hidden Markov models (HMMs) are sequence models. That is, given a sequence of inputs, such as words, an HMM will compute a sequence of outputs of the same length. An HMM model is a graph where nodes are probability distributions over labels and edges give the probability of transitioning from one node to the other. Together, these can be used to compute the probability of a label sequence given the input sequence. Fig. 6.3 illustrates a small HMM for weather given the temperatures over a number of days. According to this model, the temperature is unlikely to change dramatically from 1 day to the next. That is, it is unlikely to go directly from $<0°C$ to $>20°C$.

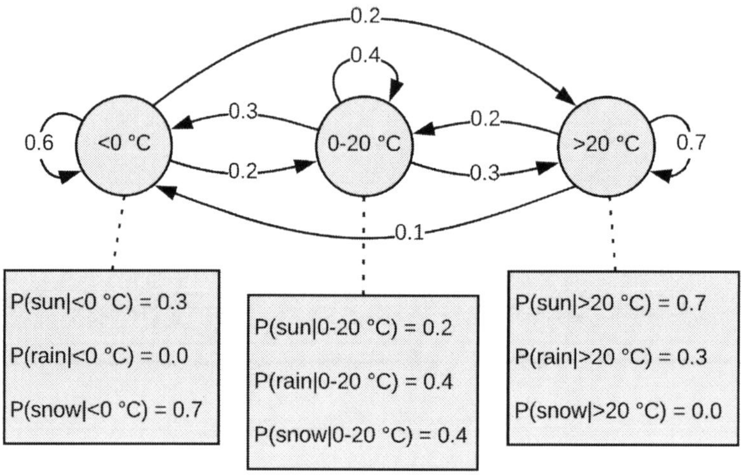

FIGURE 6.3

A simple hidden Markov model (HMM) for the weather over a number of days. Together, the transition probabilities between the states of the HMM and the emission probabilities at each state encode the probability of a sequence of temperatures and conditions. Such a model could be used to predict the next day's weather given the current weather.

HMMs are often used for tagging tasks such as POS tagging and NER where a given label depends not only on the word in question but also the sequence of labels up to that point (e.g., it is unlikely that a verb follows a preposition).

Support vector machine

The support vector machine (SVM) is a model that learns how to separate data points (e.g., words or sentences) into one of two possible classes. As opposed to *n*-gram models and HMMs, the SVM is a *discriminative* model in that it does not estimate probabilities of belonging to one class or the other but rather it finds a *decision boundary* that separates the data points in geometric space (Cortes & Vapnik, 1995). Furthermore, because there are potentially many boundaries that separate the data points, the SVM finds the boundary that has the maximum distance from the two closest data points of opposite classes (these are the support vectors). Fig. 6.4 shows a decision boundary estimated by an SVM on some toy data.

SVMs are most often used for prediction: after estimating a decision boundary on a set of data points for which we know the labels, the SVM predicts the labels of any new points by computing which side of the boundary they fall on. SVMs work quite well for a variety of NLP problems and, until recently, were used by many state-of-the-art systems for text classification.

Neural networks

The SVM in its standard form is a *linear classifier*. That is, the decision boundary it finds is always perfectly straight, as shown in Fig. 6.4. However, the classes in many

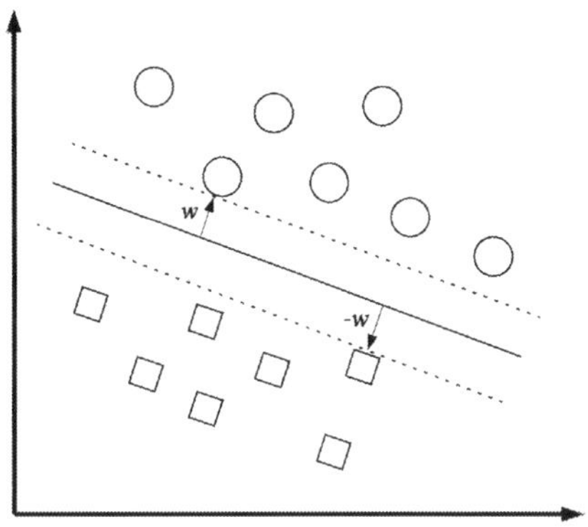

FIGURE 6.4

An illustration of the decision boundary estimated by an SVM on toy data. The data points are arranged in two dimensions and belong to one of two classes (the *circles* or *squares*). The optimal decision boundary estimated by the SVM is given by the *solid line*. The support vector is w ($-w$ in the opposite direction), which defines the distance between the boundary and the closest examples of opposite classes.

datasets are not separable by a linear boundary. Although it is possible to modify the SVM via something called a *kernel function* to allow it to estimate a nonlinear decision boundary, it is still limited in its expressivity by the exact kernel function used. Neural network models (NNs), on the other hand, can approximate any function whatsoever, meaning the nonlinear decision boundary they find can be as nuanced as desired (Lu, Pu, Wang, Hu, & Wang, 2017). Furthermore, this decision boundary can be found automatically without manual experimentation, as is necessary with SVMs and kernel functions. NNs accomplish this by transforming the data multiple times before computing the output. Each of these intermediary transformations is called a *hidden layer* and an NN that uses multiple hidden layers is called a *deep neural network*. Fig. 6.5 illustrates a simplistic neural network with a single hidden layer that predicts whether a patient is at risk for heart failure given three risk factors. The three risk factor values are first transformed into two hidden values by multiplying them by a set of weights. The result of this multiplication is in turn multiplied by a second set of weights to obtain the final values, which correspond to the decision.

NNs are realized in a wide variety of architectures. The neural network in Fig. 6.5 is an example of a *feed-forward neural network* (FFNN), as the values of each layer are fed directly into the next layer via a linear transformation. Other common NN architectures are the *convolutional neural network* (CNN) and the *recursive neural network* (RNN). Illustrations of these architectures are given in Fig. 6.6.

CNNs (Fig. 6.6A) were originally designed to mimic the human vision system to improve performance on automatic image recognition (Lu et al., 2017). Rather than apply a linear transformation at each layer as in FFNNs, CNNs use mathematical operations called *convolutions* to filter the input and highlight important

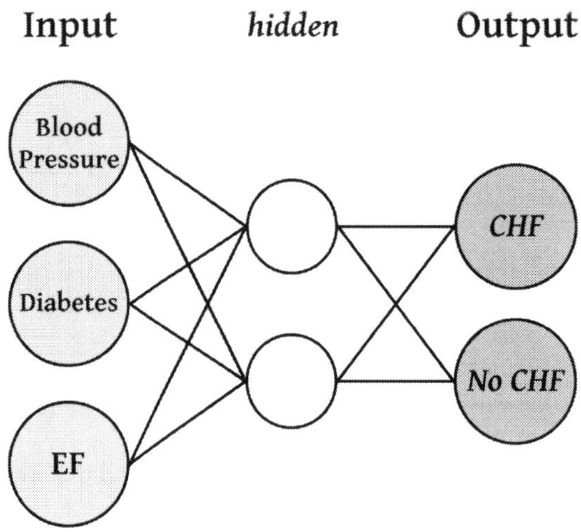

FIGURE 6.5

A feed-forward neural network with a single hidden layer for predicting whether a patient is as risk for congestive heart failure (CHF) given the values of their risk factors.

information. In the case of images, this filter can, for example, find the edges of the objects in the image. However, it was found that CNNs also achieve very good results on text classification tasks (Lecun et al., 1998)(Kim, 2014). CNNs operate on text data by, in a sense, looking at the entire document at once to distill its overall meaning, and the convolutions, rather than highlighting the edges of an image, highlight important words or phrases. For this reason, CNNs are the best for text classification tasks such as determining whether a patient is at risk for heart disease from a clinical note.

RNNs (Fig. 6.6B) are different from feed-forward neural networks and CNNs in that they operate over time (Rumelhart, Hinton, & Williams, 1986). In other words, the hidden layer at the current time step is composed of the hidden layer from the previous time step. In this way, the RNN remembers what it has previously seen. This makes RNNs well-suited to text data because they can reason about the current word given the previous words in the input. Although standard RNNs use the same operations as FFNNs to compute the hidden states, two common variations, long short-term memory (LSTM) (Sepp & Jürgen, 1997) and gated recurrent unit (GRU) (Cho et al., 2014), replace the standard hidden state with specialized operations that allow the network to selectively remember or forget certain parts of what it has seen before. This allows the network to take long-range context into account when handling a given word. LSTMs accomplish this by using a series of memory cells that selectively remember or forget information as it passes from the hidden layer at the previous time step to the current time step.

Standard RNNs are good for transforming an input sequence into an equal length output sequence, such as biomedical text input to word labels for an NER task. To

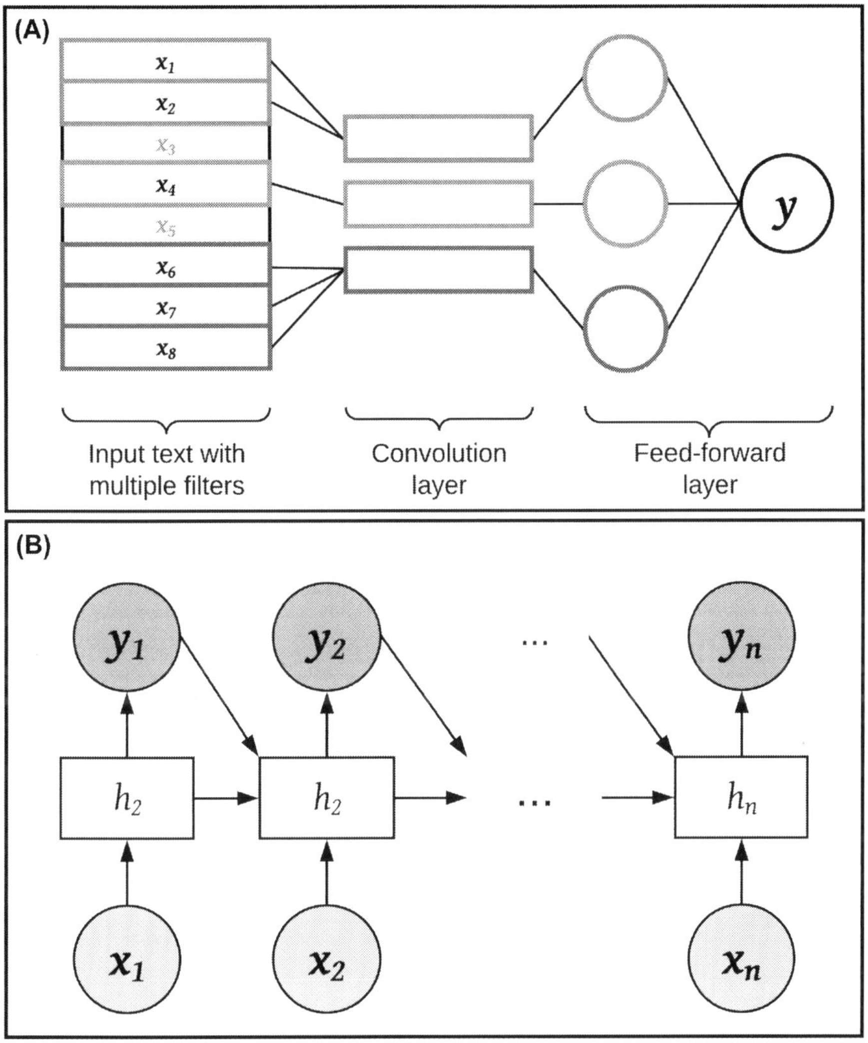

FIGURE 6.6

(A) A convolutional neural network (CNN) architecture. Each word of a given input text is represented as a vector (usually a word embedding). A series of filters (i.e., convolutions) of various sizes is applied to the input ($x_{1\text{-}2}$, x_4, and $x_{6\text{-}8}$, *respectively)* to obtain an intermediate representation, which is then passed to a feed-forward neural network for prediction. (B) A recursive neural network (RNN) architecture. Here x_t, h_t, and y_t refer to the inputs, hidden units, and outputs, respectively. The values of the hidden unit at a given time *t* are composed of the values from the previous time step and thus the previous input is used to predict the current output. Furthermore, these hidden units can be swapped out for long-short term memory (LSTM) units or gated recurrent units (GRU) to improve the processing of long-range dependencies in the input.

transform between two sequences of unequal length, on the other hand, an *encoder-decoder* network is used (Fig. 6.7A), which uses a stack of two RNNs (Sutskever, Vinyals, & Le, 2014). The first RNN (the encoder) transforms the input into a single context vector (sometimes called a thought vector). The context vector is then fed into the second RNN (the decoder), which uses it to generate the required output. Because the input and output can be of unequal length, encoder–decoder networks are often used for tasks such as question answering (where the input is a question and the output is an answer) and translation, for example, from English to Chinese or from complex clinical text to a simplified version fit for a layperson.

Nevertheless, the encoder–decoder architecture described earlier has two issues. First, the entire input sequence is encoded into a single context vector, which is not ideal especially for very long input sequences. Second, the context vector is only made available to the initial state of the decoder, meaning that later decoder states will use it less. To combat these issues, the *attention mechanism* was developed, which provides a robust way to pass information from the encoder to the decoder (Bahdanau, Cho, & Bengio, 2015). Rather than use a single context vector, the attention mechanism computes a new context vector for each decoder output state from a weighted combination of the input states (Fig. 6.7B). These weights indicate how much "attention" the decoder pays to each input word when generating each output word.

Although early uses of attention were in encoder–decoder networks using RNNs, it was later found that the RNNs themselves could be replaced by a modified attention mechanism called *self-attention* (Cheng, Dong, & Lapata, 2016). In self-attention, the hidden state for each input word is computed from a weighted combination of the other input words. By looping back on itself in this way, self-attention can compute a hidden state for each word (like RNNs) by looking at all the words surrounding it (like CNNs). The seminal paper, "Attention is All You Need" (Vaswani et al., 2017), showed that an encoder–decoder network using only attention and self-attention (an architecture called a *transformer*) significantly outperforms RNN-based encoder–decoder networks on a number of different NLP tasks.

The use of attention and self-attention culminated with the development of Google's BERT model, short for *Bidirectional Encoder Representations from Transformers* (Devlin, Chang, Lee, & Toutanova, 2019). By applying the model architecture from "Attention is All You Need" to a language modeling task, BERT can learn contextualized representations of input words. These contextualized representations are similar to word embeddings, but use the surrounding words to determine the sense of the target word and adjust its embedding accordingly. This eliminates the issue with standard word embeddings in which multiple meanings (for example, river *bank* and money *bank*) are collapsed into a single vector. Furthermore, BERT's contextualized representations can be used as a starting point for training task-specific NLP models in a process called *fine-tuning*. The BERT paper showed that by using a BERT model trained on a sufficiently large amount of text data, fine-tuning can achieve state-of-the-art performance on a variety of NLP tasks using a fraction of the training data of its competitors.

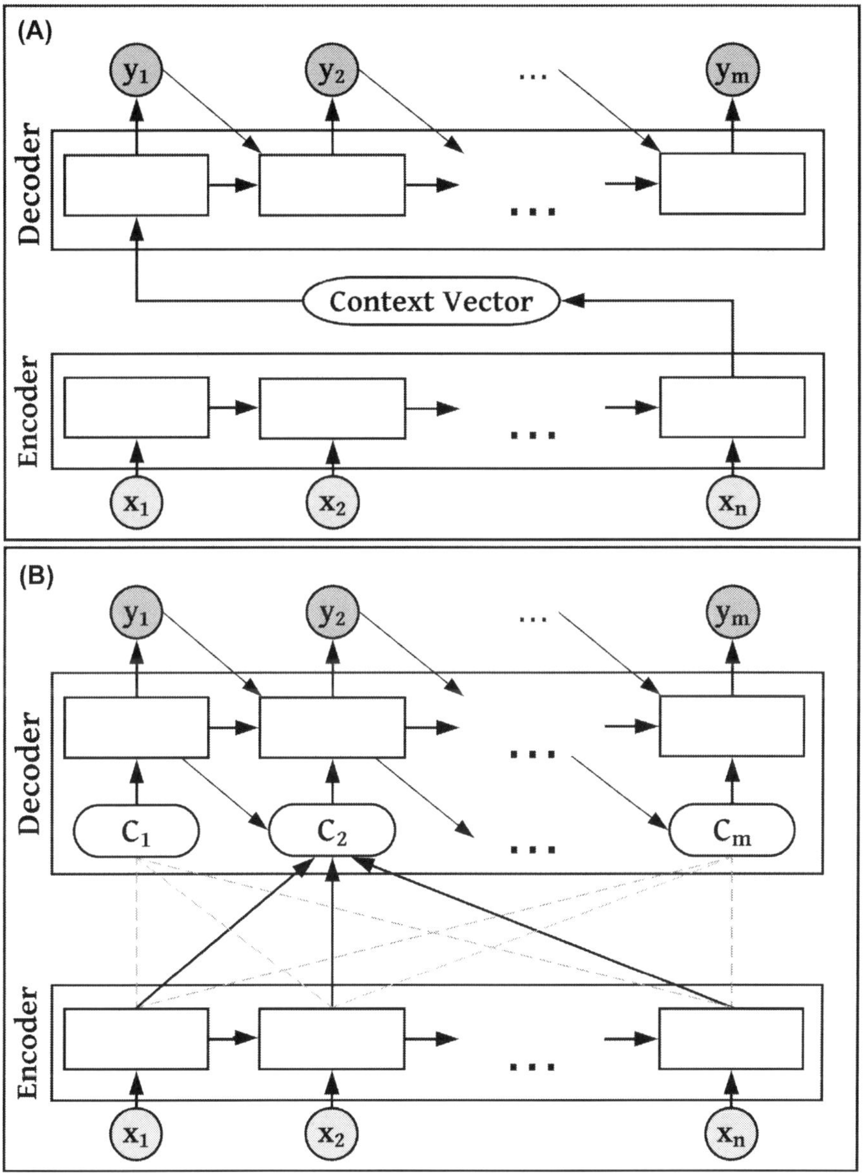

FIGURE 6.7

(A) A standard encoder–decoder network. The encoder uses an RNN to compress the input into a single context vector, which is then fed into the first layer of the decoder, which is also an RNN. The lengths of the input and output need not be the same length. (B) An encoder-decoder network with attention. As opposed to a single context vector, attention allows the network to use all the encoder hidden states when computing the output of each decoder hidden states. The encoder hidden states are weighted by parameters learned during model training.

The combination of BERT's performance and the relative ease of fine-tuning it for specific tasks shot it into the spotlight, motivated a variety of modified versions, such as AlBERT (Lan et al., 2020), RoBERTA (Liu et al., 2019), and BART (Lewis et al., 2019), and inspired researchers to apply it in various disciplines, such as relationship extraction (Chen et al., 2019). The base BERT model has been retrained on both biomedical (Lee et al., 2019) and clinical (Kexin, Jaan, & Rajesh, 2019) text, leading to significant improvements over the state-of-the-art on a number of tasks in these domains.

Biomedical NLP resources and systems

The field of biomedical NLP is growing rapidly. As such, many new tools and resources are being developed for performing a variety of NLP tasks on biomedical and clinical text. This section briefly describes a few often used, open-source resources and systems.

Biomedical terminologies and ontologies

The Unified Medical Language System

The UMLS (Bodenreider, 2004), developed by the US National Library of Medicine (NLM), collects and integrates a large number of biomedical terminologies and ontologies into a single system. The UMLS encompasses the Metathesaurus, the Semantic Network, and the SPECIALIST Suite of tools. The Metathesaurus is a collection of biomedical terms from over 200 existing vocabularies such as MeSH, RxNorm, and SNOMED-CT. The Semantic Network consists of a set of broad subject categories (i.e., semantic types) and relationships (i.e., semantic relations) between these semantic types. The SPECIALIST suite provides a lexicon and lexical tools for working with biomedical text.

RxNorm

RxNorm (Liu, Ma, Moore, Ganesan, & Nelson, 2005) is a system for normalizing drug names and supporting interoperability between health systems. It normalizes names by assigning types to each component of the name. For example, the drug name "Fluoxetine 4 MG/ML Oral Solution" comprises an ingredient (Fluoxetine), a strength (4 MG/ML), and a dose form (Oral Solution).

Biomedical NLP systems

MetaMap and SemRep

MetaMap (National Library of Medicine, 2019) is a tool developed by the US NLM for linking entities to the UMLS. The SemRep (Rindflesch & Fiszman, 2003) tool extracts relationships between UMLS concepts identified by MetMap from the biomedical literature. Both tools use a rule-based approach. The NLM regularly runs MetaMap and SemRep on the entirety of PubMed abstracts to produce a database of extracted relationships, called SemMedDB (Kilicoglu, Shin, Fiszman, Rosemblat, & Rindflesch, 2012).

cTAKES

Apache cTAKES (Savova et al., 2010) is a free information extraction system for biomedical and clinical text originally developed at the Mayo Clinic. It can extract mentions, perform entity linking, mark negated or uncertain expressions, and identify expressions of time (e.g., "follow up in 1–2 weeks"). cTAKES is built using the Apache Unstructured Information Management Architecture (UIMA) (Ferrucci, Lally, Verspoor, & Nyberg, 2009), which means that its pipelines are specified using modular components. In addition to modularity, UIMA allows for interoperability between systems built according to its framework.

The BioMedical information collection and understanding system

BioMedICUS (2019) is a free and open-source system for large-scale biomedical and clinical text analysis developed at the University of Minnesota Institute for Health Informatics. It contains a number of components for doing low-level tasks such as tokenization, sentence boundary detection, and POS, tagging as well as prespecified pipelines for processing clinical documents. These pipelines focus on areas such as social history, time, and measures.

CLAMP (clinical language annotation, modeling, and processing toolkit)

CLAMP (Soysal et al., 2018) is developed at The University of Texas Health Science Center at Houston. It incorporates state-of-the-art NLP modules that can be combined into customized pipelines and is built upon the UIMA framework, meaning it is compatible with other systems such as cTAKES. CLAMP also provides user-friendly interfaces for end-users to leverage NLP components to build customized NLP pipelines. CLAMP allows users to combine both machine learning and deep learning with rule-based approaches for performance optimization.

Summary

Natural language processing seeks to address a variety of problems, from word tokenization to syntactic parsing to question answering. In the biomedical domain, however, information extraction (IE) is the major focus, and it comprises subtasks such as named-entity recognition (NER), entity linking, and relationship extraction. In general, the NLP methods used for these tasks can be divided into two categories. The first method is rule-based, which employs regular expressions, keyword lists, and other manually defined logic to process the text. Although rule-based methods have largely been discarded by the NLP community in favor of modern machine learning approaches, they still find a home in the biomedical domain, as they are easy to implement, use, and interpret, and in many cases perform competitively with modern approaches. Some notable rule-based systems are MetaMap and SemRep for entity linking and relationship extraction, and NegEx for negation detection. The second category is machine learning methods, which automatically learn to perform NLP tasks by using statistical modeling techniques. Most notably, the explosion of deep neural networks in recent years has motivated a number of powerful models for NLP, including CNNs, RNNs, and transformers, the last of which have shown enormous potential across many tasks and domains. Nevertheless, earlier models such as HMMs and SVMs often perform well in practice.

A number of existing tools exist for NLP in the biomedical domain. The Unified Medical Language System (UMLS) provides an invaluable and expansive biomedical terminology. The UMLS is used by the aforementioned MetaMap and SemRep to perform entity linking and relationship extraction. Finally, cTAKES, CLAMP, and BioMedICUS are toolkits for biomedical NLP that include functionality for a variety of tasks, from tokenization to relationship extraction.

Applications of NLP in cardiovascular research

It is a rising trend to apply NLP in research related to cardiovascular diseases. The American College of Cardiology released the "Roadmap for Innovation" in 2017 that emphasizes the importance of NLP and machine learning in the era of digital health (Bhavnani et al., 2017). The rapid development and adoption of EHR systems documenting vast amounts of real-world patient data have created many research opportunities. Valuable clinical information (e.g., symptoms, lab values, disease histories) for cardiovascular research is found documented in free text in EHR systems, making it impractical to manually identify the information. Thus, one important topic is to accurately extract the target information from a large collection of clinical notes in an automatic way. In this section, we summarize relevant studies to demonstrate the role of NLP in current research related to cardiovascular diseases.

NLP for information extraction

Extracting information from text data is one of the primary tasks of NLP in the clinical area. The unstructured part of EHRs mainly contains three types of data: real-time data, retrospective data, and clinically irrelevant data (Bhavnani et al., 2017). The real-time data often contains information about the current encounter (e.g., diagnosis, test values, symptoms of diseases). The retrospective data provides information about past encounters (e.g., past diagnosis, treatments applied). The clinical irrelevant data includes information such as administration information, etc. Research regarding NLP applications in clinical settings has provided practical ways to extract target information from the clinical texts in the face of irrelevant information.

Rule-based NLP applications in cardiovascular research

Researchers have been using NLP for information extraction tasks in clinical settings since the 2000s. Friedlin et al. developed an NLP tool called REgenstrief data eXtraction tool (REX), with the object to extract a list of targeted congestive heart failure (CHF)-related concepts (e.g., CHF, cardiomegaly) from X-ray reports (Friedlin & McDonald, 2006). The REX uses regular expressions and a set of rules to detect target concepts and analyze their contexts. They found that the performance of REX was better than the human annotators in the evaluation, with an average sensitivity of 0.98. The shortcomings of REX include that it could only extract simple concepts and it is not good at dealing with spelling errors and variations, acronyms disambiguation, and contextual information (e.g., negation). Khalifa et al. developed an NLP application based on the Apache UIMA framework by integrating two existing NLP resources, cTAKES and Textractor, to identify cardiovascular risk factors (e.g., obesity, smoking status, diabetes) from clinical notes (Khalifa & Meystre, 2015). The raw texts are preprocessed (e.g., sentence detection, tokenization, POS tagging), which are then fed into the rule-based modules in cTAKES and Textractor to extract different risk factors. This study applies dictionary look-up, machine learning, and regular expression methods in the pipeline to identify different variables. For instance, it uses a machine learning module to identify the smoking status of patients with features like words of the sentence and their POS tags. Medications and laboratory results are detected using the rule-based pattern matching modules, while disease and risk factor terms are identified using the UMLS Metathesaurus lookup module from Textractor. This application was evaluated using the 2014 i2b2 challenge dataset and achieved an overall microaveraged F1-measure of 87.47% and a macroaveraged F1-measure of 86.99%. This study shows that it is practical to reuse and integrate existing NLP tools to solve similar tasks with minor modifications.

NLP has also been used for cardiovascular disease phenotyping in the EHR. Ye, Kalloo, Dalenberg, and Kullo (2013) developed a pipeline that identifies atherosclerotic cardiovascular disease phenotypes from both the EHR and biorepository data of patients which includes both structured data and unstructured data. Rule-based NLP methods were used to identify risk factors and medication usages, such as smoking status, diabetes, hypoglycemic agents, and antihypertensive medications. The output of the system was further combined with structured data (e.g., demographics, laboratory tests, diagnosis codes) to finally determine the phenotypes of patients.

Machine learning and deep learning with NLP applications in research

NLP applications in clinical settings have been evolving from rule-based methods to machine learning and more advanced deep learning methods. Weng, Wagholikar, McCray, Szolovits, and Chueh (2017) constructed an NLP pipeline based on cTAKES to identify the subtype of the clinical notes, such as cardiology reports. Both shallow machine learning and deep learning methods were investigated in the study. Two types of features were used. The first is the lexical features of words (e.g., bag-of-word and UMLS concepts) obtained by the NLP pipeline. The other type of feature is the distributed word (word2vec) and document representations (distributed memory model of paragraph vectors). The classifiers used include the Naïve Bayes algorithm, multinomial logistic regression, regularized SVM with linear kernel, and two ensemble algorithms, random forest and adaptive boosting. They also evaluated two deep learning methods: CNN and LSTM models. The best performance was obtained by the LSTM model with fastText word embeddings, with an AUC score of 0.98.

Zhang et al. (2018) developed both rule-based and machine learning-based NLP methods to identify the New York Heart Association (NYHA) class of patients from their clinical notes, and compared their performances. The NYHA can be used as a measure of a patient's response to cardiac resynchronization therapy (CRT), which may provide a better understanding of the progression of heart failure and to assess CRT effectiveness. For the machine learning-based method, the texts were preprocessed (e.g., remove stop words, normalize lexical variants) and bag of words and *n*-grams were extracted as features. Using these features, SVM, logistic regression, and random forest models were developed and evaluated. The random forest model with *n*-gram features obtained the best performance with an F-1 score of 93.78%, which surpassed the performance of the rule-based method. The study indicates that traditional machine learning algorithms with features created by NLP methods can be effective in clinical text classification tasks.

Chokwijitkul, Nguyen, Hassanzadeh, and Perez (2018) evaluated two deep learning architectures, CNN and RNN as well as three RNN variants, including LSTM, bidirectional long short-term memory (Bi-LSTM), and GRU, for extracting cardiac risk factors from EHRs. The dataset used in the study comes from the i2b2/UTHealth shared task. The extracted risk factors included clinical concepts related to coronary artery disease, diabetes, hypertension, hyperlipidemia, and relevant medications. The CNN model used the window approach to classify each individual word at a time instead of the entire sentence for the NER task. This approach assumed that the label of a word is dependent on its neighboring words within the window size. The clinical text was transformed into pretrained word embeddings as the input of neural network models. The evaluation showed that RNN-type networks outperformed CNN in their task. Among different RNN models, Bi-LSTM achieved the best performance with an F-measure of 0.908 on the test data. Their study shows that the deep learning approaches were comparable to highly feature-engineered hybrid systems and could obtain promising results without the help of any knowledge-driven methods. Thus, lots of human labor could be saved by deep learning approaches.

Viani et al. (2019) developed a novel application that used an RNN to extract clinical events (e.g., problem, test, treatment) from cardiology reports written in Italian. Traditional machine learning methods often rely on complicated linguistic features (e.g., lexical, morphological, and syntactic features), which usually need manually feature engineering. For neural networks, it could learn the representations starting from tokens or characters. A GRU model with pretrained word embeddings and POS tags as features was compared with a dictionary look-up method, a standard SVM model, and a set of existing Bi-LSTM-CRF models. The GRU model obtained an F-1 score of 87.4% and after integration with the dictionary look-up method, the F-1 score increased to 90.1%, indicating that combining different methods may increase the performance at low cost.

NLP for clinical decision support

Clinical decision support (CDS) aims to aid clinicians by providing them with easily accessible health-related information at the point of potential actionable advice. NLP is instrumental in extracting valuable information from unstructured data to represent clinical knowledge and drive CDS interventions that would improve the quality of healthcare (Demner-Fushman, Chapman, & McDonald, 2009).

Moon et al. (2019) developed and deployed NLP algorithms to automatically extract sudden cardiac death (SCD) factors such as syncope, family history of SCD, and hypertrophic cardiomyopathy from clinical narratives. They developed regular expression and rule-based NLP algorithms using MedTagger, which has been adopted enterprise-wide by the Mayo Clinic. The keywords for different clinical concepts used for the developed algorithms were obtained by manually reviewing and searching lexical variations and synonyms from the UMLS Metathesaurus. The assertion and negation detection module in MedTagger was integrated into the NLP pipeline to improve the results. For evaluation, they compared the NLP pipeline with the standard method of using the registry, billing codes, and patient surveys to extract syncope, family history of SCD, and hypertrophic cardiomyopathy. The NLP pipeline showed superior performance to the standard method for this task. The extracted SCD risk factors are expected to be fed into clinical decision support systems to increase the efficiency of the HCM patients' management workflow for providers, with the ultimate goal of improving the quality of care.

Meystre et al. (2016) developed a new CHF treatment performance measure information extraction system based on the UIMA framework called CHIEF. The system used a combination of rules, dictionaries, and machine learning methods to extract left ventricular function mentions and values, CHF medications, and documented reasons for a patient not receiving these medications. For instance, to extract the left ventricular ejection fraction (LVEF) mentions and corresponding values, it used both regular expressions and machine learning with morphological (e.g., prefixes and suffixes), lexical (e.g., words themselves and *n*-grams), syntactic (e.g., POS tags), and semantic features (e.g., the output of regular expression). The output information was integrated to classify the CHF treatment quality of patients. If a patient was identified with low LVEF measurements (<40%) and did not take the required medications, it indicated poor treatment quality. It may achieve the fast and scalable detection of CHF patients not receiving recommended treatment, which could help clinicians decide further treatment for patients at the point of care.

NLP for predictive models related to cardiovascular diseases

The combination of NLP and machine learning methods is promising in the development of models for predicting cardiovascular diseases. Choi, Schuetz, Stewart, and Sun (2016) explored using RNNs to predict the initial diagnosis of heart failure (HF). Compared to traditional machine learning methods, RNNs (LSTM and GRU specifically) are more suited to identify patterns in longitudinal data. The GRU model was developed to capture the relations among events (e.g., diagnosis of diseases, medication orders, and procedure orders) with time stamps. They processed the clinical events in both structured and unstructured data using a standard NLP pipeline and encoded the events using a set of one-hot vectors as the input of the GRUs. The output is the risk of HF for a patient. When comparing with logistic regression, SVM, and multilayer perceptron, the GRU model obtained an AUC of 0.883, which is significantly higher than other models. This study could help the early detection of HF, which opens new opportunities for delaying or preventing progression to the diagnosis of HF and reduce cost. Maragatham and Devi (2019) performed a similar study, developing an LSTM model that used the temporal information of patients' hospital encounters to predict their risk of HF. They applied skip-gram to encode the procedure, medication, and diagnosis events as vectors for LSTM model training. They evaluated the LSTM model with different observation windows (3–12 months) and compared its performance with several baseline models (e.g., SVM, multilayer perceptron, logistic regression). The LSTM model obtained the best performance with an AUC of 0.7969, using a 12-month observation window. Using NLP and deep learning methods to predict the risk of disease is promising and has the potential to be integrated into CDS systems to improve the quality of care.

NLP for cohort identification of cardiovascular diseases

Cohort identification is another important task in the field of clinical research. It is a vital step for other tasks such as clinical trial screening and recruitment and public health studies. Traditional cohort identification uses structured data such as the International Classification of Diseases (ICD) codes and billing codes. However, using NLP methods on unstructured clinical texts has shown to better improve cohort identification tasks. Wang et al. (2015) developed an NLP-based algorithm to find CHF cases from the EHR. In total, there are 32 CHF discriminant features included in the final pipeline. The features were extracted from both structured (e.g., demographics, vital signs) and unstructured data (e.g., ICD code of "CHF," terms like "heart failure" and "congestive heart") in the EHR. A binary classification algorithm was then developed based on the features to judge if the patient has CHF. In evaluation, the NLP-based algorithm obtained an F-measure of 0.789. In addition, the algorithm was integrated into the Health Information Exchange population exploration system in the state of Maine to conduct a real-time CHF identification task. The algorithm identified over a 1000 patients that have not been coded as CHF patients previously. Geva et al. (2017) developed a computable algorithm to improve the cohort identification of pulmonary hypertension (PH). To include text data in the EHR as features for the computable algorithm development, they applied an NLP tool, Narrative Information Linear Extraction package, to identify clinical concepts (e.g., medication mentions, symptoms of the diseases) relevant to PH in patients' clinical notes. Based on the extracted features, they fitted an adaptive least absolute shrinkage and selection operator (LASSO) penalized logistic regression model to identify if the patient has PH. The algorithm was evaluated and obtained the best performance with AUC of 0.900. The developed algorithm is promising with respect to the identification of large cohorts of PH patients for further large-scale research on this disease entity.

Conclusion

In summary, NLP applications in cardiovascular diseases have been developed for various tasks in processing and analyzing clinical texts, and a few of them have been applied in real practice. Currently, the major goals of NLP applications in clinical research are to extract and classify the target information of patients. NLP methods have evolved from rule-based methods to conventional machine learning-based methods. Most recently, advanced deep learning techniques, especially transformers such as BERT, have led to unprecedented advances in NLP, and have the potential to radically change how we approach NLP problems by removing the need for vast amounts of task-specific training data. There are still challenges and opportunities for NLP tasks in cardiovascular diseases. In the future, NLP will continue to play vital roles in both clinical research and clinical practice.

References

Bahdanau, D., Cho, K., & Bengio, Y. (2015). Neural machine translation by jointly learning to align and translate. *Proceedings of the 2015 International Conference on Learning Representations (ICLR)*.

Bhavnani, S. P., Parakh, K., Atreja, A., Druz, R., Graham, G. N., Hayek, S. S., et al. (2017). 2017 roadmap for innovation—ACC health policy statement on healthcare transformation in the era of digital health, big data, and precision health: A report of the American College of cardiology task force on health policy statements and systems of care. *Journal of the American College of Cardiology, 70*(21), 2696–2718.

Bodenreider, O. (2004). The Unified Medical Language System (UMLS): Integrating biomedical terminology. *Nucleic Acids Research, 32*(Database issue), D267–D270.

Bojanowski, P., Grave, E., Joulin, A., & Mikolov, T. (2017). Enriching word vectors with subword information. *Transactions of the Association for Computational Linguistics, 5*, 135–146.

Carrell, D. S., Schoen, R. E., Leffler, D. A., Morris, M., Rose, S., Baer, A., et al. (2017). Challenges in adapting existing clinical natural language processing systems to multiple, diverse health care settings. *Journal of the American Medical Informatics Association, 24*(5), 986–991.

Chapman, W. W., Hilert, D., Velupillai, S., Kvist, M., Skeppstedt, M., Chapman, B. E., et al. (2013). Extending the NegEx lexicon for multiple languages. *Studies in Health Technology and Informatics, 192*, 677–681.

Choi, E., Schuetz, A., Stewart, W. F., & Sun, J. (2016). Using recurrent neural network models for early detection of heart failure onset. *Journal of the American Medical Informatics Association, 24*(2), 361–370.

Chokwijitkul, T., Nguyen, A., Hassanzadeh, H., & Perez, S. (2018). Identifying risk factors for heart disease in electronic medical records: A deep learning approach. In *Proceedings of the BioNLP 2018 workshop* (pp. 18–27).

Chen, T., Wu, M., & Li, H. (2019). A general approach for improving deep learning-based medical relation extraction using a pre-trained model and fine-tuning. *Database : The Journal of Biological Databases and Curation*, baz116.

Cheng, J., Dong, L., & Lapata, M. (2016). Long short-term memory-networks for machine reading. *Proceedings of the 2016 Conference on Empirical Methods in Natural Language Processing (EMNLP)*, 551–561.

Cho, K., van Merriënboer, B., Gulcehre, C., Bahdanau, D., Bougares, F., Schwenk, H., et al. (2014). Learning phrase representations using {RNN} Encoder{–}Decoder for statistical machine translation. In *Proc EMNLP* (Vol. 2014, pp. 1724–1734).

Cortes, C., & Vapnik, V. (1995). Support-vector networks. *Machine Learning*, 273–297.

Demner-Fushman, D., Chapman, W. W., & McDonald, C. J. (2009). What can natural language processing do for clinical decision support? *Journal of Biomedical Informatics, 42*(5), 760–772.

Devlin, J., Chang, M., Lee, K., & Toutanova, K. (2019). Bert: Pre-training of deep bidirectional transformers for language understanding. *Proceedings of the 2019 Conference of the North American Chapter of the Association for Computational Linguistics: Human Language Technologies (NAACL-HLT)*, 4171–4186.

Ferrucci, D., Lally, A., Verspoor, K., & Nyberg, E. (2009). *Unstructured information management architecture (UIMA) version 1.0*. Available at https://docs.oasis-open.org/uima/v1.0/uima-v1.0.pdf. (Accessed October 2019).

Friedlin, J., & McDonald, C. J. (2006). A natural language processing system to extract and code concepts relating to congestive heart failure from chest radiology reports. In *Proc AMIA annual sym proc* (Vol. 2006, p. 269).

Friedman, C., Kra, P., & Rzhetsky, A. (2002). Two biomedical sublanguages: A description based on the theories of Zellig Harris. *Journal of Biomedical Informatics, 35*(4), 222–235.

Geva, A., Gronsbell, J. L., Cai, T., Murphy, S. N., Lyons, J. C., Heinz, M. M., et al. (2017). A computable phenotype improves cohort ascertainment in a pediatric pulmonary hypertension registry. *The Journal of Pediatrics, 188*, 224–231.

Herrero-zazo, M., Segura-Bedmar, I., Martínez, P., & Declerck, T. (2013). The DDI corpus: An annotated corpus with pharmacological substances and drug-drug interactions. *Journal of Biomedical Informatics, 46*(5), 914–920.

Kexin, H., Jaan, A., & Rajesh, R. (2019). ClinicalBERT: Modeling clinical notes and predicting hospital readmission. *arXiv preprint, 1904.05342*. https://arxiv.org/abs/1904.05342.

Khalifa, A., & Meystre, S. (2015). Adapting existing natural language processing resources for cardiovascular risk factors identification in clinical notes. *Journal of Biomedical Informatics, 58*, 128–132.

Kilicoglu, H., Shin, D., Fiszman, M., Rosemblat, G., & Rindflesch, T. C. (2012). SemMedDB: A PubMed-scale repository of biomedical semantic predications. *Bioinformatics, 28*(23), 3158–3160.

Kim, Y. (2014). Convolutional neural networks for sentence classification. *Proc EMNLP*, 1746–1751.

Lan, Z., Chen, M., Goodman, S., Gimpel, K., Sharma, P., & Soricut, R. (2020). Albert: A lite bert for self-supervised learning of language representations. *Proceedings of the 2020 International Conference on Learning Representations (ICLR)*.

Lecun, Y., Bottou, L., Bengio, Y., & Haffner, P. (1998). Gradient-based learning applied to document recognition. *Proceedings of the IEEE, 86*(11), 2278–2324.

Lee, J., Yoon, W., Kim, S., Kim, D., Kim, S., So, C. H., et al. (2019). BioBERT: A pre-trained biomedical language representation model for biomedical text mining. *Bioinformatics, 36*(4), 1234–1240.

Lewis, M., Liu, Y., Goyal, N., Ghazvininejad, M., Mohamed, A., Levy, O., et al. (2019). Bart: Denoising sequence-to-sequence pre-training for natural language generation, translation, and comprehension. *arXiv preprint, 1910.13461*. https://arxiv.org/abs/1910.13461.

Liu, S., Ma, W., Moore, R., Ganesan, V., & Nelson, S. (2005). RxNorm: Prescription for electronic drug information exchange. *IT Professional, 7*(5), 17–23.

Liu, Y., Ott, M., Goyal, N., Du, J., Joshi, M., Chen, D., et al. (2019). Roberta: A robustly optimized bert pretraining approach. *arXiv preprint, 1907.11692*. https://arxiv.org/abs/1907.11692.

Liu, S., Tang, B., Chen, Q., & Wang, X. (2016). Drug-drug interaction extraction via convolutional neural networks. *Computational and Mathematical Methods in Medicine, 2016*, 6918381.

Lu, Z., Pu, H., Wang, F., Hu, Z., & Wang, L. (2017). The expressive power of neural networks: A view from the width. *Advances in Neural Information Processing Systems, 30*, 6231–6239.

Maintenance and Support Services Organization. (2019). *Medical dictionary for regulatory Activities (MedDRA)*. Available at https://www.meddra.org.

Manning, C. D., Raghavan, P., & Schütze, H. (2008). Scoring, term weighting, and the vector space model. In *Introduction to information retrieval* (pp. 100–123). Cambridge University Press.

Maragatham, G., & Devi, S. (2019). LSTM model for prediction of heart failure in bBig data. *Journal of Medical Systems, 43*(5), 111.

Meystre, S. M., Kim, Y., Gobbel, G. T., Matheny, M. E., Redd, A., Bray, B. E., et al. (2016). Congestive heart failure information extraction framework for automated treatment performance measures assessment. *Journal of the American Medical Informatics Association, 24*(1), 40–46.

Mikolov, T., Sutskever, I., Chen, K., Corrado, G., & Dean, J. (2013). Distributed representations of words and phrases and their compositionality. *Proceedings of the 2013 Conference on Neural Information Processing Systems (NIPS)*, 3111–3119.

Moon, S., Liu, S., Scott, C. G., Samudrala, S., Abidian, M. M., Geske, J. B., et al. (2019). Automated extraction of sudden cardiac death risk factors in hypertrophic cardiomyopathy patients by natural language processing. *International Journal of Medical Informatics, 128*, 32–38.

National Library of Medicine. (2019). *MetaMap - a tool for recognizing UMLS concepts in text*. Available at https://metamap.nlm.nih.gov/.

Pennington, J., Socher, R., & Manning, C. D. (2014, October). Glove: Global vectors for word representation. In *Proceedings of the 2014 conference on empirical methods in natural language processing (EMNLP)* (pp. 1532–1543).

Pyysalo, S., Ginter, F., Moen, H., Salakoski, T., & Ananiadou, S. (2013). Distributional semantics resources for biomedical text processing. *Proc LBM*, 39–44.

Rindflesch, T. C., & Fiszman, M. (2003). The interaction of domain knowledge and linguistic structure in natural language processing: Interpreting hypernymic propositions in biomedical text. *Journal of Biomedical Informatics, 36*(6), 462–477.

Rumelhart, D., Hinton, G., & Williams, R. (1986). Learning representations by back-propagating errors. *Nature, 323*(6088), 533–536.

Savova, G. K., Masanz, J. J., Ogren, P. V., Zheng, J., Sohn, S., Kipper-Schuler, K. C., et al. (2010). Mayo clinical text analysis and knowledge extraction system (cTAKES): Architecture, component evaluation and applications. *Journal of the American Medical Informatics Association, 17*(5), 507–513.

Sepp, H., & Jürgen, S. (1997). Long short-term memory. *Neural Computation, 9*(8), 1735–1780.

Soysal, E., Wang, J., Jiang, M., Wu, Y., Pakhomov, S., Liu, H., et al. (2018). Clamp - a toolkit for efficiently building customized clinical natural language processing pipelines. *Journal of the American Medical Informatics Association, 25*(3), 331–336.

The BioMedical Information Collection and Understanding System (BioMedICUS). (2019). [online]. Available at https://nlpie.github.io/biomedicus/. [Accessed October 11, 2019].

Sutskever, I., Vinyals, O., & Le, Q. V. (2014). Sequence to sequence learning with neural networks. *Proceedings of the 2013 Conference on Neural Information Processing Systems (NIPS), 2014*, 3104–3112.

Vaswani, A., Shazeer, N., Parmar, N., Uszkoreit, J., Jones, L., Gomez, A. N., et al. (2017). Attention is all you need. *Proceedings of the 2013 Conference on Neural Information Processing Systems (NIPS), 2017*, 5998–6008.

Viani, N., Miller, T. A., Napolitano, C., Priori, S. G., Savova, G. K., Bellazzi, R., et al. (2019). Supervised methods to extract clinical events from cardiology reports in Italian. *Journal of Biomedical Informatics*, 103219.

Wang, Y., Luo, J., Hao, S., Xu, H., Shin, A. Y., Jin, B., et al. (2015). NLP based congestive heart failure case finding: A prospective analysis on statewide electronic medical records. *International Journal of Medical Informatics, 84*(12), 1039–1047.

Weng, W. H., Wagholikar, K. B., McCray, A. T., Szolovits, P., & Chueh, H. C. (2017). Medical subdomain classification of clinical notes using a machine learning-based natural language processing approach. *BMC Medical Informatics and Decision Making, 17*(1), 155.

Wu, S. T., Sohn, S., Ravikumar, K. E., Wagholikar, K., Jonnalagadda, S. R., Liu, H., et al. (2013). Automated chart review for asthma cohort identification using natural language processing: An exploratory study. *Annals of Allergy, Asthma, and Immunology, 111*(5), 364–369.

Ye, Z., Kalloo, F. S., Dalenberg, A. K., & Kullo, I. J. (2013). An electronic medical record-linked biorepository to identify novel biomarkers for atherosclerotic cardiovascular disease. *Global Cardiology Science and Practice, 2013*(1), 10.

Zhang, R., Cairelli, M. J., Fiszman, M., Rosemblat, G., Kilicoglu, H., Rindflesch, T. C., et al. (2014). Using semantic predications to uncover drug-drug interactions in clinical data. *Journal of Biomedical Informatics, 49*, 134–147.

Zhang, Y., Chen, Q., Yang, Z., Lin, H., & Lu, Z. (2019). BioWordVec, improving biomedical word embeddings with subword information and MeSH. *Scientific Data, 6*(1), 52.

Zhang, R., Ma, S., Shanahan, L., Munroe, J., Horn, S., & Speedie, S. (2018). Discovering and identifying New York heart association classification from electronic health records. *BMC Medical Informatics and Decision Making, 18*(2), 48.

CHAPTER

Contemporary advances in medical imaging

7

Alpana Senapati, Shaden Khalaf, Mouaz H. Al-Mallah

Houston Methodist DeBakey Heart and Vascular Center, Houston, TX, United States

Chapter outline

Introduction

Innovations in cardiovascular imaging have changed the field of cardiovascular medicine with a significant impact on medical diagnosis, treatment, and outcomes. Contemporary advances in echocardiography (ECHO), nuclear cardiology, cardiac computed tomography (CT), and cardiac magnetic resonance imaging (CMR) continue to evolve in the 21st century with an improved understanding of cardiac structure and pathophysiology. From the evaluation of basic cardiac anatomy and

Machine Learning in Cardiovascular Medicine. https://doi.org/10.1016/B978-0-12-820273-9.00007-5

function, valvular heart disease, coronary artery disease (CAD), and cardiomyopathies, each noninvasive imaging modality adds to the armamentarium of diagnostic tools physicians and cardiologists can utilize to effectively diagnose and treat patients. With constant advances in new therapeutics and catheter-based interventions, such as transcatheter aortic valve replacement (TAVR), cardiac imaging plays an integral part in the decision-making process for patient care. Understanding the advantages and limitations of each modality is essential in choosing the proper test and often a multimodality imaging approach is needed. In this chapter, we provide an overview of the contemporary advances in noninvasive imaging while highlighting gaps where artificial intelligence can further advance the field.

Echocardiography

History of echocardiography

Echocardiography is ubiquitous in clinical practice as it continues to be the initial noninvasive imaging test to be ordered. It provides a comprehensive assessment of cardiac structure, function, and hemodynamics becoming one of the most utilized cardiac imaging tests. The origins of ECHO date back to the 1880s with the discovery of piezoelectricity by Pierre Curie and Jacques Curie (1881) (Feigenbaum, 1996). ECHO uses piezoelectric crystals inside transducers to create ultrasound waves to provide live images of the heart. Clinical use of ECHO emerged in the 1950s by Carl Hellmuth Hertz and Inge Edler, known as the "Father of Echocardiography" (Edler and Lindstrom, 2004; Lindstrom et al., 1969; Maleki and Esmaeilzadeh, 2012). Christian Doppler first introduced the phenomenon of the Doppler effect in 1842, which measures the change in frequency or wavelength of waves in relation to an observer relative to the ultrasound wave source (Maleki and Esmaeilzadeh, 2012). Clinical use expanded worldwide with the first use of Doppler technology by Japanese investigators Satomura and Yoshida, and later followed by Edler and Harvey Feigenbaum and colleagues (Feigenbaum et al.,1965; Satomura et al., 1955; Yoshida et al., 1954). The breakthrough in the clinical application of Doppler came when Jarle Holen and Liv Hatle demonstrated that Doppler ultrasound can accurately assess hemodynamic data in the late 1970s (Hatle, Angelsen, & Tromsdal, 1979; Holen et al., 1976). Soon thereafter, ECHO became well established as the leading modality of choice for cardiac evaluation, and continued to evolve with advances in two-dimensional (2D), spectral and color Doppler, three-dimensional (3D), and strain imaging.

Contrast agents in echocardiography

The field of contrast ECHO began in 1969 with Raymond Gramiak and Pravin Shah while performing ultrasound on a patient undergoing cardiac catheterization using indocyanine green dye, which produced a cloud of echoes within the heart (Gramiak et al., 1969). Harvey Feigenbaum and the Mayo Clinic team later validated these findings for use on left ventricular endocardial border detection and cardiac shunts (Feigenbaum et al., 1970). Ultrasound enhancing agents (UEA) are a class of products that comprise microbubbles that transverse the pulmonary capillaries and enter the left cardiac chamber. The microbubbles oscillate to enhance ultrasound signals to improve visualization of the cardiac chamber borders and are especially useful in technically challenging patients (Nagueh and Quinones, 2014; Porter et al., 2018). Available UEAs that are US Food and Drug Administration (FDA) approved for use are composed of encapsulated microbubbles and include Lumason (Bracco Diagnostics, Princeton, NJ), Definity (Lantheus Medical Imaging, North Billerica, MA), and Optison (GE Healthcare, Marlborough, MA) (Porter et al., 2018). These UEA agents are small (average size between 1.1 and 4.5 μm) and are encapsulated with albumin or lipid and contain a safe high-molecular-weight gas (Porter et al., 2018). The use of UEAs has expanded in various settings with improved detection of CAD and wall motion abnormalities (Kurt et al., 2009; Mehta and Eagle, 2000; Wei, 2010; Wyrick et al., 2008). The use of UEAs is overall safe and effective for various clinical settings, both in pediatric and adult individuals, encompassing both standard ECHO and stress ECHO (Aggeli et al., 2008; Al-Mallah et al., 2008; Appis, Tracy, & Feinstein, 2015; Mulvagh et al., 2008; Muskula and Main, 2017; Porter et al., 2018; Wei et al., 2008). UEAs also are useful for the detection of intracardiac thrombi or masses, evaluation of cardiomyopathies, and enhancement of Doppler signals for the assessment of pulmonary artery pressures and aortic stenosis (Khawaja, Shaikh, & Al-Mallah, 2010; Mitchell et al., 2019; Mulvagh et al., 2008; Porter et al., 2014; Siebelink et al., 2009). Advances in ECHO with the development of UEAs allow better image quality in a wide variety of clinical settings, cardiac disease states, and in challenging patients.

Three-dimensional echocardiography

Recent advances in ECHO transducer technology and faster computer processing systems using automated algorithms allow for 3D rendered images for clinical use. Several advantages of 3D ECHO include accurate quantitation of left ventricular volumes, ejection fraction, and a full volumetric dataset that allows for infinite slice planes for direct visualization of specific cardiac anatomic structures (Hung et al., 2007; Nagueh and Quinones, 2014; Poon, Leung, & Leung, 2019). The use of 3D ECHO is preferred over that of 2D ECHO for left ventricle volume and function assessment when image quality is excellent. A 3D examination of the left ventricle in the apical window using semiautomated algorithms for endocardial border detection throughout the cardiac cycle generates a full volume dataset for left ventricular volumes and ejection fraction (Mitchell et al., 2019; Poon et al., 2019). Its use has been well validated for use in valvular heart disease, specifically for use in mitral valve disease using 3D real-time imaging, and guiding percutaneous cardiac interventions (Fig. 7.1) (Geyer et al., 2019; Little et al., 2008; Poon et al., 2019).

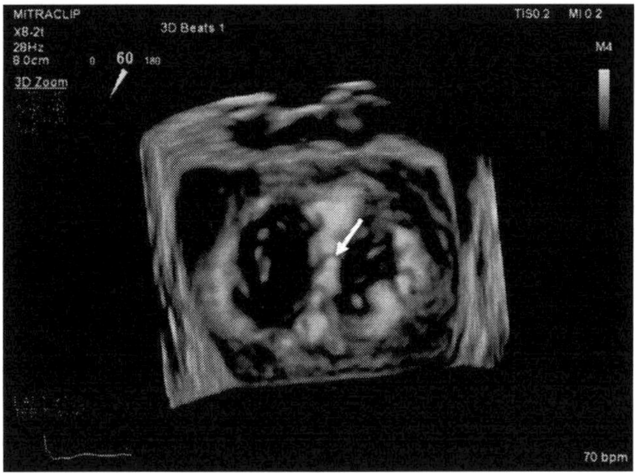

FIGURE 7.1

3D transesophageal echocardiogram view of the mitral valve post mitral valve clip (*arrow*).

Echocardiography strain imaging

Strain imaging continues to evolve as a novel and promising tool for the evaluation of cardiac function in several cardiac disease states. Speckle tracking ECHO is a recent development utilizing small speckles of reflected ultrasound throughout the cardiac cycle. Strain measures the fractional shortening of two points (speckles) within the myocardium along the circumferential, radial, or longitudinal planes in the 4-,3-, and 2-chamber views. It then calculates an average strain curve in each direction, which is then quantified and displayed as a polar map with more negative values associated with the better myocardial function (Fig. 7.2) (Donnelly and Hanna, 2017; Mitchell et al., 2019; Nagueh and Quinones, 2014). Global longitudinal strain (GLS) is the most commonly used measure of left ventricle global systolic function, with a normal value in healthy individuals of around −20% (Lang et al., 2015). Clinical applications of strain imaging include myocardial assessment of patients with hypertension, CAD, valvular heart disease, heart failure, and cardiomyopathies (Adda et al., 2012; Ishii et al., 2009). Strain can detect early myocardial dysfunction and can provide incremental prognostic information in addition to the clinical data and ejection fraction data that is obtained with a standard ECHO evaluation (Di Salvo et al., 2010; Shanks et al., 2010). In cardiac amyloidosis, a specific strain pattern of "relative apical sparing" (Fig. 7.3) using GLS has been shown to improve diagnosis and may have prognostic implications (Phelan et al., 2012; Senapati et al., 2016). In patients with aortic stenosis, worse GLS correlates with increasing disease severity and can be used to further risk stratify patients (Collier, Phelan, & Klein, 2017; Kearney et al., 2012; Lancellotti et al., 2010). Similarly, in mitral valve disease, GLS has been shown to be useful as a prognostic marker for patients after mitral valve surgery (Hiemstra et al., 2020). In cardio-oncology, strain imaging has emerged

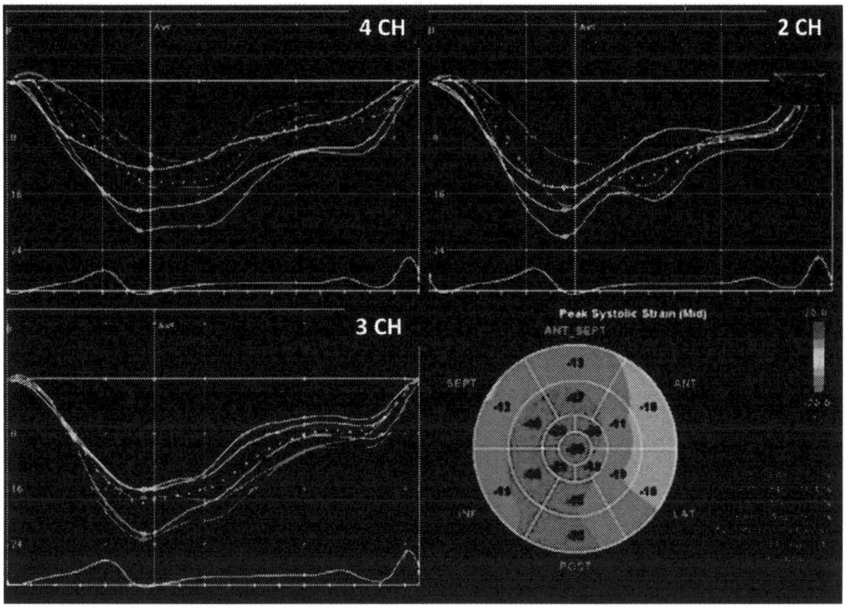

FIGURE 7.2

Average strain curves in the 4-, 2-, and 3-chamber views displayed as a 17-segment myocardial model polar map.

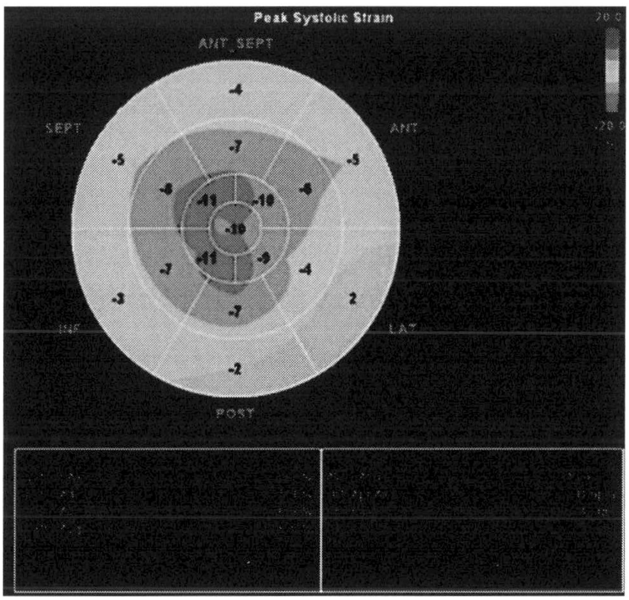

FIGURE 7.3

Strain echocardiographic pattern of "relative apical sparing" in cardiac amyloidosis.

as a useful tool to identify early left ventricular dysfunction after chemotherapy. A reduction in GLS in patients receiving chemotherapy has been shown to occur before a drop in the ejection fraction, therefore early identification of at-risk patients can help to guide clinical decision making (Plana et al., 2014; Yeh and Vejpongsa, 2015). Continued technologic advances will allow strain imaging to evolve and provide a wide array of clinical applications in cardiovascular disease.

There has been immense interest in machine learning and use of artificial intelligence in cardiovascular medicine with new technological advances and a growing population with cardiovascular disease. Noninvasive imaging remains a gatekeeper initial test in the management of patients with cardiovascular diseases. The use of machine learning in echocardiography offers a simplified diagnostic approach by utilizing complex mathematical models to integrate multiple imaging parameters from a high volume and large clinical imaging database (Narula, Shameer, Salem Omar, Dudley, & Sengupta, 2016). Machine learning algorithms are accurate and reliable in echocardiography, allowing the interpretation of a large volume of data with high diagnostic accuracy and reduced variability. This allows for less emphasis on the skill of the interpreter and enables easier serial echocardiography assessments. Further, it will open up opportunities for primary care and rural areas to use echocardiography. Studies have shown reliable results with machine learning applied to myocardial strain analysis to differentiate different forms of myocardial remodeling, chemotherapy-induced cardiotoxicity, and cardiomyopathies (Narula et al., 2016; Zhang et al., 2018). Continued advances in machine learning algorithms hold promise in cardiac imaging, both as a segmentation and image interpretation tool, and an approach for the extraction of novel prognostic markers of cardiac disease progression.

Nuclear cardiology

Single-photon emission computed tomography

Stress testing is a routinely used noninvasive modality applied to assess patients presenting with chest pain, with the aim of detecting the presence or absence of significant CAD. There are multiple types of stress testing ranging from exercise electrocardiography (ECG) treadmill, stress ECHO, stress CMR, or nuclear stress test. Single-photon emission computed tomography (SPECT) is one of the most widely utilized types of nuclear stress testing in the United States and usually uses Technetium-99m or Thallium-201 radiotracers for myocardial perfusion imaging. Combining these stress tests with exercise allows for assessment for functional capacity, chronotropic incompetence, and exercise capacity, all of which have significant clinical and prognostic implications (Al-Mallah, Hachamovitch, Dorbala, & Di Carli, 2009; Al-Mallah et al., 2016a, 2016b; Ward et al., 2007). An alternative approach to assessing changes in myocardial perfusion in individuals who are unable to exercise involves the administration of intravenous coronary validator agents. Common vasodilator pharmacologic stress agents used are regadenoson (Lexiscan,

Astellas Pharma US, Inc., Deerfield, IL) or less commonly adenosine and dipyridamole stress agents. SPECT has its limitations especially in obese patients with BMI $\geq$40 kg/m^2 due to image resolution and artifacts. In addition, it does not allow for a quantitative assessment of myocardial blood flow. Advances in SPECT imaging using new gamma camera technology with cadmium–zinc–telluride (CZT) improve image resolution, accuracy, and acquisition time and reduce radiation exposure (Budzynska, Osiecki, Mazurek, Piszczek, & Dziuk, 2019; Cantoni et al., 2019; Nudi et al., 2017). Image artifacts causing false-positive stress tests are still present for SPECT-CZT as compared to SPECT, but to a lesser extent (Oddstig, Martinsson, Jogi, Engblom, & Hindorf, 2019). SPECT has a sensitivity and specificity of 85% and 66% compared to 89% and 69% for SPECT-CZT for the detection of CAD (Cantoni et al., 2019). SPECT-CZT, like SPECT, can be combined with cardiac CT, for calcium scoring to visualize coronary vessel calcification, for added information for the diagnosis of CAD (Bailly, Thibault, Courtehoux, Metrard, & Ribeiro, 2019). Recent research in SPECT-CZT is evaluating its feasibility for myocardial blood flow quantitation in patients with CAD and heart failure (Agostini et al., 2018; Ma et al., 2019).

Position emission tomography

Positron emission tomography (PET) has gained popularity over the past several years owing to multiple factors, including shorter study duration, better image resolution, ability to perform metabolic imaging, and lower radiation exposure. Given the enhanced image quality and absolute quantification of myocardial flow reserve (MFR), PET has higher accuracy for the detection of CAD (Fig. 7.4). MFR is also important for early detection of CAD and a marker for myocardial dysfunction in the absence of obstructive CAD (Al Badarin, Aljizeeri, Almasoudi, & Al-Mallah, 2017; Al-Mallah, Sitek, Moore, Di Carli, & Dorbala, 2010; Aljizeeri et al., 2019). MFR $<$1.5 is associated with worse clinical outcomes and MFR $>$2 has an excellent ability to rule out high-risk CAD (Murthy et al., 2018).

In addition, PET metabolic imaging utilizes the native myocardial metabolic pathway that is normally dependent on fatty acid metabolism. Although inflammatory cells have higher rates of glucose metabolism, infarcted "nonviable" myocardial tissue, on the other hand, experience disturbances in native metabolism. This is the basis for PET's ability to determine myocardial viability, evaluation of inflammatory conditions, as well as infections of heart valves (Al Moudi and Sun, 2014; Dilsizian, 2016; Khalaf, Chamsi-Pasha, & Al-Mallah, 2019). Several studies have shown PET to be a useful tool in diagnosing infective endocarditis of prosthetic valves and implanted cardiac devices (Mahmood et al., 2019). Cardiac involvement in the setting of sarcoidosis, a systemic inflammatory disease, is an important clinical entity to recognize. PET imaging using radiolabeled glucose, fluorodeoxyglucose (FDG)-18, identifies areas of active inflammation in sarcoidosis with good diagnostic accuracy (Youssef et al., 2012). PET is a powerful tool in the evaluation of CAD, inflammatory disease and sarcoid (Fig. 7.5), and infected cardiac devices or valves.

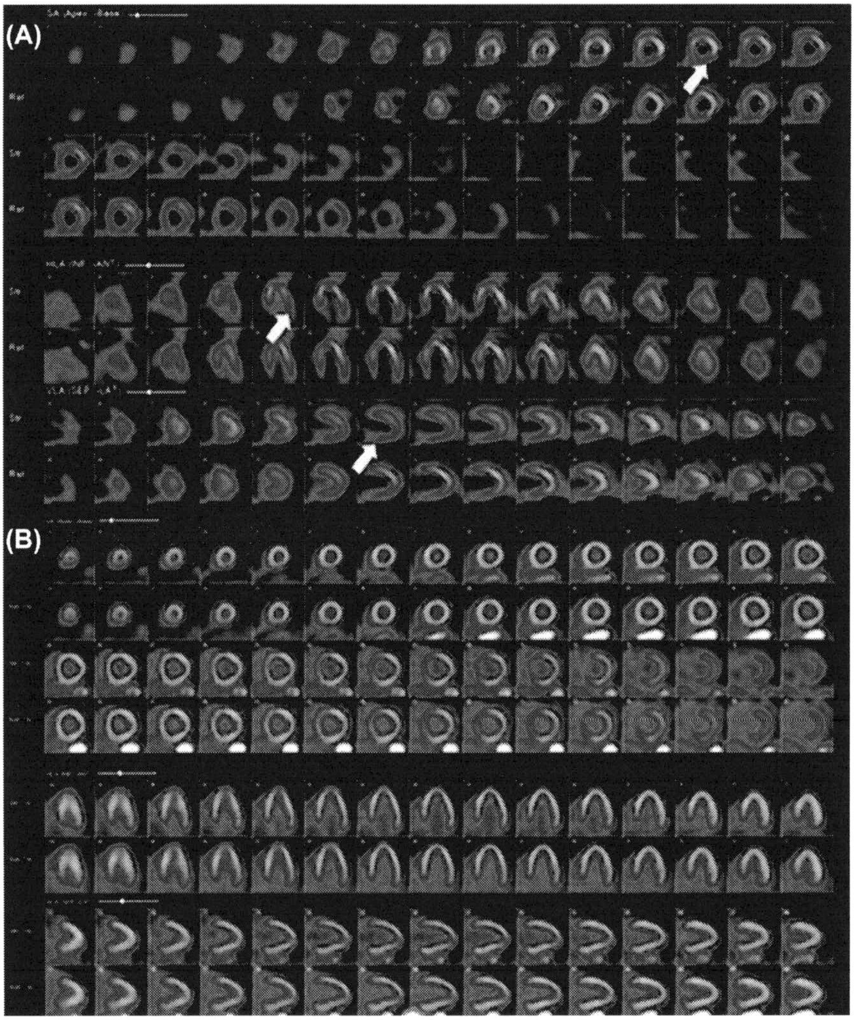

FIGURE 7.4

Image quality by SPECT (A) and PET (B) performed on the same patient 1 week apart. The PET images show no evidence of perfusion abnormalities while the SPECT imaging shows attenuation artifacts (*arrows*).

Technetium-99m pyrophosphate imaging

Cardiac amyloidosis is a type of infiltrative cardiomyopathy that results from the abnormal amyloid fibril deposition in the heart, with two main types: transthyretin (ATTR) and light-chain (AL). Diagnosis is usually made through serum blood markers along with fat pad biopsy, endomyocardial biopsy, CMR, or technetium-99m pyrophosphate (^{99m}Tc-PYP). New advances in noninvasive imaging with ^{99m}Tc-PYP imaging

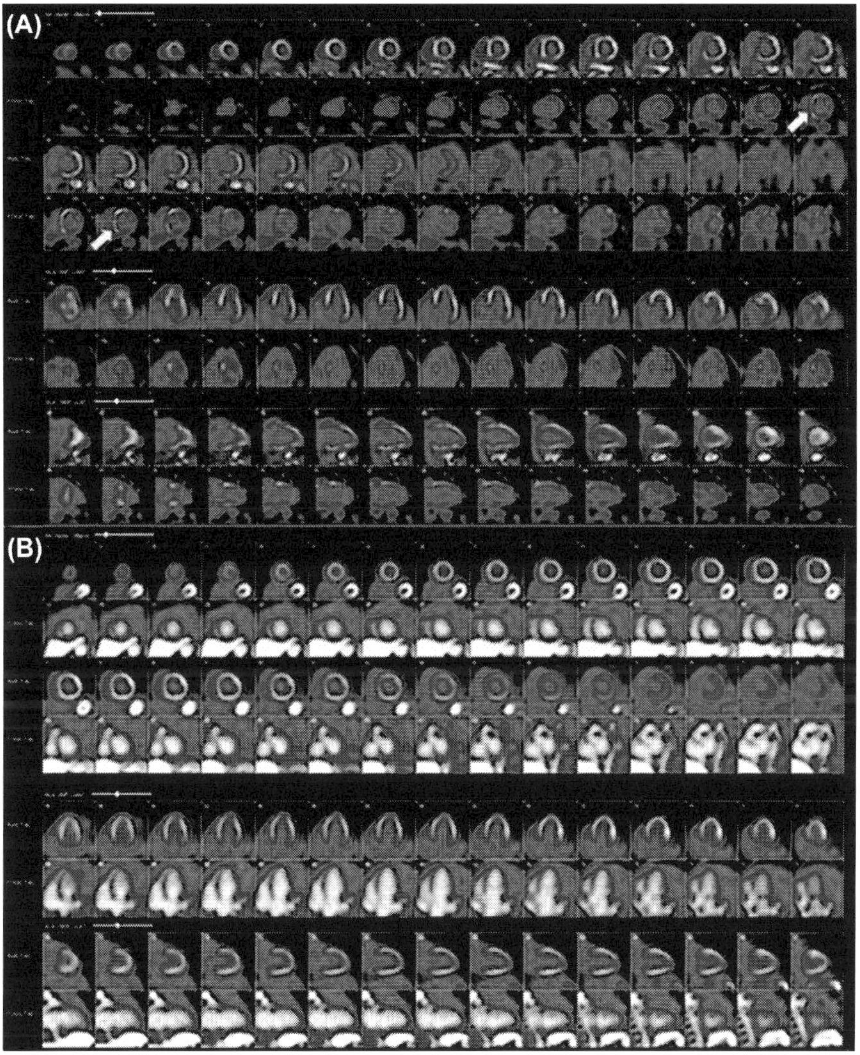

FIGURE 7.5

(A) Inflammatory cardiomyopathy with positive FDG uptake (*arrows*) and (B) normal PET perfusion in a patient with cardiac sarcoidosis.

have improved the ability to differentiate between ATTR and AL subtypes of cardiac amyloid, which is a critical step for treatment planning (Bhogal et al., 2018; Tuzovic et al., 2017; Bokhari et al., 2013). ^{99m}Tc-PYP is given as an intravenous radiotracer and myocardial uptake is seen in patients with ATTR amyloid. Semiquantitative scoring (visual score 0, no uptake to 3, uptake greater than bone) and quantitatively using the heart/contralateral lung ratio is used to determine the likelihood of ATTR amyloid (Fig. 7.6). Using a heart-to-contralateral ratio >1.5 and an intense diffuse

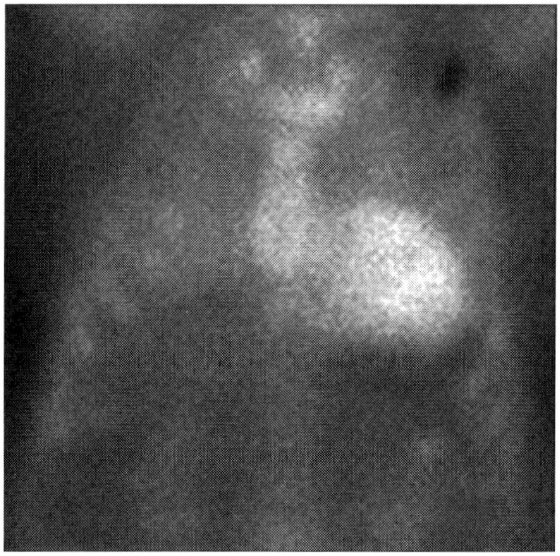

FIGURE 7.6

PYP scan in a patient with biopsy-proven ATTR cardiac amyloidosis. The heart to contralateral lung ratio is 2.1.

myocardial tracer uptake (visual score 2 or 3) was found to have a 100% specificity for ATTR amyloid (Bokhari et al., 2013; Castano et al., 2016; Harb et al., 2017; Papantoniou et al., 2015). ^{99m}Tc-PYP imaging has revolutionized the field of cardiac amyloidosis with its wide availability and ease of use for noninvasive diagnosis of ATTR cardiac amyloidosis.

Cardiac computed tomography

Cardiac CT assessment of coronary artery disease

In the last decade, cardiac CT has gained acceptance as a tool for anatomical assessment of the coronary vasculature. The rapid advances in technology using dual X-ray source and wide multidetectors (320-slice detector scanners) allow for lower radiation exposure while maintaining higher temporal and spatial resolution with isotropic voxels down to 0.3 mm in x, y, and z axes (Chaikriangkrai, Choi, Nabi, & Chang, 2014). Cardiac CT is ECG gated to the cardiac cycle, which allows for the evaluation of cardiac anatomy, coronary arteries for anomalies and CAD, and valves in different phases of the cardiac cycle. In addition, it has a high negative predictive value (to rule out obstructive CAD) (Al-Mallah, Aljizeeri, Villines, Srichai, & Alsaileek, 2015; Hulten et al., 2013) and several criteria are used for measurement of luminal stenosis as well as assessment of atherosclerotic plaque for the presence of high-risk features (Al-Mallah et al., 2014; Al-Mallah and Nasir, 2011; Nakanishi et al., 2017; Nakazato et al., 2013, 2014) that are associated with the occurrence of the incident acute coronary syndrome (Andreini et al., 2019; Ferencik et al., 2015; Liu et al., 2018). Due to its fast acquisition time and superior spatial resolution, several studies have shown that cardiac CT may be a cost-effective tool in the evaluation of patients presenting with acute and stable chest pain (Douglas et al., 2015a; Handy et al., 2016; Hoffmann et al., 2012; Sharma et al., 2019).

CT fractional flow reserve

Although cardiac CT provides an excellent anatomic assessment of the coronaries, it has limited accuracy especially in the settings of high heart rates and dense calcifications. A recent technology that allows for the noninvasive functional assessment of the coronary arteries has emerged (Al-Mallah and Ahmed, 2016). CT fractional flow reserve (FFR) is an FDA-approved method that allows for the assessment of the physiologic significance of a coronary artery lesion. CT FFR helps determine the functional significance of a specific atherosclerotic lesion through estimation of coronary flow in different segments of the coronary arteries (Chaikriangkrai, Choi, Nabi, & Chang, 2014; Han et al., 2018; Kim, Suh, Han, Kim, & Choi, 2019). Although invasive FFR can be performed during invasive coronary angiography for detection of ischemia, with reference standard for ischemia defined as an FFR of $\leq$0.80, studies have shown that CT FFR can provide a noninvasive estimate with acceptable diagnostic accuracy. Such noninvasive functional measures are extremely vital within the context of the clinical workflow as they enhance the utility of cardiac CT as a gatekeeper for downstream invasive testing (Fig. 7.7) (Chinnaiyan et al., 2017; Douglas et al., 2015b; Norgaard et al., 2014, 2018; Patel et al., 2020). Further refinement of this technology is being researched to enhance its correlation with invasive FFR utilizing newer machines and deep learning tools. Recent studies have applied machine learning-based CT-FFR in assessing coronary ischemia as compared to invasive FFR and CT coronary angiography with good diagnostic accuracy (Tesche et al., 2020; Zhou et al., 2019).

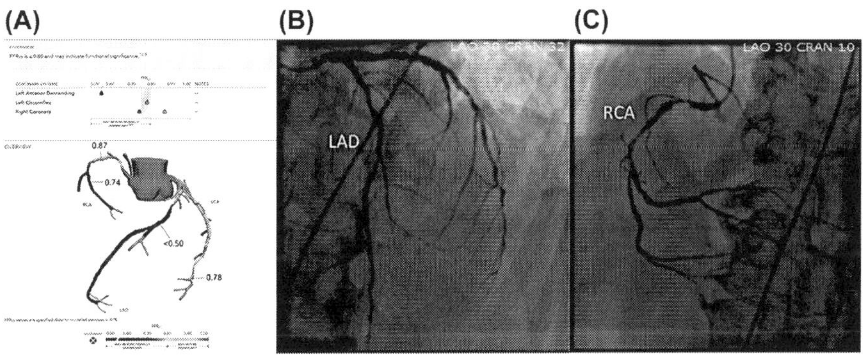

FIGURE 7.7

Cardiac CT FFR in a patient with a significant lesion (FFR $\leq$ 0.80) in left anterior descending (LAD) and right coronary arteries (RCA).

CT perfusion

The major advantage of CT perfusion is the ability to simultaneously and noninvasively evaluate coronary anatomy and physiology. Similar to nuclear cardiology, associated coronary perfusion defects during stress testing using pharmacologic agents, such as adenosine or regadenoson, allow for the identification of ischemia and functionally obstructive CAD (Cury et al., 2010; Feuchtner et al., 2012; Ko et al., 2011; Rochitte et al., 2014). With rest and stress images, infarcted or nonviable myocardium is seen as hypoattenuating areas reflecting diminished contrast content in rest and stress (Al-Mallah and Aljizeeri, 2017). Defects in stress imaging but not in rest imaging are indicative of ischemic myocardium. There are several protocols for CT perfusion and studies have shown that adding CT perfusion in addition to a standard cardiac CT increases the diagnostic accuracy of cardiac CT in identifying hemodynamically significant CAD (Becker and Becker, 2013; Dwivedi et al., 2013; Rochitte et al., 2014). Due to risks of excessive radiation and significant interobserver variability with CT perfusion, the use of CT perfusion is limited and remains mostly a research tool.

Cardiac CT in structural heart disease

Although coronary CT angiography is the most commonly used application for cardiac CT, it is not the only one. With recent advances in percutaneous therapies such as TAVR, mitral valve clip, and transcatheter mitral valve replacement, the clinical applications of cardiac CT in structural heart disease has rapidly expanded (Nasir et al., 2010). For instance, evaluation of cardiac valve anatomy, aortic annular sizing (Fig. 7.8), mitral valve annulus, mitral annular calcification, aorta, and peripheral

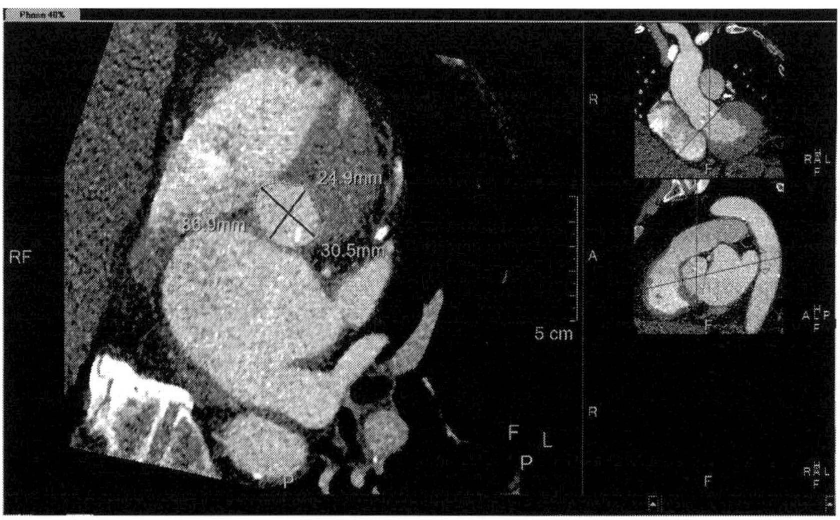

FIGURE 7.8

Cardiac CT preprocedural TAVR planning for aortic annulus sizing (aortic perimeter and major and minor axes) reconstructed in oblique axes after image reconstruction.

vascular assessment are key components for evaluation before transcatheter interventions (Blanke et al., 2019; Hell and Achenbach, 2019; Jilaihawi et al., 2012). Cardiac CT is also useful for evaluation of prosthetic para-valvular regurgitation or obstruction, as well as complications secondary to infective endocarditis (Feuchtner et al., 2009; Habets, Mali, & Budde, 2012; Kim et al., 2019; Sucha et al., 2015; Symersky, Budde, De Mol, & Prokop, 2009). Cardiac CT allows accurate evaluation of the left atrium, left atrial appendage, pulmonary veins, and assessment of atrial clots, which are important for device sizing before left atrial appendage closure and procedural planning for atrial fibrillation ablation.

Cardiac magnetic resonance

T1 mapping and late-gadolinium enhancement

Clinical application of CMR started in the 1980s with the assessment of the cardiac structure, function, and myocardial tissue using T1 and T2 relaxation properties. With the development of contrast agents, myocardial perfusion, and viability assessment with detection of scar ensued in the 1990s. CMR can also be used to assess the aorta, coronary and peripheral vessels (Pohost, 2008). It allows for accurate quantification of cardiac chamber size, function, and valvular properties (Al-Mallah and Shareef, 2011; Aljizeeri, Sulaiman, Alhulaimi, Alsaileek, & Al-Mallah, 2017; Alotaibi, Aljizeeri, Al-Mallah, & Alsaileek, 2017). Cardiac imaging is usually performed with 1.5 or 3 T scanners with gadolinium-based contrast (for scar assessment), but is contraindicated in patients with estimated global filtration rate (eGFR) <30 mL/min/1.73 m^2.

With the advent of faster imaging processing systems, CMR has the added advantage of assessing myocardial tissue characteristics using T1 and T2 mapping. Specific myocardial tissue characteristic T1, T2, and T2* magnetization relaxation times help differentiate normal tissue from myocardial interstitial fibrosis, edema, and iron deposition within the myocardium. Contrast T1 mapping performed before and after gadolinium contrast can generate values for calculation of native T1 time and extracellular volume fraction, which correlates with the amount of diffuse interstitial fibrosis or infiltration of the heart (Ghosn and Shah, 2014; Lopez-Mattei and Shah, 2013; Messroghli et al., 2017; Miller et al., 2013; Moon et al., 2013).

Contrast-enhanced CMR is well established for noninvasive evaluation of replacement fibrosis for myocardial scar assessment using late gadolinium enhancement (LGE). CMR-LGE is the only noninvasive modality to directly visualize myocardial scar, whereas previously this was only possible through direct evaluation of tissue from invasive endomyocardial biopsies or autopsies (Aljizeeri et al., 2017). In CAD, CMR-LGE has been shown to predict viable versus nonviable myocardium (transmural scar defined as extent of LGE $>50\%$ of the myocardium) and forecast recovery after revascularization (Fig. 7.9) (Shah et al., 2013). In several other cardiac diseases, T1 mapping and LGE have been shown to have a good histopathologic correlation

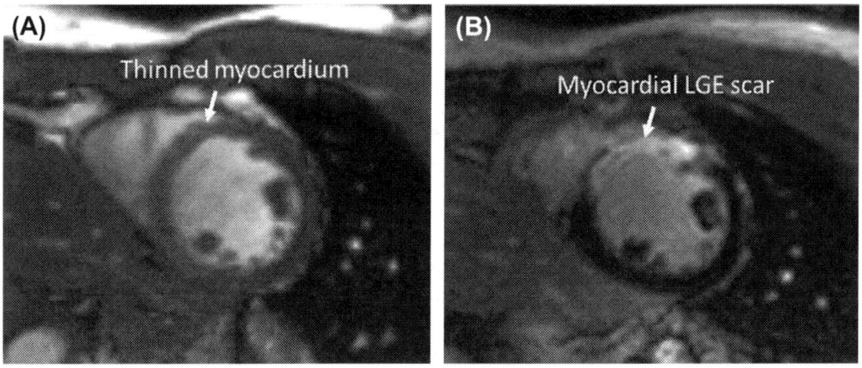

FIGURE 7.9

CMR with thinned myocardial segment and associated transmural late gadolinium enhancement (LGE) scar.

with myocardial biopsy with the added value of providing prognostically useful data, especially in cardiac sarcoidosis and amyloidosis (Bing et al., 2019; Kitkungvan et al., 2018; Kockova et al., 2016; Kwong et al., 2015; Martinez-Naharro et al., 2017).

CMR stress perfusion

Coronary artery disease remains the leading cause of morbidity of mortality worldwide. Efforts to select the most cost-effective test for detecting and managing CAD continue to evolve. With significant advances in CMR due to its superior spatial and temporal resolution, stress perfusion CMR has evolved as a novel imaging tool for the detection and quantification of CAD and is comparable to other stress imaging modalities (Multimodality Writing Group for Stable Ischemic Heart et al., 2014; Nagel et al., 1999; Nandalur, Dwamena, Choudhri, Nandalur, & Carlos, 2007; Patel et al., 2013; Raman et al., 2016; Wahl et al., 2004). In some parts of the work, it has even become one of the most utilized stress perfusion modalities. Vasodilator (regadenoson or adenosine) or dobutamine are the main pharmacologic agents used in stress CMR and less commonly exercise stress CMR. Compared to SPECT, PET, cardiac CT, and invasive angiography with FFR, stress CMR has been shown to have similar or superior diagnostic accuracy for the detection of obstructive CAD (Greenwood et al., 2014; Jaarsma et al., 2012; Nagel et al., 2019; Schwitter et al., 2013; Takx et al., 2015). Stress perfusion CMR provides prognostic information and helps with risk stratification of patients to guide revascularization strategies while being cost-effective in stable chest pain patients (Greenwood et al., 2016; Kwong et al., 2019). In addition to obstructive CAD, stress CMR has been shown to be helpful for guiding therapy and decision making in patients with myocardial infarction with nonobstructive CAD, heart failure, and hypertrophic cardiomyopathy (Crawford et al., 2014; Dastidar et al., 2019; Hernandez, 2018; Kandolin et al., 2019; Tower-Rader, Betancor, Lever, & Desai, 2017).

Valvular heart disease

Cardiac magnetic resonance imaging plays a key role in the evaluation of valvular heart disease and has several advantages including excellent signal-to-noise ratio and being able to image in any cardiac plane. Phase-contrast velocity mapping is used in CMR to assess flow through the systemic aorta and pulmonary artery systems to accurately quantify flow. CMR can accurately assess cardiac function, assess the mechanism of valvular pathology, quantify valvular regurgitation ("leaky valves"), and evaluate its effects on cardiac chamber size and function when ECHO is inconclusive or additional information is needed (Ghosn and Shah, 2014; Lopez-Mattei and Shah, 2013; Shah, 2010; Zoghbi et al., 2017). CMR is well validated for assessing flow in regurgitant lesions including aortic and mitral valve regurgitation (Kon, Myerson, Moat, & Pennel, 2004). Identifying the amount of regurgitation in prosthetic valves can be challenging on ECHO and CMR can assess complications of prosthetic valve dysfunction, such as paravalvular leak, stenosis, pseudoaneurysm, and the effect on cardiac function and clinical outcomes (Al-Najafi, Sanchez, & Lerakis, 2016; Rajiah et al., 2019; Ribeiro et al., 2016; Sherif et al., 2011). When Doppler ECHO is inconclusive, CMR can assess the degree of aortic stenosis severity by calculating anatomic valve area or using phase-contrast velocity mapping to calculate peak velocity across the aortic valve with good correlation with Doppler ECHO and transesophageal ECHO (Defrance et al., 2012; John et al., 2003; Kilner et al., 1993; Lopez-Mattei and Shah, 2013; Reant et al., 2006). In addition to quantification of native valvular and prosthetic valvular dysfunction, CMR can assess LGE and diffuse fibrosis that further adds prognostic information (Bing et al., 2019; Kitkungvan et al., 2018; Treibel et al., 2018). CMR strain-encoded imaging has recently emerged and allows evaluation of regional deformation of myocardial tissue, similar to ECHO strain. The use of CMR strain is useful in evaluating patients with CAD, heart failure, and valvular disease (Lapinskas et al., 2019; Romano et al., 2018; Scatteia, Baritussio, & Bucciarelli-Ducci, 2017; Stefek et al., 2019; Vo, Marwick, & Negishi, 2018).

CMR 4D flow

Recent advancements in CMR using 3D phase-contrast imaging, or 4D flow, allows for measurement of flow in the x, y, and z planes, instead of assessing flow only in the traditional 2D plane. With advances in image processing, 4D flow datasets that are then imported into a software program for analysis can be acquired in 8–15 min. With a 4D flow sequence, anatomy, as well as 3D flow information throughout the cardiac cycle, can be collected and temporally related to visualize complex flow patterns in normal or structural heart disease (Gatehouse et al., 2005; Shah et al., 2013). 4D flow is useful in the evaluation of complex flow patterns in venous and arterial systems, such as aortic aneurysms, aortic dissections, and congenital heart disease (Clough, Waltham, Giese, Taylor, & Schaeffter, 2012; Stankovic et al., 2010; Uribe et al., 2009).

The integration of machine learning-based algorithms within clinical CMR can further improve workflow efficiency through faster postprocessing of acquired scans resulting in faster time to diagnose. Automated algorithms assessing cardiac anatomy, cardiac volumes, flow quantification, and 4D flow can result in a comprehensive cardiovascular assessment within minutes (Avendi, Kheradvar, & Jafarkhani, 2016; Scannell et al., 2020; Tsay & Patterson, 2018). The utilization of machine learning for optimized postprocessing of scans can increase access to CMR to a wider variety of patients and populations. For more on machine learning in CMR, please refer to Chapter 10 titled "Magnetic Resonance Imaging and Artificial Intelligence."

Conclusions

Innovations in cardiovascular imaging continue to evolve with recent and ongoing advances in ECHO, nuclear cardiology, cardiac CT, and CMR. Each cardiac imaging modality has its own advantages and often a multimodality imaging approach is needed in diagnosing complex patients. As imaging technology and understanding of cardiac pathophysiology continue to improve, cardiac imaging will no longer be solely used for disease diagnosis but will expand to incorporate prognostic data that could be useful for decision making and therapy selection. The rapid incorporation of machine learning algorithms within clinical imaging workflows holds promise in improving patient access to cardiac imaging. Multimodality cardiac imaging has and will remain an essential of contemporary approaches to patient-centered comprehensive care.

References

Adda, J., Mielot, C., Giorgi, R., Cransac, F., Zirphile, X., Donal, E., et al. (2012). Low-flow, low-gradient severe aortic stenosis despite normal ejection fraction is associated with severe left ventricular dysfunction as assessed by speckle-tracking echocardiography: A multicenter study. *Circulation: Cardiovascular Imaging, 5*, 27–35.

Aggeli, C., Giannopoulos, G., Roussakis, G., Christoforatou, E., Marinos, G., Toli, C., et al. (2008). Safety of myocardial flash-contrast echocardiography in combination with dobutamine stress testing for the detection of ischaemia in 5250 studies. *Heart, 94*, 1571–1577.

Agostini, D., Roule, V., Nganoa, C., Roth, N., Baavour, R., Parienti, J. J., et al. (2018). First validation of myocardial flow reserve assessed by dynamic (99m)Tc-sestamibi CZT-SPECT camera: Head to head comparison with (15)O-water PET and fractional flow reserve in patients with suspected coronary artery disease. The WATERDAY study. *European Journal of Nuclear Medicine and Molecular Imaging, 45*, 1079–1090.

Al Badarin, F., Aljizeeri, A., Almasoudi, F., & Al-Mallah, M. H. (2017). Assessment of myocardial blood flow and coronary flow reserve with positron emission tomography in ischemic heart disease: Current state and future directions. *Heart Failure Reviews, 22*, 441–453.

Al Moudi, M., & Sun, Z. H. (2014). Diagnostic value of (18)F-FDG PET in the assessment of myocardial viability in coronary artery disease: A comparative study with (99m)Tc SPECT and echocardiography. *Journal of Geriatric Cardiology, 11*, 229–236.

Al-Mallah, M. H., & Ahmed, A. M. (2016). Controversies in the use of fractional flow reserve form computed tomography (FFRCT) vs. Coronary angiography. *Current Cardiovascular Imaging Reports, 9*, 34.

Al-Mallah, M. H., & Aljizeeri, A. (2017). Finding a gatekeeper to coronary angiography: A step in the right direction. *European Heart Journal – Cardiovascular Imaging, 18*, 978–979.

Al-Mallah, M. H., Aljizeeri, A., Villines, T. C., Srichai, M. B., & Alsaileek, A. (2015). Cardiac computed tomography in current cardiology guidelines. *Journal of Cardiovascular Computed Tomography, 9*, 514–523.

Al-Mallah, M., Alqaisi, F., Arafeh, A., Lakhdar, R., AL-Tamsheh, R., & Ananthasubramaniam, K. (2008). Long term favorable prognostic value of negative treadmill echocardiogram in the setting of abnormal treadmill electrocardiogram: A 95 month median duration follow-up study. *Journal of the American Society of Echocardiography, 21*, 1018–1022.

Al-Mallah, M. H., Hachamovitch, R., Dorbala, S., & Di Carli, M. F. (2009). Incremental prognostic value of myocardial perfusion imaging in patients referred to stress single-photon emission computed tomography with renal dysfunction. *Circulation: Cardiovascular Imaging, 2*, 429–436.

Al-Mallah, M. H., Juraschek, S. P., Whelton, S., Dardari, Z. A., Ehrman, J. K., Michos, E. D., et al. (2016a). Sex differences in cardiorespiratory fitness and all-cause mortality: The Henry ford ExercIse testing (FIT) project. *Mayo Clinic Proceedings, 91*, 755–762.

Al-Mallah, M. H., & Nasir, K. (2011). Plaque composition and stenosis severity: Is there any hope for plaque regression? *Atherosclerosis, 219*, 395–396.

Al-Mallah, M. H., Qureshi, W. T., Keteyian, S. J., Brawner, C. A., Alam, M., Dardari, Z., et al. (2016b). Racial differences in the prognostic value of cardiorespiratory fitness (results from the Henry ford exercise testing project). *The American Journal of Cardiology, 117*, 1449–1454.

Al-Mallah, M. H., Qureshi, W., Lin, F. Y., Achenbach, S., Berman, D. S., Budoff, M. J., et al. (2014). Does coronary CT angiography improve risk stratification over coronary calcium scoring in symptomatic patients with suspected coronary artery disease? Results from the prospective multicenter international confirm registry. *European Heart Journal – Cardiovascular Imaging, 15*, 267–274.

Al-Mallah, M. H., & Shareef, M. N. (2011). The role of cardiac magnetic resonance imaging in the assessment of non-ischemic cardiomyopathy. *Heart Failure Reviews, 16*, 369–380.

Al-Mallah, M. H., Sitek, A., Moore, S. C., Di Carli, M., & Dorbala, S. (2010). Assessment of myocardial perfusion and function with PET and PET/CT. *Journal of Nuclear Cardiology, 17*, 498–513.

Al-Najafi, S., Sanchez, F., & Lerakis, S. (2016). The crucial role of cardiac imaging in transcatheter aortic valve replacement (TAVR): Pre- and post-procedural assessment. *Current Treatment Options in Cardiovascular Medicine, 18*, 70.

Aljizeeri, A., Alali Alfaris, M., Ahmed, D., Farea, J., Elneama, A., Suliman, I., et al. (2019). Are patients with end-stage liver disease maximally vasodilated?: Insights from Rubidium-82 PET dilatory perfusion. *JACC: Cardiovascular Imaging, 12*, 2583–2584.

Aljizeeri, A., Sulaiman, A., Alhulaimi, N., Alsaileek, A., & Al-Mallah, M. H. (2017). Cardiac magnetic resonance imaging in heart failure: Where the alphabet begins! *Heart Failure Reviews, 22*, 385–399.

Alotaibi, A. M., Aljizeeri, A., Al-Mallah, M., & Alsaileek, A. (2017). Utility of cardiac magnetic resonance in recurrent myocarditis. *Journal of the Saudi Heart Association, 29*, 214–218.

Andreini, D., Magnoni, M., Conte, E., Masson, S., Mushtaq, S., Berti, S., et al. (2019). Coronary plaque features on CTA can identify patients at increased risk of cardiovascular events. *JACC: Cardiovascular Imaging.* S1936-878X(19)30611-4. https://doi.org/10.1016/j.jcmg.2019.06.019. Online ahead of print.

Appis, A. W., Tracy, M. J., & Feinstein, S. B. (2015). Update on the safety and efficacy of commercial ultrasound contrast agents in cardiac applications. *Echo Research and Practice, 2*, R55–R62.

Avendi, M. R., Kheradvar, A., & Jafarkhani, H. (2016). A combined deep-learning and deformable-model approach to fully automatic segmentation of the left ventricle in cardiac MRI. *Medical Image Analysis, 30*, 108–119.

Bailly, M., Thibault, F., Courtehoux, M., Metrard, G., & Ribeiro, M. J. (2019). Added value of myocardial blood flow quantification and calcium scoring during CZT SPECT myocardial perfusion imaging for coronary artery disease screening. *Clinical Nuclear Medicine, 44*, e617–e619.

Becker, A., & Becker, C. (2013). CT imaging of myocardial perfusion: Possibilities and perspectives. *Journal of Nuclear Cardiology, 20*, 289–296.

Bhogal, S., Ladia, V., Sitwala, P., Cook, E., Bajaj, K., Ramu, V., et al. (2018). Cardiac amyloidosis: An updated review with emphasis on diagnosis and future directions. *Current Problems in Cardiology, 43*, 10–34.

Bing, R., Cavalcante, J. L., Everett, R. J., Clavel, M. A., Newby, D. E., & Dweck, M. R. (2019). Imaging and impact of myocardial fibrosis in aortic stenosis. *JACC: Cardiovascular Imaging, 12*, 283–296.

Blanke, P., Weir-Mccall, J. R., Achenbach, S., Delgado, V., Hausleiter, J., Jilaihawi, H., et al. (2019). Computed tomography imaging in the context of transcatheter aortic valve implantation (TAVI)/transcatheter aortic valve replacement (TAVR): An expert consensus document of the Society of Cardiovascular Computed Tomography. *Journal of Cardiovascular Computed Tomography, 13*, 1–20.

Bokhari, S., Castano, A., Pozniakoff, T., Deslisle, S., Latif, F., & Maurer, M. S. (2013). (99m) Tc-pyrophosphate scintigraphy for differentiating light-chain cardiac amyloidosis from the transthyretin-related familial and senile cardiac amyloidoses. *Circulation: Cardiovascular Imaging, 6*, 195–201.

Budzynska, A., Osiecki, S., Mazurek, A., Piszczek, S., & Dziuk, M. (2019). Feasibility of myocardial perfusion imaging studies in morbidly obese patients with a cadmium-zinc-telluride cardiac camera. *Nuclear Medicine Review. Central and Eastern Europe, 22*, 18–22.

Cantoni, V., Green, R., Acampa, W., Zampella, E., Assante, R., Nappi, C., et al. (2019). Diagnostic performance of myocardial perfusion imaging with conventional and CZT single-photon emission computed tomography in detecting coronary artery disease: A meta-analysis. *Journal of Nuclear Cardiology.* https://doi.org/10.1007/s12350-019-01747-3. Online ahead of print.

Castano, A., Deluca, A., Weinberg, R., Pozniakoff, T., Blaner, W. S., Pirmohamed, A., et al. (2016). Serial scanning with technetium pyrophosphate ((99m)Tc-PYP) in advanced ATTR cardiac amyloidosis. *Journal of Nuclear Cardiology, 23*, 1355–1363.

Chaikriangkrai, K., Choi, S. Y., Nabi, F., & Chang, S. M. (2014). Important advances in technology and unique applications to cardiovascular computed tomography. *Methodist DeBakey Cardiovascular Journal, 10*, 152–158.

Chinnaiyan, K. M., Akasaka, T., Amano, T., Bax, J. J., Blanke, P., DE Bruyne, B., et al. (2017). Rationale, design and goals of the HeartFlow assessing diagnostic value of non-invasive FFRCT in Coronary Care (ADVANCE) registry. *Journal of Cardiovascular Computed Tomography, 11*, 62–67.

Clough, R. E., Waltham, M., Giese, D., Taylor, P. R., & Schaeffter, T. (2012). A new imaging method for assessment of aortic dissection using four-dimensional phase contrast magnetic resonance imaging. *Journal of Vascular Surgery, 55*, 914–923.

Collier, P., Phelan, D., & Klein, A. (2017). A test in context: myocardial strain measured by speckle-tracking echocardiography. *Journal of the American College of Cardiology, 69*, 1043–1056.

Crawford, T., Mueller, G., Sarsam, S., Prasitdumrong, H., Chaiyen, N., Gu, X., et al. (2014). Magnetic resonance imaging for identifying patients with cardiac sarcoidosis and preserved or mildly reduced left ventricular function at risk of ventricular arrhythmias. *Circulation: Arrhythmia and Electrophysiology, 7*, 1109–1115.

Curie, P., & Curie, J. (1881). Lois du degagement de l'electricite par pression, dans la tourmaline. *Comptes rendus de l'Académie des Sciences, 92*, 186.

Cury, R. C., Magalhaes, T. A., Borges, A. C., Shiozaki, A. A., Lemos, P. A., Junior, J. S., et al. (2010). Dipyridamole stress and rest myocardial perfusion by 64-detector row computed tomography in patients with suspected coronary artery disease. *The American Journal of Cardiology, 106*, 310–315.

Dastidar, A. G., Baritussio, A., De Garate, E., Drobni, Z., Biglino, G., Singhal, P., et al. (2019). Prognostic role of CMR and conventional risk factors in myocardial infarction with nonobstructed coronary arteries. *JACC: Cardiovascular Imaging, 12*, 1973–1982.

Defrance, C., Bollache, E., Kachenoura, N., Perdrix, L., Hrynchyshyn, N., Bruguiere, E., et al. (2012). Evaluation of aortic valve stenosis using cardiovascular magnetic resonance: Comparison of an original semiautomated analysis of phase-contrast cardiovascular magnetic resonance with Doppler echocardiography. *Circulation: Cardiovascular Imaging, 5*, 604–612.

Di Salvo, G., Pacileo, G., Limongelli, G., Baldini, L., Rea, A., Verrengia, M., et al. (2010). Non sustained ventricular tachycardia in hypertrophic cardiomyopathy and new ultrasonic derived parameters. *Journal of the American Society of Echocardiography, 23*, 581–590.

Dilsizian, V. (2016). Highlights from the updated joint ASNC/SNMMI PET myocardial perfusion and metabolism clinical imaging guidelines. *Journal of Nuclear Medicine, 57*, 1327–1328.

Donnelly, J. P., & Hanna, M. (2017). Cardiac amyloidosis: An update on diagnosis and treatment. *Cleveland Clinic Journal of Medicine, 84*, 12–26.

Douglas, P. S., Hoffmann, U., Patel, M. R., Mark, D. B., AL-Khalidi, H. R., Cavanaugh, B., et al. (2015a). Outcomes of anatomical versus functional testing for coronary artery disease. *New England Journal of Medicine, 372*, 1291–1300.

Douglas, P. S., Pontone, G., Hlatky, M. A., Patel, M. R., Norgaard, B. L., Byrne, R. A., et al. (2015b). Clinical outcomes of fractional flow reserve by computed tomographic angiography-guided diagnostic strategies vs. usual care in patients with suspected coronary artery disease: The prospective longitudinal trial of FFR(CT): Outcome and resource impacts study. *European Heart Journal, 36*, 3359–3367.

Dwivedi, G., Cocker, M., Yam, Y., Achenbach, S., Al-Mallah, M., Berman, D. S., et al. (2013). Predictive value of cardiac computed tomography and the impact of renal function on all cause mortality (from Coronary Computed Tomography Angiography Evaluation for Clinical Outcomes). *The American Journal of Cardiology, 111*, 1563–1569.

Edler, I., & Lindstrom, K. (2004). The history of echocardiography. *Ultrasound in Medicine and Biology, 30*, 1565–1644.

Feigenbaum, H. (1996). Evolution of echocardiography. *93*, 1321–1327.

Feigenbaum, H., Stone, J. M., Lee, D. A., Nasser, W. K., & Chang, S. (1970). Identification of ultrasound echoes from the left ventricle by use of intracardiac injections of indocyanine green. *Circulation, 41*, 615–621.

Feigenbaum, H., Waldhausen, J. A., & Hyde, L. P. (1965). Ultrasound diagnosis of pericardial effusion. *JAMA, 191*, 711–714.

Ferencik, M., Mayrhofer, T., Puchner, S. B., Lu, M. T., Maurovich-Horvat, P., Liu, T., et al. (2015). Computed tomography-based high-risk coronary plaque score to predict acute coronary syndrome among patients with acute chest pain—Results from the ROMICAT II trial. *Journal of Cardiovascular Computed Tomography, 9*, 538–545.

Feuchtner, G. M., Plank, F., Pena, C., Battle, J., Min, J., Leipsic, J., et al. (2012). Evaluation of myocardial CT perfusion in patients presenting with acute chest pain to the emergency department: Comparison with SPECT-myocardial perfusion imaging. *Heart, 98*, 1510–1517.

Feuchtner, G. M., Stolzmann, P., Dichtl, W., Schertler, T., Bonatti, J., Scheffel, H., et al. (2009). Multislice computed tomography in infective endocarditis: Comparison with transesophageal echocardiography and intraoperative findings. *Journal of the American College of Cardiology, 53*, 436–444.

Gatehouse, P. D., Keegan, J., Crowe, L. A., Masood, S., Mohiaddin, R. H., Kreitner, K. F., et al. (2005). Applications of phase-contrast flow and velocity imaging in cardiovascular MRI. *European Radiology, 15*, 2172–2184.

Geyer, M., Sotiriou, E., Tamm, A. R., Ruf, T. F., Kreidel, F., Yang, Y., et al. (2019). Advanced protocol for three-dimensional transesophageal echocardiography guidance implementing real-time multiplanar reconstruction for transcatheter mitral valve repair by direct annuloplasty. *Journal of the American Society of Echocardiography, 32*, 1359–1365.

Ghosn, M. G., & Shah, D. J. (2014). Important advances in technology and unique applications related to cardiac magnetic resonance imaging. *Methodist DeBakey Cardiovascular Journal, 10*, 159–162.

Gramiak, R., Shah, P. M., & Kramer, D. H. (1969). Ultrasound cardiography: Contrast studies in anatomy and function. *Radiology, 92*, 939–948.

Greenwood, J. P., Herzog, B. A., Brown, J. M., Everett, C. C., Nixon, J., Bijsterveld, P., et al. (2016). Prognostic value of cardiovascular magnetic resonance and single-photon emission computed tomography in suspected coronary heart disease: Long-term follow-up of a prospective, diagnostic accuracy cohort study. *Annals of Internal Medicine, 165*, 1–9.

Greenwood, J. P., Motwani, M., Maredia, N., Brown, J. M., Everett, C. C., Nixon, J., et al. (2014). Comparison of cardiovascular magnetic resonance and single-photon emission computed tomography in women with suspected coronary artery disease from the Clinical Evaluation of Magnetic Resonance Imaging in Coronary Heart Disease (CE-MARC) Trial. *Circulation, 129*, 1129–1138.

Habets, J., Mali, W. P., & Budde, R. P. (2012). Multidetector CT angiography in evaluation of prosthetic heart valve dysfunction. *Radiographics, 32*, 1893–1905.

Handy, C. E., Desai, C. S., Dardari, Z. A., Al-Mallah, M. H., Miedema, M. D., Ouyang, P., et al. (2016). The association of coronary artery calcium with noncardiovascular disease: The multi-ethnic study of atherosclerosis. *JACC: Cardiovascular Imaging, 9*, 568–576.

Han, D., Hartaigh, B. O., Gransar, H., Lee, J. H., Rizvi, A., Baskaran, L., et al. (2018). Incremental prognostic value of coronary computed tomography angiography over coronary calcium scoring for major adverse cardiac events in elderly asymptomatic individuals. *European Heart Journal – Cardiovascular Imaging, 19*, 675–683.

Harb, S. C., Haq, M., Flood, K., Guerrieri, A., Passerell, W., Jaber, W. A., et al. (2017). National patterns in imaging utilization for diagnosis of cardiac amyloidosis: A focus on Tc99m-pyrophosphate scintigraphy. *Journal of Nuclear Cardiology, 24*, 1094–1097.

Hatle, L., Angelsen, B. A., & Tromsdal, A. (1979). Noninvasive assessment of aortic stenosis by Doppler ultrasound. *British Heart Journal, 43*, 284–292.

Hell, M. M., & Achenbach, S. (2019). CT support of cardiac structural interventions. *British Journal of Radiology, 92*, 20180707.

Hernandez, L. E. (2018). Myocardial stress perfusion magnetic resonance in children with hypertrophic cardiomyopathy. *Cardiology in the Young, 28*, 702–708.

Hiemstra, Y. L., Tomsic, A., Van Wijngaarden, S. E., Palmen, M., Klautz, R. J. M., Bax, J. J., et al. (2020). Prognostic value of global longitudinal strain and etiology after surgery for primary mitral regurgitation. *JACC: Cardiovascular Imaging, 13*, 577–585.

Hoffmann, U., Truong, Q. A., Schoenfeld, D. A., Chou, E. T., Woodard, P. K., Nagurney, J. T., et al. (2012). Coronary CT angiography versus standard evaluation in acute chest pain. *New England Journal of Medicine, 367*, 299–308.

Holen, J., Aaslid, R., Landmark, K., & Simonsen, S. (1976). Determination of pressure gradient in mitral stenosis with a non-invasive ultrasound Doppler technique. *Acta Medica Scandinavica, 199*, 455–460.

Hulten, E., Villines, T. C., Cheezum, M. K., Berman, D. S., Dunning, A., Achenbach, S., et al. (2013). Usefulness of coronary computed tomography angiography to predict mortality and myocardial infarction among Caucasian, African and East Asian ethnicities (from the CONFIRM [coronary CT angiography evaluation for clinical outcomes: An international multicenter] registry). *The American Journal of Cardiology, 111*, 479–485.

Hung, J., Lang, R., Flachskampf, F., Shernan, S. K., Mcculloch, M. L., Adams, D. B., et al. (2007). 3D echocardiography: A review of the current status and future directions. *Journal of the American Society of Echocardiography, 20*, 213–233.

Ishii, K., Suyama, T., Imai, M., Maenaka, M., Yamanaka, A., Makino, Y., et al. (2009). Abnormal regional left ventricular systolic and diastolic function in patients with coronary artery disease undergoing percutaneous coronary intervention: Clinical significance of post-ischemic diastolic stunning. *Journal of the American College of Cardiology, 54*, 1589–1597.

Jaarsma, C., Leiner, T., Bekkers, S. C., Crijns, H. J., Wildberger, J. E., Nagel, E., et al. (2012). Diagnostic performance of noninvasive myocardial perfusion imaging using single-photon emission computed tomography, cardiac magnetic resonance, and positron emission tomography imaging for the detection of obstructive coronary artery disease: A meta-analysis. *Journal of the American College of Cardiology, 59*, 1719–1728.

Jilaihawi, H., Kashif, M., Fontana, G., Furugen, A., Shiota, T., Friede, G., et al. (2012). Cross-sectional computed tomographic assessment improves accuracy of aortic annular sizing for transcatheter aortic valve replacement and reduces the incidence of paravalvular aortic regurgitation. *Journal of the American College of Cardiology, 59*, 1275–1286.

John, A. S., Dill, T., Brandt, R. R., Rau, M., Ricken, W., Bachmann, G., et al. (2003). Magnetic resonance to assess the aortic valve area in aortic stenosis: How does it compare to current diagnostic standards? *Journal of the American College of Cardiology, 42*, 519–526.

Kandolin, R. M., Wiefels, C. C., Mesquita, C. T., Chong, A. Y., Boland, P., Glineur, D., et al. (2019). The current role of viability imaging to guide revascularization and therapy decisions in patients with heart failure and reduced left ventricular function. *Canadian Journal of Cardiology, 35*, 1015–1029.

Kearney, L. G., Lu, K., Ord, M., Patel, S. K., Profitis, K., Matalanis, G., et al. (2012). Global longitudinal strain is a strong independent predictor of all-cause mortality in patients with aortic stenosis. *European Heart Journal – Cardiovascular Imaging, 13*, 827–833.

Khalaf, S., Chamsi-Pasha, M., & Al-Mallah, M. H. (2019). Assessment of myocardial viability by PET. *Current Opinion in Cardiology, 34*, 466–472.

Khawaja, O. A., Shaikh, K. A., & Al-Mallah, M. H. (2010). Meta-analysis of adverse cardiovascular events associated with echocardiographic contrast agents. *The American Journal of Cardiology, 106*, 742–747.

Kilner, P. J., Manzara, C. C., Mohiaddin, R. H., Pennell, D. J., Sutton, M. G., Firmin, D. N., et al. (1993). Magnetic resonance jet velocity mapping in mitral and aortic valve stenosis. *Circulation, 87*, 1239–1248.

Kim, J. Y., Suh, Y. J., Han, K., Kim, Y. J., & Choi, B. W. (2019). Diagnostic value of advanced imaging modalities for the detection and differentiation of prosthetic valve obstruction: A systematic review and meta-analysis. *JACC: Cardiovascular Imaging, 12*, 2182–2192.

Kitkungvan, D., Nabi, F., Kim, R. J., Bonow, R. O., Khan, M. A., Xu, J., et al. (2018). Myocardial fibrosis in patients with primary mitral regurgitation with and without prolapse. *Journal of the American College of Cardiology, 72*, 823–834.

Ko, S. M., Choi, J. W., Song, M. G., Shin, J. K., Chee, H. K., Chung, H. W., et al. (2011). Myocardial perfusion imaging using adenosine-induced stress dual-energy computed tomography of the heart: Comparison with cardiac magnetic resonance imaging and conventional coronary angiography. *European Radiology, 21*, 26–35.

Kockova, R., Kacer, P., Pirk, J., Maly, J., Sukupova, L., Sikula, V., et al. (2016). Native T1 relaxation time and extracellular volume fraction as accurate markers of diffuse myocardial fibrosis in heart valve disease- comparison with targeted left ventricular myocardial biopsy. *Circulation Journal, 80*, 1202–1209.

Kon, M. W., Myerson, S. G., Moat, N. E., & Pennell, D. J. (2004). Quantification of regurgitant fraction in mitral regurgitation by cardiovascular magnetic resonance: Comparison of techniques. *Journal of Heart Valve Disease, 13*, 600–607.

Kurt, M., Shaikh, K. A., Peterson, L., Kurrelmeyer, K. M., Shah, G., Nagueh, S. F., et al. (2009). Impact of contrast echocardiography on evaluation of ventricular function and clinical management in a large prospective cohort. *Journal of the American College of Cardiology, 53*, 802–810.

Kwong, R. Y., Ge, Y., Steel, K., Bingham, S., Abdullah, S., Fujikura, K., et al. (2019). Cardiac magnetic resonance stress perfusion imaging for evaluation of patients with chest pain. *Journal of the American College of Cardiology, 74*, 1741–1755.

Kwong, R. Y., Heydari, B., Abbasi, S., Steel, K., Al-Mallah, M., Wu, H., et al. (2015). Characterization of cardiac amyloidosis by atrial late gadolinium enhancement using contrast-enhanced cardiac magnetic resonance imaging and correlation with left atrial conduit and contractile function. *The American Journal of Cardiology, 116*, 622−629.

Lancellotti, P., Donal, E., Magne, J., Moonen, M., O'connor, K., Daubert, J. C., et al. (2010). Risk stratification in asymptomatic moderate to severe aortic stenosis: The importance of the valvular, arterial and ventricular interplay. *Heart, 96*, 1364−1371.

Lang, R. M., Badano, L. P., Mor-Avi, V., Afilalo, J., Armstrong, A., Ernande, L., et al. (2015). Recommendations for cardiac chamber quantification by echocardiography in adults: An update from the American society of echocardiography and the European association of cardiovascular imaging. *Journal of the American Society of Echocardiography, 28*, 1-39 e14.

Lapinskas, T., Zieschang, V., Erley, J., Stoiber, L., Schnackenburg, B., Stehning, C., et al. (2019). Strain-encoded cardiac magnetic resonance imaging: A new approach for fast estimation of left ventricular function. *BMC Cardiovascular Disorders, 19*, 52.

Lindstrom, K, & Edler, I. (1969). Ultrasonic Doppler technique used in heart disease: clinical application. *Ultrasono Graphia Medica, 3*, 455-61.

Little, S. H., Pirat, B., Kumar, R., Igo, S. R., Mcculloch, M., Hartley, C. J., et al. (2008). Three-dimensional color Doppler echocardiography for direct measurement of vena contracta area in mitral regurgitation: In vitro validation and clinical experience. *JACC: Cardiovascular Imaging, 1*, 695−704.

Liu, T., Maurovich-Horvat, P., Mayrhofer, T., Puchner, S. B., Lu, M. T., Ghemigian, K., et al. (2018). Quantitative coronary plaque analysis predicts high-risk plaque morphology on coronary computed tomography angiography: Results from the ROMICAT II trial. *The International Journal of Cardiovascular Imaging, 34*, 311−319.

Lopez-Mattei, J. C., & Shah, D. J. (2013). The role of cardiac magnetic resonance in valvular heart disease. *Methodist DeBakey Cardiovascular Journal, 9*, 142−148.

Mahmood, M., Kendi, A. T., Ajmal, S., Farid, S., O'horo, J. C., Chareonthaitawee, P., et al. (2019). Meta-analysis of 18F-FDG PET/CT in the diagnosis of infective endocarditis. *Journal of Nuclear Cardiology, 26*, 922−935.

Maleki, M., & Esmaeilzadeh, M. (2012). The evolutionary development of echocardiography. *Iranian Journal of Medical Sciences, 37*, 222−232.

Martinez-Naharro, A., Treibel, T. A., Abdel-Gadir, A., Bulluck, H., Zumbo, G., Knight, D. S., et al. (2017). Magnetic resonance in transthyretin cardiac amyloidosis. *Journal of the American College of Cardiology, 70*, 466−477.

Ma, R., Wang, L., Wu, D., Wang, M., Sun, X., Hsu, B., et al. (2019). Myocardial blood flow quantitation in patients with congestive heart failure: Head-to-head comparison between rapid-rotating gantry SPECT and CZT SPECT. *Journal of Nuclear Cardiology*. https://doi.org/10.1007/s12350-019-01621-2. Online ahead of print.

Mehta, R. H., & Eagle, K. A. (2000). Missed diagnoses of acute coronary syndromes in the emergency room—continuing challenges. *New England Journal of Medicine, 342*, 1207−1210.

Messroghli, D. R., Moon, J. C., Ferreira, V. M., Grosse-Wortmann, L., He, T., Kellman, P., et al. (2017). Clinical recommendations for cardiovascular magnetic resonance mapping of T1, T2, T2* and extracellular volume: A consensus statement by the society for cardiovascular magnetic resonance (SCMR) endorsed by the European association for cardiovascular imaging (EACVI). *Journal of Cardiovascular Magnetic Resonance, 19*, 75.

Miller, C. A., Naish, J. H., Bishop, P., Coutts, G., Clark, D., Zhao, S., et al. (2013). Comprehensive validation of cardiovascular magnetic resonance techniques for the assessment of myocardial extracellular volume. *Circulation: Cardiovascular Imaging, 6*, 373−383.

Mitchell, C., Rahko, P. S., Blauwet, L. A., Canaday, B., Finstuen, J. A., Foster, M. C., et al. (2019). Guidelines for performing a comprehensive transthoracic echocardiographic examination in adults: Recommendations from the American society of echocardiography. *Journal of the American Society of Echocardiography, 32*, 1–64.

Moon, J. C., Messroghli, D. R., Kellman, P., Piechnik, S. K., Robson, M. D., Ugander, M., et al. (2013). Myocardial T1 mapping and extracellular volume quantification: A society for cardiovascular magnetic resonance (SCMR) and CMR working Group of the European society of cardiology consensus statement. *Journal of Cardiovascular Magnetic Resonance, 15*, 92.

Multimodality Writing Group For Stable Ischemic Heart, D., Wolk, M. J., Bailey, S. R., Doherty, J. U., Douglas, P. S., Hendel, R. C., et al. (2014). ACCF/AHA/ASE/ASNC/HFSA/HRS/SCAI/SCCT/SCMR/STS 2013 multimodality appropriate use criteria for the detection and risk assessment of stable ischemic heart disease: A report of the American college of cardiology foundation appropriate use criteria task force, American heart association, American society of echocardiography, American society of nuclear cardiology, heart failure society of America, heart rhythm society, society for cardiovascular angiography and interventions, society of cardiovascular computed tomography, society for cardiovascular magnetic resonance, and society of thoracic surgeons. *Journal of Cardiac Failure, 20*, 65–90.

Mulvagh, S. L., Rakowski, H., Vannan, M. A., Abdelmoneim, S. S., Becher, H., Bierig, S. M., et al. (2008). American society of echocardiography consensus statement on the clinical applications of ultrasonic contrast agents in echocardiography. *Journal of the American Society of Echocardiography, 21*, 1179–1201. Quiz 1281.

Murthy, V. L., Bateman, T. M., Beanlands, R. S., Berman, D. S., Borges-Neto, S., Chareonthaitawee, P., et al. (2018). Clinical quantification of myocardial blood flow using PET: Joint position paper of the SNMMI cardiovascular council and the ASNC. *Journal of Nuclear Cardiology, 25*, 269–297.

Muskula, P. R., & Main, M. L. (2017). Safety with echocardiographic contrast agents. *Circulation: Cardiovascular Imaging, 10*.

Nagel, E., Greenwood, J. P., Mccann, G. P., Bettencourt, N., Shah, A. M., Hussain, S. T., et al. (2019). Magnetic resonance perfusion or fractional flow reserve in coronary disease. *New England Journal of Medicine, 380*, 2418–2428.

Nagel, E., Lehmkuhl, H. B., Bocksch, W., Klein, C., Vogel, U., Frantz, E., et al. (1999). Noninvasive diagnosis of ischemia-induced wall motion abnormalities with the use of high-dose dobutamine stress MRI: Comparison with dobutamine stress echocardiography. *Circulation, 99*, 763–770.

Nagueh, S. F., & Quinones, M. A. (2014). Important advances in technology: Echocardiography. *Methodist DeBakey Cardiovascular Journal, 10*, 146–151.

Nakanishi, R., Baskaran, L., Gransar, H., Budoff, M. J., Achenbach, S., Al-Mallah, M., et al. (2017). Relationship of hypertension to coronary atherosclerosis and cardiac events in patients with coronary computed tomographic angiography. *Hypertension, 70*, 293–299.

Nakazato, R., Arsanjani, R., Achenbach, S., Gransar, H., Cheng, V. Y., Dunning, A., et al. (2014). Age-related risk of major adverse cardiac event risk and coronary artery disease extent and severity by coronary CT angiography: Results from 15 187 patients from the international multisite CONFIRM study. *European Heart Journal – Cardiovascular Imaging, 15*, 586–594.

Nakazato, R., Gransar, H., Berman, D. S., Cheng, V. Y., Lin, F. Y., Achenbach, S., et al. (2013). Relationship of low- and high-density lipoproteins to coronary artery plaque composition by CT angiography. *Journal of Cardiovascular Computed Tomography, 7*, 83–90.

Nandalur, K. R., Dwamena, B. A., Choudhri, A. F., Nandalur, M. R., & Carlos, R. C. (2007). Diagnostic performance of stress cardiac magnetic resonance imaging in the detection of coronary artery disease: A meta-analysis. *Journal of the American College of Cardiology, 50*, 1343–1353.

Narula, S., Shameer, K., Salem Omar, A. M., Dudley, J. T., & Sengupta, P. P. (2016). Machine-learning algorithms to automate morphological and functional assessments in 2D echocardiography. *Journal of the American College of Cardiology, 68*, 2287–2295.

Nasir, K., Katz, R., Al-Mallah, M., Takasu, J., Shavelle, D. M., Carr, J. J., et al. (2010). Relationship of aortic valve calcification with coronary artery calcium severity: The multi-ethnic study of atherosclerosis (MESA). *Journal of Cardiovascular Computed Tomography, 4*, 41–46.

Norgaard, B. L., Leipsic, J., Gaur, S., Seneviratne, S., Ko, B. S., Ito, H., et al. (2014). Diagnostic performance of noninvasive fractional flow reserve derived from coronary computed tomography angiography in suspected coronary artery disease: The NXT trial (analysis of coronary blood flow using CT angiography: Next steps). *Journal of the American College of Cardiology, 63*, 1145–1155.

Norgaard, B. L., Terkelsen, C. J., Mathiassen, O. N., Grove, E. L., Botker, H. E., Parner, E., et al. (2018). Coronary CT angiographic and flow reserve-guided management of patients with stable ischemic heart disease. *Journal of the American College of Cardiology, 72*, 2123–2134.

Nudi, F., Iskandrian, A. E., Schillaci, O., Peruzzi, M., Frati, G., & Biondi-Zoccai, G. (2017). Diagnostic accuracy of myocardial perfusion imaging with CZT technology: Systemic review and meta-analysis of comparison with invasive coronary angiography. *JACC: Cardiovascular Imaging, 10*, 787–794.

Oddstig, J., Martinsson, E., Jogi, J., Engblom, H., & Hindorf, C. (2019). Differences in attenuation pattern in myocardial SPECT between CZT and conventional gamma cameras. *Journal of Nuclear Cardiology, 26*, 1984–1991.

Papantoniou, V., Valsamaki, P., Kastritis, S., Tsiouris, S., Delichas, Z., Papantoniou, Y., et al. (2015). Imaging of cardiac amyloidosis by (99m)Tc-PYP scintigraphy. *Hellenic Journal of Nuclear Medicine, 18*(Suppl. 1), 42–50.

Patel, M. R., Norgaard, B. L., Fairbairn, T. A., Nieman, K., Akasaka, T., Berman, D. S., et al. (2020). 1-Year impact on medical practice and clinical outcomes of FFRCT: The ADVANCE registry. *JACC: Cardiovascular Imaging, 13*, 97–105.

Patel, M. R., White, R. D., Abbara, S., Bluemke, D. A., Herfkens, R. J., Picard, M., et al. (2013). 2013 ACCF/ACR/ASE/ASNC/SCCT/SCMR appropriate utilization of cardiovascular imaging in heart failure: A joint report of the American college of radiology appropriateness criteria committee and the American college of cardiology foundation appropriate use criteria task force. *Journal of the American College of Cardiology, 61*, 2207–2231.

Phelan, D., Collier, P., Thavendiranathan, P., Popovic, Z. B., Hanna, M., Plana, J. C., et al. (2012). Relative apical sparing of longitudinal strain using two-dimensional speckle-tracking echocardiography is both sensitive and specific for the diagnosis of cardiac amyloidosis. *Heart, 98*, 1442–1448.

Plana, J. C., Galderisi, M., Barac, A., Ewer, M. S., Ky, B., Scherrer-Crosbie, M., et al. (2014). Expert consensus for multimodality imaging evaluation of adult patients during and after cancer therapy: A report from the American society of echocardiography and the European association of cardiovascular imaging. *Journal of the American Society of Echocardiography, 27*, 911–939.

Pohost, G. M. (2008). The history of cardiovascular magnetic resonance. *JACC: Cardiovascular Imaging, 1*, 672–678.

Poon, J., Leung, J. T., & Leung, D. Y. (2019). 3D echo in routine clinical practice - state of the art in 2019. *Heart Lung and Circulation, 28*, 1400–1410.

Porter, T. R., Abdelmoneim, S., Belcik, J. T., Mcculloch, M. L., Mulvagh, S. L., Olson, J. J., et al. (2014). Guidelines for the cardiac sonographer in the performance of contrast echocardiography: A focused update from the American society of echocardiography. *Journal of the American Society of Echocardiography, 27*, 797−810.

Porter, T. R., Mulvagh, S. L., Abdelmoneim, S. S., Becher, H., Belcik, J. T., Bierig, M., et al. (2018). Clinical applications of ultrasonic enhancing agents in echocardiography: 2018 American society of echocardiography guidelines update. *Journal of the American Society of Echocardiography, 31*, 241−274.

Rajiah, P., Moore, A., Saboo, S., Goerne, H., Ranganath, P., Macnamara, J., et al. (2019). Multimodality imaging of complications of cardiac valve surgeries. *Radiographics, 39*, 932−956.

Raman, S. V., Dickerson, J. A., Mazur, W., Wong, T. C., Schelbert, E. B., Min, J. K., et al. (2016). Diagnostic performance of treadmill exercise cardiac magnetic resonance: The prospective, multicenter exercise CMR's accuracy for cardiovascular stress testing (EXACT) trial. *Journal of the American Heart Association, 5*.

Reant, P., Lederlin, M., Lafitte, S., Serri, K., Montaudon, M., Corneloup, O., et al. (2006). Absolute assessment of aortic valve stenosis by planimetry using cardiovascular magnetic resonance imaging: Comparison with transesophageal echocardiography, transthoracic echocardiography, and cardiac catheterisation. *European Journal of Radiology, 59*, 276−283.

Ribeiro, H. B., Orwat, S., Hayek, S. S., Larose, E., Babaliaros, V., Dahou, A., et al. (2016). Cardiovascular magnetic resonance to evaluate aortic regurgitation after transcatheter aortic valve replacement. *Journal of the American College of Cardiology, 68*, 577−585.

Rochitte, C. E., George, R. T., Chen, M. Y., Arbab-Zadeh, A., Dewey, M., Miller, J. M., et al. (2014). Computed tomography angiography and perfusion to assess coronary artery stenosis causing perfusion defects by single photon emission computed tomography: The CORE320 study. *European Heart Journal, 35*, 1120−1130.

Romano, S., Judd, R. M., Kim, R. J., Kim, H. W., Klem, I., Heitner, J. F., et al. (2018). Feature-tracking global longitudinal strain predicts death in a multicenter population of patients with ischemic and nonischemic dilated cardiomyopathy incremental to ejection fraction and late gadolinium enhancement. *JACC: Cardiovascular Imaging, 11*, 1419−1429.

Satomura, S., Matsubara, S., & Yoshioka, M. (1955). A new method of mechanical vibration and its applications. *Memoris of the Institute of Scientific and Industrial Research, 13*, 125.

Scannell, C. M., Veta, M., Villa, A. D. M., Sammut, E. C., Lee, J., Breeuwer, M., et al. (2020). Deep-learning-based preprocessing for quantitative myocardial perfusion MRI. *Journal of Magnetic Resonance Imaging, 51*, 1689−1696.

Scatteia, A., Baritussio, A., & Bucciarelli-Ducci, C. (2017). Strain imaging using cardiac magnetic resonance. *Heart Failure Reviews, 22*, 465−476.

Schwitter, J., Wacker, C. M., Wilke, N., AL-Saadi, N., Sauer, E., Huettle, K., et al. (2013). MR-IMPACT II: Magnetic resonance imaging for myocardial perfusion assessment in coronary artery disease trial: Perfusion-cardiac magnetic resonance vs. single-photon emission computed tomography for the detection of coronary artery disease: A comparative multicentre, multivendor trial. *European Heart Journal, 34*, 775−781.

Senapati, A., Sperry, B. W., Grodin, J. L., Kusunose, K., Thavendiranathan, P., Jaber, W., et al. (2016). Prognostic implication of relative regional strain ratio in cardiac amyloidosis. *Heart, 102*, 748−754.

Shah, D. J. (2010). Functional valve assessment: The emerging role of cardiovascular magnetic resonance. *Methodist DeBakey Cardiovascular Journal, 6*, 15−19.

Shah, D. J., Kim, H. W., James, O., Parker, M., Wu, E., Bonow, R. O., et al. (2013). Prevalence of regional myocardial thinning and relationship with myocardial scarring in patients with coronary artery disease. *Journal of the American Medical Association, 309*, 909−918.

Shanks, M., Ng, A. C., Van De Veire, N. R., Antoni, M. L., Bertini, M., Delgado, V., et al. (2010). Incremental prognostic value of novel left ventricular diastolic indexes for prediction of clinical outcome in patients with ST-elevation myocardial infarction. *The American Journal of Cardiology, 105*, 592–597.

Sharma, A., Coles, A., Sekaran, N. K., Pagidipati, N. J., Lu, M. T., Mark, D. B., et al. (2019). Stress testing versus CT angiography in patients with diabetes and suspected coronary artery disease. *Journal of the American College of Cardiology, 73*, 893–902.

Sherif, M. A., Abdel-Wahab, M., Beurich, H. W., Stocker, B., Zachow, D., Geist, V., et al. (2011). Haemodynamic evaluation of aortic regurgitation after transcatheter aortic valve implantation using cardiovascular magnetic resonance. *EuroIntervention, 7*, 57–63.

Siebelink, H. M., Scholte, A. J., Van De Veire, N. R., Holman, E. R., Nucifora, G., Van Der Wall, E. E., et al. (2009). Value of contrast echocardiography for left ventricular thrombus detection postinfarction and impact on antithrombotic therapy. *Coronary Artery Disease, 20*, 462–466.

Stankovic, Z., Frydrychowicz, A., Csatari, Z., Panther, E., Deibert, P., Euringer, W., et al. (2010). MR-based visualization and quantification of three-dimensional flow characteristics in the portal venous system. *Journal of Magnetic Resonance Imaging, 32*, 466–475.

Stefek, H. A., Berhane, H., Robinson, J. D., Reilly, B., Ruh, A., Markl, M., et al. (2019). Comprehensive MR analysis of cardiac function, aortic hemodynamics and left ventricular strain in pediatric cohort with isolated bicuspid aortic valve. *Pediatric Cardiology, 40*, 1450–1459.

Sucha, D., Symersky, P., Tanis, W., Mali, W. P., Leiner, T., Van Herwerden, L. A., et al. (2015). Multimodality imaging assessment of prosthetic heart valves. *Circulation: Cardiovascular Imaging, 8*, e003703.

Symersky, P., Budde, R. P., De Mol, B. A., & Prokop, M. (2009). Comparison of multidetector-row computed tomography to echocardiography and fluoroscopy for evaluation of patients with mechanical prosthetic valve obstruction. *The American Journal of Cardiology, 104*, 1128–1134.

Takx, R. A., Blomberg, B. A., El Aidi, H., Habets, J., DE Jong, P. A., Nagel, E., et al. (2015). Diagnostic accuracy of stress myocardial perfusion imaging compared to invasive coronary angiography with fractional flow reserve meta-analysis. *Circulation: Cardiovascular Imaging, 8*.

Tesche, C., Otani, K., De Cecco, C. N., Coenen, A., De Geer, J., Kruk, M., et al. (2020). Influence of coronary calcium on diagnostic performance of machine learning CT-FFR: Results from MACHINE registry. *JACC: Cardiovascular Imaging, 13*, 760–770.

Tower-Rader, A., Betancor, J., Lever, H. M., & Desai, M. Y. (2017). A comprehensive review of stress testing in hypertrophic cardiomyopathy: Assessment of functional capacity, identification of prognostic indicators, and detection of coronary artery disease. *Journal of the American Society of Echocardiography, 30*, 829–844.

Treibel, T. A., Kozor, R., Schofield, R., Benedetti, G., Fontana, M., Bhuva, A. N., et al. (2018). Reverse myocardial remodeling following valve replacement in patients with aortic stenosis. *Journal of the American College of Cardiology, 71*, 860–871.

Tsay, D., & Patterson, C. (2018). From machine learning to artificial intelligence applications in cardiac care. *Circulation, 138*, 2569–2575.

Tuzovic, M., Yang, E. H., Baas, A. S., Depasquale, E. C., Deng, M. C., Cruz, D., et al. (2017). Cardiac amyloidosis: Diagnosis and treatment strategies. *Current Oncology Reports, 19*, 46.

Uribe, S., Beerbaum, P., Sorensen, T. S., Rasmusson, A., Razavi, R., & Schaeffter, T. (2009). Four-dimensional (4D) flow of the whole heart and great vessels using real-time respiratory self-gating. *Magnetic Resonance in Medicine, 62*, 984–992.

Vo, H. Q., Marwick, T. H., & Negishi, K. (2018). MRI-derived myocardial strain measures in normal subjects. *JACC: Cardiovascular Imaging, 11*, 196–205.

Wahl, A., Paetsch, I., Gollesch, A., Roethemeyer, S., Foell, D., Gebker, R., et al. (2004). Safety and feasibility of high-dose dobutamine-atropine stress cardiovascular magnetic resonance for diagnosis of myocardial ischaemia: Experience in 1000 consecutive cases. *European Heart Journal, 25*, 1230–1236.

Ward, R. P., AL-Mallah, M. H., Grossman, G. B., Hansen, C. L., Hendel, R. C., Kerwin, T. C., et al. (2007). American Society of Nuclear Cardiology review of the ACCF/ASNC appropriateness criteria for single-photon emission computed tomography myocardial perfusion imaging (SPECT MPI). *Journal of Nuclear Cardiology, 14*, e26–38.

Wei, K. (2010). Utility contrast echocardiography in the emergency department. *JACC: Cardiovascular Imaging, 3*, 197–203.

Wei, K., Mulvagh, S. L., Carson, L., Davidoff, R., Gabriel, R., Grimm, R. A., et al. (2008). The safety of deFinity and Optison for ultrasound image enhancement: A retrospective analysis of 78,383 administered contrast doses. *Journal of the American Society of Echocardiography, 21*, 1202–1206.

Wyrick, J. J., Kalvaitis, S., Mcconnell, K. J., Rinkevich, D., Kaul, S., & Wei, K. (2008). Cost-efficiency of myocardial contrast echocardiography in patients presenting to the emergency department with chest pain of suspected cardiac origin and a nondiagnostic electrocardiogram. *The American Journal of Cardiology, 102*, 649–652.

Yeh, E. T., & Vejpongsa, P. (2015). Subclinical cardiotoxicity associated with cancer therapy: Early detection and future directions. *Journal of the American College of Cardiology, 65*, 2523–2525.

Yoshida, T, Mori, M., Nimura, Y, Okimura, M, Hikita, G, & Nakanishi, K (1954). Study of examining the heart with ultrasonics; IV: clinical applications. *Japanese Circulation Journal, 20*, 228.

Youssef, G., Leung, E., Mylonas, I., Nery, P., Williams, K., Wisenberg, G., et al. (2012). The use of 18F-FDG PET in the diagnosis of cardiac sarcoidosis: A systematic review and metaanalysis including the Ontario experience. *Journal of Nuclear Medicine, 53*, 241–248.

Zhang, J., Gajjala, S., Agrawal, P., Tison, G. H., Hallock, L. A., Beussink-Nelson, L., et al. (2018). Fully automated echocardiogram interpretation in clinical practice. *Circulation, 138*, 1623–1635.

Zhou, F., Wang, Y. N., Schoepf, U. J., Tesche, C., Tang, C. X., Zhou, C. S., et al. (2019). Diagnostic performance of machine learning based CT-FFR in detecting ischemia in myocardial bridging and concomitant proximal atherosclerotic disease. *Canadian Journal of Cardiology, 35*, 1523–1533.

Zoghbi, W. A., Adams, D., Bonow, R. O., Enriquez-Sarano, M., Foster, E., Grayburn, P. A., et al. (2017). Recommendations for noninvasive evaluation of native valvular regurgitation: A report from the American society of echocardiography developed in collaboration with the society for cardiovascular magnetic resonance. *Journal of the American Society of Echocardiography, 30*, 303–371.

CHAPTER

Ultrasound and artificial intelligence

8

Brian A. Telfer[1], Viksit Kumar[2], Aaron D. Aguirre[3,4,5], Anthony E. Samir[2], Laura J. Brattain[1,2]

[1] *Human Health and Performance Systems, MIT Lincoln Laboratory, Lexington, MA, United States;* [2] *Center for Ultrasound Research & Translation, Department of Radiology, Massachusetts General Hospital, Harvard Medical School, Boston, MA, United States;* [3] *Cardiology Division, Massachusetts General Hospital, Harvard Medical School, Boston, MA, United States;* [4] *Wellman Center for Photomedicine, Massachusetts General Hospital, Boston, MA, United States;* [5] *Center for Systems Biology, Massachusetts General Hospital, Boston, MA, United States*

Chapter outline

Introduction

Compared to other major medical imaging modalities such as X-ray, computed tomography (CT), and magnetic resonance imaging, medical ultrasound (US) has unique attributes that make it the preferred modality for many clinical applications. In particular, US is nonionizing, portable, and provides real-time imaging, with adequate spatial and depth resolution to visualize tissue dynamics. The ability to measure Doppler information is also important, particularly for measuring blood flows. The small size of US transducers is a key attribute for intravascular

Machine Learning in Cardiovascular Medicine. https://doi.org/10.1016/B978-0-12-820273-9.00008-7

applications. In addition, accessibility has been increased with the use of portable US, which continues to move toward a smaller footprint and lower cost. Nowadays, some US probes can even be directly connected to a phone or tablet.

On the other hand, US also has unique challenges, particularly in that image quality is highly dependent on the operator's skill in acquiring images based on the proper position, orientation, and probe pressure. Additional challenges that further require operator skill include the presence of noise, artifacts, limited field of view, difficulty in imaging structures behind bone and air, and device variability across manufacturers. Sonographers become highly proficient through extensive training and long experience, but high intra- and interobserver variability remains. This skill dependence has limited the wider use of US by healthcare providers who are not US imaging specialists.

Recent advances in machine learning (ML) have been increasingly applied to medical US (Brattain, Telfer, Dhyani, Grajo, & Samir, 2018), with a goal of reducing intra- and interobserver variability as well as interpretation time. As progress toward these goals is made, US use by nonspecialists is expected to proliferate, including nurses at the bedside or medics in the field. The acceleration in ML applications for medical US can be seen from the increasing number of publications (Fig. 8.1) and Food and Drug Administration (FDA) approvals (Table 8.1) in the past few years. Fig. 8.1 shows that cardiovascular applications (spanning the heart, brain

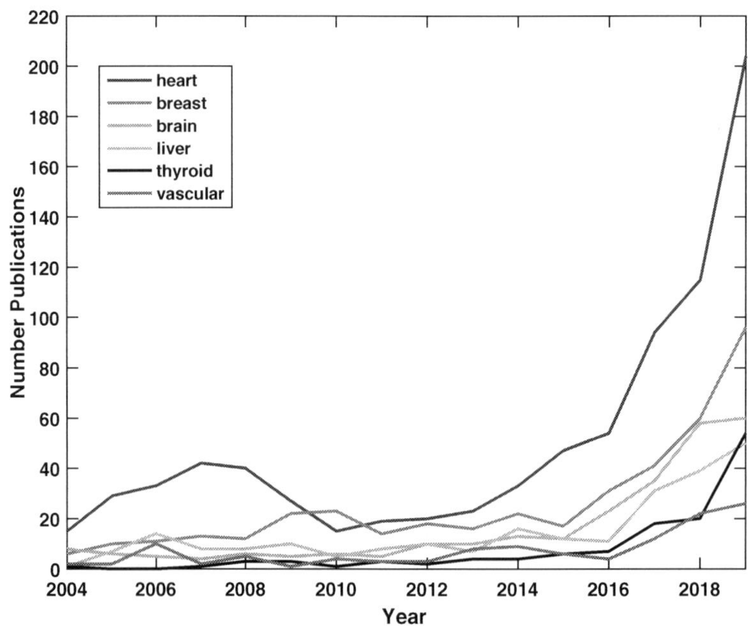

FIGURE 8.1

Number of publications per year that include keywords relating to artificial intelligence, machine learning, and ultrasound, by organ, 2004–2019.

Table 8.1 FDA-cleared AI products that combine artificial intelligence and ultrasound. Purpose is based on company press releases.

Year	Organ/ system	Company	Product name	Purpose
2018	Bladder	EchoNous (Redmond, USA)	EchoNous Bladder	Determine bladder volume
2018	Brain	Neural Analytics (Los Angeles, USA)	Lucid Robotic System	Diagnose severe stroke with transcranial ultrasound
2019	Breast	Koios Medical (Chicago, USA)	Koios DS Breast 2.0	Computer-aided diagnosis of breast cancer
2017	Breast	Qview Medical QVCAD (Los Altos, USA)	QVCAD	Computer-aided diagnosis of breast cancer
2018	Heart	Bay Labs (San Francisco, USA)	EchoMD AutoEF	Calculate ejection fraction
2013	Heart	DiA/GE (Be'er Shiva, Israel; Boston USA)	LVivo EF/ Vscan Extend	Calculate ejection fraction
2013 –18	Heart	Philips (Amsterdam, NL)	QLab, HeartModel	Identify views, analyze LV function, LA and RV longitudinal strain
2017	Heart	Siemens Healthineers (Erlangen, Germany)	TrueFusion	Fuse ultrasound with live fluoroscopy images to navigate structural heart procedures
2018	Heart	Ventripoint Diagnostics (Toronto, Canada)	VMS Plus	3D volumetric analysis of all four chambers
2019	Vascular	Sonavex (Baltimore, USA)	EchoSure	Detect blood flow changes from microvascular and vascular surgery

and vessels) have received the most attention, compared to other organs. Table 8.1 shows that pace of US FDA-cleared artificial intelligence (AI) products that combine AI and ultrasound is accelerating. Of note, many of the products have been approved over the last couple of years. Companies such as Butterfly Network (Guilford, CT) have also demonstrated AI-driven applications for portable ultrasound and more FDA clearances are expected to be published.

The goals of this chapter are to highlight the recent progress, as well as the current challenges and future opportunities. Specifically, this chapter addresses topics such as the following: (1) what is the current state of machine learning for medical US application, both in research and commercially; (2) what applications are receiving the most attention and have performance improvements been quantified;

(3) how do ML solutions fit in an overall workflow; and (4) what open-source datasets are available for the broader community to contribute to progress in this field. The focus is on cardiovascular applications (Section Cardiovascular/echocardiography), but common themes and differences for other applications for medical US are also summarized (Section Breast, liver, and thyroid ultrasound). A discussion is offered in Discussion and outlook section.

Cardiovascular/echocardiography

Echocardiography is one of the most widely used diagnostic imaging modalities in cardiology. The most common form is transthoracic echocardiography (TTE), with other forms that include transesophageal echocardiography (TEE), intracardiac echocardiography (ICE), and intravascular US (IVUS) (Fig. 8.2). It is the only imaging modality that provides real-time imaging of the heart.

Two-dimensional echocardiography is ubiquitously used to guide the diagnosis and down-stream management of numerous cardiac pathologies. Detailed analysis of echocardiography, however, is resource intensive, often resulting in visual estimation rather than precise calculation. ML proposes to automate many of these processes. There are already a number of widely adopted commercial software tools

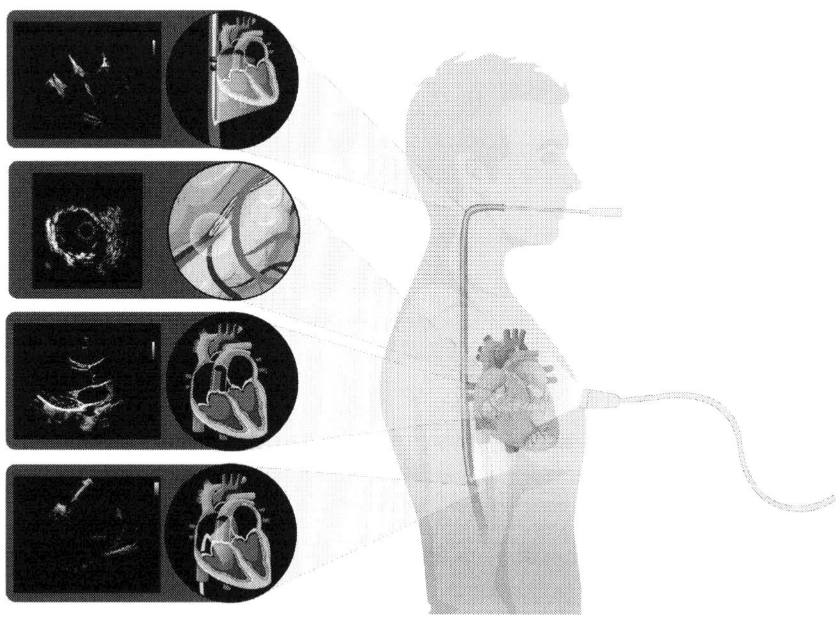

FIGURE 8.2

Illustration of the four types of echocardiography. From top to bottom: transesophageal echocardiography, intravascular ultrasound, transthoracic echocardiography, and intracardiac echocardiography.

developed for the functional analysis of 2D echocardiography data (e.g., EchoPAC by GE Healthcare, QLAB by Philips, etc.), which can perform an assortment of tasks ranging from segmentation and detection of anatomic landmarks to blood tracking. All results from such analyses, however, are wholly dependent upon careful annotation of input data, which is implicitly deferred to the clinician. That is, to use these tools, the physician evaluating the imaging study must manually select the correct study, choose an appropriate window, and specify which parameters to compute. These steps, while trivial in the evaluation of an individual echocardiogram, become limitations when analyzing significantly larger datasets (Al'Aref et al., 2019).

In this chapter, we consider recent applications of ML to transthoracic, transesophageal, intracardiac, and intravascular echocardiography. Fetal echocardiography is yet another specialty topic for which ML is being applied (Yu, Guo, Wang, Yu, & Chen, 2017), but which is not considered herein (a recent review can be found in Brattain et al. (2018)).

Transthoracic echocardiography and transesophageal echocardiography

TTE and TEE characterize both structure and function of the heart. TTE is acquired using a probe applied externally to the chest wall with modest pressure. A coupling gel is used to minimize impedance mismatch. Several standard probe locations are used including the left parasternal, apical, and subcostal positions and various cross-sectional views are obtained using tilting or rotation of the probe in these positions. For TEE, an endoscopic probe is inserted through the mouth of the patient, typically under conscious sedation, and passed through the oropharynx into the esophagus. The depth of insertion (e.g., upper esophageal, midesophageal, and transgastric) as well as the angle and tilt of the probe head provides standard views. TEE provides clearer pictures of the upper chambers and can be particularly useful on obese patients or intubated patients.

TTE is the most commonly used form of echocardiography in the assessment of structural and functional cardiovascular abnormalities. A complete TTE examination consists of a sonographer measuring nearly 30 parameters relating to physical dimensions and flow rates of the heart's chambers, valves, and aorta. These measurements are based on four primary US views, including both B-mode images and video as well as Doppler. 3D and strain imaging are additional modalities that are becoming increasingly common. Data and measurements from the exam are then reviewed by a trained cardiologist, who adds findings to the report. With the large number of measurements as well as types of data, echocardiography provides a rich problem space for ML. A key difference in echocardiography compared to other cardiovascular imaging modalities is that US image acquisition is highly interactive and operator-dependent.

The number of recent reviews (Alsharqi et al., 2018; Bizopoulos and Koutsouris, 2019; Kusunose, Haga, Abe, & Sata, 2019; Seetharam, Shrestha, & Sengupta, 2019) and editorials (D'Hooge and Fraser, 2018; Leeson, 2019; Tsay and Patterson, 2018) on ML for cardiovascular applications and echocardiography, in particular, exemplify the significant attention currently being given to ML-based solutions within the realm of echocardiography. In the rest of this section, we will focus on recent ML advances in the area of TTE, although many of the approaches discussed could be extended to TEE.

ML has been and will continue to improve TTE and TEE in several respects:

- increasing accuracy and precision while reducing bias,
- reducing image acquisition and study review times,
- enabling new types of measurements that would be impractical without automation,
- increasing accessibility of more indications,
- increasing accessibility to more patients.

Although 2D TTE remains the primary source of inputs, the increasing use of 3D TTE has augmented the accuracy of ML approaches by utilizing larger amounts of imagery. In addition, ML combined with portable US has the potential to allow aspects of TTE to be performed outside of a dedicated TTE facility, possibly even by nonexperts. Semiautomated and automated quantification in TTE has been progressing for over a decade in commercial products (Nolan and Thavendiranathan, 2019) and several companies now have products that use artificial intelligence to calculate TTE measurements (Table 8.1). Initial capabilities were primarily based on more conventional image processing approaches, but ML has been increasingly incorporated over time.

The top-level TTE workflow is summarized in Fig. 8.3. A sonographer performs an exam in real time on a patient. A typical TTE exam requires about 30 min. During

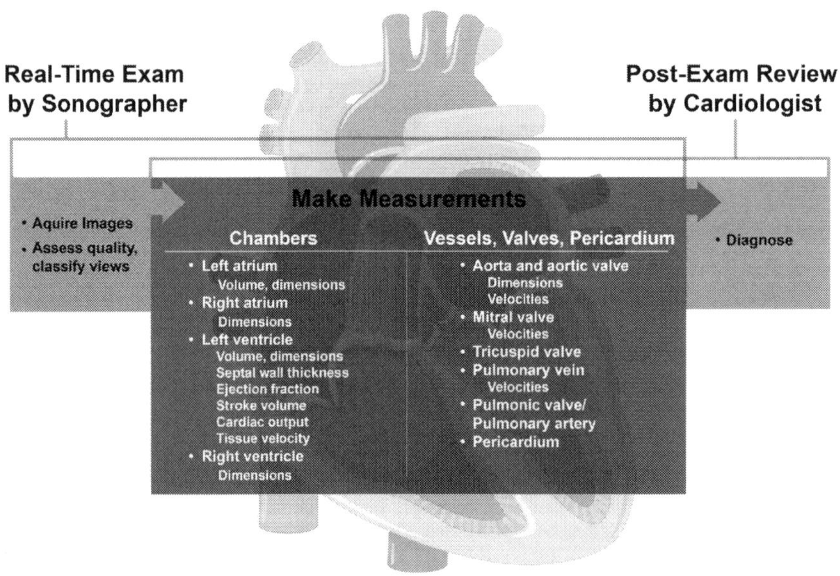

FIGURE 8.3

Echocardiogram workflow for 2D transthoracic echocardiography.

the exam, the sonographer positions the probe for a particular view and makes real-time measurements. After the exam is completed, a cardiologist reviews the images and measurements, and based on a conglomerate of specific findings establishes a particular diagnosis. Most of the published work on applying ML to TTE relates to view classification and measurements of the left ventricle. Little attention has been given to improving image acquisition, although this is a goal of Bay Labs (San Francisco, USA), as noted by Tsay and Patterson (2018).

Quality assessment and view classification

Significant research has been focused on automated view classification. Typical TTE exams collect at least 70 videos at different views (Zhang et al., 2018), including B-mode, M-mode, and Doppler. During the exam, view classification may be used to help a nonexpert in probe positioning and also to determine particular images that can be used for measurements. Following the exam, classification may be helpful to organize the TTE videos that are thereafter stored for future review.

Table 8.2 summarizes 2D view classification results from recent publications. The goal of view classification is to categorize a specific video as being from a particular view. The approach is usually to classify random images from a video and apply logic to classify the entire video. Penatti et al. (2015) provide a literature review of earlier work. The most recent papers all leverage deep convolutional neural networks (CNNs).

In addition to view classification, it is also important to assess image quality, as measurement accuracy has been shown to decrease as image quality decreases

Table 8.2 Examples of the application of machine learning for 2D echocardiography view classification.

Year	First author	#Views/ subviews	Database	Method	Accuracy
2015	Penatti	4	52 videos 1 US system	Bag of words	98%
2017	Khamis	3	309 videos, 2 sites	Spatiotemporal feature extraction, dictionary learning	95%
2017	Gao	8	432 videos, 2 sites 2 US systems	CNN, spatiotemporal fusion	92%
2018	Madani	15	267 studies 1 site >3 US systems	CNN	98% videos; 92% images
2018	Zhang	23	277 studies (7168 videos) 5 US systems	CNN	84%
2019	Ostvik	7	465 studies (7141 videos)	CNN	99%

CNN, *convolutional neural networks.*

(Medvedofsky et al., 2015). In one case, a CNN has been used to estimate a 6-level image quality score based on apical 4-chamber views graded by a human expert (Abdi et al., 2017). The CNN model resulted in a mean absolute error between the CNN and expert scores of 0.71 ± 0.58 on the 6-level scale, which was similar to an intrarater variability of 0.65. The maximum CNN output for view classification has also been assessed as a quality indicator, with the quality of the input image found to correlate modestly with a large, unambiguous CNN output value [12]. In addition to classifying 2D TTE views, machine learning has also been applied to classifying 3D TTE views [20].

Left ventricle measurements

Most automated measurement methods have been developed to measure left ventricular volume, in particular the end-diastolic and end-systolic volumes (EDV, ESV) and ejection fraction (EF). Fig. 8.4 compares a typical human process with a representative deep learning (DL) process, modeled after Leclerc et al. (2019) and Zhang et al. (2018) in which a U-Net (Ronneberger, Fischer, & Brox, 2015) segments the left ventricle rather than a human working with semiautomated software. In both cases, EF is then computed as $EF = (EDV - ESV)/EDV$.

Table 8.3 summarizes recent studies that compared ML-based automated measurement methods with criterion measurements. A few of these studies have also assessed accuracy of measuring the left atrial volume. The criterion measurement has been either manual interpretation of TTE (biplane Simpson's method (Simpson

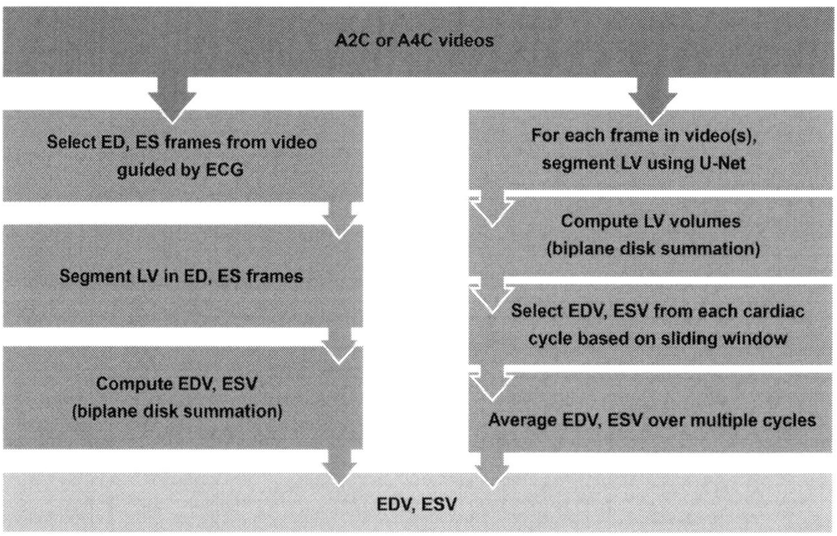

FIGURE 8.4

Typical process for human (top) and deep learning (bottom) measurement of end diastolic and systolic volumes (EDV, ESV). *LV:* left ventricle; *A2C and A4C:* apical 2-chamber and 4-chamber views; *ECG:* electrocardiogram.

Table 8.3 Summary of machine learning-based automated algorithm accuracy evaluation. Concordance measured by subtracting algorithm measurement from criterion measurement, so negative value indicates underestimate. Concordance reported as 95% limits of agreement (bias ± 1.96 standard deviations). Note: bias not reported for (Zhang et al., 2018).

Year	Author	Echo type	Algorithm/ product	Criterion method	Concordance		
					EDV (mL)	ESV (mL)	EF (%)
2015	(Knackstedt et al., 2015)	2D	TomTec	Manual			0.3 ± 18.7
2017	(Hovnanians et al., 2017)		Philips a2DQ	Manual	−4.3 ± 42.0	−2.1 ± 26.2	−0.15 ± 10.5
2018	(Zhang et al., 2018)		U-Net	Manual	±56	±39	±20
2019	(Leclerc et al., 2019)		U-Net	Manual	−6.9 ± 23.1	−3.7 ± 17.6	−1.0 ± 13.9
2019	(Miyoshi et al., 2019)		CNN	Manual			1.0 ± 11.9
2016	(Tsang et al., 2016)		Philips Heart Model	Manual	2 ± 40	10 ± 40	−6±16
2015	(Medvedofsky et al., 2015)				−3±46	−2±18	0 ± 19
2017	(Medvedofsky et al., 2017)			Manual	−14 ± 20	−6±20	−2±7
2018	(Barletta et al., 2018)			CMR	−6±63	9 ± 54	2.5 ± 13
2018	(Levy et al., 2018)			CMR	−12 ± 47	−5.9 ± 39	0.8 ± 12
2019	(Wu et al., 2019)			CMR	−5±66	−5±74	−0.7 ± 19

CMR, *cardiac magnetic resonance;* CNN, *convolutional neural networks.*

et al., 2017) for 2D), or manual interpretation (discs' summation method (Winter et al., 2008)) of cardiac magnetic resonance (CMR). The databases differ considerably, with mean EDV, ESV, and EF varying from 103 to 203 mL, 52 to 132 mL, and 35% to 58%, respectively; number of sites varying from one to six sites; number of devices varying from one to seven; number of manual annotators varying from one to three; and the number of cases (total of training and testing cases) varying from 20 to 50,000. The smaller databases were typically used in studies comparing TTE measurements with that of CMR.

The biases (mean errors) of the ML methods compared to the criterion methods in Table 8.3 are small, most often underestimates. However, the 95% limits of agreement (LoAs, equivalent to 1.96 standard deviations assuming a Gaussian distribution) are large compared to values for EDV, ESV, and EF (ranges of mean values noted earlier). The large LoAs make it very difficult to compare performance of different methods. On the other hand, these LoAs are comparable to interobserver and intersite LoAs. For example, one study (Knackstedt et al., 2015) found interobserver variability for manual EF was $-12.4 \pm 17.4\%$ and for local measurements compared to those of a reference center, $-0.9 \pm 18.1\%$. Concordance improves when manual editing of the automated measurements was performed by someone with experience in editing (EF of -3.4 ± 17 improved to 0.3 ± 10) (Medvedofsky et al., 2015). Concordance also improved when poor quality images were not processed; presumably these cases would be assigned for manual interpretation. Particular subjects, such as those with BMI>30, are more challenging cases because their images are more frequently of poor quality. The method tested by Hovnanians, Win, Makkiya, Zheng, & Taub (2017) was found to perform similarly for this population as for the rest of the cases, which has been an encouraging result. The percentage of cases in a database that are processed by the automated method is termed the feasibility. In several investigations, cases for automated measurement are chosen by a manual evaluation of image quality. In general, one would expect that as feasibility is decreased, concordance would increase. Concordance is also affected by the number and diversity of cases included in the database in the first place, by the number of sites, and by other factors. For example, databases may be limited to subjects with sinus rhythm, etc.

Kitano, Nabeshima, Otsuji, Negishi, & Takeuchi (2019) performed a large meta-analysis of semiautomated and automated methods that were evaluated in 38 studies with 1881 subjects that used CMR as the criterion method. The semiautomated methods required the user to manually initialize anatomical landmarks, followed by automatic border detection and tracking of ventricular volume through one cardiac cycle. Fully automated methods automatically detected borders and tracked volume through time, although manual corrections could then be applied. The automated software was from Philips Healthcare (Andover, MA) and Siemens AG (Munich, Germany). The review concludes that although both semiautomated and automated software underestimate EDV and ESV, underestimates from automated software were smaller. EF mean estimates from both approaches were similar to those from CMR (semiautomated, $-0.6 \pm 1.3\%$; automated, $-1.1 \pm 2.4\%$). The individual studies were published between 2011 and 2018, but the meta-analysis did not report on any possible improvement in accuracy over time.

Insight on the source of poor concordance for specific cases is not generally reported. Medvedofsky et al. (2017) quantified how for 3D TTE, accuracy significantly degrades for poor quality images. For example, for the 40% of images that were good quality versus 27% of the images that were of poor quality, EF accuracy decreased from $0.1 \pm 4.4\%$ to $0.3 \pm 16\%$. A detailed analysis of 20 outliers in EDV volume measurement by CNN (Zhang et al., 2018) found three cases where the CNN result was correct and the manual result was markedly inaccurate, three cases where the CNN and manual result were about equally incorrect (over- and under-measurement), and five cases where the CNN result was clearly wrong. Of these five significant CNN errors, two cases had used intravenous contrast, on which the CNN had not been trained, one case that involved a complex congenital heart disease with unusual anatomy, one that involved a mechanical mitral valve with shadowing and reverberation artifact, and one that included a calcified false tendon in the LV along with a pericardial effusion. More attention is needed to diagnosing error sources both in the criterion scoring and in the automated method, to improve performance.

It has been clearly established that automated methods save time. An example of time reduction for 2D TTE of 217 ± 69 to 116 ± 57 s per patient was found using Philips a2DQ software (Hovnanians et al., 2017). For 3D TTE, a time reduction of 3.6 min for a semiautomated approach to 35 s for an automated approach was found (Narang et al., 2019). It is also clear that an automated method significantly reduces or eliminates interobserver variability. Rodríguez-Zanella et al. (2019) pointed out that "in clinical practice intralaboratory, interobserver, intervendor, and test/re-test reproducibility are more important characteristics of an imaging technique than its overall accuracy versus any reference method."

3D TTE has been shown to be more accurate and reproducible than 2D TTE and is recommended by the most recent American Society of Echocardiography and European Association of Cardiovascular Imaging guidelines (Otani, Nakazono, Salgo, Lang, & Takeuchi, 2016). However, 3D TTE has not been fully adopted because of special training and expertise required, and time-consuming interpretation. Thus, automated techniques have the potential to increase the adoption of 3D TTE.

Various ML approaches have been applied, with a progression over time toward deep convolutional neural networks, with Leclerc and colleagues (Leclerc et al., 2019) evaluating the accuracy of CNN architectures against more classical ML approaches. Accuracy from U-Net architectures has been shown to be similar to accuracy from more complex architectures (Leclerc et al., 2019). Accuracy as a function of the number of patients was also evaluated, with a conclusion that accuracy plateaued after training on more than 250 patients (evaluated up to nearly 500 patients), for a database of primarily two- and four-chamber apical views (Leclerc et al., 2019). This is an important and interesting finding. Another study (Zhang et al., 2018) used data from 791 cases representing five views to train a U-Net for segmentation. In these approaches, image augmentation is generally used, but transfer learning and active learning were not used.

As a commercial product, the details of the HeartModel algorithm are not well described, but reportedly are based on ML (Narang et al., 2019). At a summary level, volume models of the left ventricle and left atrium are built by detecting the endocardial surface and matching features from the volume to shapes from a database of about 1000 3D TTEs representing a wide range of morphologies and function (Tsang et al., 2016). The volume model is then locally adapted to the patient's data. From this description, the algorithm can be considered as an adaptive artificial intelligence algorithm rather than an artificial neural network. HeartModel's accuracy for measuring regurgitant volume from 3D TTE has also been compared to CMR measurements (Levy et al., 2018), resulting in a concordance of 2 ± 24 mL.

Most methods that use CNNs have taken a standard approach of segmenting the left ventricle and measuring volumes and ejection fractions from the segmented frames. In an attempt to measure ejection fraction for all cases, even those with poor image quality, CNNs have also been used to measure ejection fraction based on estimating contraction coefficients from two images, without segmenting the chamber boundaries (Asch et al., 2019). This approach allowed the CNNs to be trained on 50,000 cases for which EF had been manually measured over the course of 10 years.

Attention is also being directed toward applying ML-based solutions for portable US. Zhang et al. (2018) prototyped a fully automated pipeline based on ML as a step toward this vision. A method was also developed for fast computation of ejection fraction on a mobile app (based on a U-Net with adversarial training), resulting in a median absolute error of 6.2% (Jafari et al., 2019). The goal of real-time TTE on a mobile app can also leverage progress in that efficient CNNs for other applications (Laube and Zell, 2019; Tan et al., 2019) as well as advances in processing chips.

Finally, although measurement of LV volume has received the most attention in the literature, detailed mitral valve analysis from 3D transesophageal echocardiography has also been reported using automated commercial software (eSie Valve Software, Siemens Healthcare, Mountain View, CA) (Jeganathan et al., 2017).

Disease classification

In addition to computing standard measurements, ML has been applied to disease classification, which is one of the most important and impactful applications of ML. Recent work has shown that ML can improve the identification of subclinical diseases from echocardiography. For example, hypertrophic cardiomyopathy was discriminated from physiological hypertrophy using three ML algorithms (support vector machines, random forests, and artificial neural networks) for TTE measurements (Narula, Shameer, Salem Omar, Dudley, & Sengupta, 2017), with the resulting 96% sensitivity and 77% specificity significantly outperforming conventional parameters. Other examples include aiding the diagnosis of myocardial infarction (Sudarshan et al., 2016), myocardial wall motion assessment (Omar et al., 2018), and mitral regurgitation (Moghaddasi and Nourian, 2016), differentiating constrictive pericarditis from restrictive cardiomyopathy (Sengupta et al., 2016), and characterizing heart failure with preserved ejection fraction (Sanchez-Martinez et al., 2018).

Nearly all recent published applications of machined learning for TTE have used supervised learning; the work by Sanchez-Martinez (Sanchez-Martinez et al., 2017) is notable for applying unsupervised clustering to characterize functional responses to stress in heart failure with preserved ejection fraction syndrome. ML has also been used to directly detect disease from TTE images. In a study from Zhang et al. (2018), hypertrophic cardiomyopathy, cardiac amyloidosis, and pulmonary arterial hypertension were detected with AUCs of 0.93 (based on parasternal long-axis and apical four-chamber (A4c) videos), 0.87, and 0.85 (A4c view), respectively.

Intracardiac echocardiography

ICE has emerged as a useful tool in the everyday practice of electrophysiology and percutaneous interventions (Narducci, Vitulano, Pazzano, & Pelargonio, 2015). Advances in catheter-based US transducers and imaging technology have made this modality integral to guiding anatomical mapping procedures as part of transcutaneous ablative procedures as well as percutaneous cardiac structural interventions (Dravid, Hope, & McKinnie, 2008).

For use, an ICE catheter is inserted into the patient's vasculature (e.g., femoral vein) and navigated to the heart, where it acquires 2D B-mode images of cardiac structures. Often in conjunction with fluoroscopy, ICE has been used for transseptal interventions (e.g., shunt closure), catheter ablation of atrial fibrillation, and ventricular tachycardia, among other applications. ICE offers direct visualization of the surrounding tissue and is often used for instrument tracking and for assessing tissue/instrument contact. Compared to external US probes, ICE can achieve higher quality views of targets in the near-field with higher acoustic frequencies, reducing aberration and attenuation due to layers of intervening muscle, fat, and other tissues. When compared to TEE, ICE is also associated with decreased procedure length while eliminating the risks of endotracheal or esophageal intubation and general anesthesia (Boccalandro, Baptista, Muench, Carter, & Smalling, 2004).

The use of ICE, however, requires highly trained operators. For example, although 2D ICE offers a wider cross-section view than a typical TTE/TEE, it has limited thickness along the axial direction. Obtaining the optimal view requires dexterity and experience in manipulating control knobs. Image interpretation also requires a substantial amount of expertise. In addition, the continuous contractile movement of the heart makes it challenging to keep up with the motion while maintaining an adequate view with the use of an ICE probe. Ablation lesion tracking and assessment as well as ablation catheter tracking also demand a high level of training.

One way to mitigate the thin viewing plane is to compound multiple 2D ICE slices together using robotic manipulation. A GPU-enabled real-time 3D ICE volume stitching and visualization method has been proposed (Brattain, 2014; Loschak, Brattain, & Howe, 2017). Robotic steering of the catheter is used to collect the 2D slices, and compensation for constrained catheter shaft motion was also considered (Degirmenci et al., 2016; Loschak, Degirmenci, & Howe 2017). A model-based ICE volume stitching approach was also implemented (Ionasec, 2015).

Commercial 3D ICE imaging systems are also emerging. Sequoia US system (Siemens Healthineers) equipped with SoundStar 3D diagnostic US catheter (10F, 90 cm, Biosense Webster, Inc., Diamond Bar, CA) is capable of acquiring US volumes from within the heart. A more recent 3D ICE system, Acuson SC2000 Prime with ACUSON AcuNav (Siemens Healthineers), can provide 2D/3D quantification using AI. The details of the AI methods used are not available.

Although numerous technical approaches have been investigated over the past decades, few have provided a reliable and reproducible advancement in clinical outcomes. Several recurring procedures are needed to raise the success rate to the 70% −80% range (Vasamreddy et al., 2004). Lacking adequate visualization of the lesion and the ablation catheter is a primary cause of the high recurrent rate of catheter ablation. There has not been much work devoted to applying ML to catheter ablation, or cardiac procedure guidance overall. However, considering the progress made in TTE/TEE, we speculate that if used properly, ML has great potential in improving ablation outcome, particularly through enhancing ablation lesion monitoring and ablation catheter tracking.

Intravascular ultrasound

Intravascular ultrasound (IVUS) is an imaging modality that has a specially designed catheter with a miniaturized US probe attached to the distal end of the catheter. The coronary arteries are the most frequent imaging target for IVUS. IVUS is used to exam the vessel walls and assess calcification and atheromatous plaque, which cannot be seen by conventional angiography.

IVUS imaging has also been proven useful in guiding coronary intervention procedures based on a comprehensive assessment of atherosclerotic plaque quantitative and qualitative characteristics, while also proving to be useful for coronary stent evaluation postimplantation (Nakamura et al., 2001). IVUS guidance is associated with a reduction in the incidence of myocardial infarction and the composite endpoint of death/myocardial infarction in the both short and long term (Casella et al., 2003; Claessen et al., 2011). Additionally, IVUS is used for guiding peripheral and endovascular intervention procedures (Clark et al., 2006; Krokidis and Hatzidakis, 2011).

Compared to intravascular optical coherence tomography (OCT), which has $\sim 10\times$ better imaging resolution, IVUS offers much better penetration ($\sim 5\times$) depth for the assessment of lipid or necrotic plaques (Fujii et al., 2015; Kubo et al., 2013). Furthermore, intravascular OCT catheters require injection of contrast for blood clearance. IVUS does not require a contrast injection as ultrasound waves can penetrate through blood (Maehara, Matsumura, Ali, Mintz, & Stone, 2017, Maehara, Mintz, & Stone, 2013). Typically, an IVUS catheter is advanced 10 mm distal to the target lesion and imaging is performed during pullback with a pull-back speed ~ 0.5 mm/s (Maehara et al., 2017). A large number of images are acquired in a very short interval. It is labor intensive to evaluate the large number of acquired images manually. As a result, automated interpretation is one of the top priorities for improving the utility of IVUS, with examples of recent work summarized in Table 8.4.

Table 8.4 Summary of ML-based algorithms for automated IVUS interpretation.

Year	References	Method	Goal/tissue target	Data size	Performance
2015	(Gao et al., 2015)	Unsupervised clustering	Detection/lumen and MA borders	337 images	Average distance error: 0.08 mm (lumen), 0.10 mm (MA)
2016	(Araki et al., 2016)	SVM	Classification/ High-risk atherosclerotic plaque	15 subjects/ 42,975 images	Accuracy 94.95%; AUC of 0.95
2016	(Banchhor et al., 2016)	Fuzzy c-Mean, k-means, HMRF	Estimate calcium volume measurement		Jaccard index: 0.7 and Dice Similarity: 0.8
2016	(Conjeti et al., 2016)	Domain adaptation of decision tree	Tissue characterization/ plaque	13 postmortem hearts/53 frames 3 hearts/7 frames	logDOR 25
2017	(Su et al., 2017)	Auto encoder	Detection/lumen and MA border	6 subjects/538 images	Detection error (lumen border: 0.0928 ± 0.0935 mm, MA border: 0.1056 ± 0.1088 mm
2017	(Hong et al., 2017)	BCN	Segmentation/ luminal and wall	50 subjects	Dice similarity score: 0.6260 (lumen) and 0.7801 (wall)
2018	(Jun et al., 2018)	FNN, KNN, RF, CNN	Detection/plaque	12,325 images	AUC of 0.911 for CNN
2018	(Sofian et al., 2018)	CNN with ResNet101	Detection/ calcification	2175 images (MICCAI 2011 challenges dataset B)	Accuracy: 0.9995
2018	(Balocco et al., 2018)	CNN with varying depth	Detection/frames of arterial calcification	8 subjects/8914 images	F1 score 0.67

Continued

Table 8.4 Summary of ML-based algorithms for automated IVUS interpretation.—*cont'd*

Year	References	Method	Goal/tissue target	Data size	Performance
2018	(Kim et al., 2018)	U-Net	Segmentation/ coronary arteries	35 subjects/ 38,478 images	Dice similarity score: 0.9 (lumen) and 0.84 (wall)
2018	(Yang et al., 2018)	FCN with multibranch architecture (ResNeXT, Inception, SegNet, U-Net), augmentation	Segmentation/ Lumen and media vessel walls	MICCAI 2011 dataset	Lumen: 0.90 JM (0.06), 0.26 (0.25) HD Media: 0.86 (0.11) JM, 0.48 (0.44) HD
2019	(Ko and Lee, 2019)	FCN w/transfer learning	**Segmentation**	70 datasets/2000 −4000 images per dataset	Dice score of 0.97
2019	(Lo Vercio, del Fresno, & Larrabide, 2019)	SVM, RF	Segmentation of vessel wall	Balocco's public dataset (Balocco et al., 2014)	AUC 0.97 and JM 0.85
2019	(Yang et al., 2019)	Dual Path U-Net	Segmentation/ arterial walls	Balocco's public dataset (Balocco et al., 2014)	0.9 JM, 0.25 mm HD

BCN, *brief convolutional network*; HD, *Hausdorff distances*; HMRF, *hidden Markov random fiel;* JM, *Jaccard measure*; logDOR, *log-compressed multiclass diagnostic odds ratio*; MA, *media-adventitia.*

Although conventional image processing remains an important approach to facilitate the interpretation of IVUS, ML has shown promise in IVUS tissue classification (TC) and anatomical component segmentation. IVUS TC is largely related to atheromatous plaque and calcification assessment. Earlier efforts focused on texture-based or radio frequency-based feature extractions, followed by using support vector machine (SVM) or random forest (RF) as a classifier for plaque or calcification assessment (Sheet, Karamalis, Eslami, Noël, Chatterjee et al., 2014; Sheet, Karamalis, Eslami, Noël, Virmani et al., 2014).

Over the past few years, DL has been increasingly applied to TC within the IVUS realm. Because of the limited annotations available, Conjeti et al. (2016) proposed domain adaption of decision trees for the detection of plaque. Moreover, 13 postmortem hearts with 53 IVUS frames were used for training of an initial model, which was then adjusted by the incorporation of seven IVUS frames from three live subjects. The key idea here is to train a model using a large amount of data obtained from a relevant domain where data are more readily available, and then refine the model with a small dataset from the domain of interest. To gain insights on the performance of various popular machine learning approaches, Jun et al. (2018) compared the results of a fully connected network (FNN), k-nearest neighbors (KNN), RF, and CNN on 12,325 IVUS images and concluded that CNN yields the best AUC (0.911). To investigate the role of the depth of CNNs, Balocco, González, Ñanculef, Radeva, & Thomas (2018) experimented with CNNs of varying depth on an 8-subject/8914-image dataset. Their goal was to down select a set of frames with calcification present and pass them for further evaluation by physicians. They concluded that a 5-convolutional layer model outperforms the shallower CNNs, although they cautioned about balancing architecture depth with size of data. Data augmentation and learning regularization were also recommended. CNNs with different DL backbones, such as ResNet101, has also been used for calcification detection (Sofian, Than, Mohammad, & Noor, 2018).

In regards to segmentation of anatomical components on IVUS, the majority of the work relates to segmenting arterial and media-adventitial (MA) walls. Kim, Lee, Hwang, & Kim (2018) used a U-Net on a fairly large dataset of 35 subjects/38,478 images. Yang, Tong, Faraji, & Basu (2018) reported on an open-source IVUS-Net for multiclass segmentation, while fully convolutional network (FCN) with transfer learning (Ko and Lee, 2019) and dual path U-Net have also been reported (Yang, Faraji, & Basu, 2019). U-Net and variants of it are the most commonly used segmentation architecture. Dice similarity metrics, Jaccard measure (JM), and Hausdorff distances (HD) are the mostly commonly used evaluation metrics. Postprocessing (e.g., ellipse filling on the segmented masks) is often necessary for improved accuracy. Further, several papers used the same publicly available IVUS dataset (Balocco et al., 2014).

Breast, liver, and thyroid ultrasound

Besides cardiovascular applications, other organs that have been frequently explored with ML algorithms include breast, liver, thyroid, prostate, brain, fetal, and skin. Different anatomical organs have their own unique requirements and challenges. In the past 3 years, as a result of rapid technical advances in DL algorithms, DL-based approaches have been by far the preferred ML approach for the detection of desired tissue, segmentation of desired tissue to perform measurements, and classification of tissue into a specific class. In this section, we will focus on breast, liver, and thyroid, where we have seen a burst of DL algorithm development.

Breast ultrasound

Breast tissue cells are prone to cancer and as such breast cancer is the leading cause of death in women aged 20–59 years (Siegel, Miller, & Jemal, 2016). Mammography is the primary modality for screening for breast cancer followed by US as a secondary modality for focused examination of the pathological tissue. Mammography is discouraged in younger women due to radiation risk, denser breast tissue, and the possibility of pregnancy. In such cases, US along with other modalities act as the primary modality for screening for breast cancer. Pathological tissue exhibits morphological and textural visual cues such as shape, margin, echo pattern, posterior features, presence of calcifications, or architectural distortion as defined by the Breast Imaging Reporting and Data System (BI-RADS) (Liberman and Menell, 2002), which serves as the basis for biopsy recommendation. In the absence of a mammographic examination, a scrupulous scanning of the entire breast by US is required to screen for breast masses. Data acquired from 2D breast US examination, both images and cine clips, is subjective due to probe orientation, angle between skin and transducer, and compression of soft tissue. 3D whole breast US removes the subjectivity associated with data acquisition; however, it offers lower contrast ratio compared to 2D US. Furthermore, focused 2D US lacks anatomical context, increasing the complexity of differentiating pathological tissue from normal tissue. The anatomical context is also present to some degree in 3D US as the probe translates from one end to other.

The application of DL in breast US can be broken down into three broad categories: detection/localization of abnormal breast tissue/mass or tumor, segmentation of breast mass, and classification of breast mass as benign or malignant. Detection or localization of breast masses have special relevance in the absence of mammogram exams and can improve the efficiency and productivity of the screening process especially when the exam is performed by less-skilled users. Segmentation of breast masses is important in identifying certain morphologic features associated with breast cancer such as irregular boundaries, spikes in boundary called as spiculations, and irregular shape. Suspicious breast masses are recommended for biopsy based on the BI-RADS score and are prone to overdiagnosis leading to undue emotional, pain, and economical stress associated with biopsy performance. Classifying breast masses as benign or malignant can reduce the number of unnecessary biopsies. Textural and morphological features extracted from sonographic images have been commonly used in the past few decades with limited success. Deep neural networks have been commonly used for detection (Cao et al., 2017; Chiang, Huang, Chen, Huang, & Chang, 2018; Shin, Lee, Yun, &, 2017; Wang et al., 2018; Yap et al., 2017), segmentation (Bian, Lee, Chou, & Cheng, 2017; Kumar et al., 2018; Lei et al., 2018; Xu et al., 2019; Yap, Goyal, Osman, Martí et al., 2018), and classification of tumors (Becker et al., 2018; Cheng et al., 2016; Han et al., 2017; Huynh, Drukker, & Giger, 2016; Shi et al., 2016; Shin et al., 2017; Xiao et al., 2018; Yap, Goyal, Osman, Martí et al., 2018; Zhang et al., 2016) as are summarized in Table 8.5. The common challenges associated with breast tissue are heterogeneity of breast tissue, lack of anatomical context, similar appearance of tissues like cyst and blood vessels, low contrast ratio, speckle noise, limited field of view, and subjectivity in data acquisition.

Liver ultrasound

Liver, the largest organ in the body, is prone to cancer and fatty liver disease. Liver cancer is the fifth and seventh most common cause of cancer death among men and women, respectively (Siegel, Miller, & Jemal, 2019). One common pathway to liver cancer is liver steatosis or fatty liver disease. Fatty liver disease can be broken down into two subtypes: Nonalcoholic fatty liver disease (NAFLD) and alcoholic fatty liver disease also called as alcoholic steatohepatitis. NAFLD is prevalent in about 25% of the world population (Perumpail et al., 2017) and can be further subclassified into simple fatty liver and nonalcoholic steatohepatitis (NASH). Simple fatty liver does not cause liver damage; whereas, NASH can lead to inflammation and liver cell damage causing fibrosis of the liver eventually leading to cirrhosis or liver cancer.

US has been extensively used as the primary imaging modality for the evaluation of liver disease. Abdominal US exam covering liver, kidney, and spleen is the most common US exam. Acquiring and annotating standard US views is a meticulous task. DL algorithms have been proposed for view classification (Cheng and Malhi, 2017; Xu et al., 2018) to reduce the operator time. US detects liver steatosis as a result of increased echogenicity of liver parenchyma (Lee and Park, 2014) compared to kidney cortex. Liver steatosis is usually graded on a four-point scale (Saadeh et al., 2002). Fibrosis of liver is also characterized by hyper echogenicity along with increased stiffness. Liver fibrosis can be graded on different scales such as METAVIR, Knodell, and the Ishak score. Table 8.5 summarizes the papers using DL algorithms for steatosis, fibrosis and cirrhosis classification (Biswas et al., 2018; Byra et al., 2018; Frid-Adar et al., 2018; Liu, Xie, Yu, Niu, & Sun, 2017; Meng et al., 2017; Schmauch et al., 2019). Abdominal US can be challenging due to the limited acoustic window, poor penetration in bariatric patients, low contrast ratio, speckle noise, limited field of view, and subjectivity in data acquisition.

Table 8.5 Application of deep learning algorithms for detection, segmentation and classification of different organs.

Year	First author	Problem type	Organ	Algorithm
2016	(Cheng et al., 2016)	Tumor classification	Breast	Stacked denoising auto-encoder
2016	(Huynh et al., 2016)	Tumor classification	Breast	Pretrained CNN
2016	(Zhang et al., 2016)	Tumor classification	Breast	Pointwise gated and restricted Boltzmann machine
2016	(Shi et al., 2016)	Tumor classification	Breast	Stacked deep polynomial network
2017	(Han et al., 2017)	Tumor classification	Breast	GoogLeNet
2017	(Yap et al., 2017)	Tumor Detection	Breast	LeNet, U-Net, FCN-AlexNet
2017	(Shin et al., 2017)	Tumor detection and classification	Breast	Faster R-CNN
2017	(Bian et al., 2017)	Tumor segmentation	Breast	VGG16 based encoder decoder
2017	(Cao et al., 2017)	Tumor detection	Breast	Single Shot MultiBox Detector
2018	(Becker et al., 2018)	Tumor classification	Breast	CNN, proprietary commercial
2018	(Kumar et al., 2018)	Tumor segmentation	Breast	U-Net variant
2018	(Wang et al., 2018)	Tumor detection	Breast	3D convolutional network
2018	(Xiao et al., 2018)	Tumor classification	Breast	InceptionV3, ResNet50, and Xception
2018	(Chiang et al., 2018)	Tumor detection	Breast	3D CNN
2018	(Lei et al., 2018)	Tumor segmentation	Breast	Convolutional encoder −decoder network
2019	(Yap, Goyal, Osman, Ahmad et al., 2018)	Tumor segmentation and classification	Breast	Fully convolutional network-AlexNet
2019	(Xu et al., 2019)	Tumor segmentation	Breast	Custom 8-layer CNN
2017	(Cheng and Malhi, 2017)	View classification	Liver	CaffeNet and VGGNet
2017	(Meng et al., 2017)	Fibrosis classification	Liver	VGGNet
2017	(Liu, Song, Wang, Zhao, Chen, 2017)	Cirrhosis classification	Liver	Pretrained 8-layer CNN
2018	(Byra et al., 2018)	Steatosis grade classification	Liver	Inception-ResNet-v2
2018	(Biswas et al., 2018)	Risk stratification	Liver	Custom 12-layer CNN

Table 8.5 Application of deep learning algorithms for detection, segmentation and classification of different organs.—*cont'd*

Year	First author	Problem type	Organ	Algorithm
2018	(Xu et al., 2018)	View classification	Liver	ResNet50 variant
2018	(Frid-Adar et al., 2018)	Lesion classification	Liver	Generative Adversarial Networks
2019	(Schmauch et al., 2019)	Detection and classification of focal lesions	Liver	ResNet50 variant
2017	(Liu, Xie et al., 2017)	Nodule classification	Thyroid	VGG-Face
2017	(Chi et al., 2017)	Nodule classification	Thyroid	Pretrained GoogLeNet
2017	(Ma, Wu, Zhu et al., 2017)	Nodule classification	Thyroid	Two fused custom CNN's
2017	(Ma, Wu, Jiang, Zhao et al., 2017)	Nodule segmentation	Thyroid	Custom 17-layer CNN
2017	(Ma, Wu, Jiang, Zhu et al., 2017)	Nodule detection	Thyroid	Cascade CNN
2018	(Song et al., 2018)	Nodule detection and classification	Thyroid	Multitask cascade CNN
2018	(Li et al., 2018)	Nodule detection and classification	Thyroid	Faster R-CNN variant
2019	(Li et al., 2019)	Nodule classification	Thyroid	Darknet-19 variant
2019	(Ko et al., 2019)	Nodule classification	Thyroid	Imagenet-vgg-verydeep16 and imagenet-vgg-f
2019	(Wang et al., 2019)	Nodule detection and classification	Thyroid	YOLOv2

CNN, *Convolutional neural network.*

Thyroid ultrasound

The increase in incidence of thyroid cancer is greater than any other cancer, standing at 4.5% per year over the last 10 years (US Preventive Services Task Force et al., 2017). Thyroid nodules are mostly benign with a low malignancy rate of 4.5% −6% (Lin et al., 2005). It is common to have multiple nodules within the thyroid gland. US Preventive Services Task Force recommends against screening (including neck palpation and US) for thyroid cancer in asymptomatic adults (US Preventive Services Task Force et al., 2017). Due to the lack of a screening process, thyroid nodules are found incidentally by palpation or by diagnostic imaging modalities. US is the primary diagnostic tool for thyroid cancer as it is inexpensive and readily available. Besides differentiating between cystic and solid nodules, ultrasonography features are related to the pathology of the nodule. The sonographic features that may increase the risk of malignancy include hypoechoic solid nodules, taller-than-wide nodules, irregular margins, extrathyroidal extension, and presence of microcalcifications. Contrarily, the presence of peripheral vascularity, round shape, isoechogenicity, spongiform appearance, smooth margins, and cystic composition are associated with benignancy (Brito et al., 2014; Frates et al., 2005; Moon et al., 2008). Subcentimeter nodules identified by US are not recommended for fine needle aspiration (Mazzaferri, 1999) as they lack the potential to be clinically significant thyroid cancers.

The applications of DL for thyroid evaluation can be broken down into the following three tasks: (1) detection of all thyroid nodules (Li et al., 2018; Ma, Wu, Jiang, Zhu, & Kong, 2017; Song et al., 2018; Wang et al., 2019), (2) segmentation of thyroid nodules to estimate the size, volume, and shape of nodules (Ma, Wu, Jiang, Zhu, & Kong, 2017), and (3) classification of nodule as benign or malignant to avoid unnecessary biopsies (Ko et al., 2019; Li et al., 2019; Lin et al., 2005; Liu, Xie et al., 2017; Ma, Wu, Zhu, Xu, & Kong, 2017). Table 8.5 summarizes the DL papers associated with the above-mentioned tasks. Thyroid US faces challenges such as subjectivity in data acquisition, and a wide range of textural differences in thyroid US features.

Discussion and outlook

Progress in applying ML to ultrasound has accelerated over the last few years, as part of an overall movement to increase automation for medical applications over at least the past decade. As is the case in other domains, DL algorithms and in particular, CNNs, have become the predominant classification methods, overtaking other classifiers operating on hand-crafted features. As in other types of biomedical image interpretation, U-Nets have been found to perform well for segmentation tasks.

In TTE, most approaches have used variants of CNNs to segment the left ventricle and then measure volumes and ejection fraction from those boundaries. However, in a few cases, CNNs have been trained to operate directly on the US images to make measurements. Most approaches have trained CNNs in standard ways and do not make full use of transfer learning (Pan and Yang, 2009) active learning (Settles, 2011), or adversarial networks (Goodfellow et al., 2014). Limited amount of work has been done in exploiting dynamics from videos (Patra and Noble, 2019). Similarly, little has been done with unsupervised learning (Hinton, Sejnowski, & Poggio, 1999) to better understand cardio-phenotypes based on echocardiography. These techniques can overcome limited annotations, reduce the amount of annotations needed, create synthetic data if necessary, or operate without the presence of truth data. They are all expected to be adopted by this community more in the future.

Progress has been made in accuracy, but there is much more to be done. Automated methods eliminate intra- and interobserver variability of the measurements themselves, although variability still exists due to variability in the imaging. Thus, automated methods have improved precision compared to manual interpretation. Automated methods have also been shown to have low biases when compared to manual measurements as the criterion standard. However, LoAs are large. Imagine telling a patient that they have an ejection fraction between 25% and 65% (due to a LoA of ±20% as reported in some results). That would not be helpful. Still, the large LoAs are partially a result of the fact that the criterion standard is manual measurement, which itself suffers from intra- and interobserver variability. In addition, much more work needs to be done to characterize and understand the outlier error cases. This has been done to some level in a couple of investigations, but needs to be performed systematically and routinely. That will lead to improved results.

Most research has focused on the aspects of the TTE system flow depicted in Fig. 8.3, and in particular on view of classification and left ventricular measurements. The vast majority of published papers have focused on B-mode video processing, rather than the Doppler information that is also available from echocardiography examinations. Electrocardiogram (ECG) data are also collected during TTE, which is particularly useful for determining the underlying rhythm and correlating a particular image to the cardiac cycle. However, ECG data are also not typically used in research on ML for TTE, in some cases for convenience (it is simpler to just analyze US imagery) and in some cases because the focus is on point-of-care US where ECG may not be included. More broadly, there is a rich research space to fuse US information with other sensor modalities.

Consideration of workflow of an ML system within a clinical environment is an important topic. Commercial products that incorporate some degree of artificial intelligence do address workflow and aspects of workflow, such as provision for manual editing as well as measuring time saved. However, workflow and human factors are typically not evaluated explicitly. In academic papers, the focus is on a particular subfunction, and workflow and the overall system that is needed are typically not considered. Published papers with more of a system view will be valuable.

A critical aspect of automated echocardiography that has been typically neglected involves aiding nonexperts in acquiring the correct views (guiding probe position, orientation) and high-quality images. For example, it is well reported that automated algorithms are less accurate for poor quality images, even when acquired by trained sonographers. Improving image quality will lead to improved accuracy. Aiding image acquisition and identifying which images to process needs to become a significant topic in future research.

Finally, approaches to explainable AI are notably missing from the literature. The AI community is increasingly giving attention on how humans gain trust in AI-enabled systems. As echocardiography is image-based, trust can be promoted as the human views segmentations or other annotations on images interpreted by the AI system. For example, the human will learn types of cases that the AI system can measure accurately, and which types of cases need to be scrutinized more closely. Still, trust is an important aspect of a human-machine system, and the research into explainable AI has much to contribute to echocardiography.

Overall, we expect that the acceleration in ML applications to echocardiography will lead to the following:

- improved accuracy and faster measurements for 2D echocardiography,
- increased adoption of 3D echocardiography (e.g., 3D TTE and 3D ICE),
- increased use of echocardiography for point of care applications by nonexperts (US at the bedside) and eventually even within the home environment,
- increased use of ML for cardiac procedure guidance,
- emerging use of ML for noninvasive, longitudinal monitoring of cardiovascular conditions with biofeedback.

More broadly speaking, ML has the potential in improving accuracy, speed, and expanding the application of medical US to the nonexpert user. Many discussion points above can also be applied to other organs such as those covered in Section 3. In general, when compared to publicly available nonmedical datasets such as ImageNet (Deng et al., 2009), which contains over 14,000,000 images, open US databases are far smaller in size. This is largely due to the high-level of domain expertise required in collecting and curating a medical dataset. Collective effort, such as MICCAI and Kaggle challenges, have proven to be helpful in promoting standardized medical datasets that can be used for algorithm benchmarking. Table 8.6 provides a snapshot of open-source US datasets, several of which were discussed in the earlier sections.

Increased automation enabled by ML will free sonographers and clinicians from making low-level measurements, which will allow them to provide more individualized care for more patients and ultimately improve the overall patient outcome.

Table 8.6 Snapshot of publically available US datasets.

Type	Size	Purpose
2D TTE (Leclerc et al., 2019)	500 patients, apical two- and four-chamber view sequences, manually annotated	Measurement of chamber volumes, ejection fraction
IVUS ("Lumen + External Elastic Border Detection in IVUS Challenge," n.d.)	Dataset A: 22 patients/77 images Dataset B: 10 patients/435 images	Segmentation of lumen and MA borders
Breast (Al-Dhabyani, Gomaa, Khaled, & Fahmy, 2020)	780 images	Classification, detection, and segmentation of malignant breast tumors
Thyroid (Wunderling et al., 2017)	Ultrasound sweeps of 8 nodules in dicom format	Segmentation of thyroid

Acknowledgments

DISTRIBUTION STATEMENT A. Approved for public release. Distribution is unlimited. This material is based upon work supported under Air Force Contract No. FA8702-15-D-0001. Any opinions, findings, conclusions, or recommendations expressed in this material are those of the author(s) and do not necessarily reflect the views of the US Air Force. The authors wish to thank Mazen S. Albaghdadi, M.D., Conor D. Barrett, M.D., Timothy Churchill, M.D., and William J. Hucker, M.D., Ph.D. for providing example echocardiography images, Coralee Lazebnikova, for the artistic rendering of the figures, and Nora Smith for her contributions as a reference librarian.

References

Abdi, A. H., Luong, C., Tsang, T., Allan, G., Nouranian, S., Jue, J., et al. (2017). Automatic quality assessment of echocardiograms using convolutional neural networks: Feasibility on the apical four-chamber view. *IEEE Transactions on Medical Imaging, 36*, 1221–1230. https://doi.org/10.1109/tmi.2017.2690836.

Al-Dhabyani, W., Gomaa, M., Khaled, H., & Fahmy, A. (2020). Dataset of breast ultrasound images. *Data Brief, 28*, 104863. https://doi.org/10.1016/j.dib.2019.104863.

Alsharqi, M., Woodward, W. J., Mumith, J. A., Markham, D. C., Upton, R., & Leeson, P. (2018). Artificial intelligence and echocardiography. *Echo Research and Practice, 5*, R115–r125. https://doi.org/10.1530/erp-18-0056.

Al'Aref, S. J., Anchouche, K., Singh, G., Slomka, P. J., Kolli, K. K., Kumar, A., et al. (2019). Clinical applications of machine learning in cardiovascular disease and its relevance to cardiac imaging. *European Heart Journal, 40*, 1975–1986. https://doi.org/10.1093/eurheartj/ehy404.

Araki, T., Ikeda, N., Shukla, D., Londhe, N. D., Shrivastava, V. K., Banchhor, S. K., et al. (2016). A new method for IVUS-based coronary artery disease risk stratification: A link between coronary & carotid ultrasound plaque burdens. *Computer Methods and Programs in Biomedicine, 124*, 161–179. https://doi.org/10.1016/j.cmpb.2015.10.022.

Asch, F. M., Poilvert, N., Abraham, T., Jankowski, M., Cleve, J., Adams, M., et al. (2019). Automated echocardiographic quantification of left ventricular ejection fraction without volume measurements using a machine learning algorithm mimicking a human expert. *Circulation: Cardiovascular Imaging, 12*. https://doi.org/10.1161/CIRCIMAGING.119.009303.

Balocco, S., Gatta, C., Ciompi, F., Wahle, A., Radeva, P., Carlier, S., et al. (2014). Standardized evaluation methodology and reference database for evaluating IVUS image segmentation. *Computerized Medical Imaging and Graphics, 38*, 70–90. https://doi.org/10.1016/j.compmedimag.2013.07.001.

Balocco, S., González, M., Ñanculef, R., Radeva, P., & Thomas, G. (2018). Calcified plaque detection in ivus sequences: Preliminary results using convolutional nets. *Lecture Notes in Computer Science*. https://doi.org/10.1007/978-3-030-01132-1_4.

Banchhor, S. K., Araki, T., Londhe, N. D., Ikeda, N., Radeva, P., Elbaz, A., et al. (2016). Five multiresolution-based calcium volume measurement techniques from coronary IVUS videos: A comparative approach. *Computer Methods and Programs in Biomedicine, 134*, 237–258. https://doi.org/10.1016/j.cmpb.2016.07.009.

Barletta, V., Hinojar, R., Carbonell, A., González-Gómez, A., Fabiani, I., Di Bello, V., et al. (2018). Three-dimensional full automated software in the evaluation of the left ventricle function: From theory to clinical practice. *The International Journal of Cardiovascular Imaging, 34*, 1205–1213.

Becker, A. S., Mueller, M., Stoffel, E., Marcon, M., Ghafoor, S., & Boss, A. (2018). Classification of breast cancer in ultrasound imaging using a generic deep learning analysis software: A pilot study. *British Journal of Radiology, 91*, 20170576.

Bian, C., Lee, R., Chou, Y.-H., & Cheng, J.-Z. (2017). Boundary regularized convolutional neural network for layer parsing of breast anatomy in automated whole breast ultrasound. In *Presented at the international conference on medical image computing and computer-assisted intervention* (pp. 259–266). Springer.

Biswas, M., Kuppili, V., Edla, D. R., Suri, H. S., Saba, L., Marinhoe, R. T., et al. (2018). Symtosis: A liver ultrasound tissue characterization and risk stratification in optimized deep learning paradigm. *Computer Methods and Programs in Biomedicine, 155*, 165–177.

Bizopoulos, P., & Koutsouris, D. (2019). Deep learning in cardiology. *IEEE Reviews in Biomedical Engineering, 12*, 168–193. https://doi.org/10.1109/RBME.2018.2885714.

Boccalandro, F., Baptista, E., Muench, A., Carter, C., & Smalling, R. W. (2004). Comparison of intracardiac echocardiography versus transesophageal echocardiography guidance for percutaneous transcatheter closure of atrial septal defect. *The American Journal of Cardiology, 93*, 437–440. https://doi.org/10.1016/j.amjcard.2003.10.037.

Brattain, L. (2014). *Enhanced ultrasound visualization for procedure guidance* (PhD Thesis).

Brattain, L. J., Telfer, B. A., Dhyani, M., Grajo, J. R., & Samir, A. E. (2018). Machine learning for medical ultrasound: Status, methods, and future opportunities. *Abdominal Radiology, 43*, 786–799. https://doi.org/10.1007/s00261-018-1517-0.

Brito, J. P., Gionfriddo, M. R., Al Nofal, A., Boehmer, K. R., Leppin, A. L., Reading, C., et al. (2014). The accuracy of thyroid nodule ultrasound to Predict thyroid cancer: Systematic review and meta-analysis. *Journal of Clinical Endocrinology & Metabolism, 99*, 1253–1263. https://doi.org/10.1210/jc.2013-2928.

Byra, M., Styczynski, G., Szmigielski, C., Kalinowski, P., Michałowski, Ł., Paluszkiewicz, R., et al. (2018). Transfer learning with deep convolutional neural network for liver steatosis assessment in ultrasound images. *International Journal of Computer Assisted Radiology and Surgery, 13*, 1895–1903.

Cao, Z., Duan, L., Yang, G., Yue, T., Chen, Q., Fu, H., et al. (2017). Breast tumor detection in ultrasound images using deep learning. In *Presented at the international workshop on patch-based techniques in medical imaging* (pp. 121–128). Springer.

Casella, G., Klauss, V., Ottani, F., Siebert, U., Sangiorgio, P., & Bracchetti, D. (2003). Impact of intravascular ultrasound-guided stenting on long-term clinical outcome: A meta-analysis of available studies comparing intravascular ultrasound-guided and angiographically guided stenting. *Catheterization Cardiovascular Interventions, 59*, 314–321. https://doi.org/10.1002/ccd.10537.

Cheng, P. M., & Malhi, H. S. (2017). Transfer learning with convolutional neural networks for classification of abdominal ultrasound images. *Journal of Digital Imaging, 30*, 234–243. https://doi.org/10.1007/s10278-016-9929-2.

Cheng, J.-Z., Ni, D., Chou, Y.-H., Qin, J., Tiu, C.-M., Chang, Y.-C., et al. (2016). Computer-aided diagnosis with deep learning architecture: Applications to breast lesions in US images and pulmonary nodules in CT scans. *Scientific Reports, 6*, 24454.

Chiang, T.-C., Huang, Y.-S., Chen, R.-T., Huang, C.-S., & Chang, R.-F. (2018). Tumor detection in automated breast ultrasound using 3-D CNN and prioritized candidate aggregation. *IEEE Transactions on Medical Imaging, 38*, 240–249.

Chi, J., Walia, E., Babyn, P., Wang, J., Groot, G., & Eramian, M. (2017). Thyroid nodule classification in ultrasound images by fine-tuning deep convolutional neural network. *Journal of Digital Imaging, 30*, 477–486.

Claessen, B. E., Mehran, R., Mintz, G. S., Weisz, G., Leon, M. B., Dogan, O., et al. (2011). Impact of intravascular ultrasound imaging on early and late clinical outcomes following percutaneous coronary intervention with drug-eluting stents. *JACC: Cardiovascular Interventions, 4*, 974–981. https://doi.org/10.1016/j.jcin.2011.07.005.

Clark, D. J., Lessio, S., O'Donoghue, M., Tsalamandris, C., Schainfeld, R., & Rosenfield, K. (2006). Mechanisms and predictors of carotid artery stent restenosis. *Journal of the American College of Cardiology, 47*, 2390–2396. https://doi.org/10.1016/j.jacc.2006.01.076.

Conjeti, S., Katouzian, A., Roy, A. G., Peter, L., Sheet, D., Carlier, S., et al. (2016). Supervised domain adaptation of decision forests: Transfer of models trained in vitro for in vivo intravascular ultrasound tissue characterization. *Medical Image Analysis, 32*, 1–17. https://doi.org/10.1016/j.media.2016.02.005.

Degirmenci, A., Loschak, P. M., Tschabrunn, C. M., Anter, E., & Howe, R. D. (2016). Compensation for unconstrained catheter shaft motion in cardiac catheters. In *Presented at the 2016 IEEE international conference on robotics and automation (ICRA)* (pp. 4436–4442). Stockholm, Sweden: IEEE. https://doi.org/10.1109/ICRA.2016.7487643.

Deng, J., Dong, W., Socher, R., Li, L.-J., Li, K., & Fei-Fei, L. (2009). Imagenet: A large-scale hierarchical image database. In *2009 IEEE conference on computer vision and pattern recognition* (pp. 248–255). Ieee.

Dravid, S. G., Hope, B., & McKinnie, J. J. (2008). Intracardiac echocardiography in electrophysiology: A review of current applications in practice: Intracardiac echocardiography in electrophysiology. *Echocardiography, 25*, 1172–1175. https://doi.org/10.1111/j.1540-8175.2008.00784.x.

D'Hooge, J., & Fraser, A. G. (2018). Learning about machine learning to create a self-driving echocardiographic laboratory. *Circulation, 138*, 1636–1638. https://doi.org/10.1161/circulationaha.118.037094.

Frates, M. C., Benson, C. B., Charboneau, J. W., Cibas, E. S., Clark, O. H., Coleman, B. G., et al. (2005). Management of thyroid nodules detected at US: Society of radiologists in ultrasound consensus conference statement. *Radiology, 237*, 794–800. https://doi.org/10.1148/radiol.2373050220.

Frid-Adar, M., Diamant, I., Klang, E., Amitai, M., Goldberger, J., & Greenspan, H. (2018). GAN-based synthetic medical image augmentation for increased CNN performance in liver lesion classification. *Neurocomputing, 321*, 321–331.

Fujii, K., Hao, H., Shibuya, M., Imanaka, T., Fukunaga, M., Miki, K., et al. (2015). Accuracy of OCT, grayscale IVUS, and their combination for the diagnosis of coronary TCFA. *JACC: Cardiovascular Imaging, 8*, 451–460. https://doi.org/10.1016/j.jcmg.2014.10.015.

Gao, Z., Hau, W. K., Lu, M., Huang, W., Zhang, H., Wu, W., et al. (2015). Automated framework for detecting lumen and media-adventitia borders in intravascular ultrasound images. *Ultrasound in Medicine and Biology, 41*, 2001–2021. https://doi.org/10.1016/j.ultrasmedbio.2015.03.022.

Goodfellow, I., Pouget-Abadie, J., Mirza, M., Xu, B., Warde-Farley, D., Ozair, S., et al. (2014). Generative adversarial nets. In *Advances in neural information processing systems* (pp. 2672–2680).

Han, S., Kang, H.-K., Jeong, J.-Y., Park, M.-H., Kim, W., Bang, W.-C., et al. (2017). A deep learning framework for supporting the classification of breast lesions in ultrasound images. *Physics in Medicine and Biology, 62*, 7714.

Hinton, G. E., Sejnowski, T. J., & Poggio, T. A. (1999). *Unsupervised learning: Foundations of neural computation*. MIT press.

Hong, Y., Hong, Y.-M., Jang, Y., Kim, S., Jeon, B., Jung, S., et al. (2017). Coronary luminal and wall mask prediction using convolutional neural network. In *Presented at the 2017 IEEE 14th international symposium on biomedical imaging (ISBI 2017)* (pp. 1049–1052). Melbourne, Australia: IEEE. https://doi.org/10.1109/ISBI.2017.7950696.

Hovnanians, N., Win, T., Makkiya, M., Zheng, Q., & Taub, C. (2017). Validity of automated measurement of left ventricular ejection fraction and volume using the Philips EPIQ system. *Echocardiography, 34*, 1575–1583. https://doi.org/10.1111/echo.13705.

Huynh, B., Drukker, K., & Giger, M. (2016). MO-DE-207B-06: Computer-aided diagnosis of breast ultrasound images using transfer learning from deep convolutional neural networks. *Medical Physics, 43*, 3705.

Ionasec, R. (2015). *Model-based positioning for intracardiac echocardiography volume stitching*. US 9,179,890 B2.

Jafari, M. H., Girgis, H., Van Woudenberg, N., Liao, Z., Rohling, R., Gin, K., et al. (2019). Automatic biplane left ventricular ejection fraction estimation with mobile point-of-care ultrasound using multi-task learning and adversarial training. *International Journal of Computer Assisted Radiology and Surgery, 14*, 1027–1037. https://doi.org/10.1007/s11548-019-01954-w.

Jeganathan, J., Knio, Z., Amador, Y., Hai, T., Khamooshian, A., Matyal, R., et al. (2017). Artificial intelligence in mitral valve analysis. *Annals of Cardiac Anaesthesia, 20*, 129–134. https://doi.org/10.4103/aca.ACA_243_16.

Jun, T. J., Kang, S. J., Lee, J. G., Kweon, J., Na, W., Kang, D., et al. (2018). Automated detection of vulnerable plaque in intravascular ultrasound images. *Medical, & Biological Engineering & Computing*. https://doi.org/10.1007/s11517-018-1925-x.

Kim, G. Y., Lee, J. H., Hwang, Y. N., & Kim, S. M. (2018). A novel intensity-based multilevel classification approach for coronary plaque characterization in intravascular ultrasound images. *BioMedical Engineering Online, 17*. https://doi.org/10.1186/s12938-018-0586-1.

Kitano, T., Nabeshima, Y., Otsuji, Y., Negishi, K., & Takeuchi, M. (2019). Accuracy of left ventricular volumes and ejection fraction measurements by contemporary three-dimensional echocardiography with semi-and fully automated software: Systematic review and meta-analysis of 1,881 subjects. *Journal of the American Society of Echocardiography, 32*, 1105–1115.

Knackstedt, C., Bekkers, S. C., Schummers, G., Schreckenberg, M., Muraru, D., Badano, L. P., et al. (2015). Fully automated versus standard tracking of left ventricular ejection fraction and longitudinal strain: The FAST-EFs multicenter study. *Journal of the American College of Cardiology, 66*, 1456–1466. https://doi.org/10.1016/j.jacc.2015.07.052.

Ko, J., & Lee, J. G. (2019). Lumen and vessel wall segmentation on intravascular ultrasound images using fully convolutional network. In *Presented at the proceedings of SPIE – the international society for optical engineering*. https://doi.org/10.1117/12.2521363.

Ko, S. Y., Lee, J. H., Yoon, J. H., Na, H., Hong, E., Han, K., et al. (2019). Deep convolutional neural network for the diagnosis of thyroid nodules on ultrasound. *Head & Neck, 41*, 885–891.

Krokidis, M., & Hatzidakis, A. (2011). The use of intravascular ultrasound (IVUS) for direct portosystemic shunt (DIPS) evaluation. *Interventional Medicine and Applied Science, 3*, 216–219. https://doi.org/10.1556/IMAS.3.2011.4.7.

Kubo, T., Akasaka, T., Shite, J., Suzuki, T., Uemura, S., Yu, B., et al. (2013). OCT compared with IVUS in a coronary lesion assessment. *JACC Cardiovascular Imaging, 6*, 1095–1104. https://doi.org/10.1016/j.jcmg.2013.04.014.

Kumar, V., Webb, J. M., Gregory, A., Denis, M., Meixner, D. D., Bayat, M., et al. (2018). Automated and real-time segmentation of suspicious breast masses using convolutional neural network. *PloS One, 13*, e0195816.

Kusunose, K., Haga, A., Abe, T., & Sata, M. (2019). Utilization of artificial intelligence in echocardiography. *Circulation Journal*. https://doi.org/10.1253/circj.CJ-19-0420.

Laube, K. A., & Zell, A. (2019). ShuffleNASNets: Efficient CNN models through modified efficient neural architecture search. In *Presented at the 2019 international joint conference on neural networks (IJCNN)* (pp. 1–6). Budapest, Hungary: IEEE. https://doi.org/10.1109/IJCNN.2019.8852294.

Leclerc, S., Smistad, E., Pedrosa, J., Ostvik, A., Cervenansky, F., Espinosa, F., et al. (2019). Deep learning for segmentation using an open large-scale dataset in 2D echocardiography. *IEEE Transactions on Medical Imaging*. https://doi.org/10.1109/tmi.2019.2900516.

Lee, S. S., & Park, S. H. (2014). Radiologic evaluation of nonalcoholic fatty liver disease. *World Journal of Gastroenterology, 20*, 7392.

Leeson, P. (2019). (Deep) learning your left from your right. *JACC Cardiovascular Imaging*. https://doi.org/10.1016/j.jcmg.2019.03.015.

Lei, B., Huang, S., Li, R., Bian, C., Li, H., Chou, Y.-H., et al. (2018). Segmentation of breast anatomy for automated whole breast ultrasound images with boundary regularized convolutional encoder–decoder network. *Neurocomputing, 321*, 178–186.

Levy, F., Marechaux, S., Iacuzio, L., Schouver, E. D., Castel, A. L., Toledano, M., et al. (2018). Quantitative assessment of primary mitral regurgitation using left ventricular volumes obtained with new automated three-dimensional transthoracic echocardiographic software: A comparison with 3-tesla cardiac magnetic resonance. *Archives of Cardiovascular Diseases, 111*, 507–517. https://doi.org/10.1016/j.acvd.2017.10.008.

Liberman, L., & Menell, J. H. (2002). Breast imaging reporting and data system (BI-RADS). *Radiologia Clinica, 40*, 409–430.

Lin, J.-D., Chao, T.-C., Huang, B.-Y., Chen, S.-T., Chang, H.-Y., & Hsueh, C. (2005). Thyroid cancer in the thyroid nodules evaluated by ultrasonography and fine-needle aspiration cytology. *Thyroid, 15*, 708–717. https://doi.org/10.1089/thy.2005.15.708.

Liu, X., Song, J., Wang, S., Zhao, J., & Chen, Y. (2017). Learning to diagnose cirrhosis with liver capsule guided ultrasound image classification. *Sensors, 17*, 149.

Liu, T., Xie, S., Yu, J., Niu, L., & Sun, W. (2017). Classification of thyroid nodules in ultrasound images using deep model based transfer learning and hybrid features. In *Presented at the 2017 IEEE international conference on acoustics, speech and signal processing (ICASSP)* (pp. 919–923). IEEE.

Li, H., Weng, J., Shi, Y., Gu, W., Mao, Y., Wang, Y., et al. (2018). An improved deep learning approach for detection of thyroid papillary cancer in ultrasound images. *Scientific Reports, 8*, 6600.

Li, X., Zhang, S., Zhang, Q., Wei, X., Pan, Y., Zhao, J., et al. (2019). Diagnosis of thyroid cancer using deep convolutional neural network models applied to sonographic images: A retrospective, multicohort, diagnostic study. *The Lancet Oncology, 20*, 193–201.

Lo Vercio, L., del Fresno, M., & Larrabide, I. (2019). Lumen-intima and media-adventitia segmentation in IVUS images using supervised classifications of arterial layers and morphological structures. *Computer Methods and Programs in Biomedicine, 177*, 113–121. https://doi.org/10.1016/j.cmpb.2019.05.021.

Loschak, P. M., Brattain, L. J., & Howe, R. D. (2017). Algorithms for automatically pointing ultrasound imaging catheters. *IEEE Transactions on Robotics, 33*, 81–91. https://doi.org/10.1109/TRO.2016.2623331.

Loschak, P. M., Degirmenci, A., & Howe, R. D. (2017). Predictive filtering in motion compensation with steerable cardiac catheters. In *Presented at the 2017 IEEE International conference on robotics and automation (ICRA)* (pp. 4830–4836). Singapore, Singapore: IEEE. https://doi.org/10.1109/ICRA.2017.7989561.

Lumen + external elastic border detection in IVUS challenge, n.d. Retrieved from http://www.cvc.uab.es/IVUSchallenge2011/index-2.html. (accessed 1 4 20).

Maehara, A., Matsumura, M., Ali, Z. A., Mintz, G. S., & Stone, G. W. (2017). IVUS-guided versus OCT-guided coronary stent implantation. *JACC: Cardiovascular Imaging, 10*, 1487–1503. https://doi.org/10.1016/j.jcmg.2017.09.008.

Maehara, A., Mintz, G. S., & Stone, G. W. (2013). OCT versus IVUS: Accuracy versus clinical utility. *JACC: Cardiovascular Imaging, 6*, 1105–1107. https://doi.org/10.1016/j.jcmg.2013.05.016.

Ma, J., Wu, F., Jiang, T., Zhao, Q., & Kong, D. (2017). Ultrasound image-based thyroid nodule automatic segmentation using convolutional neural networks. *International Journal of Computer Assisted Radiology and Surgery, 12*, 1895–1910.

Ma, J., Wu, F., Jiang, T., Zhu, J., & Kong, D. (2017). Cascade convolutional neural networks for automatic detection of thyroid nodules in ultrasound images. *Medical Physics, 44*, 1678–1691.

Ma, J., Wu, F., Zhu, J., Xu, D., & Kong, D. (2017). A pre-trained convolutional neural network based method for thyroid nodule diagnosis. *Ultrasonics, 73*, 221–230.

Mazzaferri, E. L. (1999). An overview of the management of papillary and follicular thyroid carcinoma. *Thyroid, 9*, 421–427. https://doi.org/10.1089/thy.1999.9.421.

Medvedofsky, D., Addetia, K., Patel, A. R., Sedlmeier, A., Baumann, R., Mor-Avi, V., et al. (2015). Novel approach to three-dimensional echocardiographic quantification of right ventricular volumes and function from focused views. *Journal of the American Society of Echocardiography, 28*, 1222–1231. https://doi.org/10.1016/j.echo.2015.06.013.

Medvedofsky, D., Mor-Avi, V., Byku, I., Singh, A., Weinert, L., Yamat, M., et al. (2017). Three-dimensional echocardiographic automated quantification of left heart chamber volumes using an adaptive analytics algorithm: Feasibility and impact of image quality in nonselected patients. *Journal of the American Society of Echocardiography, 30*, 879–885. https://doi.org/10.1016/j.echo.2017.05.018.

Meng, D., Zhang, L., Cao, G., Cao, W., Zhang, G., & Hu, B. (2017). Liver fibrosis classification based on transfer learning and FCNet for ultrasound images. *IEEE Access, 5*, 5804–5810.

Miyoshi, T., Addetia, K., Blitz, A., Lang, R. M., Asch, F. M., & Investigators, W. (2019). Right ventricular size and function in healthy adults: Geographical differences from the world alliance of societies of echocardiography (WASE) normal values study. *Circulation, 140*, A13056.

Moghaddasi, H., & Nourian, S. (2016). Automatic assessment of mitral regurgitation severity based on extensive textural features on 2D echocardiography videos. *Computers in Biology and Medicine, 73*, 47–55. https://doi.org/10.1016/j.compbiomed.2016.03.026.

Moon, W.-J., Jung, S. L., Lee, J. H., Na, D. G., Baek, J.-H., Lee, Y. H., et al. (2008). Benign and malignant thyroid nodules: US differentiation—multicenter retrospective study. *Radiology, 247*, 762–770. https://doi.org/10.1148/radiol.2473070944.

Nakamura, M., Nishikawa, H., Mukai, S., Setsuda, M., Nakajima, K., Tamada, H., et al. (2001). Impact of coronary artery remodeling on clinical presentation of coronary artery disease: An intravascular ultrasound study. *Journal of the American College of Cardiology, 37*, 63–69. https://doi.org/10.1016/S0735-1097(00)01097-4.

Narang, A., Mor-Avi, V., Prado, A., Volpato, V., Prater, D., Tamborini, G., et al. (2019). Machine learning based automated dynamic quantification of left heart chamber volumes. *European Heart Journal Cardiovascular Imaging, 20*, 541–549. https://doi.org/10.1093/ehjci/jey137.

Narducci, M. L., Vitulano, N., Pazzano, V., & Pelargonio, G. (2015). Technology update: Intracardiac echocardiography – a review of the literature. *Medical Devices: Evidence and Research*, 231. https://doi.org/10.2147/MDER.S49567.

Narula, S., Shameer, K., Salem Omar, A. M., Dudley, J. T., & Sengupta, P. P. (2017). Reply: Deep learning with unsupervised feature in echocardiographic imaging. *Journal of the American College of Cardiology, 69*, 2101–2102. https://doi.org/10.1016/j.jacc.2017.01.062.

Nolan, M. T., & Thavendiranathan, P. (2019). Automated quantification in echocardiography. *JACC: Cardiovascular Imaging, 12*, 1073–1092. https://doi.org/10.1016/j.jcmg.2018.11.038.

Omar, H. A., Domingos, J. S., Patra, A., Upton, R., Leeson, P., & Noble, J. A. (2018). Quantification of cardiac bull's-eye map based on principal strain analysis for myocardial wall motion assessment in stress echocardiography. In *Presented at the 2018 IEEE 15th international symposium on biomedical imaging (ISBI 2018)* (pp. 1195–1198). Washington, DC: IEEE. https://doi.org/10.1109/ISBI.2018.8363785.

Otani, K., Nakazono, A., Salgo, I. S., Lang, R. M., & Takeuchi, M. (2016). Three-dimensional echocardiographic assessment of left heart chamber size and function with fully automated quantification software in patients with atrial fibrillation. *Journal of the American Society of Echocardiography, 29*, 955–965. https://doi.org/10.1016/j.echo.2016.06.010.

Pan, S. J., & Yang, Q. (2009). A survey on transfer learning. *IEEE Transactions on Knowledge and Data Engineering, 22*, 1345–1359.

Patra, A., & Noble, J. A. (2019). Multi-anatomy localization in fetal echocardiography videos. In *2019 IEEE 16th International symposium on biomedical imaging (ISBI 2019)* (pp. 1761–1764). IEEE.

Penatti, O. A., Werneck Rde, O., de Almeida, W. R., Stein, B. V., Pazinato, D. V., Mendes Junior, P. R., et al. (2015). Mid-level image representations for real-time heart view plane classification of echocardiograms. *Computers in Biology and Medicine, 66*, 66–81. https://doi.org/10.1016/j.compbiomed.2015.08.004.

Perumpail, B. J., Khan, M. A., Yoo, E. R., Cholankeril, G., Kim, D., & Ahmed, A. (2017). Clinical epidemiology and disease burden of nonalcoholic fatty liver disease. *World Journal of Gastroenterology, 23*, 8263.

Rodríguez-Zanella, H., Muraru, D., Secco, E., Boccalini, F., Azzolina, D., Aruta, P., et al. (2019). Added value of 3- versus 2-dimensional echocardiography left ventricular ejection fraction to predict arrhythmic risk in patients with left ventricular dysfunction. *JACC: Cardiovascular Imaging, 12*, 1917–1926. https://doi.org/10.1016/j.jcmg.2018.07.011.

Ronneberger, O., Fischer, P., & Brox, T. (2015). U-Net: Convolutional networks for biomedical image segmentation. In N. Navab, J. Hornegger, W. M. Wells, & A. F. Frangi (Eds.), *Medical image computing and computer-assisted intervention – MICCAI 2015* (pp.

234–241). Cham: Springer International Publishing. https://doi.org/10.1007/978-3-319-24574-4_28.

Saadeh, S., Younossi, Z. M., Remer, E. M., Gramlich, T., Ong, J. P., Hurley, M., et al. (2002). The utility of radiological imaging in nonalcoholic fatty liver disease. *Gastroenterology, 123*, 745–750.

Sanchez-Martinez, S., Duchateau, N., Erdei, T., Fraser, A. G., Bijnens, B. H., & Piella, G. (2017). Characterization of myocardial motion patterns by unsupervised multiple kernel learning. *Medical Image Analysis, 35*, 70–82. https://doi.org/10.1016/j.media.2016.06.007.

Sanchez-Martinez, S., Duchateau, N., Erdei, T., Kunszt, G., Aakhus, S., Degiovanni, A., et al. (2018). Machine learning analysis of left ventricular function to characterize heart failure with preserved ejection fraction. *Circulation: Cardiovascular Imaging, 11*, e007138. https://doi.org/10.1161/circimaging.117.007138.

Schmauch, B., Herent, P., Jehanno, P., Dehaene, O., Saillard, C., Aubé, C., et al. (2019). Diagnosis of focal liver lesions from ultrasound using deep learning. *Diagnostic and Interventional Imaging, 100*, 227–233.

Seetharam, K., Shrestha, S., & Sengupta, P. P. (2019). Artificial intelligence in cardiovascular medicine. *Current Treatment Options in Cardiovascular Medicine, 21*, 25. https://doi.org/10.1007/s11936-019-0728-1.

Sengupta, P. P., Huang, Y. M., Bansal, M., Ashrafi, A., Fisher, M., Shameer, K., et al. (2016). Cognitive machine-learning algorithm for cardiac imaging: A pilot study for differentiating constrictive pericarditis from restrictive cardiomyopathy. *Circulation: Cardiovascular Imaging, 9*. https://doi.org/10.1161/circimaging.115.004330.

Settles, B. (2011). From theories to queries: Active learning in practice. In *Active learning and experimental design workshop in conjunction with AISTATS 2010* (pp. 1–18).

Sheet, D., Karamalis, A., Eslami, A., Noël, P., Chatterjee, J., Ray, A. K., et al. (2014a). Joint learning of ultrasonic backscattering statistical physics and signal confidence primal for characterizing atherosclerotic plaques using intravascular ultrasound. *Medical Image Analysis, 18*, 103–117. https://doi.org/10.1016/j.media.2013.10.002.

Sheet, D., Karamalis, A., Eslami, A., Noël, P., Virmani, R., Nakano, M., et al. (2014b). Hunting for necrosis in the shadows of intravascular ultrasound. *Computerized Medical Imaging and Graphics, 38*, 104–112. https://doi.org/10.1016/j.compmedimag.2013.08.002.

Shin, S. Y., Lee, S., Yun, I. D., & Lee, K. M. (2017). Joint weakly and semi-supervised deep learning for localization and classification of masses in breast ultrasound images. *ArXiv*.

Shi, J., Zhou, S., Liu, X., Zhang, Q., Lu, M., & Wang, T. (2016). Stacked deep polynomial network based representation learning for tumor classification with small ultrasound image dataset. *Neurocomputing, 194*, 87–94. https://doi.org/10.1016/j.neucom.2016.01.074.

Siegel, R. L., Miller, K. D., & Jemal, A. (2016). Cancer statistics. *CA. Cancer Journal for Clinicians, 66*, 7–30.

Siegel, R. L., Miller, K. D., & Jemal, A. (2019). Cancer statistics. *CA. Cancer Journal for Clinicians, 69*, 7–34.

Simpson, J., Lopez, L., Acar, P., Friedberg, M. K., Khoo, N. S., Ko, H. H., et al. (2017). Three-dimensional echocardiography in congenital heart disease: An expert consensus document from the European association of cardiovascular imaging and the American Society of Echocardiography. *Journal of the American Society of Echocardiography, 30*, 1–27. https://doi.org/10.1016/j.echo.2016.08.022.

Sofian, H., Than, J. C. M., Mohammad, S., & Noor, N. M. (2018). Calcification detection of coronary artery disease in intravascular ultrasound image: Deep feature learning approach. *International Journal of Integrated Engineering, 10*, 43–57.

Song, W., Li, S., Liu, J., Qin, H., Zhang, B., Zhang, S., et al. (2018). Multitask cascade convolution neural networks for automatic thyroid nodule detection and recognition. *IEEE Journal of Biomedical and Health Informatics, 23*, 1215–1224.

Sudarshan, V. K., Acharya, U. R., Ng, E. Y., Tan, R. S., Chou, S. M., & Ghista, D. N. (2016). Data mining framework for identification of myocardial infarction stages in ultrasound: A hybrid feature extraction paradigm (part 2). *Computers in Biology and Medicine, 71*, 241–251. https://doi.org/10.1016/j.compbiomed.2016.01.029.

Su, S., Gao, Z., Zhang, H., Lin, Q., Hau, W. K., & Li, S. (2017). Detection of lumen and media-adventitia borders in IVUS images using sparse auto-encoder neural network. In *Presented at the proceedings – international symposium on biomedical imaging* (pp. 1120–1124). https://doi.org/10.1109/ISBI.2017.7950713.

Tan, M., Chen, B., Pang, R., Vasudevan, V., Sandler, M., Howard, A., et al. (2019). MnasNet: Platform-Aware neural architecture search for mobile. In *The IEEE conference on computer vision and pattern recognition (CVPR)*.

Tsang, W., Salgo, I. S., Medvedofsky, D., Takeuchi, M., Prater, D., Weinert, L., et al. (2016). Transthoracic 3D echocardiographic left heart chamber quantification using an automated adaptive analytics algorithm. *JACC: Cardiovascular Imaging, 9*, 769–782. https://doi.org/10.1016/j.jcmg.2015.12.020.

Tsay, D., & Patterson, C. (2018). From machine learning to artificial intelligence applications in cardiac care. *Circulation, 138*, 2569–2575. https://doi.org/10.1161/circulationaha.118.031734.

US Preventive Services Task Force, Bibbins-Domingo, K., Grossman, D. C., Curry, S. J., Barry, M. J., Davidson, K. W., et al. (2017). Screening for thyroid cancer: US Preventive Services Task Force recommendation statement. *Journal of the American Medical Association, 317*, 1882. https://doi.org/10.1001/jama.2017.4011.

Vasamreddy, C. R., Lickfett, L., Jayam, V. K., Nasir, K., Bradley, D. J., Eldadah, Z., et al. (2004). Predictors of recurrence following catheter ablation of atrial fibrillation using an irrigated-tip ablation catheter. *Journal of Cardiovascular Electrophysiology, 15*, 692–697. https://doi.org/10.1046/j.1540-8167.2004.03538.x.

Wang, N., Bian, C., Wang, Y., Xu, M., Qin, C., Yang, X., et al. (2018). Densely deep supervised networks with threshold loss for cancer detection in automated breast ultrasound. In *Presented at the international conference on medical image computing and computer-assisted intervention* (pp. 641–648). Springer.

Wang, L., Yang, S., Yang, S., Zhao, C., Tian, G., Gao, Y., et al. (2019). Automatic thyroid nodule recognition and diagnosis in ultrasound imaging with the YOLOv2 neural network. *World Journal of Surgical Oncology, 17*, 12.

Winter, M. M., Bernink, F. J., Groenink, M., Bouma, B. J., van Dijk, A. P., Helbing, W. A., et al. (2008). Evaluating the systemic right ventricle by CMR: The importance of consistent and reproducible delineation of the cavity. *Journal of Cardiovascular Magnetic Resonance, 10*, 40. https://doi.org/10.1186/1532-429X-10-40.

Wu, V. C.-C., Kitano, T., Nabeshima, Y., Otani, K., Chu, P.-H., & Takeuchi, M. (2019). Optimal threshold of three-dimensional echocardiographic fully automated software for quantification of left ventricular volumes and ejection fraction: Comparison with cardiac magnetic resonance disk-area summation method and feature tracking method. *PloS One, 14*.

Wunderling, T., Golla, B., Poudel, P., Arens, C., Friebe, M., & Hansen, C. (2017). Comparison of thyroid segmentation techniques for 3D ultrasound. In M. A. Styner, & E. D. Angelini (Eds.), *Presented at the SPIE medical imaging, Orlando, Florida, United States* (p. 1013317). https://doi.org/10.1117/12.2254234.

Xiao, T., Liu, L., Li, K., Qin, W., Yu, S., & Li, Z. (2018). Comparison of transferred deep neural networks in ultrasonic breast masses discrimination. *BioMed Research International, 2018.*

Xu, Z., Huo, Y., Park, J., Landman, B., Milkowski, A., Grbic, S., et al. (2018). Less is more: Simultaneous view classification and landmark detection for abdominal ultrasound images. In *Presented at the international conference on medical image computing and computer-assisted intervention* (pp. 711–719). Springer.

Xu, Y., Wang, Y., Yuan, J., Cheng, Q., Wang, X., & Carson, P. L. (2019). Medical breast ultrasound image segmentation by machine learning. *Ultrasonics, 91*, 1–9.

Yang, J., Faraji, M., & Basu, A. (2019). Robust segmentation of arterial walls in intravascular ultrasound images using Dual Path U-Net. *Ultrasonics, 96*, 24–33. https://doi.org/10.1016/j.ultras.2019.03.014.

Yang, J., Tong, L., Faraji, M., & Basu, A. (2018). IVUS-Net: An intravascular ultrasound segmentation network. In *Lecture notes in computer science (including subseries lecture notes in artificial intelligence and lecture notes in bioinformatics)*. https://doi.org/10.1007/978-3-030-04375-9_31.

Yap, M. H., Goyal, M., Osman, F., Ahmad, E., Martí, R., Denton, E., et al. (2018). End-to-end breast ultrasound lesions recognition with a deep learning approach. In *Presented at the medical imaging 2018: Biomedical applications in molecular, structural, and functional imaging* (p. 1057819). International Society for Optics and Photonics.

Yap, M. H., Goyal, M., Osman, F. M., Martí, R., Denton, E., Juette, A., et al. (2018). Breast ultrasound lesions recognition: End-to-end deep learning approaches. *Journal of Medical Imaging, 6*, 011007.

Yap, M. H., Pons, G., Martí, J., Ganau, S., Sentís, M., Zwiggelaar, R., et al. (2017). Automated breast ultrasound lesions detection using convolutional neural networks. *IEEE Journal of Biomedical and Health Informatics, 22*, 1218–1226.

Yu, L., Guo, Y., Wang, Y., Yu, J., & Chen, P. (2017). Segmentation of fetal left ventricle in echocardiographic sequences based on dynamic convolutional neural networks. *IEEE Transactions on Biomedical Engineering, 64*, 1886–1895. https://doi.org/10.1109/tbme.2016.2628401.

Zhang, J., Gajjala, S., Agrawal, P., Tison, G. H., Hallock, L. A., Beussink-Nelson, L., et al. (2018). Fully automated echocardiogram interpretation in clinical practice. *Circulation, 138*, 1623–1635. https://doi.org/10.1161/circulationaha.118.034338.

Zhang, Q., Xiao, Y., Dai, W., Suo, J., Wang, C., Shi, J., et al. (2016). Deep learning based classification of breast tumors with shear-wave elastography. *Ultrasonics, 72*, 150–157. https://doi.org/10.1016/j.ultras.2016.08.004.

CHAPTER

9 Computed tomography and artificial intelligence

Damini Dey[1], Andrew Lin[1], Donghee Han[2], Piotr J. Slomka[1]

[1]*Biomedical Imaging Research Institute and Artificial Intelligence in Medicine, Cedars-Sinai Medical Center, Los Angeles, CA, United States;* [2]*Department of Imaging and the Smidt Heart Institute, Cedars-Sinai Medical Center, Los Angeles, CA, United States*

Chapter outline

Introduction

In the past 2 decades, significant advances in computed tomography (CT) hardware and software have expanded the clinical utility of CT imaging. In the cardiovascular domain, imaging of the coronary arteries is technically challenging due to their

Machine Learning in Cardiovascular Medicine. https://doi.org/10.1016/B978-0-12-820273-9.00009-9

small size and rapid constant motion. Cardiac CT requires high temporal resolution to "freeze" such motion and, therefore, has benefited from steadily increasing CT gantry rotation speeds. Imaging of the coronary arteries also requires high spatial resolution and reconstruction of submillimeter slices. Multiple CT slices need to be acquired simultaneously for complete coverage of the heart in one breath-hold. To meet these requirements, technical advances in CT hardware have focused on improved temporal and spatial resolution and more detectors for increased coverage (Halliburton et al., 2012). Dual-energy CT offers promising applications in material decomposition and energy-selective imaging, which were previously unavailable with single energy CT (McCollough, Leng, Yu, & Fletcher, 2015). Further, spectral photon-counting CT that uses energy-resolving detectors is a novel and promising modality in diagnostic radiology (Willemink, Persson, Pourmorteza, Pelc, & Fleischmann, 2018). Recently, there have been applications of artificial intelligence (AI) in all aspects of CT imaging—acquisition, reconstruction, analysis, and measurement of novel features—which may further improve the value and reduce cost. In this chapter, we summarize the recent advances in AI applied to several key aspects of CT imaging, interpretation, and analysis.

Terminology

In general, AI describes a computational program that can perform tasks that are associated with human intelligence, as such learning and problem-solving. Machine learning (ML) represents learning techniques that can give AI the ability to learn, typically progressively from large datasets, without making any prior assumptions. Finally, deep learning (DL) methods, a subset of ML, mimic human cognition by using multiple layers of convolutional neural networks (CNNs); in each layer, convolutional filters perform as feature extractors. In the past few years, the popularity and application of DL methods have expanded dramatically.

Artificial intelligence in CT image reconstruction

Given the hardware developments in CT imaging, there is a demand for image reconstruction algorithms to be more accurate and computationally efficient. In addition to well-established *analytic* and *iterative* methods for CT image reconstruction, a new wave of *data-driven* and *learning-based* methods has emerged, which draw extensively from DL techniques.

Deep learning approaches

The power of DL lies in the multilayer approach used to extract and identify important features from an input. CNNs are usually trained by first defining a loss function, which measures the discrepancy between the actual and desired outputs of the CNN for the training dataset. An optimized set of weights for the convolutional layers is then determined using an iterative optimization algorithm (Zhang & Dong, 2019). In CT image reconstruction, the input is typically an image whose quality is compromised due to data noise or artifacts, and the desired output is a higher quality image. CNNs can be applied as a pre- or postprocessing step to images reconstructed with classical filtered backprojection (Gantt, Jin, & Lu, 2019; Pelt & Batenburg, 2014), used in tandem with iterative reconstruction algorithms (Ahn et al., 2018; Bazrafkan, Van Nieuwenhove, Soons, De Beenhouwer, & Sijbers, 2019), or trained to learn a projection onto a set of desired reconstruction images (Gupta, Jin, Nguyen, McCann, & Unser, 2018). Additionally, a number of different CNN architectures are possible, including autoencoders, modularized frameworks, U-Nets, and generative adversarial networks (Zhang & Dong, 2019).

Denoising low-dose CT images

DL approaches have the potential to help reduce CT radiation doses while speeding up reconstruction times. With an estimated 80 million CT scans performed in the United States each year (Reports, 2015), there has been increasing concern over radiation exposure and the potential cancer risk (Smith-Bindman et al., 2009). It is, therefore, important to keep radiation dose "as low as reasonably achievable (ALARA)" to avoid excessive radiation exposure to the patient. Low-dose CT can now be used for many clinical indications, achieving good spatial resolution and diagnostic performance while minimizing radiation-related risk (Rampinelli, Origgi, & Bellomi, 2013). However, reduction of the CT radiation dose by lowering the tube current increases data noise, which can introduce artifacts into reconstructed images and compromise their diagnostic quality. To address this issue, numerous noise reduction techniques have been developed, using either sonogram filtration (Manduca et al., 2009; Wang, Lu, Li, & Liang, 2005), iterative reconstruction (Geyer et al., 2015; Willemink et al., 2013), or image postprocessing (Chen, Zhang, Kalra, et al. 2017; Chen, Zhang, Zhang, et al., 2017; Shan et al., 2019; Wolterink, Leiner, Viergever, & Isgum, 2017; Yang, Yan, Kalra, & Wang, 2017). Sonogram filtration methods process raw or log-transformed data before image reconstruction and often suffer from edge blurring or resolution loss. Iterative reconstruction algorithms are time consuming and require access to raw sonogram data. By contrast, image postprocessing techniques directly process an image that has already been reconstructed from raw data, and several DL frameworks have hence been applied to this step for denoising low-dose CT.

Chen et al. first trained a CNN to reduce noise in CT images by learning a patch-based neural net (Chen, Zhang, Zhang, et al., 2017), and later refined it with an autoencoder structure for end-to-end mapping of low-to normal-dose CT images (Chen, Zhang, Kalra, et al., 2017). Yang, Yan, Kalra, and Wang (2017) introduced perceptive loss as the objective function of their CNN, which compared the denoised output against the ground truth in terms of extracted features. Wolterink et al. (2017) used a generative adversarial network to estimate normal-dose CT images from low-dose CT images and hence reduce noise. A generator CNN was trained to transform low-dose CT images into normal-dose CT images, while a discriminator CNN simultaneously distinguished the output of the generator from real normal-dose CTs and provided feedback during training. This method resulted in substantial noise reduction when tested on both phantom and noncontrast cardiac CTs, and enabled more accurate coronary calcium scoring in low-dose CT images with high noise levels. Recently, Shan et al. (2019) developed a modularized CNN for end-to-process mapping whereby intermediate denoised images were obtained with associated noise reduction directions toward a final denoised image. The learned workflow-enabled radiologists in the loop to evaluate image quality and optimize the denoising depth in a task-specific fashion. This CNN was tested against commercial iterative reconstruction algorithms on an external dataset of the chest and abdominal CTs, performing favorably or comparably in terms of noise suppression, with a much faster processing time ($\sim$100 slices per second per mapping depth) than iterative reconstructions.

Improving sparse-view CT reconstruction

Low-dose CT can also be obtained by reducing the number of projections over the same acquisition arc, also known as a sparse-view protocol (Kudo, Suzuki, & Rashed, 2013). Compared to tube current reduction, sparse-view CT reconstruction does not increase image noise but instead suffers from streaking artifacts due to under-sampling in the projection domain. Many approaches have been proposed to improve sparse-view CT reconstruction, including regularized iterative reconstruction algorithms incorporating expected properties of the to-be-reconstructed image (Zhang, Liang, Dong, Xie, & Cao, 2018), compressed sensing that reconstructs images from far less sampling data than required by the Nyquist sampling theorem (Donoho, 2006), and the interpolation of missing data in sparse-view sinograms (Bertram, Wiegert, Schafer, Aach, & Rose, 2009; Brooks, Weiss, & Talbert, 1978). Recently, several DL techniques have been applied to remove streak artifacts from sparse-view CT images. Zhang, Liang, et al. (2018) applied an end-to-end neural network with deconvolution layers to sparse-view CT images that had been reconstructed using filtered backprojection. This model effectively suppressed streaking artifact and preserved detailed structures when applied to clinical datasets of the chest and hip CT images. A GoogLeNet-based CNN (Xie et al., 2018) used residual learning to study the artifacts of sparse-view CT reconstruction and then subtracted the learned artifacts from the sparse reconstructed images, finally recovering a clear correction image. This algorithm outperformed state-of-the-art sparse-view reconstruction methods in removing artifacts and achieved an intensity of image reconstruction similar to the full-view projective image. DL models based on U-Net, a large receptive field neural network, have also outperformed regularized iterative reconstruction and compressed sensing approaches in producing artifact-free sparse-view CT images, with considerably faster computational speeds (Han, Yoo, & Ye, 2016; Jin, McCann, Froustey, & Unser, 2017).

Removing image artifacts

CNNs have also been used to reduce artifacts caused by metal or beam hardening. Metallic implants in patients are highly attenuated, leading to "hardening" of the polychromatic X-ray beams and resultant streak artifacts in reconstructed CT images (Man, Nuyts, Dupont, Marchal, & Suetens, 1999). Despite the fact that a large number of metal artifact reduction (MAR) techniques have emerged over the years, no standard solution currently exists. MAR algorithms can be classified into three groups: physical effects correction, interpolation in the projection domain, and iterative reconstruction of images. Due to the high complexity of certain metal materials, sizes, and positions, it is difficult to achieve clinically satisfactory results using a single MAR strategy, and residual image artifacts often remain. DL was first applied to MAR by Park et al. who utilized a U-Net to correct metal-induced beam hardening in both the projection domain (Park, Chung, Lee, Kim, & Seo, 2017) and image domain (Park, Lee, Kim, & Seo, 2017), with promising results in simulated studies of titanium hip prostheses. Gjesteby, Yang, Xi, Shan et al. (2017) and Gjesteby, Yang, Xi, Zhou et al. (2017) used CNNs to refine the performance of a state-of-the-art MAR method known as normalized MAR, achieving additional correction of metal streaks in critical image regions of simulated titanium hip phantoms. Zhang, Chu, and Yu (2017) developed a CNN-based open MAR framework, which fused information from the original and corrected images to suppress artifacts. This method was validated on simulated and real CT datasets containing metallic hip prostheses, dental fillings, and surgical clips (Zhang & Yu, 2018). Finally, Cong and Wang (2017) specifically addressed the issue of beam hardening by using a DL-based CT image reconstruction method to transform measured projection data into a monochromatic image. This algorithm was trained and tested using a clinical dual-energy CT dataset and achieved high accuracy in projection correction compared to ground truth images, with a relative error of less than 0.2%.

In summary, the described AI techniques can optimize or even replace modern analytic and iterative reconstruction methods with algorithms learned from big data. They can improve the quality of images reconstructed from insufficient data, as occurring in low-dose or sparse-view CT, or overcome artifacts due to adverse object conditions such as metal in a patient. Although not yet ready for clinical implementation, these algorithms have the potential to enhance CT workflow by providing rapid and computationally efficient image reconstruction.

Artificial intelligence in cardiac CT image analysis

The big data generated by precision cardiovascular imaging is heterogeneous and rapidly changing, requiring increasingly sophisticated and efficient computational models for their analysis and interpretation. Noninvasive imaging with coronary artery calcium (CAC) scoring noncontrast cardiac CT and coronary CT angiography (CTA) plays a key role in the diagnosis, risk stratification, and management of patients with cardiovascular disease. By facilitating CT image segmentation, measurements, and subsequent clinical pathways, AI may assist physicians in delivering more efficient and personalized care with improved accuracy and reduced costs.

Image segmentation

One of the first DL algorithms developed for cardiac CT used marginal space learning—which incrementally learns classifiers in marginal spaces of lower dimensions—to automatically segment all four heart chambers in 3D CT volumes. This method was tested on a large annotated dataset, achieving accurate boundary delineation with an average speed of 4 s per volume (Zheng, Barbu, Georgescu, Scheuering, & Comaniciu, 2008). Subsequently, several CNN-based cardiac chamber segmentation methods have been applied to noncontrast cardiac CT (Shahzad et al., 2017; Xu, Wu, & Feng, 2018; Ye, Wang, Zhang, & Wang, 2019); and CTA (Bruns et al., 2019; Singh et al., 2019; Zreik et al., 2016). The majority of these studies trained a U-Net using 3D volume as input; the advantages of U-Net being its highly symmetric encoding and decoding modules, and the skip connections between the modules that maintain spatial information lost in the decoding process (Ronneberger, Fischer, & Brox, 2015, pp. 234–241).

Coronary artery segmentation from coronary CTA enables luminal stenosis grading, plaque assessment, and blood flow simulation. The two major steps in this process are coronary centerline extraction and lumen segmentation. Centerline extraction has typically been performed using CNNs as a pre- or postprocessing step for traditional methods (Gülsün, Funka-Lea, Sharma, Rapaka, & Zheng, 2016; Guo et al., 2019). By contrast, Wolterink, Leiner, and Išgum (2019) recently trained a 3D dilated CNN to predict the most likely direction and radius of an artery at any given point in a coronary CTA image based on a local image patch. This algorithm was tested on multiple datasets and successfully performed automated centerline extraction on 92% of coronary artery segments, with each centerline being processed in approximately 0.4 s. For lumen and vessel wall segmentation, most CNNs are trained end to end to predict dense segmentation probability maps (Merkow, Marsden, Kriegman, & Tu, 2016, pp. 371–379; Shen et al., 2019). Alternatively, graph convolutional networks can be applied, whereby vertices on the lumen surface are considered as graph nodes and the spatial locations of these tubular surface mesh vertices are directly optimized (Wolterink et al., 2019).

Automated segmentation of heart valves has been of recent interest due to the surge in transcatheter valve replacement procedures, which require valve modeling and precise sizing of the implants. Marginal space learning can provide patient-specific modeling and segmentation of all four valves from cardiac CT; this framework incorporates a constrained multilinear shape method to efficiently estimate the patient-specific valve parameters from cine images (Grbic et al., 2012). A regression tree-based ML algorithm has been shown to accurately localize eight aortic valve landmarks within 12 ms on preprocedural 3D CTA images of patients undergoing transcatheter aortic valve implantation (TAVI) (Al et al., 2018).

Further, a fuzzy classification method was able to automatically segment the aortic root from pre-TAVI CTA datasets, with a high agreement between automated radius measurements and expert manual delineations (Elattar et al., 2014). Rotational C-arm CT can provide intraprocedural imaging guidance for TAVI, by acquiring 3D datasets immediately before valve deployment. Zheng et al., (2012) used marginal space learning to automatically segment the aorta and detect eight key aortic valve landmarks in a C-arm CT volume, extracting a real-time 3D aorta model that was superimposed on 2D fluoroscopic images during real TAVI cases.

Semiautomated plaque analysis

Beyond the detection of coronary stenosis, coronary CTA enables a comprehensive assessment of atherosclerotic plaque, with certain visually determined plaque characteristics conferring an increased risk of future cardiovascular events (Motoyama et al., 2015; Nerlekar et al., 2018; Williams et al., 2019). However, such qualitative measures are prone to interreader variability and a significant rate of false-positive findings (Min, Shaw, & Berman, 2010). The recent development of semiautomated plaque analysis software has enabled the accurate quantification and characterization of coronary plaque from CTA. Dey et al. (2009) used a multistep adaptive algorithm to segment the coronary arterial wall, followed by an iterative algorithm to classify noncalcified and calcified plaque components based on scan-specific voxel attenuation values. This method was validated against intravascular ultrasound, with excellent correlation for noncalcified plaque volumes ($R = 0.94$, $P < .001$) (Dey et al., 2010) as well as total plaque volume ($R = 0.943$, $P = .001$) (Matsumoto et al., 2019). Fig. 9.1 shows a case example of semiautomated plaque quantification that employs these algorithms.

Automated measurements

CAC assessed via CT is a direct marker of coronary atherosclerosis and a strong predictor of adverse cardiovascular events (Budoff et al., 2007; Budoff et al., 2018). Current clinical measurement of the CAC score is tedious and involves manually placing regions of interest around all coronary calcified lesions, defined as contiguous voxels above a CT attenuation threshold of 130 Hounsfield units (Agatston et al., 1990). In scans not dedicated to calcium scoring, this can be complicated by high noise, low resolution, and motion artifacts. AI algorithms applied to large datasets of ungated, noncontrast chest CTs acquired for lung cancer screening have been shown to accurately and automatically calculate the Agatston CAC score compared to the manual reference standard (Isgum, Prokop, Niemeijer, Viergever, & van Ginneken, 2012; Lessmann, Ginneken, JoZreikng, De Vos, & Viergever, 2017). Isgum et al. (2012) introduced the first ML method in which CAC lesions were described with features and subsequently classified using a k-nearest neighbor and support vector machine approach. Lessmann, Ginneken, JoZreikng, De Vos, & Viergever, (2017) applied DL with two consecutive CNNs: the first CNN used a

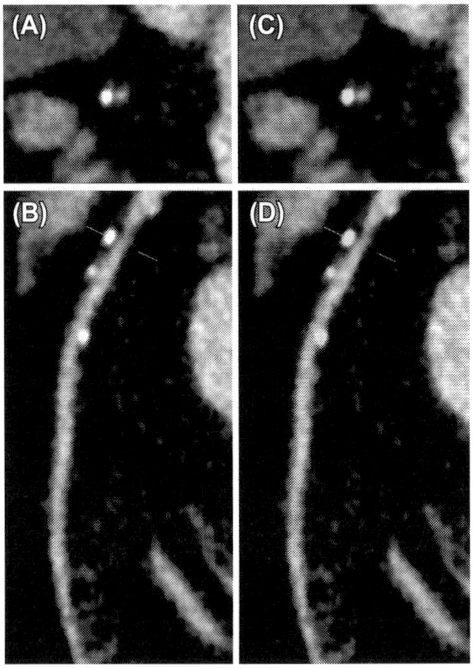

FIGURE 9.1 Automated coronary plaque quantification on coronary CT angiography.

Case example of plaque quantification using automated research software (Autoplaque). (A) Cross-section and (B) curved multiplanar view of mixed plaque in the proximal-to-mid left anterior descending artery. (C, D) Corresponding views following automated plaque analysis, whereby an iterative, recursive 3D region-growing algorithm classifies the noncalcified (red overlay) and calcified (yellow overlay) plaque components based on voxel attenuation values.

large receptive field that identified and labeled potential calcifications, and second CNN discarded false positive detections of the first network. More recent ML and DL approaches have enabled reliable and rapid automated Agatston score measurements from dedicated noncontrast CAC scoring CT scans (Martin et al., 2019; Wolterink, Leiner, Takx, Viergever, & Isgum, 2015). Wolterink et al. (2015) used randomized trees classifiers to identify calcified lesions and allocate them to a specific coronary artery; lesions that could not be classified with high certainty were presented for expert review. When tested on 530 CT images, automated quantification of CAC volume using this ML model demonstrated high agreement with expert readers (intraclass correlation coefficient = 0.95). Martin et al., (2019) trained a CNN with ResNet architecture to perform automated CAC scoring using 2000 labeled datasets, achieving high accuracy compared to expert manual measurements ($R = 0.97$) in 511 test scans. Several studies have also demonstrated the feasibility of automated CAC measurements on coronary CTA acquisitions (Mylonas et al., 2014;

Wolterink et al., 2016; Yang et al., 2016), the use of CTA images obviating the need for a dedicated noncontrast scan and hence reducing radiation dose.

The noninvasive detection and quantification of coronary stenosis is the most widely used clinical application of coronary CTA, with current reporting based on a subjective visual assessment by clinicians. Kelm et al. (2011) used an ML approach to automatically identify, grade, and classify coronary stenosis caused by all types of plaques. Following centerline extraction and lumen segmentation, their random forest model trained on 229 CTA volumes performed accurate stenosis identification and lumen cross-sectional area estimation, with an average processing time of 1.8 s per case. Kang et al. (2015) utilized a two-stage ML algorithm incorporating a support vector machine to automatically classify lesions as obstructive ($\geq$50% luminal narrowing) and nonobstructive (<50% luminal narrowing), with an accuracy of 94% compared to expert readers. With respect to DL approaches, Zreik, van Hamersvelt, Wolterink et al. (2019) trained a multitask recurrent CNN to reliably detect coronary plaque and determine its anatomic significance (no stenosis, obstructive, or nonobstructive). Further, Hong et al. (2019) used an M-Net-based CNN to automatically quantify percent diameter stenosis, minimal luminal area, and contrast density difference for 716 coronary lesions, demonstrating a strong correlation with an expert reader for all measurements.

The hemodynamic significance of a given coronary stenosis can also be assessed using AI techniques. A novel DL algorithm to calculate noninvasive fractional flow reserve (FFR), trained on a large dataset of synthetically generated coronary models, demonstrated similar accuracy and faster execution times (~2.4 s per case) compared to computationally demanding 3D flow simulations, with invasive FFR as the reference standard (Itu et al., 2016). This DL model was further validated against invasive FFR in a multicenter registry of 351 patients and 525 vessels (Coenen et al., 2018) and has the potential to deliver real-time, on-site FFR computations with high predictive power. Recently, an unsupervised DL technique to analyze a complete coronary artery from a coronary CTA-derived 3D volume and extract lower-dimensional encoding was shown to reliably detect functionally significant stenosis as defined by invasive FFR (Zreik et al., 2020).

Epicardial adipose tissue (EAT) is a metabolically active fat depot within the visceral pericardium that directly encases the coronary arteries. There is evidence that EAT exerts a local pathological effect on the coronary vasculature (Mazurek et al., 2003) and associates with adverse cardiovascular events (Cheng et al., 2010). Manual quantification of EAT volume from noncontrast CT is technically challenging and requires 10–15 min per case (Dey et al., 2008). Commandeur et al. (2018) used a multitask DL algorithm to perform rapid (<2 s per case), fully automated quantification of EAT volume from CAC scoring CT datasets (Fig. 9.2), with high accuracy compared to an expert reader ($R = 0.979$, $P < .001$). This model was validated in a large multicenter study of 850 patients scanned using different CT vendors and protocols (Commandeur, Goeller, et al., 2019), and integrated into research software for deployment (Fig. 9.3). Rapid, automated quantification of CAC and EAT has the potential for integration into routine clinical reporting of CAC scoring CT and may improve workflow and accuracy in cardiovascular risk assessment.

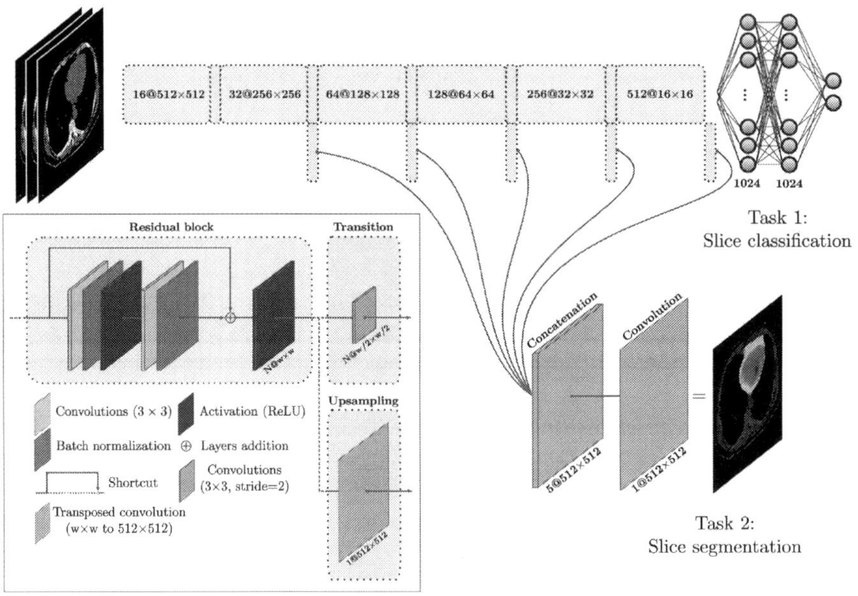

FIGURE 9.2 Deep learning-based quantification of epicardial adipose tissue.

In this deep learning model, a convolutional neural network is trained to perform intrapericardial segmentation. Each slice of a CT scan is given as input along with the immediately previous and following slices. The model performs a first task of classification to determine whether the input is located between the superior and inferior limits of the heart and then segments the pericardium as a second task. The epicardial adipose tissue in a whole scan is quantified by repeating this procedure for each slice and applying Hounsfield unit thresholding. *ReLU*: rectified linear unit.

Reproduced with permission from Commandeur, F., Goeller, M., Razipour, A., Cadet, S., Hell, M.M., Kwiecinski, J., et al 2019a, 'Fully automated CT quantification of epicardial adipose tissue by deep learning: A multicenter study', Radiology: Artificial Intelligence, vol. 1, no. 6, p. e190045.

Outcome prediction

ML integration of clinical and cardiac CT data has been shown to improve the prediction of cardiovascular outcomes. In 6814 participants from the MESA (Multi-Ethnic Study of Atherosclerosis), Ambale-Venkatesh et al. (2017) used a random survival forest technique to identify the top 20 predictors of cardiovascular outcomes among 735 variables from questionnaires, noncontrast cardiac CT scans, and serum biomarkers. They demonstrated the CAC score to be the most important predictor of all atherosclerotic cardiovascular disease (ASCVD) combined outcomes, and their ML model performed better than established risk scores with increased prediction accuracy. Han et al. (2019) showed an ML model combining clinical, laboratory,

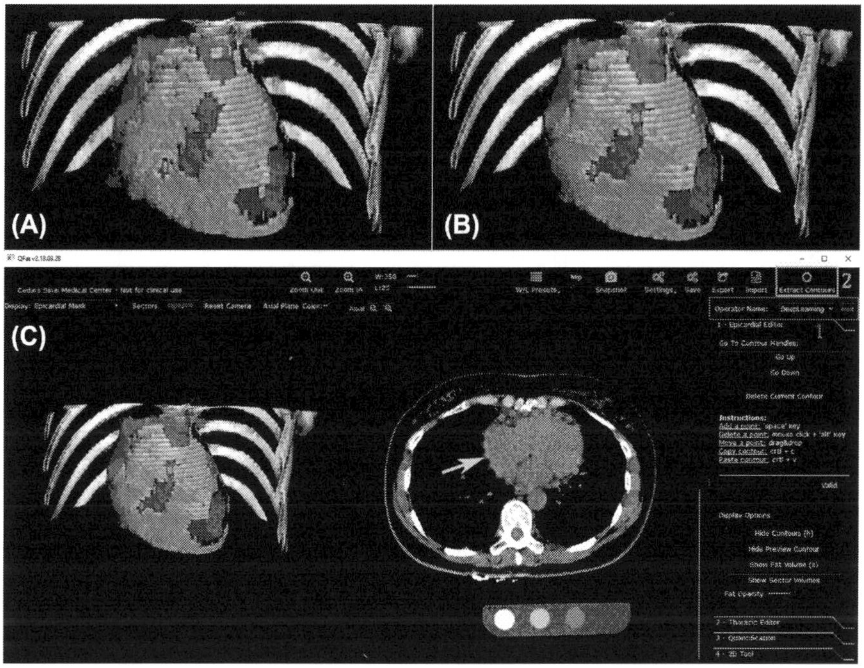

FIGURE 9.3 Deep learning algorithm integrated into research software.

Three-dimensional representations of epicardial adipose tissue (pink overlay), (A), as manually identified by an expert reader and, (B), as automatically identified by the algorithm. (C) Screenshot of QFAT software with the embedded deep learning approach. The pericardium is automatically identified by the algorithm (yellow *arrow*) by selecting a new operator (1, red) and by using the "extract contours" option (2, green).

Reproduced with permission from Commandeur, F., Goeller, M., Razipour, A., Cadet, S., Hell, M.M., Kwiecinski, J., et al 2019a, 'Fully automated CT quantification of epicardial adipose tissue by deep learning: A multicenter study', Radiology: Artificial Intelligence, vol. 1, no. 6, p. e190045.

and CAC score parameters to improve risk reclassification for all-cause mortality over conventional risk prediction approaches in an asymptomatic population undergoing health checkups. Commandeur et al. (2019) used a gradient-boosted ML algorithm for prognostication in 1912 asymptomatic subjects from the EISNER (Early Identification of Subclinical Atherosclerosis by Noninvasive Imaging Research) trial who underwent CAC scoring CT scans and 14-year follow-up. Their ML risk score integrating clinical variables, CAC score, and automated EAT measurements outperformed the ASCVD risk score and CAC score alone for long-term prediction of cardiac events. Fig. 9.4 demonstrates explainable, individualized risk prediction using this ML model in two EISNER subjects; such prediction breakdown can help clinical practice by personalizing therapeutic strategies (Commandeur et al., 2019).

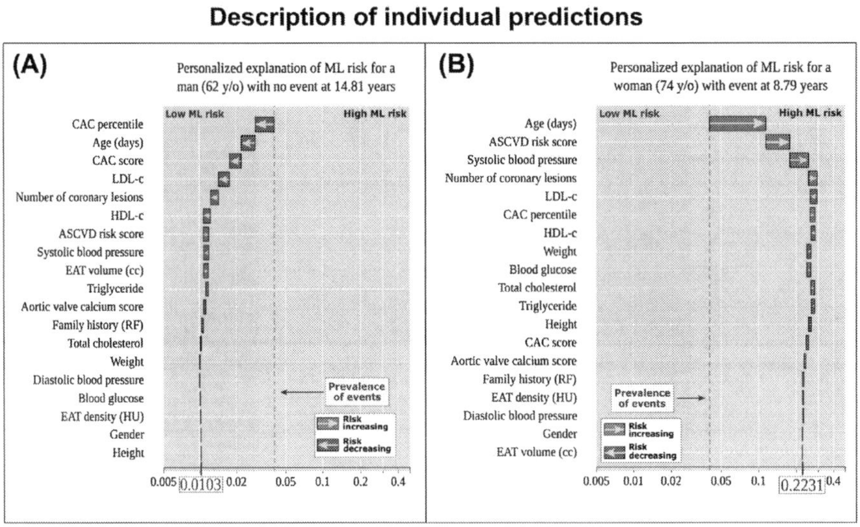

FIGURE 9.4 Explanations of individualized machine learning risk prediction.

Machine learning (ML) model for the prediction of myocardial infarction (MI) and cardiac death with subject-specific feature importance for a 62-year-old man with no event (A) and a 74-year-old woman with an MI (B). The X-axis corresponds to the ML risk score. The arrows represent the influence of each covariate on the overall prediction, either decreasing (blue) or increasing (red) the risk of future events. The combination of all covariates' influence provides the final ML risk score. Subject (A) has a low ML risk score (0.0105), while subject (B) has a high ML risk score (0.2206). The blue and red colors provide the separation between low and high ML risk as determined by Youden's index. The gray dashed lines represent the overall prevalence of events in the population (~4%). *ASCVD*: atherosclerotic cardiovascular risk; *CAC*: coronary artery calcium; *EAT*: epicardial adipose tissue; *HDL-c*: high-density lipoprotein cholesterol; *HU*: Hounsfield units; *LDL-c*: low-density lipoprotein cholesterol; *RF*: risk factor.

Reproduced with permission from Commandeur, F., Slomka, P.J., Goeller, M., Chen, X., Cadet, S., Razipour, A., et al 2019b, 'Machine learning to predict the long-term risk of myocardial infarction and cardiac death based on clinical risk, coronary calcium, and epicardial adipose tissue: A prospective study', Cardiovascular Research.

Coronary CTA measures can also enhance ML risk prediction models. In an analysis of 10,030 patients from the CONFIRM (Coronary CT Angiography EvaluatioN For Clinical Outcomes: An InteRnational Multicenter) registry, an ML model combining 25 clinical and 44 visually assessed coronary CTA parameters (based on lesion location, composition, and stenosis severity) predicted 5-year all-cause mortality with an area under the curve (AUC) of 0.79, significantly better than existing clinical or CTA metrics alone (Motwani et al., 2017) (Fig. 9.5). Further, a substudy of the NXT (Analysis of Coronary Blood Flow Using CT Angiography: Next Steps) trial showed ML risk score combining clinical variables and CTA-derived quantitative plaque metrics to improve the prediction of lesion-specific

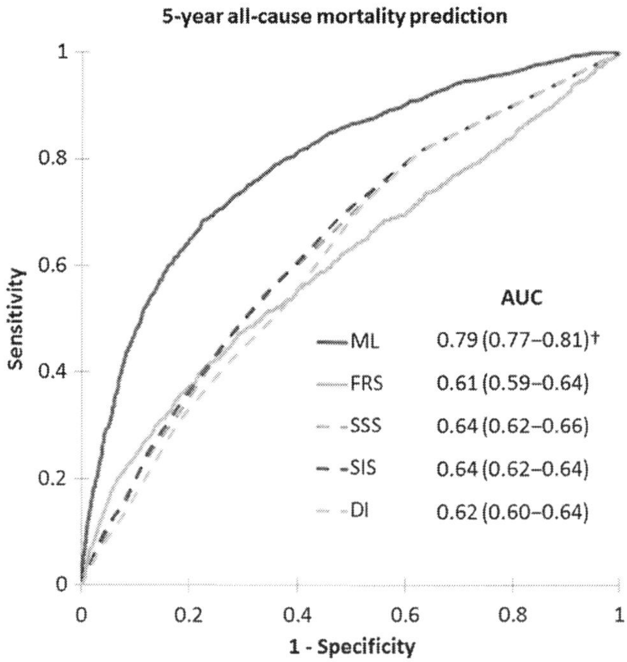

FIGURE 9.5 Machine learning prediction of all-cause mortality.

Receiver-operating characteristics curves for prediction of 5-year all-cause mortality: an ML risk score integrating clinical and coronary CTA data had a higher AUC than the Framingham Risk Score (FRS) and CTA severity scores (Segment Stenosis Score [SSS], Segment Involvement Score [SIS], Duke Index [DI]) ($P < .001$†). *AUC*: area under the curve; *CTA*: computed tomography angiography; *ML*: machine learning.

Reproduced with permission from Motwani, M., Dey, D., Berman, D.S., Germano, G., Achenbach, S., Al-Mallah, M.H., et al 2017, 'Machine learning for prediction of all-cause mortality in patients with suspected coronary artery disease: A 5-year multicentre prospective registry analysis', European Heart Journal, vol. 38, no. 7, pp. 500–507.

ischemia by invasive FFR with an AUC of 0.84, over pretest likelihood of coronary artery disease, stenosis severity (Fig. 9.6), or individual plaque measures (Dey et al., 2018).

Cardiac CT radiomics

Radiomics is a new field of quantitative image analysis in which digital medical images are converted into mineable high-dimensional data (Gillies, Kinahan & Hricak, 2016). From a given imaging region of interest, dedicated software is used to extract large numbers (typically >1000) of radiomic features not discernible to the human eye, which can be classified as shape-based, intensity-based, or texture-based (Kolossvary, Kellermayer, Merkely, & Maurovich-Horvat, 2018). ML techniques

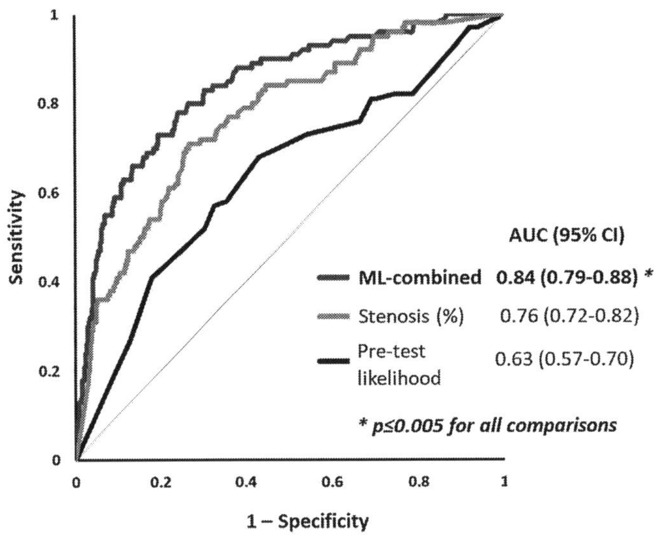

FIGURE 9.6 Machine learning prediction of lesion-specific ischemia.

The machine learning risk score (ML-combined) integrating clinical features and quantitative CTA measures had a significantly higher area under the curve (AUC) than quantitative percent stenosis and pretest likelihood of coronary artery disease for the prediction of lesion-specific ischemia by invasive fractional flow reserve.

Reproduced with permission from: Dey, D., Gaur, S., Ovrehus, K.A., Slomka, P.J., Betancur, J., & Goeller, M., et al. (2018). Integrated prediction of lesion-specific ischaemia from quantitative coronary CT angiography using machine learning: A multicentre study. European Radiology, 28(6), 2655–2664.

such as regression, random forests, and artificial neural networks can then be used to identify associations between the features and clinical or outcome data. Although the initial applications of radiomics were for texture analysis in cardiac magnetic resonance imaging, recent studies have applied this process to cardiac CT, specifically, coronary plaque characterization.

Coronary atherosclerotic plaques consist of distinct histological components with different CT attenuation values; each voxel is a separate measurement of the amount of radiation absorbed in the given volume. This has enabled the noninvasive assessment of plaque characteristics in vivo with coronary CTA. However, as previously discussed, the visual interpretation of plaque features is subject to interobserver variability, and current plaque quantification methods are based on absolute voxel attenuation values without considering the relationship between voxels. Radiomics can provide an objective, mathematical, or AI-based description of plaque that incorporates the spatial distribution of voxels. This approach has been shown to outperform conventional qualitative and quantitative parameters for the detection of high-risk plaque (Kolossváry et al., 2017; Kolossváry, Park, et al.,

2019). Moreover, in an ex vivo study, a radiomics-based ML model improved the discriminatory power of coronary CTA in identifying histologically determined advanced atherosclerotic lesions (Kolossváry, Karády, et al., 2019).

In summary, AI applications have shown great promise in facilitating cardiac CT image analysis, by automating image segmentation and quantification and assisting physician interpretation. Furthermore, the integration of clinical and imaging data using AI models improves cardiovascular risk stratification over conventional methods. Radiomic analysis is a new field that quantifies spatial and textural features from images and utilizes AI approaches to identify novel imaging biomarkers from big data. In the foreseeable future, AI may work in the background of cardiac CT image analysis software and routine clinical reporting, automatically collecting data and enabling real-time diagnosis and risk stratification.

Artificial intelligence in CT brain for stroke

Stroke is one of the major causes of morbidity and mortality worldwide (Koton et al., 2014). Rapid diagnosis and early intervention are critical for improving outcomes in patients with stroke (Jauch et al., 2013). In patients with stroke symptoms, imaging of the brain and craniocervical vessels using a CT or MRI scan is crucial for making diagnosis and treatment decisions (Latchaw et al., 2009; Wintermark et al., 2013). Compared to MRI, CT imaging has the advantage of being widely available and rapidly accessible in emergency settings. CT scanning of all patients with acute stroke as an initial imaging evaluation is reliable and cost-effective (Gonzalez, 2013; Young, Benesch, & Jahromi, 2010). Primary CT-based imaging methods for acute ischemic stroke include noncontrast CT, CT perfusion (CTP), and CTA. Recently, several AI techniques to analyze CT images have been used for the timely diagnosis and treatment of stroke.

Noncontrast CT

The noncontrast CT scan is an effective and rapid method of CT examination without contrast injection. It is the current standard for the initial imaging of patients who are suspected of having a stroke. The two different types of stroke, hemorrhagic and ischemic, have different treatment strategies. The biggest advantage of noncontrast CT is its ability to detect an intracranial hemorrhage to discriminate hemorrhagic stroke. DL-based AI applications can improve the detection of intracranial hemorrhage. The largest contemporary study consisting of 313,318 noncontrast CT scans evaluated a fully automated (DL) algorithm designed to detect time-sensitive abnormal findings, including intracranial hemorrhage (Chilamkurthy et al., 2018). The DL algorithm performed well in identifying abnormal findings of CT scans with sensitivity values of 82%–91%, specificity values of 90%–91%, and AUC ranging from 0.86 to 0.97. In another study that developed a DL-based prediction model to optimize noncontrast CT interpretation workflow, application of the prediction model demonstrated a 96% reduction in time to diagnosis of intracranial hemorrhage (Arbabshirani et al., 2018).

In the case of ischemic stroke, CT findings are often subtle compared to hemorrhagic stroke; thus, training and experience are required for accurate noncontrast CT evaluations. The Alberta Stroke Program Early CT Score (ASPECTS) is a 10-point quantitative score that measures the extent of early ischemic changes in brain noncontrast CT images (Barber, Demchuk, Zhang, & Buchan, 2000) and has been widely used diagnosis and prognosis of ischemic stroke. Current guidelines recommend ASPECTS as an imaging selection criterion for thrombectomy in patients who are suspected of having suffered an acute ischemic stroke (Powers et al., 2018). However, there is a limited interrater agreement, with several factors such as past stroke history, brain atrophy, leukoaraiosis, and reader experiences influencing the accurate estimation of the ASPECTS (Barber et al., 2000; Gupta et al., 2012). One potential utility of AI in noncontrast CT images for stroke diagnosis is its ability to perform rapid and standardized measurements of the ASPECTS. In an initial study, which assessed the capability of an AI algorithm to estimate the ASPECTS in 34 patients, the algorithm performed similar to an expert radiologist, with a sensitivity of 46% and specificity of 94% (Herweh et al., 2016). These findings are consistent in more recent studies, in which DL-based software showed improved efficiency, accuracy, and better agreement with radiologists (Goebel et al., 2018; Maegerlein et al., 2019). Additionally, in a recent study of 184 patients who underwent a mechanical thrombectomy, AI-derived ASPECTS demonstrated good correlation with quantitative assessments of infarct core and infarct volume by perfusion imaging (Olive-Gadea et al., 2019).

The hyperdense sign in noncontrast CT represents a marker of thromboembolus in patients with acute ischemic stroke. AI-based automated detection of the hyperdense vessel sign in emergency imaging may accelerate the identification of patients with ischemic stroke. This was demonstrated in a study that used an AI algorithm to detect the hyperdense middle cerebral artery (MCA) dot sign (Takahashi et al., 2014). The investigators employed an ML technique that consisted of (1) extraction of the Sylvian fissure region where the MAC dot sign was located, (2) feature extraction, and (3) classification using a support vector machine classifier that assessed a total 27 CT images in seven patients with the MCA dot sign verified by two neuroradiologists. The ML algorithm showed a noticeable sensitivity of 97.5% (39/40) and a low false-positive rate of 0.5 per image in the CT dataset.

CT angiography

CTA can effectively evaluate the presence and extent of intracranial vessel occlusion in patients with acute ischemic stroke. In addition, CTA enables the visualization of the anatomy of the proximal extracranial vessel and collateral flows, providing information to guide endovascular treatment. Although it is not challenging for expert readers to interpret CTA, the automated analysis of CTA imaging can be used to assist physicians by improving the accuracy and efficiency of their readings. In a study of 875 patients, where 46% of the subjects had large artery occlusions, automated CTA analysis based on a DL algorithm showed an accuracy of 86%, sensitivity of 90%, and specificity of 83% compared to the interpretations of experienced stroke neurologists (Barreira et al., 2018). In another study that analyzed 650 CTA scans that were performed to evaluate acute ischemic stroke, a DL algorithm using a convolutional neural network demonstrated a sensitivity of 82% and a specificity of 94% (Chatterjee, Somayaji, & Kabakis, 2019). Notably, the maximal running time of the automated CTA analysis was under 5–8 min (Barreira et al., 2018; Chatterjee, et al., 2019). As early diagnosis and treatment are crucial for reducing morbidity and mortality, automated CTA analysis can potentially be used for triaging or flagging patients with critical findings immediately after acquisition. The system can support physicians by decreasing the scan-to-notification time and accelerating rapid decision-making in stroke treatment.

Another application of AI algorithms to CTA images is to assess collateral flow. Collateral flow is defined as blood flow through a network of preexistent vascular anastomoses. Assessments of collateral flow based on CTA imaging can identify patients who may benefit from an endovascular thrombectomy (Berkhemer et al., 2016). The most commonly used tool for quantifying collateral flow is the Tan score (Tan et al., 2007). As is the case for many image-based scoring systems, the challenge to the implementation of the Tan score is interobserver variability and limited reproducibility. The application of the automated AI-based method to evaluate collateral scores may help overcome this limitation. Grunwald et al. tested a fully automated approach to collateral scoring using an ML algorithm (Grunwald et al., 2019). Automated software was used to generate e-CTA scores that were compared to manual collateral score assessments quantified by three experienced neuroradiologists. This ML approach provided real-time analysis of collateral scores, which agreed in 90% of the scans, with the remaining 10% falling within one point of the consensus (ICC 0.93, 0.90–0.95) compared to the radiologist's assessment. The sensitivity and specificity values of the automated algorithm for identifying favorable collateral flow (collateral score range 2–3) were 0.99 (0.93–1.00) and 0.94 (0.70–1.00), respectively. Automated rapid assessment of the collateral capacity of CTA has the potential to augment the triage of stroke patients for endovascular treatment (Boers et al., 2018; Grunwald et al., 2019).

CT perfusion

CTP is a series of 3D CT scans acquired after intravenous contrast injection, which demonstrates blood perfusion in the brain. Compared to noncontrast CT, CTP can identify earlier ischemic changes. Prior research evaluated estimates of CTP post-processing quantitative parameters including cerebral blood flow, cerebral blood volume, and mean transit time. These perfusion parameters help identify brain tissue characterizations in patients with acute ischemic stroke. Automated and operator-independent imaging processing tools have been used to quantify CTP parameters, providing rapid processing of quantitative CTP parameters within several minutes (Dehkharghani et al., 2015).

CTP enables the distinction of potentially salvageable ischemic tissue (penumbra) from irreversibly infarcted tissue (core). The accurate estimation of core and penumbra volume is an important point in CTP image analysis that helps determine the optimal treatment strategy. Core and penumbra volumes evolve over time: the core becomes larger, as the penumbral tissue becomes irreversibly damaged core. These changes depend on various clinical and imaging factors, such as the location of the infarct and the baseline size of the core and penumbral tissues. Prior research has demonstrated that ML algorithms can predict not only core and penumbra volumes, but also how the core volume changes over time (Lucas et al., 2018). In a study of 188 subjects with acute ischemic stroke, a fully convolutional neural network consisting of four inputs—(1) the CTP, (2) the CTP downsampled, (3) the arterial input function, and (4) metadata—learned to predict the final infarct volume with −2.3 mL mean error (Robben et al., 2020). This information may potentially help clinicians in predicting brain tissue status over time following an acute stroke event.

In summary, AI applications for CT-based stroke evaluation have performed well in rapid, automated, and standardized analysis of noncontrast CT, CTA, and CTP scans. This information can provide real-time support for clinical decision-making and individualized treatment strategies for patients with acute ischemic stroke. Future studies will need to determine the feasibility and efficacy of integrating these AI algorithms into the clinical workflow of stroke management.

Artificial intelligence in CT for cancer screening

Cancer is a group of diseases involving the uncontrolled growth and spread of abnormal cells in the body. Cancer remains the most prominent cause of morbidity and mortality, accounting for approximately 1.8 million new cancer cases in 2019; a number that is projected to rise to more than 2.3 million by 2030 (Siegel, Miller, & Jemal, 2019; Smith, Smith, Hurria, Hortobagyi, & Buchholz, 2009). As early detection of cancer can increase the chances of successful treatment and reduce mortality, there are many screening strategies available for the various types of cancer. Imaging with CT or MRI is widely available and has been shown to be an effective cancer screening method. However, most imaging findings rely on radiologists' subjective visual assessment. In addition, the clinical significance of screening findings is often uncertain, and differentiating between malignant tumors and benign mimics remains challenging. For the last two decades, computer-assisted diagnosis (CAD) has been used to assist physicians in improving the efficacy of imaging interpretation (Shiraishi, Li, Appelbaum, & Doi, 2011). Recently, AI algorithms have been applied to CAD for various cancer imaging such as lung, CNS, breast, and prostate cancer. Studies of CNS, breast, and prostate cancers are mostly performed using other imaging tools, such as MRI or mammography; however, CT tools are used exclusively for lung cancer screening. The following summary focuses on the role of AI in CT for lung cancer screening.

Lung nodule detection

Early detection of malignant lung nodules improves prognosis, and CT is a useful modality for detecting lung nodules. Current guidelines suggest that a low-dose CT scan (LDCT) is an appropriate and effective screening tool for selected high-risk populations (Tanoue, Tanner, Gould, & Silvestri, 2015). However, identifying lung nodules on CT can be difficult and time consuming. In addition, although a decrease in slice thickness of CT images increases the sensitivity of the lung nodule detection, the amount of imaging data is large (up to 500 slices in single scan) significantly increases the radiologist's workload. In this situation, AI-based CAD may be able to reduce workload and increase efficacy. The use of DL algorithms for lung nodule identification has shown the ability to detect lung nodules with high accuracy, sensitivity, and specificity. For example, a recent study of 12,754 CT scans demonstrated that a DL technique using CNNs showed increased sensitivity and specificity compared to manual reviews, regardless of patient age, gender, or radiation dose (Liu et al., 2019). The application of AI algorithms can enhance the manual identification of pulmonary nodules and reduce reading time.

Identification of high-risk imaging features

Pulmonary nodules are incidentally detected in more than 1.5 million adults in the United States each year (Gould et al. 2015). However, 96% of the pulmonary nodules identified in LDCT screens are not malignant lesions (Gould et al., 2015). Therefore, the identification of high-risk characteristics in LDCT is an important dimension for lung cancer screening. Recently, the American College of Radiology proposed the Lung Imaging Reporting and Data System (Lung-RADS) for standardizing CT interpretation based on nodule size, number, and morphological characteristics (Pinsky et al., 2015). Although recent studies using Lung-RADS have demonstrated its benefit in reducing the false positive rate compared to prior evaluation criteria (Kaminetzky et al., 2019), this system remains prone to interreader variability due to its reliance on visual assessment of nodule type and manual measurement of nodule size. To overcome this limitation, AI applications can potentially enhance the identification of high-risk nodules.

Radiomics is the process of extracting a large number of quantitative features from medical images and can be used to identify subtle imaging features of malignant lung nodules that cannot be detected by visual interpretation. Once the features have been extracted using dedicated software, a potential challenge for radiomics is to identify and prioritize stable radiomics features. Traditional ML methods can be used for feature selection from a large radiomic dataset (Parmar, Grossmann, Bussink, Lambin, & Aerts, 2015). In a study of 600 low-dose lung CT scans, a radiomics model with 23 textural features extracted from each lung nodule could predict potential malignant nodules with an accuracy of 80% (Hawkins et al., 2016). This study used ML algorithms, including a random forest classifier and correlation-based feature subset selection, to identify 23 stable and important radiomics features. The radiomics model showed better performance compared to other risk assessment models, including the Lung-RADS, in predicting malignant lesions. Another recent study that evaluated textural features of 209 lung nodules based on LDCT images also showed that a model based on 11 radiomics features demonstrated superior accuracy compared to Lung-RADS (89.9% vs. 76.5%, respectively) (Mao et al., 2019).

Recent advances in DL algorithms have demonstrated promising results for identifying, classifying, and quantifying patterns in medical imaging analysis (Shen, Wu, & Suk, 2017). CNNs are actively used for various purposes of medical imaging analysis, including the malignant risk prediction of lung nodules. For example, in a recent study of 42,290 CT scans from 14,851 patients wherein 578 subjects had biopsy-confirmed cancer within the 1-year follow-up period, researchers developed a CNN cancer risk prediction model (Ardila et al., 2019). The investigators constructed a 3D CNN full-volume model that was applied to whole LDCT volumes and a CNN region-of-interest (ROI) detection model to detect cancerous lesions in the LDCT volume. Following this, a CNN risk prediction model was developed that incorporated the full-volume model and the cancer ROI detection model to predict patient-level malignancy risk with cancer localization. The CNN risk prediction model outperformed radiologists' assessments with an absolute reduction of 11% in false positives and 5% in false negatives.

To summarize, a variety of AI applications have been used to enhance accuracy and efficiency in lung cancer CT screening. The application of such algorithms can assist in the manual identification of pulmonary nodules and reduce physician interpretation time.

Summary

In summary, with the increasing sophistication of ML and deep-learning techniques, AI-based algorithms can be utilized to improve medical imaging. When testing the performance of AI algorithms, the test data must be always kept separate from training data, to enable realistic estimates of performance. Further, external validation in data from unseen centers, not just hold-out data from the same center, can be expected to provide the most robust performance. Individualized risk prediction, which explains the AI findings for a specific patient, will be increasingly useful because they can personalize medical therapy. They will also be important for the acceptance of new AI approaches by physicians by overcoming the "black box" perception of AI. From the current data, AI-based algorithms can be expected to improve most aspects of CT imaging—personalized acquisition, reconstruction, image interpretation, and analysis and measurement of novel image features.

References

Agatston, A. S., Janowitz, W. R., Hildner, F. J., Zusmer, N. R., Viamonte, M., & Detrano, R. (1990). Quantification of coronary artery calcium using ultrafast computed tomography. *Journal of the American College of Cardiology, 15*(4), 827–832.

Ahn, C. K., Yang, Z., Heo, C., Jin, H., Park, B., & Kim, J. H. (2018). *A deep learning-enabled iterative reconstruction of ultra-low-dose CT: Use of synthetic sinogram-based noise simulation technique* (Vol. 10573). SPIE.

Al, W. A., Jung, H. Y., Yun, I. D., Jang, Y., Park, H.-B., & Chang, H.-J. (2018). Automatic aortic valve landmark localization in coronary CT angiography using colonial walk. *PloS One, 13*(7), e0200317 (e).

Ambale-Venkatesh, B., Yang, X., Wu, C. O., Liu, K., Hundley, W. G., McClelland, R., et al. (2017). Cardiovascular event prediction by machine learning: The multi-ethnic study of atherosclerosis. *Circulation Research, 121*(9), 1092–1101.

Arbabshirani, M. R., Fornwalt, B. K., Mongelluzzo, G. J., Suever, J. D., Geise, B. D., Patel, A. A., et al. (2018). Advanced machine learning in action: Identification of intracranial hemorrhage on computed tomography scans of the head with clinical workflow integration. *NPJ Digital Medicine, 1*, 9.

Ardila, D., Kiraly, A. P., Bharadwaj, S., Choi, B., Reicher, J. J., Peng, L., et al. (2019). End-to-end lung cancer screening with three-dimensional deep learning on low-dose chest computed tomography. *Nature Medicine, 25*(6), 954–961.

Barber, P. A., Demchuk, A. M., Zhang, J., & Buchan, A. M. (2000). Validity and reliability of a quantitative computed tomography score in predicting outcome of hyperacute stroke before thrombolytic therapy. ASPECTS Study Group. Alberta Stroke Programme Early CT Score. *Lancet, 355*(9216), 1670–1674.

Barreira, C., Bouslama, M., Lim, J., Al-Bayati, A., Saleem, Y., Devlin, T., et al. (2018). E-108 Aladin study: Automated large artery occlusion detection in stroke imaging study – a multicenter analysis. *Journal of Neurointerventional Surgery, 10*(Suppl. 2), A101–A102.

Bazrafkan, S., Van Nieuwenhove, V., Soons, J., De Beenhouwer, J., & Sijbers, J. (2019). *Deep neural network assisted iterative reconstruction method for low dose CT.*

Berkhemer, O. A., Jansen, I. G., Beumer, D., Fransen, P. S., van den Berg, L. A., Yoo, A. J., et al. (2016). Collateral status on baseline computed tomographic angiography and intra-arterial treatment effect in patients with proximal anterior circulation stroke. *Stroke, 47*(3), 768–776.

Bertram, M., Wiegert, J., Schafer, D., Aach, T., & Rose, G. (2009). Directional view interpolation for compensation of sparse angular sampling in cone-beam CT. *IEEE Transactions on Medical Imaging, 28*(7), 1011–1022.

Boers, A. M. M., Sales Barros, R., Jansen, I. G. H., Berkhemer, O. A., Beenen, L. F. M., Menon, B. K., et al. (2018). Value of quantitative collateral scoring on CT angiography in patients with acute ischemic stroke. *American Journal of Neuroradiology, 39*(6), 1074–1082.

Brooks, R. A., Weiss, G. H., & Talbert, A. J. (1978). A new approach to interpolation in computed tomography. *Journal of Computer Assisted Tomography, 2*(5), 577–585.

Bruns, S., Wolterink, J., van Hamersvelt, R., Zreik, M., Leiner, T., & Išgum, I. (2019). *Improving myocardium segmentation in cardiac CT angiography using spectral information* (Vol. 10949). SPIE.

Budoff, M. J., Shaw, L. J., Liu, S. T., Weinstein, S. R., Mosler, T. P., Tseng, P. H., et al. (2007). Long-term prognosis associated with coronary calcification: Observations from a registry of 25,253 patients. *Journal of the American College of Cardiology, 49*(18), 1860–1870.

Budoff, M. J., Young, R., Burke, G., Jeffrey Carr, J., Detrano, R. C., Folsom, A. R., et al. (2018). Ten-year association of coronary artery calcium with atherosclerotic cardiovascular disease (ASCVD) events: The multi-ethnic study of atherosclerosis (MESA)'. *European Heart Journal, 39*(25), 2401–2408.

Chatterjee, A., Somayaji, N. R., & Kabakis, I. M. (2019). Abstract WMP16: Artificial intelligence detection of cerebrovascular large vessel occlusion - nine month, 650 patient evaluation of the diagnostic accuracy and performance of the Viz.ai LVO algorithm. *Stroke, 50*(Suppl. 1_1). AWMP16-AWMP.

Cheng, V. Y., Dey, D., Tamarappoo, B., Nakazato, R., Gransar, H., Miranda-Peats, R., et al. (2010). Pericardial fat burden on ECG-gated noncontrast CT in asymptomatic patients who subsequently experience adverse cardiovascular events. *JACC Cardiovasc Imaging, 3*(4), 352–360.

Chen, H., Zhang, Y., Kalra, M., Lin, F., Chen, Y., Liao, P., et al. (2017a). Low-dose CT with a residual encoder-decoder convolutional neural network. *IEEE Transactions on Medical Imaging, 36*, 2524–2535.

Chen, H., Zhang, Y., Zhang, W., Liao, P., Li, K., Zhou, J., et al. (2017b). Low-dose CT via convolutional neural network. *Biomedical Optics Express, 8*(2), 679–694.

Chilamkurthy, S., Ghosh, R., Tanamala, S., Biviji, M., Campeau, N. G., Venugopal, V. K., et al. (2018). Deep learning algorithms for detection of critical findings in head CT scans: A retrospective study. *Lancet, 392*(10162), 2388–2396.

Coenen, A., Kim, Y. H., Kruk, M., Tesche, C., De Geer, J., Kurata, A., et al. (2018). Diagnostic accuracy of a machine-learning approach to coronary computed tomographic angiography-based fractional flow reserve: Result from the MACHINE consortium. *Circulation Cardiovascular Imaging, 11*(6), e007217.

Commandeur, F., Goeller, M., Betancur, J., Cadet, S., Doris, M., Chen, X., et al. (2018). Deep learning for quantification of epicardial and thoracic adipose tissue from non-contrast CT. *IEEE Transactions on Medical Imaging, 37*(8), 1835–1846.

Commandeur, F., Goeller, M., Razipour, A., Cadet, S., Hell, M. M., Kwiecinski, J., et al. (2019a). Fully automated CT quantification of epicardial adipose tissue by deep learning: A multicenter study. *Radiology: Artificial Intelligence, 1*(6), e190045.

Commandeur, F., Slomka, P. J., Goeller, M., Chen, X., Cadet, S., Razipour, A., et al. (2019b). Machine learning to predict the long-term risk of myocardial infarction and cardiac death based on clinical risk, coronary calcium, and epicardial adipose tissue: A prospective study. *Cardiovascular Research*. Dec 19:cvz321. https://doi.org/10.1093/cvr/cvz321

Cong, W., & Wang, G. (2017). *Monochromatic CT image reconstruction from current-integrating data via deep learning*.

Dehkharghani, S., Bammer, R., Straka, M., Albin, L. S., Kass-Hout, O., Allen, J. W., et al. (2015). Performance and predictive value of a user-independent platform for CT perfusion analysis: Threshold-derived automated systems outperform examiner-driven approaches in outcome prediction of acute ischemic stroke. *American Journal of Neuroradiology, 36*(8), 1419–1425.

Dey, D., Cheng, V. Y., Slomka, P. J., Nakazato, R., Ramesh, A., Gurudevan, S., et al. (2009). Automated 3-dimensional quantification of noncalcified and calcified coronary plaque from coronary CT angiography. *Journal of Cardiovascular Computed Tomography, 3*(6), 372–382.

Dey, D., Gaur, S., Ovrehus, K. A., Slomka, P. J., Betancur, J., Goeller, M., et al. (2018). Integrated prediction of lesion-specific ischaemia from quantitative coronary CT angiography using machine learning: A multicentre study. *European Radiology, 28*(6), 2655–2664.

Dey, D., Schepis, T., Marwan, M., Slomka, P. J., Berman, D. S., & Achenbach, S. (2010). Automated three-dimensional quantification of noncalcified coronary plaque from coronary CT angiography: Comparison with intravascular US. *Radiology, 257*(2), 516–522.

Dey, D., Suzuki, Y., Suzuki, S., Ohba, M., Slomka, P. J., Polk, D., et al. (2008). Automated quantitation of pericardiac fat from noncontrast CT. *Investigative Radiology, 43*(2), 145–153.

Donoho, D. L. (2006). Compressed sensing. *IEEE Transactions on Information Theory, 52*(4), 1289–1306.

Elattar, M. A., Wiegerinck, E. M., Planken, R. N., Vanbavel, E., van Assen, H. C., Baan, J., Jr., et al. (2014). Automatic segmentation of the aortic root in CT angiography of candidate patients for transcatheter aortic valve implantation. *Medical, Biological Engineering and Computing, 52*(7), 611–618.

Gantt, C., Jin, Y., & Lu, E. (2019). *Deep neural networks for sparse-view filtered backprojection imaging* (Vol. 10990). SPIE.

Geyer, L. L., Schoepf, U. J., Meinel, F. G., Nance, J. W., Jr., Bastarrika, G., Leipsic, J. A., et al. (2015). State of the art: Iterative CT reconstruction techniques. *Radiology, 276*(2), 339–357.

Gillies, R. J., Kinahan, P. E., & Hricak, H. (2016). Radiomics: Images are more than pictures, they are data. *Radiology, 278*(2), 563–577.

Gjesteby, L., Yang, Q., Xi, Y., Shan, H., Claus, B., Jin, Y., et al. (2017a). *Deep learning methods for CT image-domain metal artifact reduction* (Vol. 10391). SPIE.

Gjesteby, L., Yang, Q., Xi, Y., Zhou, Y., Zhang, J., & Wang, G. (2017b). *Deep learning methods to guide CT image reconstruction and reduce metal artifacts* (Vol. 10132). SPIE.

Goebel, J., Stenzel, E., Guberina, N., Wanke, I., Koehrmann, M., Kleinschnitz, C., et al. (2018). Automated ASPECT rating: Comparison between the frontier ASPECT score software and the brainomix software. *Neuroradiology, 60*(12), 1267–1272.

Gonzalez, R. G. (2013). Current state of acute stroke imaging. *Stroke, 44*(11), 3260–3264.

Gould, M. K., Tang, T., Liu, I. L., Lee, J., Zheng, C., Danforth, K. N., et al. (2015). Recent trends in the identification of incidental pulmonary nodules. *American Journal of Respiratory and Critical Care Medicine, 192*(10), 1208–1214.

Grbic, S., Ionasec, R., Vitanovski, D., Voigt, I., Wang, Y., Georgescu, B., et al. (2012). Complete valvular heart apparatus model from 4D cardiac CT. *Medical Image Analysis, 16*(5), 1003–1014.

Grunwald, I. Q., Kulikovski, J., Reith, W., Gerry, S., Namias, R., Politi, M., et al. (2019). Collateral automation for triage in stroke: Evaluating automated scoring of collaterals in acute stroke on computed tomography scans. *Cerebrovascular Diseases, 47*(5–6), 217–222.

Gülsün, M., Funka-Lea, G., Sharma, P., Rapaka, S., & Zheng, Y. (2016). *Coronary centerline extraction via optimal flow paths and CNN path pruning*.

Guo, Z., Bai, J., Lu, Y., Wang, X., Cao, K., Song, Q., et al. (2019). *DeepCenterline: A multi-task fully convolutional network for centerline extraction*. arXiv:1903.10481.

Gupta, H., Jin, K. H., Nguyen, H. Q., McCann, M. T., & Unser, M. (2018). CNN-based projected gradient descent for consistent CT image reconstruction. *IEEE Transactions on Medical Imaging, 37*(6), 1440–1453.

Gupta, A. C., Schaefer, P. W., Chaudhry, Z. A., Leslie-Mazwi, T. M., Chandra, R. V., Gonzalez, R. G., et al. (2012). Interobserver reliability of baseline noncontrast CT Alberta Stroke Program Early CT Score for intra-arterial stroke treatment selection. *American Journal of Neuroradiology, 33*(6), 1046–1049.

Halliburton, S., Arbab-Zadeh, A., Dey, D., Einstein, A. J., Gentry, R., George, R. T., et al. (2012). State-of-the-art in CT hardware and scan modes for cardiovascular CT. *Journal of Cardiovascular Computed Tomography, 6*(3), 154–163.

Han, D., Kolli, K. K., Gransar, H., Lee, J. H., Choi, S.-Y., Chun, E. J., et al. (2020). Machine learning based risk prediction model for asymptomatic individuals who underwent coronary artery calcium score: Comparison with traditional risk prediction approaches. *Journal of Cardiovascular Computed Tomography, 14*(2), 168–176.

Han, Y., Yoo, J., & Ye, J. C. (2016). *Deep residual learning for compressed sensing CT reconstruction via persistent homology analysis.*

Hawkins, S., Wang, H., Liu, Y., Garcia, A., Stringfield, O., Krewer, H., et al. (2016). Predicting malignant nodules from screening CT scans. *Journal of Thoracic Oncology, 11*(12), 2120–2128.

Herweh, C., Ringleb, P. A., Rauch, G., Gerry, S., Behrens, L., Mohlenbruch, M., et al. (2016). Performance of e-ASPECTS software in comparison to that of stroke physicians on assessing CT scans of acute ischemic stroke patients. *International Journal of Stroke, 11*(4), 438–445.

Hong, Y., Commandeur, F., Cadet, S., Goeller, M., Doris, M. K., Chen, X., et al. (2019). Deep learning-based stenosis quantification from coronary CT Angiography. *Proceedings of SPIE-Tthe International Society for Optical Engineering, 10949*, 109492I.

Isgum, I., Prokop, M., Niemeijer, M., Viergever, M. A., & van Ginneken, B. (2012). Automatic coronary calcium scoring in low-dose chest computed tomography. *IEEE Transactions on Medical Imaging, 31*(12), 2322–2334.

Itu, L., Rapaka, S., Passerini, T., Georgescu, B., Schwemmer, C., Schoebinger, M., et al. (2016). A machine-learning approach for computation of fractional flow reserve from coronary computed tomography'. *Journal of Applied Physiology, 121*(1), 42–52.

Jauch, E. C., Saver, J. L., Adams, H. P., Jr., Bruno, A., Connors, J. J., Demaerschalk, B. M., et al. (2013). Guidelines for the early management of patients with acute ischemic stroke: A guideline for healthcare professionals from the American heart association/American stroke association. *Stroke, 44*(3), 870–947.

Jin, K. H., McCann, M. T., Froustey, E., & Unser, M. (2017). Deep convolutional neural network for inverse problems in imaging. *IEEE Transactions on Image Processing, 26*(9), 4509–4522.

Kaminetzky, M., Milch, H. S., Shmukler, A., Kessler, A., Peng, R., Mardakhaev, E., et al. (2019). Effectiveness of lung-RADS in reducing false-positive results in a diverse, underserved, urban lung cancer screening cohort. *Journal of the American College of Radiology, 16*(4 Pt A), 419–426.

Kang, D., Dey, D., Slomka, P. J., Arsanjani, R., Nakazato, R., Ko, H., et al. (2015). Structured learning algorithm for detection of nonobstructive and obstructive coronary plaque lesions from computed tomography angiography. *Journal of Medical Imaging, 2*(1), 014003.

Kelm, B. M., Mittal, S., Zheng, Y., Tsymbal, A., Bernhardt, D., Vega-Higuera, F., et al. (2011). Detection, grading and classification of coronary stenoses in computed tomography angiography. *Medical Image Computing and Computer Assisted Intervention, 14*(Pt 3), 25–32.

Kolossváry, M., Karády, J., Kikuchi, Y., Ivanov, A., Schlett, C. L., Lu, M. T., et al. (2019a). Radiomics versus visual and histogram-based assessment to identify atheromatous lesions at coronary CT angiography: An ex vivo study. *Radiology, 293*(1), 89–96.

Kolossváry, M., Karády, J., Szilveszter, B., Kitslaar, P., Hoffmann, U., Merkely, B., et al. (2017). Radiomic features are superior to conventional quantitative computed tomographic metrics to identify coronary plaques with napkin-ring sign', *circulation. Cardiovascular imaging, 10*(12), e006843 (e).

Kolossvary, M., Kellermayer, M., Merkely, B., & Maurovich-Horvat, P. (2018). Cardiac computed tomography radiomics: A comprehensive review on radiomic techniques. *Journal of Thoracic Imaging, 33*(1), 26–34.

Kolossváry, M., Park, J., Bang, J.-I., Zhang, J., Paeng, J. C., Merkely, B., et al. (2019b). Identification of invasive and radionuclide imaging markers of coronary plaque vulnerability using radiomic analysis of coronary computed tomography angiography. *European heart journal cardiovascular Imaging, 20.*

Koton, S., Schneider, A. L., Rosamond, W. D., Shahar, E., Sang, Y., Gottesman, R. F., et al. (2014). Stroke incidence and mortality trends in US communities, 1987 to 2011. *Jama, 312*(3), 259–268.

Kudo, H., Suzuki, T., & Rashed, E. A. (2013). Image reconstruction for sparse-view CT and interior CT-introduction to compressed sensing and differentiated backprojection. *Quantitative Imaging in Medicine and Surgery, 3*(3), 147–161.

Latchaw, R. E., Alberts, M. J., Lev, M. H., Connors, J. J., Harbaugh, R. E., Higashida, R. T., et al. (2009). Recommendations for imaging of acute ischemic stroke: A scientific statement from the American heart association. *Stroke, 40*(11), 3646–3678.

Lessmann, N., Ginneken, B. M., JoZreikng, P., De Vos, B., Viergever, M., et al. (2017). Automatic calcium scoring in low-dose chest CT using deep neural networks with dilated convolutions. *IEEE Transactions on Medical Imaging*. arXiv:1711.00349.

Liu, K., Li, Q., Ma, J., Zhou, Z., Sun, M., Deng, Y., et al. (2019). Evaluating a fully automated pulmonary nodule detection approach and its impact on radiologist performance. *Radiology: Artificial Intelligence, 1*(3), e180084.

Lucas, C., Kemmling, A., Bouteldja, N., Aulmann, L. F., Madany Mamlouk, A., & Heinrich, M. P. (2018). Learning to predict ischemic stroke growth on acute CT perfusion data by interpolating low-dimensional shape representations. *Frontiers in Neurology, 9*, 989.

Maegerlein, C., Fischer, J., Monch, S., Berndt, M., Wunderlich, S., Seifert, C. L., et al. (2019). Automated calculation of the alberta stroke program early CT score: Feasibility and reliability. *Radiology, 291*(1), 141–148.

Manduca, A., Yu, L., Trzasko, J. D., Khaylova, N., Kofler, J. M., McCollough, C. M., et al. (2009). Projection space denoising with bilateral filtering and CT noise modeling for dose reduction in CT. *Medical Physics, 36*(11), 4911–4919.

Man, B. D., Nuyts, J., Dupont, P., Marchal, G., & Suetens, P. (1999). Metal streak artifacts in X-ray computed tomography: A simulation study. *IEEE Transactions on Nuclear Science, 46*(3), 691–696.

Mao, L., Chen, H., Liang, M., Li, K., Gao, J., Qin, P., et al. (2019). Quantitative radiomic model for predicting malignancy of small solid pulmonary nodules detected by low-dose CT screening. *Quantitative Imaging in Medicine and Surgery, 9*(2), 263–272.

Martin, S. S., van Assen, M., Rapaka, S., Hudson, H. T., Fischer, A. M., Varga-Szemes, A., et al. (2019). Evaluation of a deep learning–based automated CT coronary artery calcium scoring algorithm. *Journal of the American College of Cardiology: Cardiovascular Imaging*, 3233.

Matsumoto, H., Watanabe, S., Kyo, E., Tsuji, T., Ando, Y., Otaki, Y., et al. (2019). Standardized volumetric plaque quantification and characterization from coronary CT angiography: A head-to-head comparison with invasive intravascular ultrasound. *European Radiology, 26*(10), 019−06219.

Mazurek, T., Zhang, L., Zalewski, A., Mannion, J. D., Diehl, J. T., Arafat, H., et al. (2003). Human epicardial adipose tissue is a source of inflammatory mediators. *Circulation, 108*(20), 2460−2466.

McCollough, C. H., Leng, S., Yu, L., & Fletcher, J. G. (2015). Dual- and multi-energy CT: Principles, technical approaches, and clinical applications. *Radiology, 276*(3), 637−653.

Merkow, J., Marsden, A., Kriegman, D., & Tu, Z. (2016). *Dense volume-to-volume vascular boundary detection*. Springer International Publishing.

Min, J. K., Shaw, L. J., & Berman, D. S. (2010). The present state of coronary computed tomography angiography a process in evolution'. *Journal of the American College of Cardiology, 55*(10), 957−965.

Motoyama, S., Ito, H., Sarai, M., Kondo, T., Kawai, H., Nagahara, Y., et al. (2015). Plaque characterization by coronary computed tomography angiography and the likelihood of acute coronary events in mid-term follow-up. *Journal of the American College of Cardiology, 66*(4), 337−346.

Motwani, M., Dey, D., Berman, D. S., Germano, G., Achenbach, S., Al-Mallah, M. H., et al. (2017). Machine learning for prediction of all-cause mortality in patients with suspected coronary artery disease: A 5-year multicentre prospective registry analysis. *European Heart Journal, 38*(7), 500−507.

Mylonas, I., Alam, M., Amily, N., Small, G., Chen, L., Yam, Y., et al. (2014). Quantifying coronary artery calcification from a contrast-enhanced cardiac computed tomography angiography study'. *European Heart Journal Cardiovascular Imaging, 15*(2), 210−215.

Nerlekar, N., Ha, F. J., Cheshire, C., Rashid, H., Cameron, J. D., Wong, D. T., et al. (2018). Computed tomographic coronary angiography-derived plaque characteristics predict major adverse cardiovascular events: A systematic review and meta-analysis. *Circulation Cardiovascular Imaging, 11*(1), e006973.

Olive-Gadea, M., Martins, N., Boned, S., Carvajal, J., Moreno, M. J., Muchada, M., et al. (2019). Baseline ASPECTS and e-ASPECTS correlation with infarct volume and functional outcome in patients undergoing mechanical thrombectomy. *Journal of Neuroimaging, 29*(2), 198−202.

Park, H., Chung, Y., Lee, S. M., Kim, H., & Seo, J. (2017a). *Sinogram-consistency learning in CT for metal artifact reduction*.

Park, H., Lee, S. M., Kim, H., & Seo, J. (2017b). *Machine-learning-based nonlinear decomposition of CT images for metal artifact reduction*.

Parmar, C., Grossmann, P., Bussink, J., Lambin, P., & Aerts, H. (2015). Machine learning methods for quantitative radiomic biomarkers. *Scientific Reports, 5*, 13087.

Pelt, D. M., & Batenburg, K. J. (2014). Improving filtered backprojection reconstruction by data-dependent filtering. *IEEE Transactions on Image Processing, 23*(11), 4750−4762.

Pinsky, P. F., Gierada, D. S., Black, W., Munden, R., Nath, H., Aberle, D., et al. (2015). Performance of lung-RADS in the national lung screening trial: A retrospective assessment'. *Annals of Internal Medicine, 162*(7), 485−491.

Powers, W. J., Rabinstein, A. A., Ackerson, T., Adeoye, O. M., Bambakidis, N. C., Becker, K., et al. (2018). 2018 guidelines for the early management of patients with acute ischemic stroke: A guideline for healthcare professionals from the American heart association/ American stroke association. *Stroke, 49*(3), e46–e110.

Rampinelli, C., Origgi, D., & Bellomi, M. (2013). Low-dose CT: Technique, reading methods and image interpretation. Cancer Imaging *: The Official Publication of the International Cancer Imaging Society, 12*(3), 548–556.

Reports, C. (2015). *The surprising dangers of CT scans and X-rays.* viewed December 27th 2019 https://www.consumerreports.org/cro/magazine/2015/01/the-surprising-dangers-of-ct-sans-and-x-rays/index.htm.

Robben, D., Boers, A. M. M., Marquering, H. A., Langezaal, L., Roos, Y., van Oostenbrugge, R. J., et al. (2020). Prediction of final infarct volume from native CT perfusion and treatment parameters using deep learning. *Medical Image Analysis, 59*, 101589.

Ronneberger, O., Fischer, P., & Brox, T. (2015). *U-Net: Convolutional networks for biomedical image segmentation* (pp. 234–241). Springer International Publishing.

Shahzad, R., Bos, D., Budde, R. P., Pellikaan, K., Niessen, W. J., van der Lugt, A., et al. (2017). Automatic segmentation and quantification of the cardiac structures from non-contrast-enhanced cardiac CT scans. *Physics in Medicine and Biology, 62*(9), 3798–3813.

Shan, H., Padole, A., Homayounieh, F., Kruger, U., Khera, R. D., Nitiwarangkul, C., et al. (2019). Competitive performance of a modularized deep neural network compared to commercial algorithms for low-dose CT image reconstruction. *Nature Machine Intelligence, 1*(6), 269–276.

Shen, Y., Fang, Z., Gao, Y., Xiong, N., Zhong, C., & Tang, X. (2019). Coronary arteries segmentation based on 3D FCN with attention gate and level set function. *IEEE Access, 7*, 42826–42835.

Shen, D., Wu, G., & Suk, H. I. (2017). Deep learning in medical image analysis. *Annual Review of Biomedical Engineering, 19*, 221–248.

Shiraishi, J., Li, Q., Appelbaum, D., & Doi, K. (2011). Computer-aided diagnosis and artificial intelligence in clinical imaging. *Seminars in Nuclear Medicine, 41*(6), 449–462.

Siegel, R. L., Miller, K. D., & Jemal, A. (2019). Cancer statistics, 2019. *CA Cancer Journal of Clinicians, 69*(1), 7–34.

Singh, G., Al'Aref, S., Maliakal, G., van Rosendael, A., Lee, B., Wang, J., et al. (2019). Deep learning based automatic segmentation of cardiac computed tomography. *Journal of the American College of Cardiology, 73*, 1643.

Smith-Bindman, R., Lipson, J., Marcus, R., Kim, K. P., Mahesh, M., Gould, R., et al. (2009). Radiation dose associated with common computed tomography examinations and the associated lifetime attributable risk of cancer'. *Archives of Internal Medicine, 169*(22), 2078–2086.

Smith, B. D., Smith, G. L., Hurria, A., Hortobagyi, G. N., & Buchholz, T. A. (2009). Future of cancer incidence in the United States: Burdens upon an aging, changing nation. *Journal of Clinical Oncology, 27*(17), 2758–2765.

Takahashi, N., Lee, Y., Tsai, D. Y., Matsuyama, E., Kinoshita, T., & Ishii, K. (2014). An automated detection method for the MCA dot sign of acute stroke in unenhanced CT'. *Radiological Physics and Technology, 7*(1), 79–88.

Tan, J. C., Dillon, W. P., Liu, S., Adler, F., Smith, W. S., & Wintermark, M. (2007). Systematic comparison of perfusion-CT and CT-angiography in acute stroke patients. *Annals of Neurology, 61*(6), 533–543.

Tanoue, L. T., Tanner, N. T., Gould, M. K., & Silvestri, G. A. (2015). Lung cancer screening. *American Journal of Respiratory and Critical Care Medicine, 191*(1), 19–33.

Wang, J., Lu, H., Li, T., & Liang, Z. (2005). *Sinogram noise reduction for low-dose CT by statistics-based nonlinear filters* (Vol. 5747). SPIE.

Willemink, M. J., Leiner, T., de Jong, P. A., de Heer, L. M., Nievelstein, R. A., Schilham, A. M., et al. (2013). Iterative reconstruction techniques for computed tomography part 2: Initial results in dose reduction and image quality. *European Radiology, 23*(6), 1632–1642.

Willemink, M. J., Persson, M., Pourmorteza, A., Pelc, N. J., & Fleischmann, D. (2018). Photon-counting CT: Technical principles and clinical prospects. *Radiology, 289*(2), 293–312.

Williams, M. C., Moss, A. J., Dweck, M., Adamson, P. D., Alam, S., Hunter, A., et al. (2019). Coronary artery plaque characteristics associated with adverse outcomes in the SCOT-heart study. *Journal of the American College of Cardiology, 73*(3), 291.

Wintermark, M., Sanelli, P. C., Albers, G. W., Bello, J., Derdeyn, C., Hetts, S. W., et al. (2013). Imaging recommendations for acute stroke and transient ischemic attack patients: A joint statement by the American society of neuroradiology, the American College of radiology, and the society of NeuroInterventional surgery. *American Journal of Neuroradiology, 34*(11), E117–E127.

Wolterink, J. M., Leiner, T., de Vos, B. D., van Hamersvelt, R. W., Viergever, M. A., & Isgum, I. (2016). Automatic coronary artery calcium scoring in cardiac CT angiography using paired convolutional neural networks. *Medical Image Analysis, 34*, 123–136.

Wolterink, J. M., Leiner, T., & Išgum, I. (2019). *Graph convolutional networks for coronary artery segmentation in cardiac CT angiography*. Cham: Springer International Publishing.

Wolterink, J. M., Leiner, T., Takx, R. A., Viergever, M. A., & Isgum, I. (2015). Automatic coronary calcium scoring in non-contrast-enhanced ECG-triggered cardiac CT with ambiguity detection. *IEEE Transactions on Medical Imaging, 34*(9), 1867–1878.

Wolterink, J. M., Leiner, T., Viergever, M. A., & Isgum, I. (2017). Generative adversarial networks for noise reduction in low-dose CT'. *IEEE Transactions on Medical Imaging, 36*(12), 2536–2545.

Xie, S., Zheng, X., Chen, Y., Xie, L., Liu, J., Zhang, Y., et al. (2018). Artifact removal using improved GoogLeNet for sparse-view CT reconstruction. *Scientific Reports, 8*(1), 6700.

Xu, Z., Wu, Z., & Feng, J. (2018). *CFUN: Combining faster R-CNN and U-net network for efficient whole heart segmentation*.

Yang, G., Chen, Y., Ning, X., Sun, Q., Shu, H., & Coatrieux, J. L. (2016). Automatic coronary calcium scoring using noncontrast and contrast CT images. *Medical Physics, 43*(5), 2174.

Yang, Q., Yan, P., Kalra, M., & Wang, G. (2017). *CT image denoising with perceptive deep neural networks*.

Ye, C., Wang, W., Zhang, S., & Wang, K. (2019). Multi-depth fusion network for whole-heart CT image segmentation. *IEEE Access*, 1. vol. PP.

Young, K. C., Benesch, C. G., & Jahromi, B. S. (2010). Cost-effectiveness of multimodal CT for evaluating acute stroke. *Neurology, 75*(19), 1678–1685.

Zhang, Y., Chu, Y., & Yu, H. (2017). *Reduction of metal artifacts in x-ray CT images using a convolutional neural network* (Vol. 10391). SPIE.

Zhang, H., & Dong, B. (2019). *A review on deep learning in medical image reconstruction*.

Zhang, Z., Liang, X., Dong, X., Xie, Y., & Cao, G. (2018a). A sparse-view CT reconstruction method based on combination of DenseNet and deconvolution. *IEEE Transactions on Medical Imaging, 37*, 1.

Zhang, Y., & Yu, H. (2018). Convolutional neural network based metal artifact reduction in X-ray computed tomography. *IEEE Transactions on Medical Imaging, 37*(6), 1370–1381.

Zheng, Y., Barbu, A., Georgescu, B., Scheuering, M., & Comaniciu, D. (2008). Four-chamber heart modeling and automatic segmentation for 3-D cardiac CT volumes using marginal space learning and steerable features. *IEEE Transactions on Medical Imaging, 27*(11), 1668–1681.

Zheng, Y., John, M., Liao, R., Nottling, A., Boese, J., Kempfert, J., et al. (2012). Automatic aorta segmentation and valve landmark detection in C-arm CT for transcatheter aortic valve implantation. *IEEE Transactions on Medical Imaging, 31*.

Zreik, M., Leiner, T., De Vos, B., van Hamersvelt, R., Viergever, M., & Isgum, I. (2016). *Automatic segmentation of the left ventricle in cardiac CT angiography using convolutional neural networks*.

Zreik, M., van Hamersvelt, R. W., Wolterink, J. M., Leiner, T., Viergever, M. A., & Isgum, I. (2019). A recurrent CNN for automatic detection and classification of coronary artery plaque and stenosis in coronary CT angiography'. *IEEE Transactions on Medical Imaging, 38*(7), 1588–1598.

Zreik, M., van Hamersvelt, R. W., Khalili, N., Wolterink, J. M., Voskuil, M., Viergever, M. A., et al. (2020). Deep learning analysis of coronary arteries in cardiac CT angiography for detection of patients requiring invasive coronary angiography. *IEEE Transactions on Medical Imaging, 39*(5), 1545–1557.

Further reading

Wolterink, J. M., van Hamersvelt, R. W., Viergever, M. A., Leiner, T., & Isgum, I. (2019b). Coronary artery centerline extraction in cardiac CT angiography using a CNN-based orientation classifier. *Medical Image Analysis, 51*, 46–60.

Zhang, H., Wang, J., Zeng, D., Tao, X., & Ma, J. (2018b). Regularization strategies in statistical image reconstruction of low-dose x-ray CT: A review. *Medical Physics, 45*(10), e886–e907.

CHAPTER

Magnetic resonance imaging and artificial intelligence

10

Giuseppe Muscogiuri[1], Marco Guglielmo[1], Andrea Baggiano[1], Laura Fusini[1], Francesca Ricci[2], Gloria Cicala[3], Mark G. Rabbat[4,5], Andrea I. Guaricci[6], Gianluca Pontone[1]

[1]*Centro Cardiologico Monzino, IRCCS, Milan, Italy;* [2]*Department of Biomedicine and Prevention Division of Diagnostic Imaging, University of Rome "Tor Vergata," Rome, Italy;* [3]*Department of Medicine and Surgery, Azienda Ospedaliero-Universitaria di Parma, Università degli Studi, Parma, Italy;* [4]*Loyola University of Chicago, Chicago, IL, United States;* [5]*Edward Hines Jr. VA Hospital, Hines, IL, United States;* [6]*Institute of Cardiovascular Disease, Department of Emergency and Organ Transplantation, University Hospital Policlinico of Bari, Bari, Italy*

Chapter outline

Introduction

The diagnostic and prognostic role of artificial intelligence (AI) is rapidly increasing in medicine (Mintz & Brodie, 2019). Despite the increasingly prevalent notion that the application of AI in radiologic image interpretation will serve as a surrogate for physicians' diagnostic and prognostic role, AI should be considered as a decision-support tool (Tang et al., 2018). Indeed, AI-based algorithms can detect subtle changes within imaging datasets not depictable by human eyes and in a short time period (Tang et al., 2018). In the field of Radiology, AI has been applied to a multitude of organs ranging from neurovascular, abdominal, thoracic, and cardiovascular conditions (Mintz & Brodie, 2019). AI in advanced cardiovascular imaging is used in both computed tomography (CT) and magnetic resonance (MR). In cardiac CT, AI has been predominantly used for the evaluation of the coronary calcium

Machine Learning in Cardiovascular Medicine. https://doi.org/10.1016/B978-0-12-820273-9.00010-5

score and patient-level risk assessment (Singh et al., 2018; Varga-Szemes, Jacobs, & Schoepf, 2018). In cardiac MR (CMR), the application of AI has ranged from image reconstruction, image segmentation and interpretation, and thereafter prognostic stratification (Leiner, 2019; O'regan, 2020).

The aforementioned applications of AI in CMR can be helpful from numerous aspects, especially in relation to workflow optimization by shortening acquisition periods, image analysis and reporting of findings that would help with the institution of patient-specific and disease-specific therapy. In the next section of this chapter, we will provide an overview of the application of AI in CMR.

Image reconstruction with AI

In this era, CMR represents the most accurate and reproducible noninvasive cardiovascular imaging technique that allows for the evaluation of chamber volumes and function in addition to tissue characterization of ventricular walls (De Cecco, Muscogiuri, Varga-Szemes, & Schoepf, 2017). Although CMR can provide information with both high reproducibility and accuracy, it is characterized by long acquisition times and elaborate postprocessing, which pose unique challenges in certain cohorts such as pediatrics and uncooperative individuals (Muscogiuri, Suranyi et al., 2019b).

To overcome such a limitation, several strategies have been developed by different vendors. In particular, new sequences have been used for different acquisition and reconstruction of the k-space (Jeong et al., 2015; Lin et al., 2017; Okuda et al., 2015), to decrease acquisition time while preserving image quality compared to standard cine images. However, despite the quicker acquisition times with these new sequences, diagnostic image quality has been less than adequate (Greil et al., 2008; Muscogiuri et al., 2020; Tsao, Boesiger, & Pruessmann, 2003). Further, despite advances in sequencing techniques with improvements in acquisition times, the rate-limiting step has been image reconstruction. Here, the power of AI has been leveraged in an attempt to streamline this step. Schlemper, Caballero, Hajnal, Price, and Rueckert (2018) have applied a convolutional neural network (CNN) and used the data redundancy in undersampled MR imaging to shorten the time of reconstruction. In particular, the authors trained the CNN model using only 10 sampled short-axis loops that were progressively augmented using rigid transformation and elastic deformation. The authors showed that it was possible to reconstruct two-dimensional (2D) cine images nearly in real time with a time lapse of 23 ms (Schlemper et al., 2018). Quin et al. (2019) also explored the advantages of applying a CNN algorithm to image reconstruction combining the latter with iterative reconstruction allowing a potential reduction of reconstruction time. This CNN model evolves on two dimensions represented by time and iteration and could potentially denoise the images in a short time providing good images (Qin et al., 2019). Fig. 10.1 demonstrates how AI can contribute to image reconstruction in CMR.

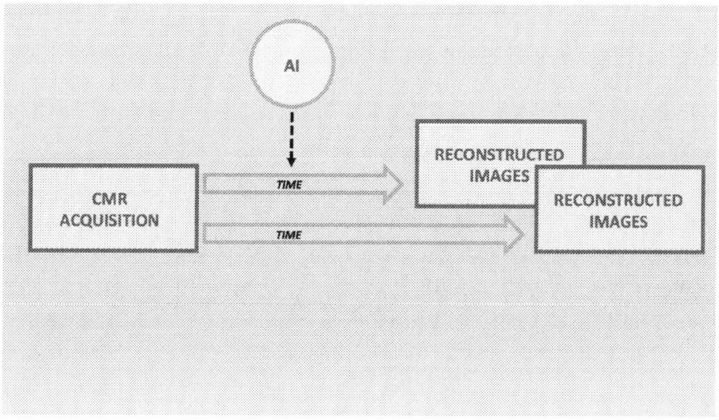

FIGURE 10.1

Illustration to highlight the advantage of artificial intelligence for image reconstruction. The application of artificial intelligence reduces the time of reconstruction as demonstrated in the *upper arrow*. The *second arrow* represents a significantly longer time in the absence of an artificial intelligence algorithm used for image reconstruction. *AI*, artificial intelligence; *CMR*, cardiac magnetic resonance.

Role of AI in cardiac MR analysis

Left and right ventricular function

Cine images represent an important aspect of CMR examinations. Acquiring cine images in long-axis and short-axis planes provides the ability to evaluate wall motion, chamber volumes and function, anatomy, and valvular disease with high reproducibility (Muscogiuri, Suranyi et al., 2019). Typically, the evaluation of volumes and function can be assessed after transferring the images to an off-line workstation and manually segmenting the images. Despite this approach being the main workhorse in clinical practice, it is fundamental to consider that this process is time consuming and even experienced cardiovascular imaging readers can spend somewhere between 20 and 30 min analyzing each case (Bai et al., 2018). New approaches using AI have been developed to optimize this part of the workflow, allowing faster analysis of biventricular volumes and function (Avendi, Kheradvar, & Jafarkhani, 2016; Curiale, Colavecchia, & Mato, 2019; Liao, Chen, Hu, & Song, 2019; Ma et al., 2016; Ngo, Lu, & Carneiro, 2017; Tan, Liew, Lim, & Mclaughlin, 2017).

The neural networks used for CMR image analysis had been developed on small but heterogeneous populations and comparisons have often been performed to establish the network with the best performance (Bernard et al., 2018). In particular, datasets used from the following challenges have been utilized for CMR-specific model development: MICCAI 2009, MICCAI 2011, MICCAI 2012, and the 2015 Kaggle Second Annual Data Science Bowl (Bernard et al., 2018). In these competitions, several networks were evaluated for the accuracy of delineating contours (Bernard et al., 2018). A network developed by Isensee et al. (2017) won the award for the best network during the joint conferences of the "Automatic Cardiac Diagnosis Challenge" (ACDC) and the 20th International Conference on Medical Image Computing and Computer-Assisted Intervention (MICCAI). The authors developed a 2D + three-dimensional (3D) U-net extracting a series of instants and dynamic features from the segmented maps while using an ensemble of 50 multilayer perceptrons and a random forest to perform classification (Isensee et al., 2017). The constructed network was shown to have good DICE scores for both end-diastolic volume and end-systolic volume in the right and left ventricles. In particular, the DICE score observed for the left ventricular cavity, right ventricular cavity, and myocardium in the end-diastolic phase was 0.96, 0.94, and 0.90, respectively; whereas, in end-systole, the same cardiac structures showed the following results: 0.93, 0.89, and 0.91, respectively (Bernard et al., 2018). From a diagnostic accuracy perspective, Khened, Varghese, and Krishnamurthi (2017) developed a DENSE U-Net model that showed a diagnostic accuracy of 0.96 and was superior to the neural network of Isensee, which comprised a 2D+3D U-net (Bernard et al., 2018). Despite the few cases analyzed in ACDC, the take-home message was stipulated by the ability to analyze biventricular volumes in both cardiac phases with good diagnostic accuracy (Bernard et al., 2018).

The aforementioned limitation of a small sample size used for the development of a neural network for image analysis was outdated by Bai and colleagues, who used the UK-biobank as the source of data (Bai et al., 2018). In particular, the authors obtained a dataset of 4875 patients containing short-axis and long-axis cine images in which a fully convolutional network (FCN) was trained (Bai et al., 2018). During the training phase, images were first annotated, subsequently exported in XML (eXtensible Markup Language) files and then the FCN was applied using pixelwise image segmentation. When defined, the final model of the FCN was applied and tested on 600 cases (Bai et al., 2018). Using this approach, the authors found a DICE metric on short-axis for the left ventricular cavity, left ventricular myocardium, and right ventricular cavity of 0.94, 0.88, and 0.90, respectively. Focusing on the evaluation of biventricular volumes while comparing FCN with manual segmentation, the authors showed a difference in terms of left end-diastolic volume, left end-systolic volume, left ventricular mass, right end-diastolic volume, and end-systolic volume of 6.1 mL, 5.3 mL, 6.9 g, 8.5mL, and 6.2 mL, respectively (Bai et al., 2018). These results highlight the utility of AI for cardiac chamber segmentation and evaluation with results comparable to that of manual segmentation (Bai et al., 2018). Even though various networks have been used for such a task, the main advantage of using AI is the optimization of workflow through a major reduction in processing times without compromising diagnostic accuracy. Fig. 10.2 represents a case example of the application of AI, compared to manual segmentation, for segmentation of ventricular wall and cavity on CMR.

Role of AI in tissue characterization

One of the main advantages of CMR is the high spatial resolution that allows for the ability to characterize myocardial tissue (De Cecco et al., 2017). In the past, tissue characterization was strictly confined to the evaluation of late gadolinium enhancement (LGE), which is acquired 10–15 min after the administration of a gadolinium-based contrast agent (Abdula et al., 2018; Kim et al., 2000), and precontrast images acquiring T1 or T2 weighted black blood images (Srichai et al., 2013; Stehling et al., 1996) (Fig. 10.3). Nevertheless, T1 or T2 weighted sequences have several limitations: T1 weighted images allow for a qualitative evaluation of tissue while T2 images may be influenced by surface coil position. Furthermore, it is not possible with T2 weighted black blood images to obtain a quantitative, specific measurement for the identification of edematous myocardium (De Cecco et al., 2017).

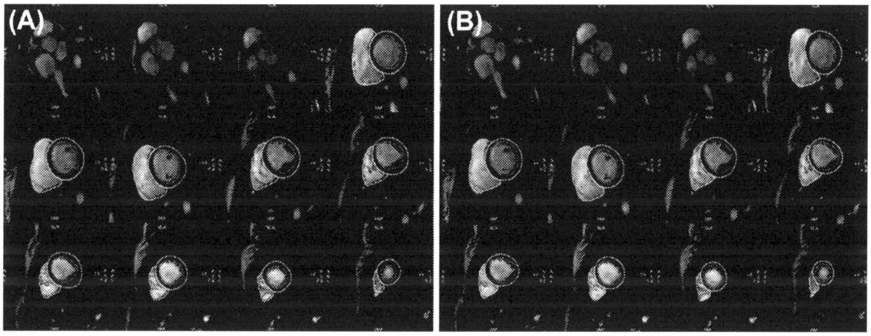

FIGURE 10.2

Right ventricular and left ventricular end-diastolic volume evaluation using a deep learning algorithm (A) and manual segmentation (B) in a 21-year-old patient who underwent cardiac magnetic resonance for ventricular arrhythmia. Both endocardial and epicardial contours are similar in systole and diastole using a deep learning protocol (A) and manual segmentation (B).

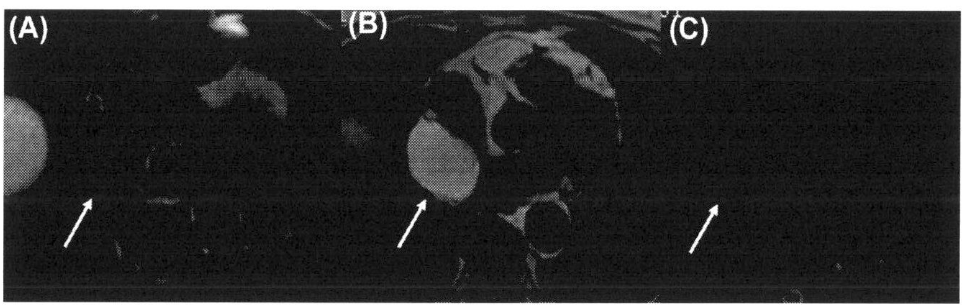

FIGURE 10.3

53 years old patient with cardiac lipoma. T1 (A) and T2 (C) sequences with fat suppression show hypointense signal intensity of cardiac lipoma (*arrow*) while T1 sequence shows the hyperintense signal intensity of lipoma (B).

LGE is typically acquired in clinical practice for the evaluation of replacement fibrosis (Muscogiuri et al., 2018). To acquire LGE, it is mandatory to administer a gadolinium-based contrast agent (Simonetti et al., 2001). Furthermore, image quality and evaluation of LGE can be influenced by several factors such as field inhomogeneity, inadequate nulling of remote myocardium, and choice of quantification method (Kim, Shah, & Judd, 2003). Recently, several sequences have been developed for the evaluation of extracellular volume to assess for diffuse or replacement fibrosis, edema, and infiltrative or storage diseases (Muscogiuri et al., 2018). In particular, T1 mapping and T2 mapping have been introduced in clinical practice for the evaluation of edema, fibrosis, or increased extracellular volume without the use of gadolinium-based contrast agents (De Cecco et al., 2017). In particular, T2 mapping is specific for the diagnosis of edema, while T1 mapping can depict edema as well as replacement or interstitial fibrosis (De Cecco et al., 2017). Based on the limitations described earlier, AI-driven approaches have aimed at increasing the diagnostic accuracy of sequences used for tissue characterization (Baessler, Mannil, Maintz, Alkadhi, & Manka, 2018; Zabihollahy, White, & Ukwatta, 2019; Zhang et al., 2019). For example, the application of AI algorithms in LGE is mainly focused on the identification and quantification of LGE. Zabihollay et al. (2019) developed a software that uses a 3D CNN-based approach to segment myocardial scar from a 3D LGE scam of the left ventricle. The authors analyzed a complete dataset of 34 patients, in whom 10 patients were used as a training set (Zabihollahy et al., 2019). They used the dice similarity coefficient (DSC) and the Jaccard index (JI) to evaluate the similarity between manual and AI segmentation. The DSC and JI were 93.63 ± 2.61% and 88.13 ± 4.70%, respectively, indicating an excellent overlap between manual and automatic segmentation. However, the evaluation of fibrosis volumes using CNN compared to manual full-width half max or standard deviation from normal tissue was statistically and significantly different (Zabihollahy et al., 2019).

Moccia et al. developed two algorithms based on a fully convolutional neural network (FCNN) for the detection of LGE in an ischemic cohort (Moccia et al., 2019). Compared to manual segmentation, the FCNN model, which functions through the detection of LGE in a predefined myocardial region, showed a sensitivity, specificity, accuracy, and DSC of 88.07%, 97.89%, 96.83%, and 71.25%, respectively (Moccia et al., 2019). Further, a recent investigation explored the possibility of using AI algorithms for the assessment of scar from noncontrast cine images on a selected population with previous myocardial infarction (Zhang et al., 2019). The authors developed a deep learning approach that evaluated for the presence of LGE on a single end-diastolic cine image and on an entire dataset of cine images. Compared to human segmentation, the best sensitivity, specificity, and accuracy were observed if the whole dataset was analyzed. In particular, the sensitivity, specificity, and area under the curve for the whole dataset of nonenhanced cine images in the overall population was 89.8%, 99.1%, and 0.94 in full nonenhanced cardiac MR versus 38.6%, 77.4%, and 0.58 if calculated in single nonenhanced cardiac MRI (Zhang et al., 2019). Focusing on quantitative analysis, the amount of infarct detected using the deep learning approach was not statistically significant from that of manual annotation: deep learning (myocardial infarcted area: $6.2 \pm 2.8\ cm^2$, myocardial infarcted area percentage: $21.5 \pm 17.3\%$) and manual segmentation (myocardial infarcted area: $5.5 \pm 2.3\ cm^2$, myocardial infarcted area percentage: $18.5 \pm 15.4\%$) (Zhang et al., 2019). Such an approach is intriguing and could have important clinical implications, especially when considering the possibility of fibrosis evaluation without the need to administer gadolinium while overcoming the T1 mapping limitations in terms of acquisition and reconstruction (Taylor, Salerno, Dharmakumar, & Jerosch-Herold, 2016).

Another interesting application of AI in terms of tissue characterization was developed by Baessler et al. using a combined approach of T1 weighted images, texture analysis, and machine learning (ML) on a cohort of hypertrophic cardiomyopathy patients (Baessler et al., 2018). The authors acquired T1 weighted images and applied a machine-learning algorithm to distinguish between patients with hypertrophic cardiomyopathy (HCM) and healthy controls (Baessler et al., 2018). Utilizing the gray level nonuniformity (GLevNonU) architecture, the authors found that using a cut off value ≥ 46 identified patients with HCM compared to normal patients with a sensitivity and specificity of 94% and 90%, respectively (Baessler et al., 2018). Considering that the GLevNonU would be biased by the presence of LGE, a possible abnormality of tissue characterization in patients with hypertrophic cardiomyopathy, the authors evaluated the performance of a combined approach with ML and texture analysis showing that in patients without LGE, it was possible to identify subtle changes with a sensitivity and specificity of 100% and 90%, respectively (Baessler et al., 2018).

AI for prognostic assessment

CMR is a useful noninvasive imaging tool that provides valuable information for prognosis and risk stratification (Chan et al., 2015; Halliday et al., 2019; Neilan et al., 2013). The combined evaluation of ejection fraction and tissue characterization provides information that can be useful for optimal therapy selection (Pontone et al., 2016). In patients with dilated cardiomyopathy, severe reduction of ejection fraction that is associated with the presence of LGE has been linked to worse outcomes, and as such the placement of an implantable cardioverter defibrillator is often considered (Disertori et al., 2016; Halliday et al., 2017; Pontone et al., 2016).

The application of ML algorithms for risk stratification has been seen in cohorts undergoing cardiac CT. Motwani and colleagues showed that an ML approach, built with a boosted ensemble algorithm and 10-fold stratified cross-validation, combining clinical data and quantitative atherosclerotic plaque measures, was able to better predict 5-year all-cause mortality compared to existing clinical or cardiac CT metrics alone (Motwani et al., 2017). In CMR, ML algorithms have also demonstrated incremental prognostic value in regards to risk assessment (Chen et al., 2019; Dawes et al., 2017; Guaricci, Masci, Lorenzoni, Schwitter, & Pontone,

2018). Dawes et al. (2017) evaluated the impact of an ML algorithm on patients with pulmonary hypertension. In particular, the algorithm used for the analysis was based on multiatlas approach with the entire dataset of labeled atlases (Aljabar, , Heckemann, Hammers, Hajnal, & Rueckert, 2009). The cohort comprised pulmonary hypertension patients who underwent CMR, right heart catheterization, and a 6-min walk test and was followed for up to 4 years (Dawes et al., 2017). The authors observed that an incorporated ML model was able to better identify patients with better outcomes (Dawes et al., 2017). The performance of an ML algorithm with an area under the curve (AUC) of 0.73 was able to identify patients with low and high risk in terms of survival (Dawes et al., 2017).

Recently, an article by Chen et al. studied the role of ML in predicting 1-year cardiovascular events in patients with dilated cardiomyopathy (Chen et al., 2019). The ML algorithm utilized a combination of metrics ranging from clinical data, pharmacotherapy, as well as imaging data, including tissue characterization (Chen et al., 2019). The authors compared the ML model with the standardized Meta-Analysis Global Group in Chronic Heart Failure (MAGGIC) score (Pocock et al., 2013) and ejection fraction. The ML approach was found to have an AUC of 0.88 compared to an AUC of 0.59 for the MAGGIC score and 0.50 for ejection fraction (Chen et al., 2019). This analysis was limited by a small sample size, while the ongoing DERIVATE study is recruiting a larger cohort of ischemic and nonischemic cardiomyopathy patients.

However, randomized trials are the next logical step with this innovative approach that could prove the effectiveness of an AI-based method on risk assessment. The ideal workflow for establishing the prognostic trajectory of individuals using ML is summarized in Fig. 10.4.

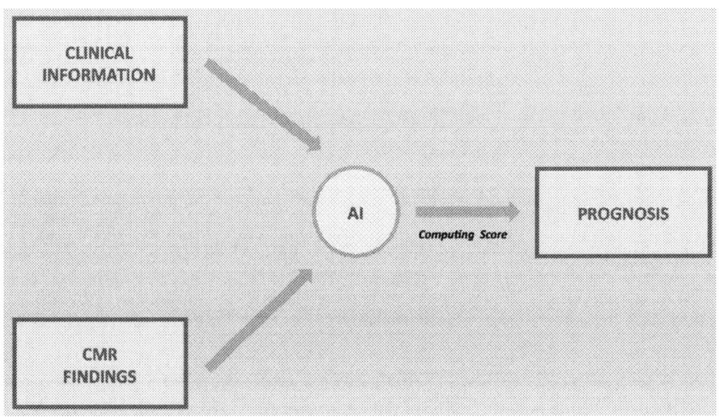

FIGURE 10.4

Illustration of the application of artificial intelligence for risk assessment. Clinical data (*upper box*) can be combined with cardiac magnetic resonance (CMR) findings to improve risk assessment.

Limitation and future directions

The major limitation of AI in cardiac MR is represented by the limited insight that is offered to clinicians regarding the inner working of various AI models, traditionally called as the "black box" aspect of AI. Furthermore it is well known that the development of a powerful AI algorithm requires large datasets for training and validation purposes. This aspect is quite challenging as each institution, site, and/or company has its own regulations and policies on data sharing and privacy. Nevertheless, the potential implications for an expanded role for AI within healthcare in general, and cardiac magnetic resonance in particular, promises to revolutionize the field, and future multi-disciplinary collaborations are likely to focus efforts and attempt at reaching solutions for current existent limitations.

Conclusions

The application of AI promises to revolutionize and advance the utility of CMR in clinical medicine and across healthcare systems. AI in CMR adds value in regards to image acquisition, reporting of findings, and contextualization of data (Fig. 10.5). At the time being, the application of AI in CMR includes the following:

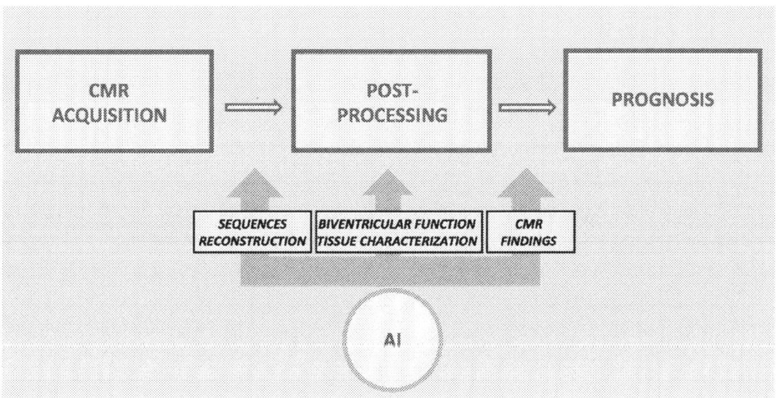

FIGURE 10.5

Application of AI during the process of image acquisition and reporting of cardiac magnetic resonance (CMR). The first application of AI could be observed between the CMR acquisition and postprocessing boxes; in this case, AI can be useful to speed up the time of image reconstruction. During image postprocessing AI can be useful for image analysis and in the evaluation of volumes and tissue characterization as can be observed in the middle box. Finally, the analysis of CMR findings with AI can be useful for prognostication of cardiovascular outcomes.

(1) Optimization of the workflow involving CMR image reconstruction and reporting.
(2) Extraction of tissue characteristics without the need for additional contrast agents.
(3) Improvement in prognostication of outcomes in select high-risk cohorts.

The ensuing few years promise to result in significant advances in the application of AI in the field of cardiac MR, with the ultimate goal of providing physicians with tools to optimize daily workflow while providing patients with specific and precise risk assessment.

References

Abdula, G., Nickander, J., Sorensson, P., Lundin, M., Kellman, P., Sigfridsson, A., et al. (2018). Synthetic late gadolinium enhancement cardiac magnetic resonance for diagnosing myocardial scar. *Scandinavian Cardiovascular Journal, 52*, 127–132.

Aljabar, P., Heckemann, R. A., Hammers, A., Hajnal, J. V., & Rueckert, D. (2009). Multi-atlas based segmentation of brain images: Atlas selection and its effect on accuracy. *NeuroImage, 46*, 726–738.

Avendi, M. R., Kheradvar, A., & Jafarkhani, H. (2016). A combined deep-learning and deformable-model approach to fully automatic segmentation of the left ventricle in cardiac MRI. *Medical Image Analysis, 30*, 108–119.

Baessler, B., Mannil, M., Maintz, D., Alkadhi, H., & Manka, R. (2018). Texture analysis and machine learning of non-contrast T1-weighted MR images in patients with hypertrophic cardiomyopathy-preliminary results. *European Journal of Radiology, 102*, 61–67.

Bai, W., Sinclair, M., Tarroni, G., Oktay, O., Rajchl, M., Vaillant, G., et al. (2018). Automated cardiovascular magnetic resonance image analysis with fully convolutional networks. *Journal of Cardiovascular Magnetic Resonance, 20*, 65.

Bernard, O., Lalande, A., Zotti, C., Cervenansky, F., Yang, X., Heng, P. A., et al. (2018). Deep learning techniques for automatic MRI cardiac multi-structures segmentation and diagnosis: Is the problem solved? *IEEE Transactions on Medical Imaging, 37*, 2514–2525.

Chan, R. H., Maron, B. J., Olivotto, I., Assenza, G. E., Haas, T. S., Lesser, J. R., et al. (2015). Significance of late gadolinium enhancement at right ventricular attachment to ventricular septum in patients with hypertrophic cardiomyopathy. *The American Journal of Cardiology, 116*, 436–441.

Chen, R., Lu, A., Wang, J., Ma, X., Zhao, L., Wu, W., et al. (2019). Using machine learning to predict one-year cardiovascular events in patients with severe dilated cardiomyopathy. *European Journal of Radiology, 117*, 178–183.

Curiale, A. H., Colavecchia, F. D., & Mato, G. (2019). Automatic quantification of the LV function and mass: A deep learning approach for cardiovascular MRI. *Computer Methods and Programs in Biomedicine, 169*, 37–50.

Dawes, T. J. W., De Marvao, A., Shi, W., Fletcher, T., Watson, G. M. J., Wharton, J., et al. (2017). Machine learning of three-dimensional right ventricular motion enables outcome prediction in pulmonary hypertension: A cardiac MR imaging study. *Radiology, 283*, 381–390.

De Cecco, C. N., Muscogiuri, G., Varga-Szemes, A., & Schoepf, U. J. (2017). Cutting edge clinical applications in cardiovascular magnetic resonance. *World Journal of Radiology, 9*, 1–4.

Disertori, M., Rigoni, M., Pace, N., Casolo, G., Mase, M., Gonzini, L., et al. (2016). Myocardial fibrosis assessment by LGE is a powerful predictor of ventricular tachyarrhythmias in ischemic and nonischemic LV dysfunction: A meta-analysis. *JACC Cardiovasc Imaging, 9*, 1046–1055.

Greil, G. F., Germann, S., Kozerke, S., Baltes, C., Tsao, J., Urschitz, M. S., et al. (2008). Assessment of left ventricular volumes and mass with fast 3D cine steady-state free precession k-t space broad-use linear acquisition speed-up technique (k-t BLAST). *Journal of Magnetic Resonance Imaging, 27*, 510–515.

Guaricci, A. I., Masci, P. G., Lorenzoni, V., Schwitter, J., & Pontone, G. (2018). CarDiac MagnEtic resonance for primary prevention implantable CardioVerter DebrillAtor ThErapy international registry: Design and rationale of the DERIVATE study. *International Journal of Cardiology, 261*, 223–227.

Halliday, B. P., Baksi, A. J., Gulati, A., Ali, A., Newsome, S., Izgi, C., et al. (2019). Outcome in dilated cardiomyopathy related to the extent, location, and pattern of late gadolinium enhancement. *JACC Cardiovasc Imaging, 12*(8(2)), 1645–1655. https://doi.org/10.1016/j.jcmg.2018.07.015.

Halliday, B. P., Gulati, A., Ali, A., Guha, K., Newsome, S., Arzanauskaite, M., et al. (2017). Association between midwall late gadolinium enhancement and sudden cardiac death in patients with dilated cardiomyopathy and mild and moderate left ventricular systolic dysfunction. *Circulation, 135*, 2106–2115.

Isensee, F., Jaeger, P., Full, P., Wolf, I., Engelhardt, S., & Maier-Hein, K. H. (2017). Automatic cardiac disease assessment on cine-MRI via time-series segmentation and domain specific features. In , *Lncs: 10663. STACOM-Miccai* (pp. 120–129).

Jeong, D., Schiebler, M. L., Lai, P., Wang, K., Vigen, K. K., & Francois, C. J. (2015). Single breath hold 3D cardiac cine MRI using kat-ARC: Preliminary results at 1.5T. *The International Journal of Cardiovascular Imaging, 31*, 851–857.

Khened, M., Varghese, A., & Krishnamurthi, G. (2017). Densely connected fully convolutional network for short-axis cardiac cine mr image segmentation and heart diagnosis using random fores. In , *Lncs: 10663. STACOM-Miccai* (pp. 140–151).

Kim, R. J., Shah, D. J., & Judd, R. M. (2003). How we perform delayed enhancement imaging. *Journal of Cardiovascular Magnetic Resonance, 5*, 505–514.

Kim, R. J., Wu, E., Rafael, A., Chen, E. L., Parker, M. A., Simonetti, O., et al. (2000). The use of contrast-enhanced magnetic resonance imaging to identify reversible myocardial dysfunction. *New England Journal of Medicine, 343*, 1445–1453.

Leiner, T. (2019). Deep learning for detection of myocardial scar tissue: Goodbye to gadolinium? *Radiology, 291*, 618–619.

Liao, F., Chen, X., Hu, X., & Song, S. (2019). Estimation of the volume of the left ventricle from MRI images using deep neural networks. *IEEE Transactions on Cybernetics, 49*, 495–504.

Lin, A. C. W., Strugnell, W., Riley, R., Schmitt, B., Zenge, M., Schmidt, M., et al. (2017). Higher resolution cine imaging with compressed sensing for accelerated clinical left ventricular evaluation. *Journal of Magnetic Resonance Imaging, 45*, 1693–1699.

Ma, Y., Wang, L., Ma, Y., Dong, M., Du, S., & Sun, X. (2016). An SPCNN-GVF-based approach for the automatic segmentation of left ventricle in cardiac cine MR images. *International Journal of Computer Assisted Radiology and Surgery, 11*, 1951–1964.

Mintz, Y., & Brodie, R. (2019). Introduction to artificial intelligence in medicine. *Minimally Invasive Therapy and Allied Technologies, 28*, 73–81.

Moccia, S., Banali, R., Martini, C., Muscogiuri, G., Pontone, G., Pepi, M., et al. (2019). Development and testing of a deep learning-based strategy for scar segmentation on CMR-LGE images. *Magma, 32*, 187–195.

Motwani, M., Dey, D., Berman, D. S., Germano, G., Achenbach, S., Al-Mallah, M. H., et al. (2017). Machine learning for prediction of all-cause mortality in patients with suspected coronary artery disease: A 5-year multicentre prospective registry analysis. *European Heart Journal, 38*, 500–507.

Muscogiuri, G., Gatti, M., DELL'Aversana, S., Andreini, D., Guaricci, A. I., Guglielmo, M., et al. (2020). Diagnostic accuracy of single-shot 2-dimensional multisegment late gadolinium enhancement in ischemic and nonischemic cardiomyopathy. *Journal of Thoracic Imaging, 35*(1), 56–63. https://doi.org/10.1097/RTI.0000000000000402.

Muscogiuri, G., Suranyi, P., Eid, M., Varga-Szemes, A., Griffith, L., Pontone, G., et al. (2019). Pediatric cardiac MR imaging: Practical preoperative assessment. *Magnetic Resonance Imaging Clinics of North America, 27*, 243–262.

Muscogiuri, G., Suranyi, P., Schoepf, U. J., De Cecco, C. N., Secinaro, A., Wichmann, J. L., et al. (2018). Cardiac magnetic resonance T1-mapping of the myocardium: Technical background and clinical relevance. *Journal of Thoracic Imaging, 33*, 71–80.

Neilan, T. G., Coelho-Filho, O. R., Danik, S. B., Shah, R. V., Dodson, J. A., Verdini, D. J., et al. (2013). CMR quantification of myocardial scar provides additive prognostic information in nonischemic cardiomyopathy. *JACC Cardiovasc Imaging, 6*, 944–954.

Ngo, T. A., Lu, Z., & Carneiro, G. (2017). Combining deep learning and level set for the automated segmentation of the left ventricle of the heart from cardiac cine magnetic resonance. *Medical Image Analysis, 35*, 159–171.

Okuda, S., Yamada, Y., Tanimoto, A., Fujita, J., Sano, M., Fukuda, K., et al. (2015). Three-dimensional cardiac cine imaging using the kat ARC acceleration: Initial experience in clinical adult patients at 3T. *Magnetic Resonance in Imaging, 33*, 911–917.

O'regan, D. P. (2020). Putting machine learning into motion: Applications in cardiovascular imaging. *Clinical Radiology, 75*(1), 33–37. https://doi.org/10.1016/j.crad.2019.04.00.

Pocock, S. J., Ariti, C. A., Mcmurray, J. J., Maggioni, A., Kober, L., Squire, I. B., et al. (2013). Predicting survival in heart failure: A risk score based on 39 372 patients from 30 studies. *European Heart Journal, 34*, 1404–1413.

Pontone, G., Guaricci, A. I., Andreini, D., Solbiati, A., Guglielmo, M., Mushtaq, S., et al. (2016). Prognostic benefit of cardiac magnetic resonance over transthoracic echocardiography for the assessment of ischemic and nonischemic dilated cardiomyopathy patients referred for the evaluation of primary prevention implantable cardioverter-defibrillator therapy. *Circ Cardiovasc Imaging, 9*.

Qin, C., Schlemper, J., Caballero, J., Price, A. N., Hajnal, J. V., & Rueckert, D. (2019). Convolutional recurrent neural networks for dynamic MR image reconstruction. *IEEE Transactions on Medical Imaging, 38*, 280–290.

Schlemper, J., Caballero, J., Hajnal, J. V., Price, A. N., & Rueckert, D. (2018). A deep cascade of convolutional neural networks for dynamic MR image reconstruction. *IEEE Transactions on Medical Imaging, 37*, 491–503.

Simonetti, O. P., Kim, R. J., Fieno, D. S., Hillenbrand, H. B., Wu, E., Bundy, J. M., et al. (2001). An improved MR imaging technique for the visualization of myocardial infarction. *Radiology, 218*, 215–223.

Singh, G., AL'Aref, S. J., Van Assen, M., Kim, T. S., Van Rosendael, A., Kolli, K. K., et al. (2018). Machine learning in cardiac CT: Basic concepts and contemporary data. *Journal of Cardiovascular Computed Tomography, 12*, 192–201.

Srichai, M. B., Lim, R. P., Lath, N., Babb, J., Axel, L., & Kim, D. (2013). Diagnostic performance of dark-blood T2-weighted CMR for evaluation of acute myocardial injury. *Investigative Radiology, 48*, 24–31.

Stehling, M. K., Holzknecht, N. G., Laub, G., Bohm, D., Von Smekal, A., & Reiser, M. (1996). Single-shot T1- and T2-weighted magnetic resonance imaging of the heart with black blood: Preliminary experience. *Magma, 4*, 231–240.

Tang, A., Tam, R., Cadrin-Chenevert, A., Guest, W., Chong, J., Barfett, J., et al. (2018). Canadian association of radiologists white paper on artificial intelligence in radiology. *Canadian Association of Radiologists Journal, 69*, 120–135.

Tan, L. K., Liew, Y. M., Lim, E., & Mclaughlin, R. A. (2017). Convolutional neural network regression for short-axis left ventricle segmentation in cardiac cine MR sequences. *Medical Image Analysis, 39*, 78–86.

Taylor, A. J., Salerno, M., Dharmakumar, R., & Jerosch-Herold, M. (2016). T1 mapping: Basic techniques and clinical applications. *JACC Cardiovasc Imaging, 9*, 67–81.

Tsao, J., Boesiger, P., & Pruessmann, K. P. (2003). k-t BLAST and k-t SENSE: Dynamic MRI with high frame rate exploiting spatiotemporal correlations. *Magnetic Resonance in Medicine, 50*, 1031–1042.

Varga-Szemes, A., Jacobs, B. E., & Schoepf, U. J. (2018). The power and limitations of machine learning and artificial intelligence in cardiac CT. *Journal of Cardiovascular Computed Tomography, 12*, 202–203.

Zabihollahy, F., White, J. A., & Ukwatta, E. (2019). Convolutional neural network-based approach for segmentation of left ventricle myocardial scar from 3D late gadolinium enhancement MR images. *Medical Physics, 46*, 1740–1751.

Zhang, N., Yang, G., Gao, Z., Xu, C., Zhang, Y., Shi, R., et al. (2019). Deep learning for diagnosis of chronic myocardial infarction on nonenhanced cardiac cine MRI. *Radiology, 291*, 606–617.

CHAPTER

Nuclear imaging and artificial intelligence

11

Benjamin C. Lee[1], Felix Y.J. Keng[2], Subhi J. Al'Aref[3], Lohendran Baskaran[1,2]

[1]*Dalio Institute of Cardiovascular Imaging, Weill Cornell Medicine/NewYork-Presbyterian, New York, NY, United States;* [2]*Department of Cardiovascular Medicine, National Heart Centre, Singapore;* [3]*Division of Cardiology, Department of Medicine, University of Arkansas for Medical Sciences, Little Rock, AR, United States*

Chapter outline

Myocardial perfusion imaging (MPI) has aided in diagnosing coronary artery disease (CAD) particularly in assessing its severity, manifest as ischemic heart disease (IHD). Single-photon emission computed tomography (SPECT) is the most widely available and validated nuclear medicine imaging modality (Fig. 11.1D). Positron emission tomography (PET) on the other hand produces higher resolution images (Fig. 11.1E) and has a higher diagnostic accuracy than SPECT and cardiovascular magnetic resonance imaging (MRI) (Jaarsma et al., 2012). Both nuclear medicine imaging modalities begin with injecting the patient with a radiopharmaceutical that is perfused into or taken up by healthy myocardium, from which the emitted radiation is measured by the sensors of the imaging camera.

Image reconstruction of nuclear medicine images was first performed by the filtered backprojection (FBP) algorithm, an analytical formulation of inverting the projection data in the camera-collected sinograms (Fig. 11.1A and B) to the original image. However, FBP does not account for the inherent Poisson noise model and therefore iterative reconstruction methods were developed to better model the

Machine Learning in Cardiovascular Medicine. https://doi.org/10.1016/B978-0-12-820273-9.00011-7

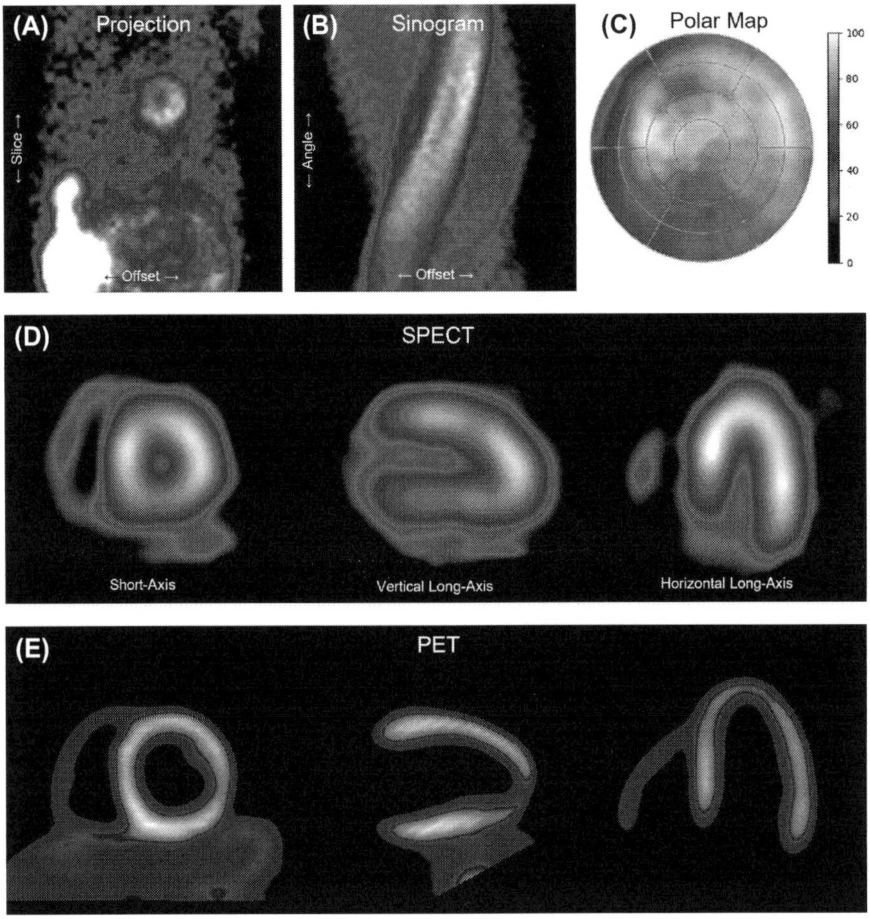

FIGURE 11.1 Nuclear medicine images

Example measurement data from a SPECT camera shows (A) a single angular projection of the torso and (B) a single slice sinogram through the heart showing the perfusion activity in the left ventricular (LV) myocardium. Filtered backprojection reconstructed ^{99m}Tc-sestamibi SPECT images of the LV are shown in (D) three multiplanar reformation cardiac views and (C) the LV myocardium segmented polar map representation with a 17-segment overlay. (E) 3D iterative reconstructed ^{13}N-ammonia PET images also in 3 cardiac views show the LV myocardium at a higher resolution.

Images from each modality are not necessarily from the same patient.

physical process and camera system thus reducing noise and artifacts in the reconstructed image.

Image analysis begins with the segmentation of the left ventricular (LV) myocardium using polar sampling to get the maximum count circumferential profile of each short-axis slice (Garcia et al., 1985). The result is a bulls eye-shaped 2D polar map image (Fig. 11.1C). For diagnosing abnormal perfusion quantitatively, the standard method is a statistical test comparing the polar map perfusion values against a distribution derived from a database of normal perfusion images (Van Train et al., 1993). Existing nuclear cardiology quantitative software packages (Ficaro, Lee, Kritzman & Corbett, 2007; Garcia et al., 2007; Germano et al., 2007) provide this and other measurements of perfusion and function that can provide a second opinion to a clinician but can also be further be used as inputs in artificial intelligence (AI) algorithms (Garcia, Klein & Taylor, 2014).

Artificial intelligence in cardiovascular nuclear medicine imaging has been applied as early as 1985 using reason-based semantic nets and fuzzy logic (Niemann et al., 1985). Various other AI methods such as rule-based expert systems, artificial neural networks, and structured machine learning (ML) have been applied in parallel timelines for the diagnosis of CAD. Image analysis using AI has mostly been performed on SPECT images in nuclear medicine likely due to the larger availability of images required by machine learning techniques.

Recent developments of deep learning algorithms such as the convolutional neural network (CNN) and availability of faster hardware such as graphics processing units have contributed to AI-based image reconstruction of nuclear medicine images, mostly for PET images due to it being a newer and higher resolution modality. These deep learning-based techniques aim to decrease reconstruction time, improve image quality, and reduce radiation dose to the patient.

Artificial intelligence in cardiac SPECT and PET image analysis

Application of AI methods for cardiovascular SPECT and PET image analysis to diagnose IHD began with expert systems using a knowledge base and associated rules. Learning-based methods soon followed with artificial neural networks. Due to heavy computational requirements to train a neural network, other machine learning methods such as support vector machines and decision tree-based methods gained more popularity. Attention to neural networks was later renewed due to hardware improvements and new deep learning algorithms in the form of convolutional neural networks. As most of the research in AI for medical image analysis has been to diagnose or predict similar disease indicators or clinical endpoints, the following topics are organized by AI algorithms: nonlearning-based expert systems, structured traditional machine learning, shallow artificial neural networks, and deep convolutional neural networks. The methods reviewed focused on cardiovascular imaging, which were mostly from SPECT systems.

Nonlearning expert systems

Expert systems are programmed with known facts and their inferences to emulate the decision-making process and performance of a human expert. The first type is case-based reasoning, which is a process of making decisions based on past experiences. It finds the most similar prior problem in the knowledge base, based on a similarity metric on relevant features, and applies that solution to the current problem. Haddad, Moertl and Porenta (1995), Haddad, Adlassnig and Porenta (1997) detected obstructive CAD, as determined by more than 70% diameter stenosis or luminal narrowing of the artery from invasive coronary angiography (ICA), by feature matching ^{201}Thallium (Tl) SPECT MPI polar map perfusion values from a 100-patient case library. The tolerances for polar map value similarity were determined beforehand by expert physician readers. The diagnostic accuracy (sensitivity = 98% and specificity = 70%) of their system showed the feasibility of mimicking the decision process of humans based on experience. In a follow-up study that increased the number of patients to 240, the performance of the case-based reasoning method was not superior to the quantitative normal database method nor expert visual interpretation and in fact performed worse in detecting CAD in the left circumflex (Khorsand et al., 2001).

A more common type of expert system employs rule-based reasoning, which encodes knowledge from experts into "if-then" association rules. Although rule-based reasoning can be more difficult to implement compared to case-based reasoning, rule-based expert systems are typically more accurate. The earliest SPECT application of a rule-based expert system was introduced in 1985, where Niemann et al. (1985) extracted LV myocardial contours and pixel values from gated blood pool SPECT images, input them into a complex semantic network, a knowledge base that connects concepts through a graph structure, and detected abnormal wall motion.

Later, Cios, Goodenday, Shah and Serpen (1996) employed a combination of rule-based reasoning and a decision tree algorithm to detect ischemia, a reduction of blood flow to the myocardium, from LV segmented values from 99 SPECT images with a resulting performance equivalent to those of an expert physician. Kurgan, Cios, Tadeusiewicz, Ogiela and Goodenday (2001) approached the same problem on 267 SPECT images by utilizing data mining, the process of finding patterns in a large dataset, to generate heuristic rules for the diagnosis of CAD. Sacha, Goodenday and Cios (2002) also approached this same problem on 728 SPECT images using a Bayesian learning data mining technique, which applies conditional probabilities from one variable to the next.

Garcia et al. (2001) reported the diagnostic performance of a rule-based expert system, called PERFEX, part of the Emory Cardiac Toolbox software, in detecting the presence and location of CAD as determined by ICA in 655 rest/stress SPECT myocardial perfusion images. PERFEX was trained on the location, size, shape, and reversibility from 70% of the data to develop 253 heuristic rules. Testing for the presence of CAD on the remaining 30% of the data resulted in equivalent performance (sensitivity = 83% and specificity = 73%) to visual interpretation by expert physicians.

To validate the association rules in PERFEX and discover new rules to predict heart disease, Ordonez et al. (2001) used data mining in 655 patients with 113 attributes that included perfusion estimates and high-risk factors. Applying constraints on the results of the data mining algorithm to limit complexity, irrelevancy, and redundancy reduced the number of rules by tenfold (Ordonez, Ezquerra & Santana, 2006). Garcia et al. (2014) highlighted one of the discovered rules as: "If the patient has a perfusion defect and the patient has had previous carotid surgery, then the patient has a high probability of CAD." In 2018, Garcia et al. (2018) validated an artificial intelligence-driven structured reporting system for diagnostic accuracy of detecting CAD and ischemia in rest/stress SPECT myocardial perfusion images from 1000 patients. The inputs into this heuristic-rule based expert system were perfusion and functional parameters over the 17 LV segments that were converted to nonparametric certainty factors instead of parametric comparisons to a normal database. Comparing the final natural language structured report to expert physicians' diagnoses showed equivalent performance.

Traditional machine learning

In contrast to expert systems, learning-based systems, such as supervised machine learning, develop a model of the data only guided by the labeled output as ground truth. The learnable weights or coefficients of a model are iteratively updated in small steps that eventually reduce the error between the model's prediction and the ground truth output. Any machine learning method that inputs features, hand-engineered by a domain expert from raw data, is considered traditional machine learning. As such, models that input structured or tabular data fall into this category. Specifically, models that are not deep neural networks also fall into this category, like support vector machines (SVMs), decision tree-based ensemble methods, and shallow artificial neural networks. Linear regression and logistic regression, borrowed from statistics, are also part of machine learning because they learn their coefficients from data.

An SVM is a classifier that separates two classes with a linear boundary that maximizes the margin between the boundary line and the closest data points from each class, known as support vectors (Fig. 11.2A). SVMs can also generate nonlinear class boundaries by warping the feature space into higher dimensions to make the two classes separable by a linear boundary. An SVM with the kernel trick replaces the expensive coordinate transformation with a simpler distance computation to the decision boundary.

Decision tree-based ensemble methods combine multiple decision trees and are categorized into either bagging or boosting ensembles. Bagging, a contraction of bootstrap aggregating, either averages or majority votes on the outputs of individual models that were trained in parallel on different subsets of the training data. Random Forest is an example that uses bagging on individual decision trees (Fig. 11.2B).

Boosting is a method that sequences individual weak learning models that perform slightly better than random guessing, with increased weights iteratively placed on the previously incorrectly predicted training cases, thus converting the whole series of weak learning models to a stronger ensembled learning model.

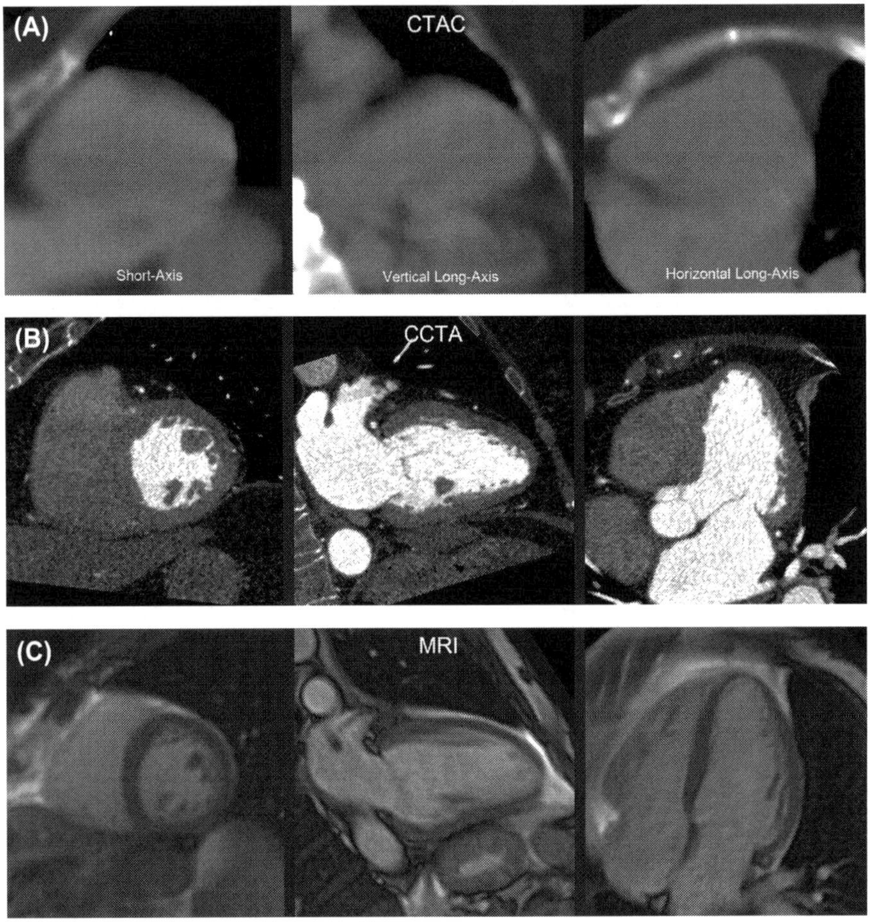

FIGURE 11.2 Nonnuclear medicine modality images

(A) Computed tomography (CT) attenuation correction images (CTAC), shown in the 3 cardiac views, are lower resolution CT images used to correct attenuation artifacts during PET and some SPECT image reconstructions. Deep learning algorithms can generate synthetic CTAC images from (C) cardiac magnetic resonance images (MRIs) in PET/MR systems. Due to the higher resolution, anatomical structures measured from cardiovascular CT angiography (CCTA) images are used as the ground truth for machine learning-based basal valve plane placement in SPECT images.

Images from each modality are not necessarily from the same patient.

AdaBoost, short for adaptive boosting, was the first boosting algorithm to use weak learners like decision stumps, decision trees with a single split. LogitBoost reframed boosting probabilistically using the logistic loss function (Fig. 11.2C) instead of the exponential loss of AdaBoost. Therefore, LogitBoost is less sensitive to noisy data and outliers compared to AdaBoost. Gradient Boosting, also called gradient boosting machine or gradient tree boosting, is a generalization of AdaBoost, where each new weak learning model added reduces the overall model error using gradients to provide the next optimization direction. XGBoost, which stands for Extreme Gradient Boosting, is a gradient boosting algorithm with many algorithmic improvements that reduce overfitting, run an order of magnitude faster than its predecessors, and are well suited on large datasets.

Both bagging and boosting reduce the variance of the output, producing more stable models. They differ in that bagging is better suited for reducing overfitting while boosting can reduce bias and increase accuracy.

In the early 2010s, preliminary work on SPECT images used SVMs to detect abnormal perfusion (Gaspar-Cunha, 2010), to diagnose ischemia more accurately than a hybrid decision tree and rule-based algorithm (Ciecholewski, 2011, 2012), and to assess cardiac motion (Marin et al., 2011). Then, Arsanjani, Xu, Dey, Fish et al. (2013) showed that using an SVM with inputs of (a) total perfusion deficit (TPD), which quantifies both the total perfusion defect extent and severity, (b) stress and rest changes in ischemia, and (c) ejection fraction (EF) derived from 957 SPECT myocardial perfusion images, performed equivalently or better in diagnosing obstructive CAD compared to physician readers using ICA as the ground truth. Betancur et al. (2017) used an SVM to estimate the mitral valve plane location based on the 3D LV surfaces from 392 SPECT images. These results were comparable against physician readers using expert placement of the valve plane on the paired coronary computed tomography (CT) angiography (CCTA) images as the ground truth (Fig. 11.3B). Haro Alonso et al. (2019) used an SVM, along with tree-based AdaBoost and Random Forest, and sparse solution-encouraging LASSO, which stands for least absolute shrinkage and selection operator, to predict cardiac death given up to 122 clinical predictors from 8321 patients with SPECT images. They concluded that the ML algorithms were more effective than a baseline logistic regression model, with SVM having the highest area under the receiver operating characteristic curve (AUC) of 0.83 compared to the baseline logistic regression AUC of 0.76. Ricciardi et al. (2020) employed data mining techniques with various machine learning algorithms on stress SPECT images from 10,265 patients to estimate abnormal perfusion via the summed stress score (SSS) and LV function via ejection fraction. Using SSS and EF computed from a quantitative software package as the ground truth, the Random Forest and Gradient Boosting algorithms achieved the highest performances (AUCs > 0.99), compared to nonensemble learning models of a decision tree, K-nearest neighbor, and Naïve Bayes.

Arsanjani, Xu, Dey, Vahistha et al. (2013) used LogitBoost on 1181 SPECT images with TPD, stress and rest changes in perfusion, transient ischemic dilation, age, sex, and other features to detect obstructive CAD determined by ICA. The LogitBoost model achieved an AUC of 0.94 and performed no worse than using only TPD or visual assessment of CAD by expert readers. Juarez-Orozco et al. (2020) also used LogitBoost but on 16 demographic, clinical, and function variables derived from 1234 ^{13}N-ammonia PET images to detect myocardial ischemia (AUC $= 0.72$) and elevated risk of major adverse cardiovascular events (MACE) (AUC $= 0.71$).

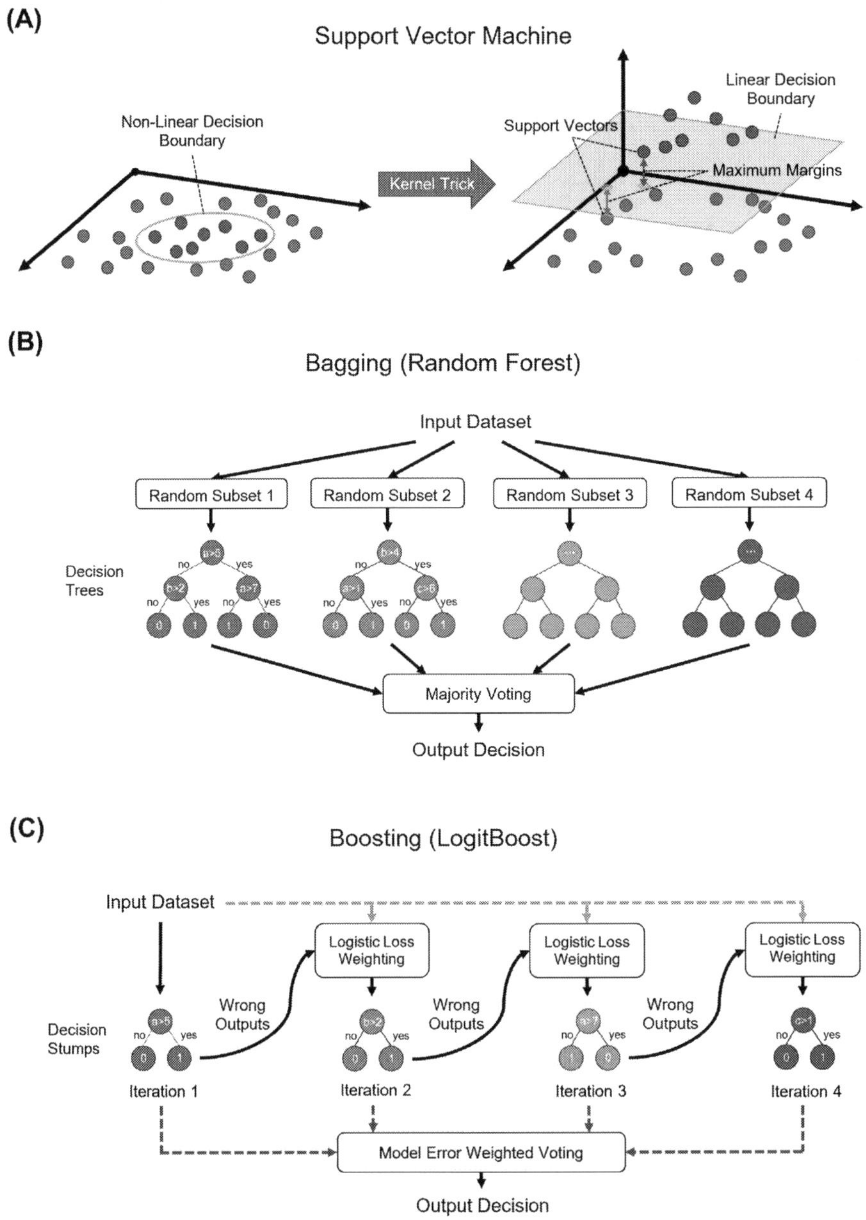

FIGURE 11.3 Traditional machine learning algorithms used in SPECT

(A) Support vector machines, (B) bagging ensemble learning, and (C) boosting ensemble learning are structured-data machine learning algorithms used for automated diagnosis of coronary artery disease using features derived from SPECT images. (A) Support vector machines using the kernel trick can convert a nonlinear decision boundary such as the

LogitBoost has also been used to predict outcomes in other studies. Arsanjani et al. (2015) again used LogitBoost to predict a revascularization event based on 33 clinical, quantitative, and imaging features derived 713 rest/stress SPECT images. The LogitBoost model was equivalent to or better than ischemic TPD and expert readers. Betancur, Otaki et al. (2018) also used LogitBoost to predict MACE based on 28 clinical, 17 stress test, and 25 imaging variables derived from 2619 stress SPECT images. The LogitBoost model performed better than the prediction from just TPD or expert readers.

REFINE SPECT (REgistry of Fast Myocardial Perfusion Imaging with NExt generation SPECT) is a registry of prognostic data with 20,418 scans from 5 sites and diagnostic data with 2079 scans from 9 sites (Slomka et al., 2018) with several outcomes predictive results using traditional statistical analysis (Han et al., 2020; Otaki et al., 2019). However, using data from REFINE SPECT, Hu et al. (2019) used LogitBoost to predict early coronary revascularization based on 18 clinical, 9 stress tests, and 28 imaging features derived from 1980 rest/stress SPECT images. The LogitBoost model performed better than TPD and expert readers at both the per-patient and per-vessel levels.

Shallow artificial neural networks

Artificial neural networks (ANNs) are learning algorithms with interconnected artificial neurons called perceptrons that are loosely based on biological neurons (Fig. 11.4A). Each perceptron or node sums multiple weighted input values and based on a threshold, outputs a single value between 0 and 1, called an activation. This nonlinear operation of limiting the output to a range is performed by a sigmoid or an S-shaped function and allows for input to output mappings that are more complicated than a simple linear combination. A feed-forward neural network propagates the weighted and combined activation signals from one layer of nodes to the next in one direction without cyclical connections. These are also called fully connected layers or multilayer perceptrons. The output at the last layer can output continuous-valued regression results or be binarized for classification. Although current deep artificial neural networks have up to hundreds of layers (Fig. 11.4B), the earliest artificial neural networks consisted of just an input layer, one hidden layer, and an output layer because a single hidden layer was found to be sufficient

circle shown to a linear hyper-plane by transforming the data in a higher dimension. Support vectors are the data points closest to the decision boundary with maximum margins. (B) Random Forest is a bagging method where multiple decision trees are trained on random subsets of the data and the output decision is a majority vote of each individual decision. (C) LogitBoost is a boosting method where multiple decision stumps, decision trees with a single spit, sequentially improve the accuracy of the prior decision stump by weighting the data according to the logistic loss function. The output decision is a vote of each model inversely weighted by the training error.

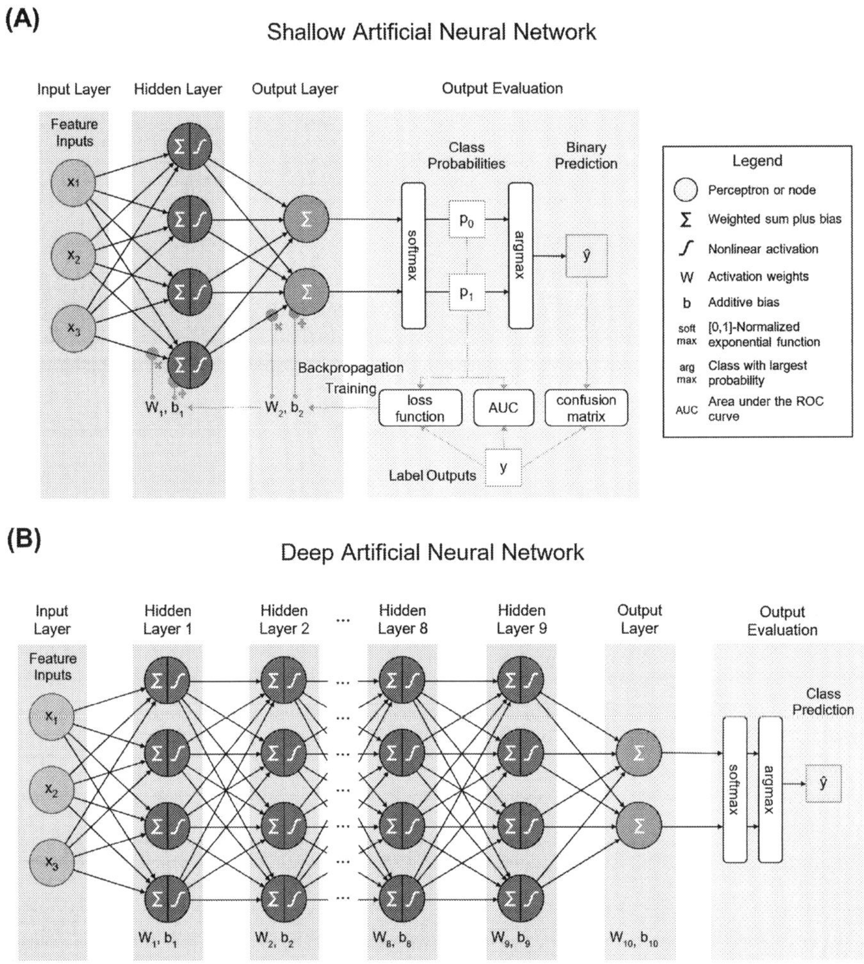

FIGURE 11.4 Shallow and deep artificial neural networks

(A) A shallow artificial neural network has only one hidden layer. Weighted feature inputs move from the input layer to the hidden layer to be summed with biases and sent through a nonlinear activation function that maps that sum to an activation value between 0 and 1. This activation is fed through another weighted sum in the output layer. In the output evaluation, the softmax function normalizes the unbounded logit outputs to class probabilities that sum to 1. The argmax function selects the most probable class as the final output of this binary classifier. Given the label outputs as ground truths, the confusion matrix and the area-under-the ROC curve evaluate the predictive performance. The derivative or gradient of the cross-entropy loss function is used in the backpropagation technique to update the two sets of weights during the training step. (B) A neural network is considered deep if there many layers and nodes, specifically if there are 2 or more hidden layers. This deep artificial neural network example has 10 trainable layers consisting of 9 hidden layers and 1 output layer, excluding the input layer, which has no weights.

to approximate any continuous function (Cybenko, 1989). Historically, these shallow neural networks were often called three-layer neural networks, counting the 3 types of layers (input, hidden, output) in the network structures. However, the contemporary definition used for deep networks counts the layers with trainable weights of the hidden layers and the output layer but not the input layer. By this definition, shallow neural networks have 2 trainable layers. The following studies discussed utilized shallow ANNs.

The perceptron was introduced in the 1950s but an efficient algorithm for training neural networks called backpropagation was not developed until 1986. In backpropagation, the gradients of the output errors generated from a loss function propagate backwards through the layers correcting their weights iteratively (Fig. 11.4A). Shallow ANNs were the earliest traditional machine learning models studied on SPECT images, 2 decades earlier than SVMs and tree-based ensemble methods noted in the prior section. Fujita, Katafuchi, Uehara and Nishimura (1992) first applied a 16 × 16 LV polar map, sampled from a ^{201}Tl SPECT myocardial perfusion image, into an ANN consisting of 256 unwrapped input image values, 100 nodes in the hidden layer, and 8 output classes. The output classes were composed of one normal diagnosis and seven other combinations of abnormal diagnoses in the 3 vascular territories. Using ICA as the ground truth, the test performance on 20% of the 74 total patients had an accuracy of 77% that ranked in between an inexperienced and an experienced radiologist. In a similar study with more data of 410 patients and limited to a single output of only detecting the presence of abnormal perfusion using normal databases as the ground truth, an accuracy of 92% (AUC = 0.96) was achieved (Hamilton, Riley, Miola & Amro, 1995). On the other hand, using 159 ^{201}Tl planar perfusion images, where the third-dimensional information is collapsed down to 2D planar projections, ANNs tasked to diagnose the presence, severity, and localization in two vascular territories of CAD did not exceed the performances of an expert physician (Porenta et al., 1994).

Later that decade, Lindahl et al. (1997) applied ANNs on ^{99m}Technetium(Tc)-sestamibi rest/stress SPECT myocardial perfusion images from 135 patients for the detection and localization of CAD in two vascular territories using ICA as the ground truth. In prior studies, limited computational resources required down-sampled polar map images as inputs into the ANN. Lindahl et al. instead performed a Fourier transform on the polar map samples and selected the 30 most significant values both as a data reduction technique and as a feature extraction step. The resulting ANN had 30 input nodes, 3 nodes in the hidden layer, and 1 output node. The ANN performed better than an expert physician in the right coronary artery (RCA) and left circumflex artery (LCX) territories (AUC = 0.82) and equivalently in the LCA territory (AUC = 0.76). Lindahl, Palmer, & Edenbrandt, 1999b also trained an ANN for reversibility, severity, extent, and localization in the three vascular territories from 338 patients and concluded its performance was equivalent to expert physicians. To test generalizability, detection of CAD trained on data from one hospital in Sweden was also shown to be noninferior when tested on data from another hospital in Denmark (Lindahl et al., 2000). However, rather than replacing the physician reader, ANNs were evaluated on their effectiveness as decision support systems. The use of ANNs to aid physicians improved their AUC for detection of CAD by 3%–5% points and reduced intra and interobserver variabilities (Lindahl, Lanke, Lundin, Palmer, & Edenbrandt, 1999a). In a later study of 243 SPECT images, Guner et al. (2010) corroborated that a combination of ANNs and expert physicians performed better than without the aid of ANNs.

Other input data reduction ideas were explored by Bagher-Ebadian, Soltanian-Zadeh, Setayeshi and Smith (2004), where a full 3D SPECT image volume was analyzed. Instead of using the maximal counts sampled on a fixed 2D polar map grid as done in past studies, they extracted 12 optimal features that maximized their separation in the feature space. Another improvement in utilizing ANNs for CAD detection was made by Allison, Heo and Iskandrian (2005) by training three separate ANNs using the same input data but customized for each of the left anterior descending, LCX, and RCA vascular territories, as opposed having one multilabel ANN making three predictions at the same time.

As previously mentioned, ensemble learning is commonly used in other traditional structured machine learning algorithms such as Random Forest and Gradient Boosting to prevent overfitting the model to the training data. However, Lomsky et al. (2008) first applied ensemble learning to ANNs for gated rest/stress SPECT MPI interpretation by training 100 separate ANNs with different initialized weights and then thresholding the average of 100 binary outputs to produce a single more stable binary prediction. The ANN ensemble model was trained on ^{99m}Tc-tetrofosmin SPECT images from 418 patients from a hospital in the United Kingdom and tested on ^{99m}Tc-sestamibi SPECT images from 532 patients from a different hospital in Sweden. Their study employed an active shape algorithm for segmenting the LV myocardium from which size, severity, and wall thickening of stress and rest defects along with the end-diastolic volume ejection fraction were extracted as input features. The ANNs each had 8 input nodes, 5 hidden nodes, and 1 output node. The output labels of ischemia or infarction, in five myocardial segments and globally, were interpreted by expert physicians as ground truths. Reported by Johansson et al. (Johansson et al., 2014; Lomsky et al., 2008), this ensemble ANN algorithm, called EXINI, performed similarly to experienced physicians and outperformed PERFEX, the rule-based expert system. A similar ensemble learning ANN was also proven to be an effective quality control tool that identified 87% of the cases in which a second evaluation of suspected ischemia differed from the initial interpretation (Tägil, Marving, Lomsky, Hesse and Edenbrandt, 2008).

Nakajima et al. (2015) evaluated the ensemble ANN that was developed by Lomsky et al. (Johansson et al., 2014; Lomsky et al., 2008) and trained on 1051 gated SPECT images from a hospital in Sweden, by testing it on 106 gated SPECT images from hospitals in Japan. The ensemble ANN detected stress defects and ischemia with performance equivalent to an expert physician. The ANN was then retrained on images from hospitals in Japan to ensure the training and test data would share the same data distributions. The training set had 1001 gated SPECT images and the test set consisted of 364 new gated SPECT images plus the earlier 106 gated

SPECT images. Drawing data from the same population proved successful with improved performance on detecting perfusion defects and ischemia (Nakajima et al., 2017, 2018). One follow-up study evaluating the accuracy of the ANN in detecting regional stress defect and ischemia (Shibutani et al., 2019) and another follow-up study using hybrid CCTA and SPECT images to detect culprit coronary arteries (Yoneyama et al., 2019) both led to equivalent or higher performances compared to expert physicians.

Deep learning and convolutional neural networks

Deep learning is a subset of machine learning in which raw data such as images or time sequences are inputted into a learning model instead of hand-engineered features. Precisely, a deep learning model is a neural network with more than one hidden layer and can be as deep as several hundred layers (Fig. 11.4B). The multiple hidden layers afford the neural network to automate the feature extraction process and create higher-order decision boundaries.

Convolutional neural networks or CNNs (Fig. 11.5A) are deep learning models that automatically extract features from large 1D, 2D, 3D, or even 4D inputs in the first part of the network consisting of convolutional layers by compressing the data into a set of important features. This is performed by a mathematical operation called a convolution that involves sliding a filter across an image. A moving average filter or a smoothing filter is an example. Any kind of filter may be learned in the training process. Input images filtered over several convolutional layers are transformed into features and are then sent to the final part of the network consisting of fully connected layers like a regular ANN, for the classification or regression prediction step. The convolutional layers allow CNNs to be efficient because only spatially local network connections are considered. The unbiased nature of the CNN's feature extraction step may lead to never-before-considered and better output-predictive features than hand-engineered ones.

CNNs gained widespread popularity in the computer vision community starting in 2012 when the AlexNet CNN outperformed state-of-the-art image classification algorithms by wide margins (Krizhevsky, Sutskever & Hinton, 2012). It was in 2018 when CNNs were first used to improve the automatic interpretation of SPECT myocardial perfusion images to predict obstructive disease. In this study with 1638 patients from multiple centers, Betancur, Commandeur et al. (2018) develop a seven-layer 2D CNN with raw and quantitative 2D LV polar maps sampled from static stress SPECT images as inputs and obstructive disease defined as ICA diameter stenosis $\geq$70% as the labeled output. The training and test splits were performed using a 10-fold cross-validation. When compared to TPD as the baseline, the CNN achieved higher AUCs (0.80 vs. 0.78 per-patient and 0.76 vs. 0.73 per-vessel). In a follow up study, Betancur et al. (2019) included both supine and semiupright SPECT images from 1160 patients from four sites in the REFINE SPECT study but performed a leave-one-center-out cross-validation. Prediction performances were similar to their prior study.

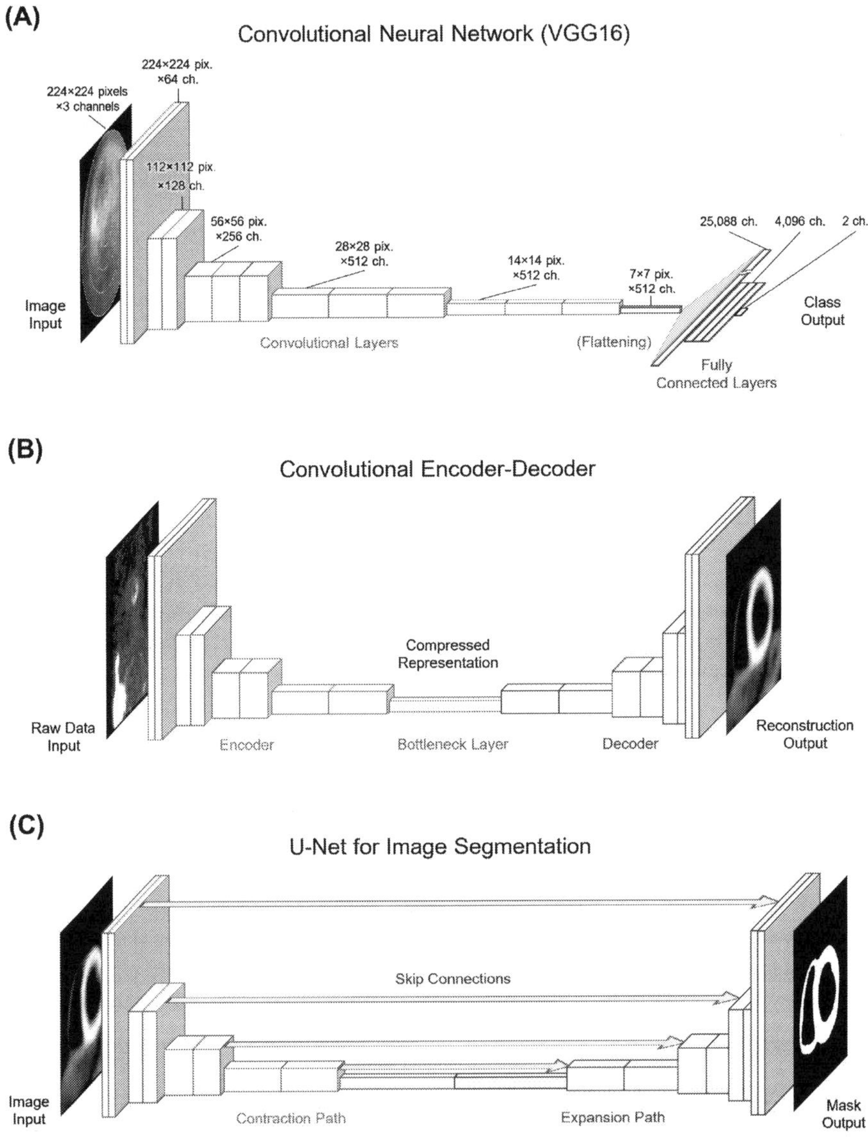

FIGURE 11.5 Convolutional neural networks and variants

(A) The convolutional neural network (CNN) in this example is the VGG16 network with a slight modification of a two-channel output layer for binary classification. The 2D input image is convolved with 64 filters with trainable weights to produce a 64-channel activation image in the first hidden layers. The successive hidden layers with half the image size but twice the channel depth progressively convert spatial information into more complex feature information. There are 16 total trainable layers consisting of a set of 13 convolutional layers and a set of 3 fully connected layers, with a flattening step in

CNNs have been applied to other conditions in nuclear medicine beyond coronary artery disease. Cardiac sarcoidosis (CS) is an inflammatory disease in which small clumps of white blood cells called granulomas form in the heart. Togo, Hirata et al. (2019) employed a 48-layer 2D CNN on ^{18}F-FDG PET images to detect patients with CS from those without CS. LV polar map images from 85 patients were first input to the convolutional layers of the Inception-v3 CNN for feature extraction. Version 3 of the Inception neural network is a computationally efficient CNN that applies various parallel convolutional filters with sizes other than the typical 3×3 size, such as 1×1 or 5×5, increasing its flexibility. After the image features have been extracted, they are weighted based on their importance and finally the adjusted features are sent into an SVM for the final classification step. This deep learning method achieved higher performance (sensitivity = 84% and specificity = 87%) compared to the standard uptake value (SUV)-based method (sensitivity = 47% and specificity = 71%).

One criticism of machine learning is that it is a "black box" where the nonlinearity of the machine learning algorithms makes it difficult to know how and why the algorithm arrived at its decision. It is desirable to understand the algorithm's feature visualization and feature importance. Togo, Ogawa et al. (2019) expanded upon their prior study to identify that 17-segment polar map regions are related to cardiac sarcoidosis by masking out each region and then computing confidence factors to visualize their importance. Three CNNs (VGG16, ResNet50, and Inception-v3) were compared and had mixed excellent and poor results against a ground truth of visual assessment by an expert physician that warrants further exploration.

In another study using PET images, Juarez-Orozco, Martinez-Manzanera, van der Zant, Knol and Knuuti (2020) employed a 50-layer 2D CNN to predict outcomes of MACE given inputs of myocardial blood flow (MBF) values, which are measures of absolute perfusion. Rest/stress ^{13}N-ammonia PET myocardial perfusion images from 1185 patients were recruited. The inputs were three 2D polar map samples of stress MBF, rest MBF, and myocardial flow reserve, which is calculated by subtracting the rest MBF from the stress MBF. Data augmentation of image rotations was used, which is a method to increase the size of the training data by generating modifications of the original data. This study used a 50-layer residual neural network (ResNet) that employs residual connections or skip connections between convolutional layers to prevent the vanishing gradient problem during the backpropagation step, which slows down training. The feature-extracting convolutional layers of the ResNet-50 neural network were replicated three times in parallel with different

between to unravel 3D array of values to a long 1D array. (B) The convolutional encoder–decoder borrows the convolutional layers of a CNN to perform encoding of the input image into a compressed representation or set of features in the bottleneck layer, from which the decoder reconstructs an output image through a series of convolutions and upsampling. (C) A U-Net is an encoder–decoder used for image segmentation but with resolution-preserving skip connections that copy and concatenate activation images in the contraction path to similar-sized activation images in the expansion path.

weights before merging their outputs into the same fully connected output layers. This CNN model achieved an AUC of 0.90, which was higher than non-CNN models with clinical, functional, or absolute perfusion inputs.

CNNs were originally designed for inputting rectangular natural images of the outside world; however, changing the network architecture of CNNs to better adapt to the desired input data is possible. Projecting the capped cylindrically 3D shape of the LV myocardium to a 2D polar map to input into a CNN introduces spatial distortions where the apical region is compressed, and the basal regions are expanded. Additionally, a 2D polar map image leaves blank regions in the corner of the circumscribed square input image. To avoid these issues, Spier et al. (2019) applied a graph-convolutional neural network, which samples the 2D myocardial surface in 3D space as an irregular mesh network graph, to 946 datasets. Their three-layer graph-CNN performed better at localizing perfusion defects than a 1D fully connected neural network and a standard 2D CNN.

CNNs have also been adapted to produce output images. This new architecture is called a fully convolutional network (FCN), which leaves out the fully connected layers of the second half of the CNN and consists of only convolutional layers. The U-Net is a special case of an FCN that mirrors the downsampling convolutional layers of a CNN with succeeding deconvolutional layers that upsample the compressed features to a segmented image of the original size (Fig. 11.5C) (Ronneberger, Fischer & Brox, 2015). The network architecture resembles the shape of the letter U. A U-Net can be trained for semantic segmentation that maps each image pixel to a class label. Although the original U-Net study segmented 2D microscopy images, the V-Net, where the V stands for volumetric and represents the shape of the network, was developed for 3D image volumes for biomedical image segmentation (Milletari, Navab & Ahmadi, 2016). Wang et al. (2019) utilized a deep 32-layer V-Net to automatically segment the enclosed volumes of the LV endocardial and epicardial surfaces from gated SPECT perfusion images. This was a feasibility study consisting of 56 patients with expert physician annotated segmentations used as the ground truth. The automated segmentation was successful with the machine learning Dice similarity coefficient of 0.90 out of a maximum score of 1.00, which is a complete overlap of segmentation masks. The LV ejection fractions were also estimated from gated endocardial surfaces from both the V-Net and the ground truth with high correlation ($r = 0.91$).

Artificial intelligence in cardiac SPECT and PET image reconstruction

Artificial intelligence in image reconstruction for nuclear medicine has been a recent development due to the relatively recent advent of the convolutional neural network. As most image reconstruction deep learning methods are CNN based, the topics are organized by the goals of the reconstruction. Some of these objectives include directly reconstructing the image from the measurement data without an explicit image measurement model in an end-to-end approach, while others include using the traditional image reconstruction algorithms but generating attenuation maps from non-CT imaging modalities or correcting for motion artifacts. Deep learning has also been applied to the last step of the image reconstruction pipeline that includes noise reduction techniques that allow for either improving image quality or lowering the radiopharmaceutical dosage while retaining the same image quality. Although much of this research in nuclear medicine has been focused on brain imaging, the techniques described here are applicable to cardiovascular imaging.

End-to-end reconstruction

In the machine learning context, end-to-end reconstruction refers to estimating an image directly from the raw measurement data of the imaging camera without using conventional filtered backprojection or iterative reconstruction methods like ordered subsets expectation-maximization (OSEM). The measurement data that SPECT and PET cameras ultimately produce are sinogram data (Fig. 11.1B), which are sets of angular projections (Fig. 11.1A). In 2018, Zhu, Liu, Cauley, Rosen and Rosen (2018) introduced a general framework called Automated Transform by Manifold Approximation (AUTOMAP), which trains a supervised deep neural network to map sinogram input data to an output image that has been conventionally reconstructed. The examples used were mostly for MRI data. Häggström, Schmidtlein, Campanella and Fuchs (2019) on the other hand developed a PET-tailored end-to-end reconstruction network called DeepPET. DeepPET is a special case of a convolutional neural network called an encoder–decoder network (Fig. 11.5B), which is a two-stage network that maps an input image to a compressed representation, also called variables in a latent space, and then maps those variables back up to the desired output image. An encoder–decoder is like a U-Net but without the skip connections and the input and output images may be of different sizes. DeepPET was trained on 291,000 pairs of sinogram inputs and reconstructed image outputs, and their data augmentations. DeepPET reconstructed higher quality images than FBP and OSEM and with reconstruction speeds 3 and 108 times faster, respectively.

Attenuation correction

For PET, varying radiopaque areas of the body can attenuate an emitted positron from reaching its intended sensor causing attenuation artifacts where some parts of the myocardium may incorrectly appear to have lower counts. Incorporating a CT attenuation correction (CTAC) map (Fig. 11.2A) from a PET/CT imaging system into the reconstruction process can correct for this. For PET/MR imaging systems, Liu, Jang, Kijowski, Bradshaw and McMillan (2017) and Leynes et al. (2018) have each developed deep convolutional encoder–decoder networks with MRI image inputs (Fig. 11.2C) and true CTAC image outputs to generate pseudo-CTAC images from MRI images. Even more remarkable is Liu et al. (2018) went a step further to generate attenuation maps by training with PET image inputs and CTAC image outputs requiring neither MRI nor CT images. Similarly, Shi, Onofrey, Liu, Liu and Liu (2020) trained a deep fully convolutional neural network and a generative adversarial network (GAN) with SPECT myocardial perfusion image inputs and CTAC image outputs to generate attenuation maps with only SPECT images.

Motion correction

Motion of the patient during imaging is unavoidable, whether it is cardiac, respiratory, or patient motion. The impact of motion is blurred imaging. Öktem, Pouchol and Verdier (2019) incorporated deep learning-based motion registration into the iterative reconstruction step to decrease noise in time-varying PET images. Li, Zhang, Qi, Asma and Qi (2019) similarly implemented unsupervised deep learning-based nonrigid registration and integrated it into the iterative reconstruction step to reduce the effects of respiratory motion.

Denoising and image quality improvement

Autoencoders are convolutional neural networks that extract the most important information in images by forcing the successive image activations of the network through a bottleneck. Autoencoders are special cases of encoder—decoder networks where the input and outputs are the same. The impact of the bottleneck retaining the most important information is a smoothing or denoising effect. Cui, Liu, Wang and Liu (2017) employed this through a stacked autoencoder to improve image quality, where the stacking implies each hidden layer is trained one at a time then combined or stacked together. Dietze, Branderhorst., Kunnen, Viergever and de Jong (2019) used an encoder—decoder network but with FBP reconstructed images as inputs and iterative reconstructed images as the outputs for enhancing the quality of SPECT FBP images but with much lower total reconstruction times on the order of seconds instead of minutes. Song et al. (2019a) used a 3D residual convolutional neural network with no downsampling layers on gated SPECT image reconstruction also with results of improved image quality and faster reconstruction.

Another wide area of deep learning reconstruction research is in denoising low-dose nuclear medicine scans. Many of these approaches generally train an encoder—decoder network with low-dose image inputs and standard-dose image outputs. During testing or running of the model, the low-dose images are transformed into ones with reduced noise. There are many examples of this using different networks. Wang et al. (2018) performed this using a 3D conditional GAN with a U-Net generator. Gong, Guan, Liu and Qi (2019) used transfer learning by training a deep CNN that was pretrained on a VGG network with simulation data. Kim et al. (2018) employed a denoising CNN with linear fitting. Ramon et al. (2018) used a 3D CNN in the form of a stacked convolutional autoencoder on SPECT MPI with doses 1/8 and 1/16 of the standard dose with success. Song et al. (2019b) implemented the aforementioned 3D residual CNN with no downsampling layers on gated SPECT MPI with 1/4 of the standard dose and results were improved spatial resolution in the LV.

A recent promising method called Deep Image Prior developed by Ulyanov et al. (Lempitsky, Vedaldi & Ulyanov, 2018; Ulyanov, Vedaldi & Lempitsky, 2020) does not require prior training inputs and associated labels to improve image quality. The U-Net used performs a regularization on the images when the degraded or noisy images are treated as outputs and random noise or an initialization of the image is set as the input. The standard supervised learning methodology is training a universal model only once using many training examples, which then is evaluated on new images using the same model. On the other hand, the deep image prior method requires training many models, each one specific to each evaluation image. This strategy requires tuning the early-stopping hyperparameter to halt the reconstruction before the undesired noise and artifacts begin to contribute to the final image. Cheng, Gadelha, Maji and Sheldon (2019) found the deep image prior was equivalent to a Gaussian prior and their new framework no longer depended on an early-stopping strategy. Many researchers have applied the deep image prior technique to PET denoising problems successfully (Cui et al., 2018, 2019; Gong, Catana, Qi & Li, 2019, Gong et al., 2019; Hashimoto, Ohba, Ote, Teramoto & Tsukada, 2019; Lim, Chun, Dewaraja & Fessler, 2020). As a contrast to the conventional supervised learning methods, these approaches, also described as unsupervised, without prior training, and regularizing methods, have promising results.

Summary

Artificial intelligence methods have had a long history in nuclear medicine imaging beginning with expert systems using mostly rule-based reasoning, next shallow artificial neural networks, and then traditional machine learning algorithms. Recent deep learning algorithms include convolutional neural networks for classification, U-Net and V-Net for segmentation, encoder–decoder networks for image reconstruction, and generative adversarial networks for image generation. One of the challenges that remain for machine learning in nuclear medicine imaging is training models with sufficiently large numbers of training images on the order of thousands to hundreds of thousands of images to avoid overfitting models. Some strategies to address the small data problem include: (1) data augmentation to increase the training set, (2) employing unsupervised, semisupervised or self-supervised learning to reduce the dependence on expensive manual data labeling, and (3) using a hybrid approach of rules and machine learning to extend the generalizability of these models. The other broader challenge is instilling confidence in machine-learned predictions by adding interpretability to address the "how" and explainability to address the "why." These are yet unsolved but critical ongoing areas of machine learning research.

References

Allison, J. S., Heo, J., & Iskandrian, A. E. (2005). Artificial neural network modeling of stress single-photon emission computed tomographic imaging for detecting extensive coronary artery disease. *The American Journal of Cardiology, 95*, 178–181. https://doi.org/10.1016/j.amjcard.2004.09.003

Arsanjani, R., Dey, D., Khachatryan, T., Shalev, A., Hayes, S. W., Fish, M., et al. (2015). Prediction of revascularization after myocardial perfusion SPECT by machine learning in a large population. *Journal of Nuclear Cardiology, 22*, 877–884. https://doi.org/10.1007/s12350-014-0027-x

Arsanjani, R., Xu, Y., Dey, D., Fish, M., Dorbala, S., Hayes, S., et al. (2013). Improved accuracy of myocardial perfusion SPECT for the detection of coronary artery disease using a support vector machine algorithm. *Journal of Nuclear Medicine, 54*, 549–555. https://doi.org/10.2967/jnumed.112.111542

Arsanjani, R., Xu, Y., Dey, D., Vahistha, V., Shalev, A., Nakanishi, R., et al. (2013). Improved accuracy of myocardial perfusion SPECT for detection of coronary artery disease by machine learning in a large population. *Journal of Nuclear Cardiology, 20*, 553–562. https://doi.org/10.1007/s12350-013-9706-2

Bagher-Ebadian, H., Soltanian-Zadeh, H., Setayeshi, S., & Smith, S. T. (2004). Neural network and fuzzy clustering approach for automatic diagnosis of coronary artery disease in nuclear medicine. *IEEE Transactions on Nuclear Science, 51*, 184–192. https://doi.org/10.1109/TNS.2003.823047

Betancur, J., Commandeur, F., Motlagh, M., Sharir, T., Einstein, A. J., Bokhari, S., et al. (2018). Deep learning for prediction of obstructive disease from fast myocardial perfusion SPECT: A multicenter study. *Journal of the American College of Cardiology: Cardiovascular Imaging, 11*, 1654–1663. https://doi.org/10.1016/j.jcmg.2018.01.020

Betancur, J., Hu, L.-H., Commandeur, F., Sharir, T., Einstein, A. J., Fish, M. B., et al. (2019). Deep learning analysis of upright-supine high-efficiency SPECT myocardial perfusion imaging for prediction of obstructive coronary artery disease: A multicenter study. *Journal of Nuclear Medicine, 60*, 664–670. https://doi.org/10.2967/jnumed.118.213538

Betancur, J., Otaki, Y., Motwani, M., Fish, M. B., Lemley, M., Dey, D., et al. (2018). Prognostic value of combined clinical and myocardial perfusion imaging data using machine learning. *Journal of the American College of Cardiology: Cardiovascular Imaging, 11*, 1000–1009. https://doi.org/10.1016/j.jcmg.2017.07.024

Betancur, J., Rubeaux, M., Fuchs, T. A., Otaki, Y., Arnson, Y., Slipczuk, L., et al. (2017). Automatic valve plane localization in myocardial perfusion SPECT/CT by machine learning: Anatomic and clinical validation. *Journal of Nuclear Medicine, 58*, 961–967. https://doi.org/10.2967/jnumed.116.179911

Cheng, Z., Gadelha, M., Maji, S., & Sheldon, D. (2019). *A bayesian perspective on the deep image prior*. arXiv:1904.07457 [cs, stat].

Ciecholewski, M. (2011). Support vector machine approach to cardiac SPECT diagnosis. In J. K. Aggarwal, R. P. Barneva, V. E. Brimkov, K. N. Koroutchev, & E. R. Korutcheva (Eds.), *Combinatorial image analysis, lecture notes in computer science* (pp. 432–443). Berlin, Heidelberg: Springer. https://doi.org/10.1007/978-3-642-21073-0_38.

Ciecholewski, M. (2012). Ischemic heart disease detection using selected machine learning methods. *International Journal of Computer Mathematics, 90*. https://doi.org/10.1080/00207160.2012.742189

Cios, K. J., Goodenday, L. S., Shah, K. K., & Serpen, G. (1996). A novel algorithm for classification of SPECT images of a human heart. In *Presented at the proceedings ninth IEEE symposium on computer-based medical systems* (pp. 1–5). https://doi.org/10.1109/CBMS.1996.507116

Cui, J., Gong, K., Guo, N., Meng, X., Kim, K., Liu, H., et al. (2018). CT-guided PET image denoising using deep neural network without prior training data. In *Presented at the 2018 IEEE nuclear science symposium and medical imaging conference proceedings (NSS/MIC)* (pp. 1−3). https://doi.org/10.1109/NSSMIC.2018.8824397

Cui, J., Gong, K., Guo, N., Wu, C., Meng, X., Kim, K., et al. (2019). PET image denoising using unsupervised deep learning. *European Journal of Nuclear Medicine and Molecular Imaging, 46*, 2780−2789. https://doi.org/10.1007/s00259-019-04468-4

Cui, J., Liu, X., Wang, Y., & Liu, H. (2017). Deep reconstruction model for dynamic PET images. *PLoS One, 12*. https://doi.org/10.1371/journal.pone.0184667

Cybenko, G. (1989). Approximation by superpositions of a sigmoidal function. *Mathematics Control Signal Systems, 2*, 303−314. https://doi.org/10.1007/BF02551274

Dietze, M. M. A., Branderhorst, W., Kunnen, B., Viergever, M. A., & de Jong, H. W. A. M. (2019). Accelerated SPECT image reconstruction with FBP and an image enhancement convolutional neural network. *EJNMMI Physics, 6*, 14. https://doi.org/10.1186/s40658-019-0252-0

Ficaro, E., Lee, B., Kritzman, J., & Corbett, J. (2007). Corridor4DM: The Michigan method for quantitative nuclear cardiology. *Journal of Nuclear Cardiology, 14*, 455−465. https://doi.org/10.1016/j.nuclcard.2007.06.006

Fujita, H., Katafuchi, T., Uehara, T., & Nishimura, T. (1992). Application of artificial neural network to computer-aided diagnosis of coronary artery disease in myocardial SPECT bull's-eye images. *Journal of Nuclear Medicine, 33*, 272−276.

Garcia, E. V., Cooke, C. D., Folks, R. D., Santana, C. A., Krawczynska, E. G., Braal, L. D., et al. (2001). Diagnostic performance of an expert system for the interpretation of myocardial perfusion SPECT studies. *Journal of Nuclear Medicine, 42*, 1185−1191.

Garcia, E. V., Faber, T. L., Cooke, C. D., Folks, R. D., Chen, J., & Santana, C. (2007). The increasing role of quantification in clinical nuclear cardiology: The Emory approach. *Journal of Nuclear Cardiology, 14*, 420−432. https://doi.org/10.1016/j.nuclcard.2007.06.009

Garcia, E. V., Klein, J. L., Moncayo, V., Cooke, C. D., Del'Aune, C., Folks, R., et al. (2018). Diagnostic performance of an artificial intelligence-driven cardiac-structured reporting system for myocardial perfusion SPECT imaging. *Journal of Nuclear Cardiology*. https://doi.org/10.1007/s12350-018-1432-3

Garcia, E. V., Klein, J. L., & Taylor, A. T. (2014). Clinical decision support systems in myocardial perfusion imaging. *Journal of Nuclear Cardiology, 21*, 427−439. https://doi.org/10.1007/s12350-014-9857-9

Garcia, E. V., Train, K. V., Maddahi, J., Prigent, F., Friedman, J., Areeda, J., et al. (1985). Quantification of rotational thallium-201 myocardial tomography. *Journal of Nuclear Medicine, 26*, 17−26.

Gaspar-Cunha, A. (2010). Feature selection using multi-objective evolutionary algorithms: Application to cardiac SPECT diagnosis. In M. P. Rocha, F. F. Riverola, H. Shatkay, & J. M. Corchado (Eds.), *Advances in bioinformatics, advances in intelligent and soft computing* (pp. 85−92). Berlin, Heidelberg: Springer. https://doi.org/10.1007/978-3-642-13214-8_11.

Germano, G., Kavanagh, P. B., Slomka, P. J., Van Kriekinge, S. D., Pollard, G., & Berman, D. S. (2007). Quantitation in gated perfusion SPECT imaging: The Cedars-Sinai approach. *Journal of Nuclear Cardiology, 14*, 433−454. https://doi.org/10.1016/j.nuclcard.2007.06.008. Abstracts of Original Contributions, ASNC 2007, 12th Annual Scientific Session.

Gong, K., Catana, C., Qi, J., & Li, Q. (2019). PET image reconstruction using deep image prior. *IEEE Transactions on Medical Imaging, 38*, 1655–1665. https://doi.org/10.1109/TMI.2018.2888491

Gong, K., Guan, J., Kim, K., Zhang, X., Yang, J., Seo, Y., et al. (2019). Iterative PET image reconstruction using convolutional neural network representation. *IEEE Transactions on Medical Imaging, 38*, 675–685. https://doi.org/10.1109/TMI.2018.2869871

Gong, K., Guan, J., Liu, C.-C., & Qi, J. (2019). PET image denoising using a deep neural network through fine tuning. *IEEE Transactions on Radiation and Plasma Medical Sciences, 3*, 153–161. https://doi.org/10.1109/TRPMS.2018.2877644

Guner, L. A., Karabacak, N. I., Akdemir, O. U., Karagoz, P. S., Kocaman, S. A., Cengel, A., et al. (2010). An open-source framework of neural networks for diagnosis of coronary artery disease from myocardial perfusion SPECT. *Journal of Nuclear Cardiology, 17*, 405–413. https://doi.org/10.1007/s12350-010-9207-5

Haddad, M., Adlassnig, K.-P., & Porenta, G. (1997). Feasibility analysis of a case-based reasoning system for automated detection of coronary heart disease from myocardial scintigrams. *Artificial Intelligence in Medicine, Case-based Reasoning: Opportunities and Applications in Health Care, 9*, 61–78. https://doi.org/10.1016/S0933-3657(96)00361-2

Haddad, M., Moertl, D., & Porenta, G. (1995). SCINA: A case-based reasoning system for the interpretation of myocardial perfusion scintigrams. In *Presented at the computers in cardiology 1995* (pp. 761–764). https://doi.org/10.1109/CIC.1995.482776

Häggström, I., Schmidtlein, C. R., Campanella, G., & Fuchs, T. J. (2019). DeepPET: A deep encoder–decoder network for directly solving the PET image reconstruction inverse problem. *Medical Image Analysis, 54*, 253–262. https://doi.org/10.1016/j.media.2019.03.013

Hamilton, D., Riley, P. J., Miola, U. J., & Amro, A. A. (1995). A feed forward neural network for classification of bull's-eye myocardial perfusion images. *European Journal of Nuclear Medicine, 22*, 108–115. https://doi.org/10.1007/BF00838939

Han, D., Rozanski, A., Gransar, H., Sharir, T., Einstein, A. J., Fish, M. B., et al. (2020). Myocardial ischemic burden and differences in prognosis among patients with and without diabetes: Results from the multicenter international REFINE SPECT registry. *Diabetes Care, 43*, 453–459. https://doi.org/10.2337/dc19-1360

Haro Alonso, D., Wernick, M. N., Yang, Y., Germano, G., Berman, D. S., & Slomka, P. (2019). Prediction of cardiac death after adenosine myocardial perfusion SPECT based on machine learning. *Journal of Nuclear Cardiology, 26*, 1746–1754. https://doi.org/10.1007/s12350-018-1250-7

Hashimoto, F., Ohba, H., Ote, K., Teramoto, A., & Tsukada, H. (2019). Dynamic PET image denoising using deep convolutional neural networks without prior training datasets. *IEEE Access, 7*, 96594–96603. https://doi.org/10.1109/ACCESS.2019.2929230

Hu, L.-H., Betancur, J., Sharir, T., Einstein, A. J., Bokhari, S., Fish, M. B., et al. (2019). Machine learning predicts per-vessel early coronary revascularization after fast myocardial perfusion SPECT: Results from multicentre REFINE SPECT registry. *European Heart Journal Cardiovascular Imaging*. https://doi.org/10.1093/ehjci/jez177

Jaarsma, C., Leiner, T., Bekkers, S. C., Crijns, H. J., Wildberger, J. E., Nagel, E., et al. (2012). Diagnostic performance of noninvasive myocardial perfusion imaging using single-photon emission computed tomography, cardiac magnetic resonance, and positron emission tomography imaging for the detection of obstructive coronary artery disease: A meta-analysis. *Journal of the American College of Cardiology, 59*, 1719–1728. https://doi.org/10.1016/j.jacc.2011.12.040

Johansson, L., Edenbrandt, L., Nakajima, K., Lomsky, M., Svensson, S.-E., & Trägårdh, E. (2014). Computer-aided diagnosis system outperforms scoring analysis in myocardial perfusion imaging. *Journal of Nuclear Cardiology, 21*, 416–423. https://doi.org/10.1007/s12350-013-9815-y

Juarez-Orozco, L. E., Knol, R. J. J., Sanchez-Catasus, C. A., Martinez-Manzanera, O., van der Zant, F. M., & Knuuti, J. (2020). Machine learning in the integration of simple variables for identifying patients with myocardial ischemia. *Journal of Nuclear Cardiology, 27*, 147–155. https://doi.org/10.1007/s12350-018-1304-x

Juarez-Orozco, L. E., Martinez-Manzanera, O., van der Zant, F. M., Knol, R. J. J., & Knuuti, J. (2020). Deep learning in quantitative PET myocardial perfusion imaging: A study on cardiovascular event prediction. *Journal of the American College of Cardiology: Cardiovascular Imaging, 13*, 180–182. https://doi.org/10.1016/j.jcmg.2019.08.009

Khorsand, A., Haddad, M., Graf, S., Moertl, D., Sochor, H., & Porenta, G. (2001). Automated assessment of dipyridamole 201Tl myocardial SPECT perfusion scintigraphy by case-based reasoning. *Journal of Nuclear Medicine, 42*, 189–193.

Kim, K., Wu, D., Gong, K., Dutta, J., Kim, J. H., Son, Y. D., et al. (2018). Penalized PET reconstruction using deep learning prior and local linear fitting. *IEEE Transactions on Medical Imaging, 37*, 1478–1487. https://doi.org/10.1109/TMI.2018.2832613

Krizhevsky, A., Sutskever, I., & Hinton, G. E. (2012). ImageNet classification with deep convolutional neural networks. In F. Pereira, C. J. C. Burges, L. Bottou, & K. Q. Weinberger (Eds.), *Advances in neural information processing systems 25* (pp. 1097–1105). Curran Associates, Inc.

Kurgan, L. A., Cios, K. J., Tadeusiewicz, R., Ogiela, M., & Goodenday, L. S. (2001). Knowledge discovery approach to automated cardiac SPECT diagnosis. *Artificial Intelligence in Medicine, 23*, 149–169. https://doi.org/10.1016/S0933-3657(01)00082-3

Lempitsky, V., Vedaldi, A., & Ulyanov, D. (2018). Deep image prior. In *Presented at the 2018 IEEE/CVF conference on computer vision and pattern recognition* (pp. 9446–9454). https://doi.org/10.1109/CVPR.2018.00984

Leynes, A. P., Yang, J., Wiesinger, F., Kaushik, S. S., Shanbhag, D. D., Seo, Y., et al. (2018). Zero-echo-time and dixon deep pseudo-CT (ZeDD CT): Direct generation of pseudo-CT images for pelvic PET/MRI attenuation correction using deep convolutional neural networks with multiparametric MRI. *Journal of Nuclear Medicine, 59*, 852–858. https://doi.org/10.2967/jnumed.117.198051

Lim, H., Chun, I. Y., Dewaraja, Y. K., & Fessler, J. A. (2020). *Improved low-count quantitative PET reconstruction with an iterative neural network*. arXiv:1906.02327 [physics, stat].

Lindahl, D., Lanke, J., Lundin, A., Palmer, J., & Edenbrandt, L. (1999a). Improved classifications of myocardial bull's-eye scintigrams with computer-based decision support system. *Journal of Nuclear Medicine, 40*, 96–101.

Lindahl, D., Palmer, J., & Edenbrandt, L. (1999b). Myocardial SPET: Artificial neural networks describe extent and severity of perfusion defects. *Clinical Physiology, 19*, 497–503. https://doi.org/10.1046/j.1365-2281.1999.00203.x

Lindahl, D., Palmer, J., Ohlsson, M., Peterson, C., Lundin, A., & Edenbrandt, L. (1997). Automated interpretation of myocardial SPECT perfusion images using artificial neural networks. *Journal of Nuclear Medicine, 38*, 1870–1875.

Lindahl, D., Toft, J., Hesse, B., Palmer, J., Ali, S., Lundin, A., et al. (2000). Scandinavian test of artificial neural network for classification of myocardial perfusion images. *Clinical Physiology, 20*, 253–261. https://doi.org/10.1046/j.1365-2281.2000.00255.x

Liu, F., Jang, H., Kijowski, R., Bradshaw, T., & McMillan, A. B. (2017). Deep learning MR imaging–based attenuation correction for PET/MR imaging. *Radiology, 286*, 676–684. https://doi.org/10.1148/radiol.2017170700

Liu, F., Jang, H., Kijowski, R., Zhao, G., Bradshaw, T., & McMillan, A. B. (2018). A deep learning approach for 18F-FDG PET attenuation correction. *EJNMMI Physics, 5*, 24. https://doi.org/10.1186/s40658-018-0225-8

Li, T., Zhang, M., Qi, W., Asma, E., & Qi, J. (2019). Motion correction of respiratory-gated PET image using deep learning based image registration framework. In *Presented at the 15th international meeting on fully three-dimensional image reconstruction in radiology and nuclear medicine* (p. 110720Q). International Society for Optics and Photonics. https://doi.org/10.1117/12.2534851.

Lomsky, M., Gjertsson, P., Johansson, L., Richter, J., Ohlsson, M., Tout, D., et al. (2008). Evaluation of a decision support system for interpretation of myocardial perfusion gated SPECT. *European Journal of Nuclear Medicine and Molecular Imaging, 35*, 1523–1529. https://doi.org/10.1007/s00259-008-0746-9

Marin, T., Kalayeh, M. M., Pretorius, P. H., Wernick, M. N., Yang, Y., & Brankov, J. G. (2011). Numerical observer for cardiac motion assessment using machine learning. In *Presented at the medical imaging 2011: Image perception, observer performance, and technology assessment, international society for optics and photonics* (p. 79660G). https://doi.org/10.1117/12.878186

Milletari, F., Navab, N., & Ahmadi, S.-A. (2016). V-Net: Fully convolutional neural networks for volumetric medical image segmentation. In *Presented at the 2016 fourth international conference on 3D vision (3DV)* (pp. 565–571). https://doi.org/10.1109/3DV.2016.79

Nakajima, K., Kudo, T., Nakata, T., Kiso, K., Kasai, T., Taniguchi, Y., et al. (2017). Diagnostic accuracy of an artificial neural network compared with statistical quantitation of myocardial perfusion images: A Japanese multicenter study. *European Journal of Nuclear Medicine and Molecular Imaging, 44*, 2280–2289. https://doi.org/10.1007/s00259-017-3834-x

Nakajima, K., Matsuo, S., Wakabayashi, H., Yokoyama, K., Bunko, H., Okuda, K., et al. (2015). Diagnostic performance of artificial neural network for detecting ischemia in myocardial perfusion imaging. *Circulation Journal, 79*, 1549–1556. https://doi.org/10.1253/circj.CJ-15-0079

Nakajima, K., Okuda, K., Watanabe, S., Matsuo, S., Kinuya, S., Toth, K., et al. (2018). Artificial neural network retrained to detect myocardial ischemia using a Japanese multicenter database. *Annals of Nuclear Medicine, 32*, 303–310. https://doi.org/10.1007/s12149-018-1247-y

Niemann, H., Bunke, H., Hofmann, I., Sagerer, G., Wolf, F., & Feistel, H. (1985). A knowledge based system for analysis of gated blood pool studies. In *IEEE transactions on pattern analysis and machine intelligence PAMI-7* (pp. 246–259). https://doi.org/10.1109/TPAMI.1985.4767655

Öktem, O., Pouchol, C., & Verdier, O. (2019). Spatiotemporal PET reconstruction using ML-EM with learned diffeomorphic deformation. In F. Knoll, A. Maier, D. Rueckert, & J. C. Ye (Eds.), *Machine learning for medical image reconstruction, lecture notes in computer science* (pp. 151–162). Cham: Springer International Publishing. https://doi.org/10.1007/978-3-030-33843-5_14.

Ordonez, C., Ezquerra, N., & Santana, C. A. (2006). Constraining and summarizing association rules in medical data. *Knowledge and Information Systems, 9*, 1–2. https://doi.org/10.1007/s10115-005-0226-5

Ordonez, C., Omiecinski, E., de Braal, L., Santana, C. A., Ezquerra, N., Taboada, J. A., et al. (2001). Mining constrained association rules to predict heart disease. In *Presented at the proceedings 2001 IEEE international conference on data mining* (pp. 433–440). https://doi.org/10.1109/ICDM.2001.989549

Otaki, Y., Betancur, J., Sharir, T., Hu, L.-H., Gransar, H., Liang, J. X., et al. (2019). 5-Year prognostic value of quantitative versus visual MPI in subtle perfusion defects: Results from REFINE SPECT. *Journal of the American College of Cardiology: Cardiovascular Imaging*. https://doi.org/10.1016/j.jcmg.2019.02.028

Porenta, G., Dorffner, G., Kundrat, S., Petta, P., Duit-Schedlmayer, J., & Sochor, H. (1994). Automated interpretation of planar thallium-201-dipyridamole stress-redistribution scintigrams using artificial neural networks. *Journal of Nuclear Medicine, 35*, 2041–2047.

Ramon, A. J., Yang, Y., Pretorius, P. H., Johnson, K. L., King, M. A., & Wernick, M. N. (2018). Initial investigation of low-dose SPECT-MPI via deep learning. In *Presented at the 2018 IEEE nuclear science symposium and medical imaging conference proceedings (NSS/MIC)* (pp. 1–3). https://doi.org/10.1109/NSSMIC.2018.8824548

Ricciardi, C., Cantoni, V., Improta, G., Iuppariello, L., Latessa, I., Cesarelli, M., et al. (2020). Application of data mining in a cohort of Italian subjects undergoing myocardial perfusion imaging at an academic medical center. *Computer Methods and Programs in Biomedicine, 189*, 105343. https://doi.org/10.1016/j.cmpb.2020.105343

Ronneberger, O., Fischer, P., & Brox, T. (2015). U-Net: Convolutional networks for biomedical image segmentation. In N. Navab, J. Hornegger, W. M. Wells, & A. F. Frangi (Eds.), *Medical image computing and computer-assisted intervention – MICCAI 2015, lecture notes in computer science* (pp. 234–241). Cham: Springer International Publishing. https://doi.org/10.1007/978-3-319-24574-4_28.

Sacha, J. P., Goodenday, L. S., & Cios, K. J. (2002). Bayesian learning for cardiac SPECT image interpretation. *Artificial Intelligence in Medicine, 26*, 109–143. https://doi.org/10.1016/S0933-3657(02)00055-6

Shibutani, T., Nakajima, K., Wakabayashi, H., Mori, H., Matsuo, S., Yoneyama, H., et al. (2019). Accuracy of an artificial neural network for detecting a regional abnormality in myocardial perfusion SPECT. *Annals of Nuclear Medicine, 33*, 86–92. https://doi.org/10.1007/s12149-018-1306-4

Shi, L., Onofrey, J. A., Liu, H., Liu, Y.-H., & Liu, C. (2020). Deep learning-based attenuation map generation for myocardial perfusion SPECT. *European Journal of Nuclear Medicine and Molecular Imaging*. https://doi.org/10.1007/s00259-020-04746-6

Slomka, P. J., Betancur, J., Liang, J. X., Otaki, Y., Hu, L.-H., Sharir, T., et al. (2018). Rationale and design of the REgistry of fast myocardial perfusion imaging with NExt generation SPECT (REFINE SPECT). *Journal of Nuclear Cardiology*. https://doi.org/10.1007/s12350-018-1326-4

Song, C., Yang, Y., Wernick, M. N., Hendrik Pretorius, P., & King, M. A. (2019a). Approximate 4D reconstruction of cardiac-gated spect images using a residual convolutional neural network. In *Presented at the 2019 IEEE international conference on image processing (ICIP)* (pp. 1262–1266). https://doi.org/10.1109/ICIP.2019.8803772

Song, C., Yang, Y., Wernick, M. N., Pretorius, P. H., & King, M. A. (2019b). Low-dose cardiac-gated spect studies using a residual convolutional neural network. In *Presented at the 2019 IEEE 16th international symposium on biomedical imaging (ISBI 2019)* (pp. 653–656). https://doi.org/10.1109/ISBI.2019.8759586

Spier, N., Nekolla, S., Rupprecht, C., Mustafa, M., Navab, N., & Baust, M. (2019). Classification of polar maps from cardiac perfusion imaging with graph-convolutional neural networks. *Scientific Reports, 9*, 1–8. https://doi.org/10.1038/s41598-019-43951-8

Tägil, K., Marving, J., Lomsky, M., Hesse, B., & Edenbrandt, L. (2008). Use of neural networks to improve quality control of interpretations in myocardial perfusion imaging. *The International Journal of Cardiovascular Imaging, 24*, 841–848. https://doi.org/10.1007/s10554-008-9329-x

Togo, R., Hirata, K., Manabe, O., Ohira, H., Tsujino, I., Magota, K., et al. (2019). Cardiac sarcoidosis classification with deep convolutional neural network-based features using polar maps. *Computers in Biology and Medicine, 104*, 81–86. https://doi.org/10.1016/j.compbiomed.2018.11.008

Togo, R., Ogawa, T., Manabe, O., Hirata, K., Shiga, T., & Haseyama, M. (2019). Extraction of regions related to cardiac sarcoidosis in polar map images. In *Presented at the 2019 IEEE 1st global conference on life sciences and technologies (LifeTech)* (pp. 237–238). https://doi.org/10.1109/LifeTech.2019.8883975

Ulyanov, D., Vedaldi, A., & Lempitsky, V. (2020). Deep image prior. *International Journal of Computer Vision*. https://doi.org/10.1007/s11263-020-01303-4

Van Train, K. F., Areeda, J., Garcia, E. V., Cooke, C. D., Maddahi, J., Kiat, H., et al. (1993). Quantitative same-day rest-stress technetium-99m-sestamibi SPECT: Definition and validation of stress normal limits and criteria for abnormality. *Journal of Nuclear Medicine, 34*, 1494–1502.

Wang, T., Lei, Y., Tang, H., He, Z., Castillo, R., Wang, C., et al. (2019). A learning-based automatic segmentation and quantification method on left ventricle in gated myocardial perfusion SPECT imaging: A feasibility study. *Journal of Nuclear Cardiology*. https://doi.org/10.1007/s12350-019-01594-2

Wang, Y., Yu, B., Wang, L., Zu, C., Lalush, D. S., Lin, W., et al. (2018). 3D conditional generative adversarial networks for high-quality PET image estimation at low dose. *NeuroImage, 174*, 550–562. https://doi.org/10.1016/j.neuroimage.2018.03.045

Yoneyama, H., Nakajima, K., Taki, J., Wakabayashi, H., Matsuo, S., Konishi, T., et al. (2019). Ability of artificial intelligence to diagnose coronary artery stenosis using hybrid images of coronary computed tomography angiography and myocardial perfusion SPECT. *European Journal Hybrid Imaging, 3*, 4. https://doi.org/10.1186/s41824-019-0052-8

Zhu, B., Liu, J. Z., Cauley, S. F., Rosen, B. R., & Rosen, M. S. (2018). Image reconstruction by domain-transform manifold learning. *Nature, 555*, 487–492. https://doi.org/10.1038/nature25988

CHAPTER

Radiomics in cardiovascular imaging: principles and clinical implications

12

Márton Tokodi[1], Attila Kovács[1], Pál Maurovich-Horvat[2]

[1]*Heart and Vascular Center, Semmelweis University, Budapest, Hungary;* [2]*Department of Radiology, Medical Imaging Center, Semmelweis University, Budapest, Hungary*

Chapter outline

Machine Learning in Cardiovascular Medicine. https://doi.org/10.1016/B978-0-12-820273-9.00012-9

Introduction

Over the past decades, the medical community has witnessed an unprecedented expansion in the arsenal of noninvasive imaging technologies (Thomas and Banerjee, 2013). By providing exhaustive, high-resolution information about the structure and function of the heart and vasculature, imaging has become the pillar of modern cardiovascular (CV) care (Zoghbi, 2012). It has a pivotal role in setting up accurate diagnosis, quantification of disease traits, and tracking response to therapy. In addition, it has enabled remarkable improvements in disease phenotyping and prognostication, and has conspicuously advanced CV research (Carli, Geva and Davidoff, 2016).

Massive amounts of medical images are acquired along the diagnostic and therapeutic pathway that a CV patient follows. The conventional interpretation of these images is reliant on the compound of basic, guideline-defined geometric quantifiers and qualitative, often subjective, descriptors. Although this approach is sufficient for clinical judgment in most cases, it has substantial shortcomings, and there is much space to improve. Images encompass an abundance of meaningful information that is challenging to quantify, interpret, and link with diagnosis or prognosis using solely conventional metrics. Moreover, visually assessed qualitative characteristics rely on a priori knowledge, have limited capabilities to discern subtle differences, and are hard to reproduce. Therefore, medical professionals are in the pursuit of novel technologies that can augment the interpretation of images allowing more personalized patient management.

Advanced data extraction and analytic tools might offer a panacea for the limitations above. One of these emerging technologies is *radiomics*, which can be defined as the high-throughput extraction of quantitative features enabling the conversion of images into structured, mineable datasets. These high-dimensional databases can be supplied to machine learning (ML) algorithms that are capable of simultaneously interpreting myriads of variables to generate and test hypotheses, and ultimately to support clinical decisions. Radiomic features can describe the texture and spatial complexity in detail through the uncovering of patterns that are invisible to the naked eye. Thus, it provides a nearly limitless supply of imaging biomarkers with potential incremental value over conventional descriptors.

The utility of radiomics-based predictive models was first demonstrated in the field of oncology, where notable success was achieved in tumor classification, prediction of treatment response, and prognostication in multiple cohorts (Gillies, Kinahan and Hricak, 2016; Lambin et al., 2017). CV imaging is another clinical discipline that could be remarkably transformed by radiomics (Kolossvary, Kellermayer, Merkely, & Maurovich-Horvat, 2018; Oikonomou, Siddique, & Antoniades, 2020; Raisi-Estabragh et al., 2020). Radiomics analysis can be performed on standard-of-care images acquired using echocardiography, computed tomography (CT), or magnetic resonance imaging (MRI). In echocardiography, it has been reported to be suitable for the identification of cardiac amyloidosis (Pinamonti et al., 1989), myocardial fibrosis (Kagiyama et al., 2020), different stages of myocardial infarction (MI) (Sudarshan et al., 2016a), or left atrial thrombi (Sun et al., 2014). There has been an increasing interest in the application of radiomics analysis to cardiac CT images, as it has achieved promising results in the characterization of coronary plaques (Kolossvary et al., 2017; Kolossváry, Karády et al., 2019; Kolossvary, Park et al., 2019) and perivascular fat (Oikonomou et al., 2019). Proof-of-concept studies have also demonstrated the feasibility and potential clinical value of cardiac MRI radiomics in the diagnosis of hypertrophic cardiomyopathy (HCM) (Baessler, Mannil, Maintz et al., 2018) and the assessment of MI or myocardial viability (Baessler, Mannil, Oebel et al., 2018; Larroza et al., 2017; Larroza et al., 2018).

In this chapter, we provide an overview of basic radiomics concepts and workflow, along with its potential role in CV imaging. Besides its strength, we will also review some of the pitfalls that should be avoided while utilizing this state-of-the-art analytic approach.

Concept of radiomics

Radiomics is a precision phenotyping toolbox designed to break down radiological images into a multitude of features (Gillies, Kinahan, & Hricak, 2016). In contrast to the conventional practice of treating medical images as pictures intended solely for visual interpretation, radiomics aims to represent pictorial data using complex mathematical formulae, enabling a unique way of analysis. Its name expresses its similarities to other "omics" approaches (e.g., genomics, transcriptomics, proteomics) as radiomics is a field of study in life sciences that focuses on generating large-scale data from single objects or samples. One desirable attribute of "omics" data is that they are mineable with robust data analysis techniques such as ML; thus, they can be used for exploration and hypothesis generation.

Process of radiomics

The practice of radiomics comprises discrete steps (Fig. 12.1): (1) image acquisition, (2) identification of the region of interest (ROI) and segmentation, (3) extraction of descriptive features from segmented ROI, (4) generation of searchable databases, and (5) mining these databases to develop ML-based models.

Image acquisition

Images suitable for radiomics analysis can be assessed using different imaging modalities such as cardiac ultrasound, CT, MRI, or nuclear medicine imaging techniques. It should be highlighted that radiomics can be applied on standard, routinely acquired clinical images without requiring dedicated acquisition protocols

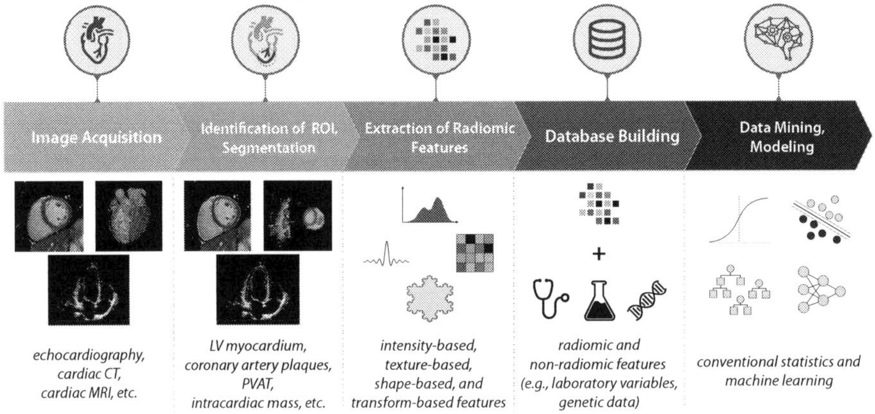

FIGURE 12.1

Pipeline of radiomics-based analysis. The radiomics workflow comprises five discrete steps: (1) image acquisition, (2) identification of the ROI and segmentation, (3) extraction of descriptive features from segmented ROI, (4) generation of searchable databases, and (5) mining these databases to develop machine learning–based models. Abbreviations: *CT*, computed tomography; *LV*, left ventricle; *MRI*, magnetic resonance imaging; *PVAT*, perivascular adipose tissue; *ROI*, region of interest.

or additional imaging. Accordingly, radiomics has a real potential for being integrated into the routine clinical workflow as an adjunct to conventional measures.

Nevertheless, the standardization of imaging protocols across different centers is still lacking. This usually does not hamper the identification and measurement of conventional features used in current clinical practice. However, when images are analyzed in-depth in order to extract pixel-based data, differences in image acquisition, reconstruction, and postprocessing parameters can introduce significant variability that is not caused by the underlying pathological conditions (Altazi et al., 2017; Berenguer et al., 2018; Kolossvary, Szilveszter et al., 2019; Mackin et al., 2015; Meyer et al., 2019; Zhao et al., 2016). Therefore, the standardization of image acquisition, reconstruction, and postprocessing should be promoted to eliminate this unnecessary confounding variability (Ger et al., 2018).

Identification of the region of interest and segmentation

Following the selection of the images to be analyzed, the next critical and challenging step is the identification and segmentation of image regions (i.e., ROI) containing essential prognostic information. As the radiomic features are extracted from the delineated areas or volumes, variations in contouring that alter the ROI will modify the values of the radiomic features as well. Therefore, it is indispensable to have a consistent contouring technique defined through standard operating procedures. There are various contouring methods ranging from manual to fully automated techniques (Chen et al., 2020). A consensus is emerging that the most reproducible results in radiomics are achievable with computer-assisted contouring followed by limited manual corrections (Gillies et al., 2016). Numerous methods exist that can ensure the robustness of feature extraction and limit the extent of bias such as evaluation by multiple operators, perturbing segmentation with noise, combining diverse algorithms, or utilizing images from different stages of the breathing cycle.

Extraction of radiomic features

After the segmentation of the ROI, radiomic features can be extracted through dedicated pipelines developed by individual manufacturers or using open-source packages. Radiomic features can be categorized into four major groups (Fig. 12.2): (1) intensity-based metrics, (2) texture-based metrics, (3) shape-based measures, and (4) transform-based features.

Intensity-based metrics

Intensity-based metrics (or first-order statistics) are calculated from individual pixel values without considering any additional information that might be obtained from analyzing the spatial relationship between pixels. The pixel values can be analyzed using the tools of histogram analysis. These metrics are used to quantify different aspects of the distribution, such as central tendency, variation, shape, and diversity.

The basic mathematical descriptors of central tendency and variation, such as *mean*, *median*, *minimum*, *maximum*, *percentiles*, and *interquartile range* are well known among medical professionals. Although these metrics are simple to calculate and interpret, they do not describe precisely the size and shape of the distribution. As several different distributions may exist with the same central tendency but different shapes, these statistics are not sufficient to characterize a given image.

The shape of the distribution is most frequently described by *variance*— a measure of sample variation around the mean— and *standard deviation (SD),* which is the square root of variance. In normally distributed datasets, the SD quantifies the band around the mean where approximately 68% of the data points are located. *Skewness* and *kurtosis* are less frequently used by medical professionals; however, they capture other meaningful characteristics of the distribution. Skewness expresses how asymmetric the distribution is. Negative skewness indicates that the bulk of the data is located to the right of the mean, whereas positive skewness means the opposite. Kurtosis quantifies how close the data points are to the mean. Small values indicate that most values are situated within one SD of the mean, and the dataset contains only a few outliers, whereas higher values imply that a large proportion of data points can be found beyond one SD. In normally distributed datasets, the kurtosis is equal to 3. It is common to compare the kurtosis of a distribution to this value to assess how data points are dispersed around the mean.

The statistics mentioned above provide information about the average, the variation, and the distribution of the sample but not about the diversity of pixel values.

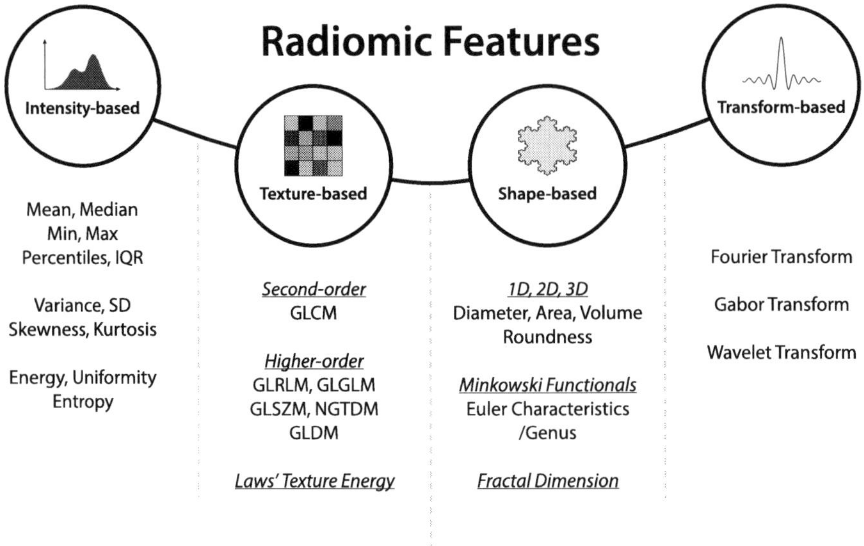

FIGURE 12.2

Classes of radiomic features. *Intensity-based metrics* (or first-order statistics) are calculated by analyzing individual pixel values using the tools of histogram analysis. *Texture-based features* attempt to describe patterns using mathematical formulae based on the spatial arrangement of pixels. The texture analytic toolbar includes diverse approaches such as second- and higher-order metrics, and Laws' texture energy approach, to name a few. *Shape-based metrics* aim to quantify the size, the localization, the spatial complexity, and the self-symmetry of the given lesion or abnormality. *Transform-based methods* convert the images into other domains, such as the frequency domain, to extract additional radiomic features. It is important to note that we are providing only a few examples of each feature class; however, thousands of other radiomic parameters can be extracted from radiological images. Abbreviations: *GLCM*, gray-level co-occurrence matrix; *GLDM*, gray-level dependence matrix; *GLGLM*, gray-level gap-length matrix; *GLRLM*, gray-level run-length matrix; *GLSZM*, gray-level size zone matrix; *IQR*, interquartile range; *NGTDM*, neighborhood gray-tone difference matrix; SD, standard deviation.

Concepts of information theory can be applied to quantify heterogeneity. *Energy* measures the overall magnitude of the values, and it is calculated by squaring and then summing them. *Uniformity* quantifies the similarity of the values, and it is computed by squaring the relative frequency of the given values and then summing them. *Entropy* measures the information content of the dataset (Shannon, 1948). Events with higher probabilities carry less information because their outcomes can be guessed more easily, whereas unlikely events bear more information as their occurrence highlights specific instances. Entropy quantifies uncertainty by weighting the information content of each event with its probability. The higher the entropy, the more heterogeneous the dataset is.

Texture-based metrics

The preceding parameters neglect all spatial information and use the absolute values of individual pixels solely. In order to consider spatial relationships, more sophisticated mathematical approaches are required. As a promising solution, the field of texture analysis was born in the 1970s, and it has been continuously evolving ever since (Davies, 2018). Texture analysis attempts to describe patterns— the systematic, spatial variation of some physical characteristics such as intensity, shape, or color—using mathematical formulae based on the spatial arrangement of pixels. The texture analytic toolbar includes diverse approaches such as second- and higher-order metrics (Fig. 12.3), and Laws' texture energy approach, to name a few.

Second-order statistics

In 1973, Haralick et al. proposed the idea of *gray-level co-occurrence matrix (GLCM)* for the classification of materials and terrains based on their texture (Haralick, Shanmugam and Dinstein, 1973). GLCMs allow the derivation of second-order metrics that are calculated from the spatial relationship of two pixels and not from the values themselves. These matrices aim to quantify how frequently different combinations of gray levels occur in pairs of pixels that are separated by a predefined distance along a given direction (Fig. 12.3).

Prior to matrix generation, pixels values are required to be discretized into a given number of bins, as pixels with exactly the same values occur very scarcely. The matrix will have the same number of rows as columns, which equals the number of bins. Next, a direction and a distance are needed to be defined to analyze the texture. The direction is usually described with an angle (Fig. 12.4). Statistics are calculated in four directions only (0, 45, 90, 135 degrees), as the remaining directions are the exact counterparts of the above. As the next step, the transposed matrix is added to the original raw GLCM to receive a symmetrical matrix. Because relative frequencies are more informative than absolute numbers, the matrix should be normalized by dividing each of its elements by the sum of all values in the GLCM. Due to the fact that texture is an intrinsic property of an image, the same results should be obtained even if the image is rotated. Therefore, to achieve rotationally invariant results, statistics are calculated on the four GLCMs and then averaged.

Haralick et al. (1973) introduced 14 different statistics that can be derived from the GLCMs; however, many more exist. To illustrate the significance of these features, we will briefly discuss energy and contrast. Energy (also known as angular second moment or uniformity) is a measure of homogeneity, and it is equal to the sum of the squared elements of the normalized GLCM. A more homogenous image (containing only a few dominant gray-tone transitions) has higher values of energy. Contrast is computed by the elementwise multiplication of the normalized GLCM and the difference in the gray-level values squared for that given row and column, and then adding up all the numbers. Greater values indicate a higher amount of local variation (i.e., a bigger difference between the values of the neighboring pixels).

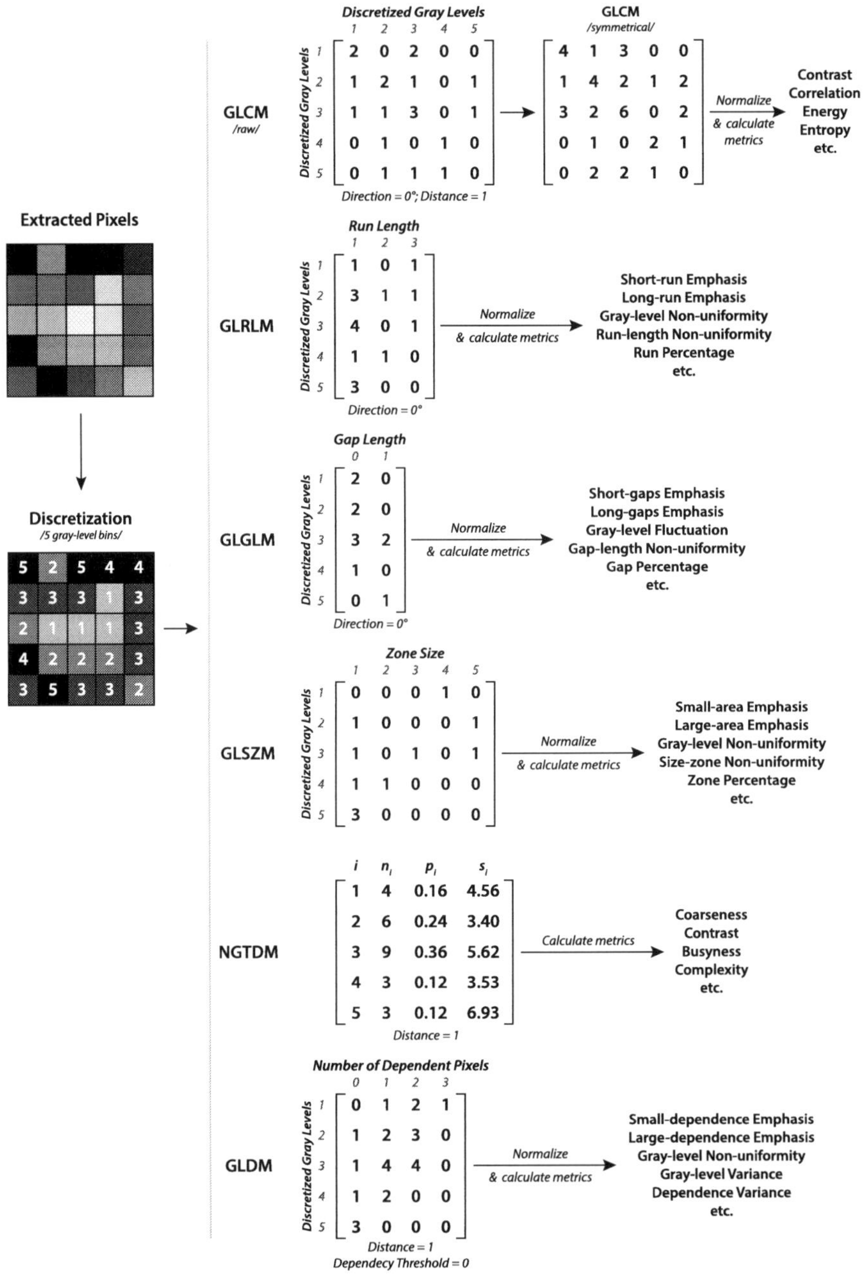

FIGURE 12.3

Characterization of texture using second- and higher-order radiomic features. After the segmentation of the ROI, pixels are discretized into a given number of bins. *GLCMs* aim to

Higher-order statistics

Whereas second-order statistics examine the relationship between pairs of pixels, higher-order metrics attempt to describe the link between three or more pixels. Hereunder, we review some of the most widely used concepts (Fig. 12.3).

Gray-level run-length matrix (GLRLM) quantifies how many consecutive pixels have the same value along a predefined direction (Galloway, 1975). The rows of the matrix represent the discretized gray levels and the columns the run lengths. Galloway introduced five descriptors— namely short-run emphasis, long-run emphasis, gray-level nonuniformity, run-length nonuniformity, and run percentage—to emphasize different properties of these matrices (Galloway, 1975). Similar to those derived from GLCMs, these statistics can be calculated in all four directions; therefore, to obtain rotationally invariant results, they should be averaged.

Gray-level gap-length matrix (GLGLM) measures the distribution of gap lengths for each gray level in an image (Xinli, Albregtsen and Foyn, 1995). The gap length is defined as the distance between two pixels having the same gray level minus one. Thus, two neighboring pixels with identical gray levels have zero gap length. If no pixel with the same gray level is found along the searching direction, the gap length is considered as infinite and is omitted. Several texture features can be extracted from GLGLMs such as short-gaps emphasis, long-gaps emphasis, gray-level fluctuation, gap-length nonuniformity, gap percentage, low gray-level gaps emphasis, high gray-level gaps emphasis.

Gray-level size zone matrix (GLSZM) quantifies the number of continuous pixels with the same gray level (Thibault et al., 2009). This matrix has a fixed number of lines equal to the number of gray levels and a dynamic number of columns

quantify how frequently different combinations of gray levels occur in pairs of pixels that are separated by a predefined distance along a given direction. *GLRLMs* quantify how many consecutive pixels have the same value along a predefined direction. *GLGLMs* measure the distribution of gap lengths for each gray level in an image. To achieve rotationally invariant results, matrices (GLCM, GLRLM, GLGLM) and statistics should be calculated in all four directions (0, 45, 90, and 135 degrees) and then averaged. *GLSZMs* quantify the number of continuous pixels with the same gray level. *NGTDMs* are generated by the summation of differences between the gray level of each pixel and its neighborhood within a predefined distance. *GLDMs* determine the number of connected pixels within a given distance that are dependent on the center pixel. A neighboring pixel is considered dependent on the center pixel if the difference between their gray-level values does not exceed a predefined threshold. The GLSZM, the NGTDM, and the GLDM are rotation independent; thus, only one of each is calculated from the given ROI. Abbreviations: *GLCM*, gray-level co-occurrence matrix; *GLDM*, gray-level dependence matrix; *GLGLM*, gray-level gap-length matrix; *GLRLM*, gray-level run-length matrix; *GLSZM*, gray-level size zone matrix; *i*, the discretized gray levels; *NGTDM*, neighborhood gray-tone difference matrix; n_i, the number of pixels with gray level i; p_i, the probability of gray level i (calculated as $n_i/n_{all\ pixels}$); *ROI*, region of interest; s_i, the sum of absolute differences for gray level i.

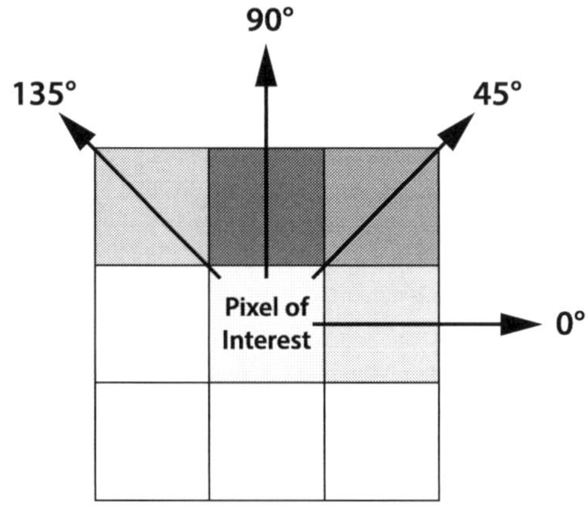

FIGURE 12.4

Schematic illustration of directions. To calculate some of the second- and higher-order radiomic features, a direction is required to be defined. By convention, the direction is described with an angle. Pixels to the east of the reference pixel are at 0 degrees, the ones to the northeast are at 45 degrees, the ones to the north are at 90 degrees, and the ones to the northwest are at 135 degrees. Statistics are calculated in four directions only (0, 45, 90, 135 degrees), as the remaining ones are the exact counterparts of the above.

determined by the size of the largest area. The more homogeneous the texture, the wider the matrix will be. Contrary to GLCM, GLRLM, and GLGLM, the GLSZM is rotation independent; thus, only one matrix is calculated from the given ROI. In an investigation focusing on the accurate classification of skin fibroblast nuclei using GLGLM, Thibault et al. suggested the utilization of 13 indices (many of them are analogical to those of the GLRLM) (Thibault et al., 2009).

A *neighborhood gray-tone difference matrix (NGTDM)* is generated by the summation of differences between the gray level of each pixel and its neighborhood within a predefined distance (Amadasun and King, 1989). Various statistics can be computed from this matrix, e.g., coarseness, contrast, and complexity. Coarseness—an indicator of the spatial rate of change—is a measure of the average difference between the center pixel and its neighborhood. A higher value implies a lower spatial change rate and a locally more uniform texture. Contrast measures the spatial intensity change, but it is also dependent on the overall gray-level dynamic range. Higher values will be observed if both the dynamic range and the spatial change rate are high (i.e., the image has a wide variety of gray levels and there are large changes between pixels and their neighborhood). Complexity quantifies the number of primitive components in the image (i.e., the image is nonuniform, and there are many rapid changes in gray-level values).

Gray-level dependence matrix (GLDM) determines the number of connected pixels within a given distance that are dependent on the center pixel (Sun and Wee, 1983). A neighboring pixel is considered dependent on the center pixel if the difference between their gray-level values does not exceed a predefined threshold. The GLDM can be used to calculate myriads of radiomic features such as small-dependence emphasis, large-dependence emphasis, gray-level nonuniformity, gray-level variance, or dependence variance.

Shape-based metrics

Beyond its texture, the spatial distribution, localization, and size of a given abnormality carry additional information with potential diagnostic and prognostic value. Some of these metrics are simple and routinely assessed in clinical practice, whereas others are less known and more challenging to understand.

One-, two-, and three-dimensional metrics

One-dimensional (1D) metrics are based on measuring the distance between two points. These parameters are commonly used in daily clinical practice to describe the size of an abnormality. Diameters measured in different directions can be used to derive new statistics that can characterize additional properties (e.g., the ratio of the longest and the shortest diameters describes the roundness or ellipticity of a lesion).

Two-dimensional (2D) metrics are calculated based on cross-sectional planes and are used to compute different parameters that are based on areas. These parameters are often used to approximate the three-dimensional (3D) properties of the abnormality (e.g., the 1D diameter and the 2D area are all considered approximations of the 3D volume).

Three-dimensional metrics aim to quantify different aspects of a volumetric shape. The geometrical properties of 3D objects have been thoroughly examined in the field of rigid-body mechanics. All objects have so-called principal axes or eigenvectors. These mutually perpendicular axes intersect at the center of mass. The force applied to the principal axes acts independently, meaning that if we push or rotate the object along or around any of these axes, the object will not move or rotate in any other direction. These eigenvectors also have eigenvalues that can be interpreted as weights proportional to the amount of mass located along that given principal axis. The eigenvectors can be used to quantify different shape-based metrics— for example, roundness, which is the difference between the largest eigenvalues of the smallest enclosing and the largest enclosed ellipse.

Fractal dimension

Fractal dimension measures the spatial complexity of an object and quantifies self-symmetry by analyzing repeating patterns at different scales (Mandelbrot, 1967). Objects with no fractal properties scale their characteristics exponentially depending on the dimension. On the other hand, fractals act differently. If a line with fractal properties is enlarged by two, it would be longer than double of its original length. This is caused by the fact that when an object with fractal properties is magnified, smaller details become visible and measurable, which affects the length and size of the given object. This phenomenon can also be observed when we try to measure the length of a coastline. With the gradual magnification of the image, more and more details emerge, leading to an increase in the total length of the coastline. Fractal patterns can be found everywhere in nature, such as the Romanesco broccoli, snowflakes, or the rings of Saturn. Another famous and illustrative example is the Mandelbrot set, which can be easily programmed into a simple code to create an infinite stream of changing, self-similar patterns.

The most simplistic concept of quantifying fractal dimension is the box-counting method. In this analysis, the boxes of a grid occupied by the object are counted repeatedly at different scales. Then, the number of occupied boxes is plotted against the reciprocal of the scale on a log-log plot, and the box-counting dimension will be equal to the slope of the line connecting the data points (Lopes and Betrouni, 2009).

Transform-based metrics

Medical images comprise an immense number of pixel values distributed in the spatial domain. These images can be transformed into the frequency domain without losing any data as spatial- and frequency-domain representations of an image carry identical information. However, each emphasizes different aspects of an image; therefore, each enables different analytic options. In the frequency domain, the pattern and rate at which the pixel values change along spatial directions are used to represent the image instead of assigning gray-level values to spatial coordinates.

Fourier and Gabor transforms

Fourier transform decomposes the spatial-domain dataset into a fundamental sine wave and its harmonics along the two dimensions of an image. By converting the image into the frequency domain using this transform, spatial information is discarded. The Gabor transform, named after the Hungarian-born physicist Dénes Gábor, is a special case of the short-time Fourier transform, which enables the simultaneous evaluation of spatial and frequency information (Gabor, 1946). Fast Fourier transform allows the identification and hence the filtration of specific frequencies (e.g., ones related to noise), whereas Gabor transform provides access to image features such as edges, texture, and blobs.

Wavelet transforms

Similar to the Fourier transform, wavelet transform also converts the image into the frequency domain, but spatial information is preserved (Dettori and Semler, 2007). When the wavelet transform is performed, a window function is translated across the original image. The width of the window determines both spatial and frequency resolutions: the narrower the window, the higher the localization precision (lower the frequency resolution). By gradually changing the width of the window, a series of transformed images is generated that can be analyzed with various statistical methods. Wavelet transformation decomposes the data into high- and low-frequency components, which describe the pattern and rate of changes along spatial directions. At high frequencies, wavelets can capture discontinuities, ruptures, and singularities in the original data. At low frequencies, the wavelet characterizes the coarse structure of the data to identify long-term trends. Thus, wavelet analysis allows the extraction of hidden but significant temporal features of the original data, while it also improves the signal-to-noise ratio.

Database building

The power of predictive models heavily depends on having well-curated, sufficiently large datasets. Radiomics-based modeling can be performed with a few hundred patients as well; however, more extensive datasets would significantly enhance the predictive power and generalizability. Moreover, holistic models incorporating additional, nonradiomic features (e.g., clinical and genomic covariates) tend to be more robust; therefore, researchers should endeavor to capture several other domains of variables as well.

Modeling

A detailed description of modeling goes beyond the focus of this chapter; however, a few aspects should be briefly discussed. The process of radiomic feature extraction will yield a virtually unlimited number of parameters, which often exceeds the sample size of the cohort used for modeling. Utilizing all features would inevitably result in overfitting; therefore, one needs to identify an optimal subset of input variables. Feature selection should be performed in a data-driven fashion and should be reported in a transparent manner. Various techniques (filtering, wrapper, and embedded methods) exist to pinpoint features that are robust, informative, and nonredundant (Devi and Sabrigiriraj, 2018; Saeys, Inza, & Larrañaga, 2007). Another solution to overcome the curse of dimensionality is to generate a new, smaller set of features from the original ones. This approach is termed dimensionality reduction and can be accomplished using several different algorithms, such as principal component analysis or manifold learning methods (Nguyen and Holmes, 2019).

Once the final set of input features is identified, one can proceed with building a specific predictive model. Different ML algorithms are associated with distinct inherent strengths and limitations; therefore, the choice of modeling definitely affects prediction performance (Parmar, Grossmann, Bussink, Lambin, & Aerts, 2015). Multiple-modeling implementations are desirable, and researchers should consider making the source-code available (e.g., via online repositories) to enable the wide-spread utilization of the models and the independent replication of results.

Emphasis should also be placed on the thorough validation of models as nonvalidated predictive systems are of limited value (Steyerberg and Harrell, 2016; Steyerberg and Vergouwe, 2014). The optimal method for validation would be the evaluation of models' performance in external cohorts. However, this is not feasible in all cases; therefore, validation can be performed internally by partitioning the entire cohort into training (used exclusively for model derivation) and test sets (used solely for assessing the performance of fitted models). Several techniques are available for internal validation, such as the hold-out method, k-fold cross-validation, leave-one-out cross-validation, random subsampling, or bootstrapping (Steyerberg et al., 2001). Nevertheless, the results of internal validation tend to be more optimistic, as they often overestimate the effectiveness of the proposed models (Siontis, Tzoulaki, Castaldi, & Ioannidis, 2015; Steyerberg, Bleeker, Moll, Grobbee, &Moons, 2003). Accordingly, externally validated models have more credibility than internally validated ones, as the former is considered more independent, which reinforces the models' generalizability.

Noting the high inconsistency in reporting of multivariate prediction tasks in medicine, researchers are advised to follow the recommendations set forth by the TRIPOD statement and the Radiomics Quality score (Collins, Reitsma, Altman, & Moons, 2015; Lambin et al., 2017).

Clinical implications

The feasibility and potential utility of radiomics have been illustrated in various clinical scenarios. Significant success has been achieved in myocardial tissue characterization, coronary artery calcium quantification, coronary plaque phenotyping, and risk stratification of CV patients (Kolossvary, Kellermayer, Merkely, & Maurovich-Horvat, 2018; Oikonomou, Siddique, & Antoniades, 2020; Raisi-Estabragh et al., 2020).

Characterization of myocardial texture

A considerable number of proof-of-concept studies have demonstrated the capabilities of texture analysis to aid clinicians in various diagnostic dilemmas related to patients with suspected hypertrophic cardiomyopathy (HCM). In a small cohort, Baessler, Mannil, Maintz, Alkadhi and Manka (2018) observed significant differences in four texture features (extracted from noncontrast T1-weighted magnetic resonance (MR) images) between HCM patients and healthy controls. They identified gray-level nonuniformity—a marker of texture inhomogeneity—as the best discriminator between the two groups (logistic regression, cross-validated area under the receiver operating characteristic curve [AUC]: 0.95). Other investigators illustrated that the texture analysis of noncontrast cine MR images could accurately classify the different etiologies of left ventricular hypertrophy such as HCM, cardiac amyloidosis, aortic stenosis, and hypertension-induced hypertrophy (Schofield et al., 2019). The differential diagnostic potential of radiomics was also exemplified by Neisius et al. (2019), whose support vectors machine (SVM)-based model (using radiomic features extracted from native T1 maps) achieved an AUC of 0.89 in the distinction between HCM and hypertensive heart disease. All these findings imply that the texture analysis of noncontrast cardiac MR images is a desirable, gadolinium-free technique that enables the detection of tissue-level myocardial alterations in HCM.

In support of the presumptions above, radiomics seems to be capable of capturing subclinical changes in noncontrast cine cardiac MR images of hypertensive patients well beyond the efficiency of conventional image interpretation techniques (Cetin et al., 2019). It is also important to highlight from this study that all features selected for inclusion in the final model were intensity- and texture-based metrics indicating that the main changes can be found in the cardiac texture and tissue. This might explain the inability of conventional structural and functional imaging indices to identify these often subclinical alterations.

The precise assessment of MI as well as myocardial viability are two other strengths of cardiac MR, and they account for a substantial proportion of referrals. Although late gadolinium enhancement (LGE) cardiac MR is a well-established and accurate method for the detection of ischemic myocardial scar (Goetti, Kozerke et al., 2011; von Knobelsdorff-Brenkenhoff and Schulz-Menger, 2012), an improvement in the detection of scarred myocardium was reported when texture- and intensity-based features were also considered (Engan, Eftestol, Orn, Kvaloy, & Woie, 2010; Kotu, Engan, Eftestøl, Ørn, & Woie, 2011). Not only the detection but also the differentiation between acute and chronic MI is essential for the appropriate patient treatment and follow-up (Roffi et al., 2016). However, there is no clear-cut difference between the hyperenhancement of the acutely and chronically infarcted myocardium on LGE images (Choi, 2006). Radiomics can facilitate this differentiation, as it was found to distinguish acute (within 1 week) from chronic MI (more than 6 months before imaging) correctly using either noncontrast cine or LGE MR images (cross-validated AUC $>$ 0.80) (Larroza et al., 2017). In cases when gadolinium-based contrast cannot be administered (e.g., chronic kidney disease), alternative methods for the characterization of myocardial scar would be desirable. As histological properties of the myocardium are reflected by texture features extracted from noncontrast cine MR images, radiomics could be applied to depict tissue-level differences between nonviable and remote segments in a gadolinium-free manner. This supposition was proved by Larroza et al. (2018) as they could detect nonviable, viable, and remote segments (as per LGE) in noncontrast enhanced cine MR images with cross-validated AUC values of 0.935, 0.819, and 0.794, respectively. In addition, by harnessing the imperceptible differences in the myocardial texture of noncontrast cine cardiac MR images, Baessler, Mannil, Oebel et al. (2018) could effectively diagnose small (volume of enhancement at LGE imaging $\leq$20%) and large (enhanced volume $>$20%) myocardial scars, with cross-validated AUCs of 0.93 and 0.92, respectively. Thus, myocardial viability could be assessed using cine MRI sequences solely without the need for gadolinium administration for further LGE acquisitions.

With its unique capability for tissue characterization, cardiac MRI has become the primary tool for the noninvasive assessment of myocardial inflammation in patients with suspected myocarditis (Biesbroek, Beek, Germans, Niessen, & van Rossum, 2015; Caforio et al., 2013). Nevertheless, many studies have reported substantial variations in sensitivity and specificity of the currently applied diagnostic tools such as the Lake Louise Criteria (Lurz et al., 2012; Lurz et al., 2016); therefore, more accurate techniques are warranted. This has offered radiomics another perfect opportunity to outperform existent methods: the diagnostic performance of Lake Louise Criteria, which accounts for mean T1 and T2 values (all AUCs <0.70), has been significantly exceeded by a predictive model combining two features (T2 run-length nonuniformity and T2 gray-level nonuniformity, cross-validated AUC: 0.88) in the identification of patients with biopsy-proven infarctlike acute myocarditis (Baessler, Luecke et al., 2018).

Despite being the most appropriate choice for myocardial tissue characterization, the availability of cardiac MR might be limited, and more accessible screening tools are warranted. Accordingly, Kagiyama et al. (2020) proposed a radiomic-based cardiac ultrasound fingerprinting pipeline that can achieve an AUC of 0.84 in diagnosing the presence of myocardial fibrosis (detected by LGE cardiac MRI). The utilization of their approach may allow cardiac ultrasound to be a cheap and widely available screening tool for myocardial fibrosis; however, further research and validation are required to overthrow the barriers in the adoption of ultrasound techniques for myocardial tissue characterization.

Other investigators have attempted to detect infarcted myocardium from radiomics-based analysis of 2D echocardiograms as various radiomics descriptors such as first-order statistics (Tak, Visser, Rahimtoola, & Chandraratna, 1992), wavelet transform (Mojsilovic, Popovic, Neskovic, & Popovic, 1997; Nesković et al., 1998), and higher-order statistics (Sudarshan et al., 2015) can be extracted from these images and can be used to distinguish normal and infarcted myocardium in a noninvasive, radiation-free, and cost-efficient manner. Sudarshan et al. demonstrated that normal and infarcted myocardium could be discriminated with 94% accuracy (Sudarshan et al., 2016b) and the extent of myocardial damage (normal, moderately and severely infarcted myocardium) can be detected with 99% accuracy (Sudarshan et al., 2016a). To enhance clinical feasibility, they combined texture metrics to create single-number indices. Another group of investigators extracted a multitude of gray-scale features (fractal dimension, entropies based on the higher-order spectra, features based on image texture and local binary patterns, and wavelet-based features) from echocardiographic images belonging to a vast database of 400 normal cases and 400 patients with coronary artery disease (Acharya et al., 2013). In the final model, they used a subset of nine features to classify the images of normal and coronary artery disease affected cases with 100% accuracy. A novel, highly discriminative "HeartIndex" was also calculated from the combination of these features to enable an objective and a quick diagnosis of coronary artery disease in the daily clinical practice.

If we are looking for alternatives to LGE MR for the detection of myocardial scar, late iodine enhancement CT scans might be an ideal choice as iodine contrast agents have similar wash-in and wash-out kinetics as gadolinium (Gerber et al., 2006). Although CT has a superior spatial resolution, the greatest limitation when compared to MR is its poor contrast to noise ratio, which hampers the confident detection and segmentation of scar based on the visual assessment (Goetti, Feuchtner et al., 2011). Radiomics might overcome this limitation, as it is capable of detecting tissue-level alterations, and there is evidence that they can be used to effectively differentiate normal from scarred myocardium using late iodine enhancement CT images (Antunes et al., 2016). In another study, including patients with proven acute MI and controls with no cardiac abnormalities, Hinzpeter et al. (2017) investigated the feasibility and accuracy of texture analysis to distinguish healthy and infarcted myocardium using contrast-enhanced cardiac CT images. The best multivariate logistic regression model (AUC: 0.90) included kurtosis and short-run high gray-level emphasis, accentuating the complementary nature of first- and higher-order radiomic features in tissue characterization. Moreover, as demonstrated by Mannil, von Spiczak, Manka and Alkadhi (2018), the identification of acute and chronic MI is also possible using exclusively noncontrast enhanced cardiac CT images in combination with texture analysis and ML algorithms.

The detection of myocardial scars is fundamental to therapeutic decision-making, risk stratification, and therapeutic monitoring; however, the assessment of diffuse remodeling of the extracellular matrix is equally important. Multiple diagnostic strategies have been tested for the noninvasive assessment of diffuse extracellular matrix remodeling as an expanding body of evidence indicates that diffuse myocardial fibrosis is a causative factor of myocardial dysfunction and ventricular arrhythmias (Mewton, Liu, Croisille, Bluemke, & Lima, 2011). One such strategy is the radiomics-based analysis of late iodine enhancement cardiac CT images, as the distinct signatures revealed by this analysis were found to be associated with left ventricular dilatation and systolic and diastolic function (Esposito et al., 2018). In addition, the findings of Esposito et al. also suggest that radiomics bears the potential to identify the different patterns of left ventricular structural remodeling characterizing patients with recurrent ventricular tachycardia.

Automated coronary artery calcium scoring

Coronary artery calcifications can be visualized and quantified with CT. The extent of coronary calcification is routinely expressed using the Agatston score, which has been shown to be an independent and robust predictor of adverse CV events (Budoff et al., 2007; Budoff et al., 2017; Shaw, Raggi, Schisterman, Berman, & Callister, 2003). Such scoring is usually performed using prospectively electrocardiography (ECG)-triggered noncontrast enhanced cardiac CT. However, coronary calcification can be visualized with native, non-ECG-gated chest CT as well. Several studies have demonstrated that low-dose non-ECG-synchronized chest CT scans allow for the quantification of coronary calcium and that there is a good agreement between scores determined using non-ECG-gated and ECG-gated CT scans (Budoff et al., 2011; Kim et al., 2008). This process can be further evolved with the help of radiomics as calcium scoring can be performed in an automated manner by feeding texture, size, and spatial features (extracted from non-ECG-gated chest CT images) into a pattern recognition system (Isgum, Prokop, Niemeijer, Viergever, & van Ginneken, 2012).

Coronary plaque phenotyping

Although the detection of coronary plaques relies on conventional coronary CT angiography-derived metrics, the comprehensive assessment of the plaque microenvironment, histology, and biology requires a more profound radiomic analysis. The radiomics-based phenotyping of a given plaque can identify high-risk plaque features—the radiological manifestations of complex pathological processes—as alterations in its attenuation histogram and texture.

Out of the four main CT signatures of high-risk coronary atherosclerotic plaques, the napkin-ring sign (NRS)—a plaque cross-section with a central area of low CT attenuation apparently in contact with the lumen, surrounded by ring-shaped higher attenuation plaque tissue (Maurovich-Horvat et al., 2010)—is the only qualitative feature (Maurovich-Horvat, Ferencik, Voros, Merkely, & Hoffmann, 2014). Therefore, the identification of NRS is significantly affected by interreader variability. Radiomics can objectively characterize NRS lesions as demonstrated by Kolossváry et al. (2017), who reported 916 radiomic features to be significantly different between NRS lesions and non-NRS plaques with similar degrees of calcium, luminal obstruction, localization, and imaging parameters. According to their analysis, GLRLM short- and long-run gray-level emphasis had the highest AUCs (0.92 and 0.89, respectively). In contrast, none of the conventional coronary CT angiography metrics discriminated between the two groups.

Visually detectable adverse plaque characteristics based on coronary CT angiography show only a modest correlation with intravascular ultrasound (IVUS), optical coherence tomography (OCT), or sodium-fluoride positron emission tomography (NaF18-PET)-derived features (Lee et al., 2017). On the other hand, coronary CT angiography radiomics is capable of identifying invasive and radionuclide imaging markers of plaque vulnerability with good to excellent diagnostic accuracy, and radiomic parameters significantly outperform conventional metrics in the detection of high-risk plaque features assessed by IVUS, OCT, and NaF18-PET (Kolossvary, Park et al., 2019). These findings imply that coronary CT angiography radiomics may provide the opportunity to detect vulnerable plaques noninvasively in daily clinical practice.

Advanced atherosclerotic lesions (early and late fibroatheroma or thin-cap atheroma) are associated with a higher risk of MI (Stary et al., 1995; Virmani, Burke, Kolodgie, & Farb, 2002); therefore, the ability to noninvasively identify these lesions could substantially improve risk stratification and prediction of future CV events. In a study involving the coronary CT angiography images of ex vivo hearts, a radiomics-based ML model outperformed expert visual assessment (cross-validated AUC: 0.73 vs. 0.65, $P < .05$), area of low attenuation (cross-validated AUC: 0.73 vs. 0.55, $P < .05$), and average Hounsfield unit (cross-validated AUC = 0.73 vs. 0.53, $P < .05$) in the identification of advanced atheromatous lesions (Kolossváry et al., 2019). These results suggest that radiomics-based ML analysis significantly improves the discriminatory power of coronary CT angiography in the detection of advanced atherosclerotic lesions.

Radiomic phenotyping of perivascular fat

There is an expanding body of evidence that perivascular adipose tissue (PVAT) plays a pivotal role in the regulation of vascular function in both homeostasis and disease (Oikonomou and Antoniades, 2019). The presence of vascular inflammation in the coronary arteries—through the release of proinflammatory cytokines—inhibits adipogenesis in the adjacent perivascular fat resulting in altered attenuation values that are also detectable on standard coronary CT angiography scans (Antonopoulos et al., 2017). Recently, in the post-hoc analysis of the CRISP-CT study, Fat Attenuation Index—a novel imaging biomarker—was found to have incremental prognostic value for the prediction of future adverse cardiac events (independent of traditional risk factors, the extent of coronary atherosclerosis, and presence of high-risk plaque features) highlighting a residual cardiac risk hidden in the PVAT radiome (Oikonomou et al., 2018). The inflammation-induced spatial changes in the PVAT texture could be captured using the tools of radiomics as well. Oikonomou et al. (2019) linked the CT radiomic profile of PVAT to the expression of genes characterizing inflammation (e.g., tumor necrosis factor α), fibrosis (e.g., COL1A1), and vascularity (e.g., endothelial-marker platelet and endothelial cell adhesion marker CD31) within the visualized tissue. Using this radiotranscriptomic approach, they proposed an integrated CT signature of pericoronary fat that captures the permanent fibrotic and microvascular changes, thus functioning as a surrogate marker of cumulative coronary injury and aging. When tested in the participants of the SCOT-HEART study, this integrated signature (i.e., the Fat Radiomic Profile) predicted a significant residual cardiac risk not captured by traditional risk factors, high-risk plaque features, or coronary calcium score.

Valvular heart disease

The assessment of mechanical prosthetic valve obstruction is challenging as the clinical and radiological signs are much alike, while the etiologies and the subsequent management are different. As part of a multimodality approach, cardiac CT has been suggested to facilitate the differentiation between thrombus, pannus, and vegetation (Nishimura et al., 2017). CT markers of shape, size, and attenuation have been proposed as indicators of specific etiologies; however, these metrics are largely subjective as they are dependent on the observer's experience and show a considerable amount of variability (Tanis et al., 2014). The potential role of radiomics has been investigated in this differential diagnostic process by Nam et al. (2019), who reported that the combination of a radiomic score (calculated from five radiomic features) and visual assessment exhibited superior performance than visual analysis alone. These results unveiled the incremental value of CT radiomic features to differentiate pannus from thrombus or vegetation in patients with suspected prosthetic valve obstruction.

Another interesting approach is the automated assessment of mitral regurgitation severity using radiomics-based ML classifiers. In a study, an SVM model utilizing texture features extracted from a rectangular ROI covering the regurgitant area of echocardiographic images exhibited an accuracy of over 95% in apical two-chamber, apical four-chamber, and parasternal long-axis views as well (Balodi, Dewal, Anand, & Rawat, 2016). These findings imply that the proposed computer-aided diagnostic system may effectively assist in establishing and confirming the stages of mitral regurgitation when conventional echocardiographic parameters cannot be assessed reliably.

Intracardiac mass

Echocardiography is the primary modality of choice for the differential diagnosis of intracardiac masses (Saric et al., 2016). During the echocardiographic assessment, the cardiologist takes predominantly two major characteristics into consideration: the movement of the mass (motion features) and the length of its base (boundary features). In addition, different masses can have distinct echogenicity and texture characteristics; nevertheless, they are challenging to quantify and interpret using conventional metrics. Radiomics was designed exactly for this purpose, and its diagnostic value was demonstrated by Guo, Wnag, Kong and Shu (2015), who combined texture, motion, and boundary features to classify intracardiac tumors and thrombi with an accuracy of 97%.

In patients with atrial fibrillation, transesophageal echocardiography is the gold standard imaging modality for the diagnosis and exclusion of left atrial (LA) and left atrial appendage (LAA) thrombi (Donal et al., 2016; Pepi et al., 2010; Saric et al., 2016). Although it exhibits excellent sensitivity and specificity in this clinical scenario, there might be cases when the presence of a thrombus cannot be ruled out completely. Radiomics might come in handy in these situations as Sun et al. (2014) demonstrated that the diagnostic accuracy of transesophageal echocardiography for LA/LAA thrombi could be significantly enhanced with a computer-aided diagnostic system combining GLCM features and an artificial neural network (AUC: 0.83 vs. 0.93, $P < .001$).

Risk stratification

The accurate identification of high-risk patients who might benefit from more intensive therapeutic strategies represents a crucial mission in CV medicine. Radiomics can be helpful in such a scenario, as the extracted quantitative features can be supplied to ML-based predictive models to determine the risk of life-threatening cardiac events, hence providing a window of opportunity for treatment and prevention.

Coronary artery calcium scoring only represents a small fraction of all information embedded in noncontrast cardiac CT scans. Therefore, a screening method that could integrate the assessment of coronary calcium and other phenotypic elements such as left ventricular hypertrophy would be highly desirable in preventive CV imaging. A study involving almost two thousand participants of the Dallas Heart Study has demonstrated that an automated pipeline using radiomic features extracted from noncontrast cardiac CT scans can achieve an AUC above 0.70 for detecting high-risk left ventricular hypertrophy (Kay et al., 2020). By enabling the consistent and cost-efficient recognition of high-risk phenotypic patterns without the need for additional imaging or radiation exposure, this automated system could further leverage the already established value of noncontrast cardiac CT in CV risk stratification.

Although numerous studies have attempted to explore the potentials of radiomics in plaque detection and phenotyping, there is only limited evidence of the prognostic value of coronary CT-derived quantitative intensity-based parameters. Nevertheless, it seems that the addition of quantitative plaque characteristics to clinical risk profile (as assessed by the Framingham risk score) and qualitative expert opinion increases the accuracy of predicting acute coronary syndrome (AUC: 0.64 vs. 0.79, $P < .05$) (Versteylen et al., 2013).

Both post-MI and HCM patients are susceptible to ventricular tachyarrhythmias that represent a well-established risk factor of sudden cardiac death. This unfavorable outcome can be prevented by the implantation of implantable cardioverter defibrillators (Al-Khatib et al., 2018). Therefore, accurate arrhythmic risk stratification is essential to enable appropriate candidate selection for implantation. Using the compound of features capturing size, location, and heterogeneity of myocardial scar on LGE MRI scans, an SVM classifier exhibited a cross-validated AUC over 0.90 for the classification of post-MI patients into high- and low-risk subgroups (Kotu et al., 2015). In HCM patients, Amano, Suzuki, Yanagisawa, Omori and Matsumoto (2018) noted significant differences in the texture features of LGE images between participants with and without a history of ventricular tachycardia. These findings suggest that radiomics might have a substantial role in the prognostication of patients who are at risk of ventricular tachyarrhythmias.

It is a well-known fact that both the presence and the greater extent of LGE are associated with worse prognosis in HCM (Weng et al., 2016); however, patients with a similar extent of LGE may experience different outcomes. Thus, novel LGE markers capturing the subtle differences are still warranted. Radiomics seems to be a promising candidate in this scenario as Cheng et al. (2018) demonstrated that increased LGE heterogeneity (higher GLRLM energy, lower skewness of intensity histogram, and lower GLCM cluster tendency) is strongly associated with worse clinical outcomes (CV death, aborted sudden cardiac death, heart transplantation, and unscheduled heart failure hospitalization) in HCM patients.

The future direction of radiomic research

Radiomics is an appealing novel technology with the potential to expand medical knowledge and reform the field of CV imaging. It has achieved promising results in various clinical scenarios; nonetheless, there is still a long way to go before it transitions to routine clinical practice.

An inevitable step toward this aim would be the standardization of image acquisition, reconstruction, and postprocessing protocols, as it can be hard to determine whether differences in radiomic features reflect biological alterations or variations in the technical details of the imaging process (Altazi et al., 2017; Berenguer et al., 2018; Kolossvary, Szilveszter et al., 2019; Mackin et al., 2015; Meyer et al., 2019; Zhao et al., 2016). Another source of heterogeneity is the questionable test-retest reliability of radiomic features (Baessler, Weiss, & Pinto Dos Santos, 2019; Kickingereder et al., 2018). As this issue poses a serious challenge to the wide-spread utilization of radiomics, there is a pressing need to establish which features are stable and robust enough to be considered for inclusion in predictive models.

It should also be highlighted that most of the published results were derived from proof-of-concept studies with relatively small sample size. Accordingly, for the verification of the proposed radiomics-based models and computer-aided diagnostic systems, multicenter and multivendor studies in larger cohorts are warranted. In addition, future investigations should also evaluate the clinical impact of radiomics analysis, as it is still unknown how much the utilization of such models improves patient care and outcomes.

After proving the generalizability of the models, it should be ensured that radiomics-based decision support tools can be used conveniently without increasing clinical workload. In the CV imaging laboratory of the future, radiomics-based ML models will be directly linked to picture archiving and communication systems, and electronic medical records. Consequently, the extracted radiomic features can be easily compared with the results of previous visits and can be merged with information from different domains of clinical data to enable individualized decision-making.

Conclusions

Radiomics represents an emerging image analysis technology with the potential to enhance the accuracy and precision of CV imaging. By unveiling imperceptible patterns in medical images, radiomics is capable of providing a detailed description of disease burden and linking distinct phenotypes to outcomes. In addition, radiomics offers a unique insight into tissue-level pathophysiology, which might facilitate the understanding of disease mechanisms. The utilization of radiomic features also holds the promise to objectify image interpretation, to tailor diagnostic and therapeutic strategies in a more individualized manner, and ultimately, to bring us one step closer to personalized patient management. If current efforts are able to resolve the existing technical challenges, radiomics will become a valuable tool in the clinical arena and will leave a permanent footprint in CV imaging.

Acknowledgments

We are grateful to Márton Kolossváry, M.D. for his insightful comments and suggestions that helped to improve this book chapter. We would like to also express our gratitude to Judit Simon, M.D. and Liliána Szabó, M.D. for their help in the preparation of illustrations.

References

Acharya, U. R., Sree, S. V., Muthu Rama Krishnan, M., Krishnananda, N., Ranjan, S., Umesh, P., et al. (2013). Automated classification of patients with coronary artery disease using grayscale features from left ventricle echocardiographic images. *Computer Methods and Programs in Biomedicine, 112*(3), 624–632.

Al-Khatib, S. M., Stevenson, W. G., Ackerman, M. J., Bryant, W. J., Callans, D. J., Curtis, A. B., et al. (2018). 2017 AHA/ACC/HRS guideline for management of patients with ventricular arrhythmias and the prevention of sudden cardiac death: A report of the American college of cardiology/American heart association task force on clinical practice guidelines and the heart rhythm society. *Circulation, 138*(13), e272–e391.

Altazi, B. A., Zhang, G. G., Fernandez, D. C., Montejo, M. E., Hunt, D., Werner, J., et al. (2017). Reproducibility of F18-FDG PET radiomic features for different cervical tumor segmentation methods, gray-level discretization, and reconstruction algorithms. *Journal of Applied Clinical Medical Physics, 18*(6), 32–48.

Amadasun, M., & King, R. (1989). Textural features corresponding to textural properties. *IEEE Transactions on Systems, Man, and Cybernetics, 19*(5), 1264–1274.

Amano, Y., Suzuki, Y., Yanagisawa, F., Omori, Y., & Matsumoto, N. (2018). Relationship between extension or texture features of late gadolinium enhancement and ventricular tachyarrhythmias in hypertrophic cardiomyopathy. *BioMed Research International, 2018*, 4092469.

Antonopoulos, A. S., Sanna, F., Sabharwal, N., Thomas, S., Oikonomou, E. K., Herdman, L., et al. (2017). Detecting human coronary inflammation by imaging perivascular fat'. *Science Translational Medicine, 9*(398).

Antunes, S., Esposito, A., Palmisanov, A., Colantoni, C., de Cobelli, F., & Del Maschio, A. (2016). Characterization of normal and scarred myocardium based on texture analysis of cardiac computed tomography images. *Conference Proceedings in IEEE Engineering in Medicine and Biology Society, 2016*, 4161–4164.

Baessler, B., Luecke, C., Lurz, J., Klingel, K., Roeder, M.v., Waha, S.d., et al. (2018). Cardiac MRI texture analysis of T1 and T2 maps in patients with infarctlike acute myocarditis. *Radiology, 289*(2), 357–365.

Baessler, B., Mannil, M., Maintz, D., Alkadhi, H., & Manka, R. (2018). Texture analysis and machine learning of non-contrast T1-weighted MR images in patients with hypertrophic cardiomyopathy-Preliminary results. *European Journal of Radiology, 102*, 61–67.

Baessler, B., Mannil, M., Oebel, S., Maintz, D., Alkadhi, H., & Manka, R. (2018). Subacute and chronic left ventricular myocardial scar: Accuracy of texture analysis on nonenhanced cine MR images. *Radiology, 286*(1), 103–112.

Baessler, B., Weiss, K., & Pinto Dos Santos, D. (2019). Robustness and reproducibility of radiomics in magnetic resonance imaging: A phantom study. *Investigative Radiology, 54*(4), 221–228.

Balodi, A., Dewal, M. L., Anand, R. S., & Rawat, A. (2016). Texture based classification of the severity of mitral regurgitation. *Computers in Biology and Medicine, 73*, 157–164.

Berenguer, R., Pastor-Juan, M.d. R., Canales-Vázquez, J., Castro-García, M., Villas, M. V., Legorburo, F. M., et al. (2018). Radiomics of CT features may Be nonreproducible and redundant: Influence of CT acquisition parameters. *Radiology, 288*(2), 407–415.

Biesbroek, P. S., Beek, A. M., Germans, T., Niessen, H. W., & van Rossum, A. C. (2015). Diagnosis of myocarditis: Current state and future perspectives. *International Journal of Cardiology, 191*, 211–219.

Budoff, M. J., Mayrhofer, T., Ferencik, M., Bittner, D., Lee, K. L., Lu, M. T., et al. (2017). Prognostic value of coronary artery calcium in the PROMISE study (prospective multicenter imaging study for evaluation of chest pain). *Circulation, 136*(21), 1993–2005.

Budoff, M. J., Nasir, K., Kinney, G. L., Hokanson, J. E., Barr, R. G., Steiner, R., et al. (2011). Coronary artery and thoracic calcium on noncontrast thoracic CT scans: Comparison of ungated and gated examinations in patients from the COPD gene cohort. *Journal of Cardiovascular Computed Tomography, 5*(2), 113–118.

Budoff, M. J., Shaw, L. J., Liu, S. T., Weinstein, S. R., Mosler, T. P., Tseng, P. H., et al. (2007). Long-term prognosis associated with coronary calcification: Observations from a registry of 25,253 patients. *Journal of the American College of Cardiology, 49*(18), 1860–1870.

Caforio, A. L., Pankuweit, S., Arbustini, E., Basso, C., Gimeno-Blanes, J., Felix, S. B., et al. (2013). Current state of knowledge on aetiology, diagnosis, management, and therapy of myocarditis: A position statement of the European society of cardiology working group on myocardial and pericardial diseases. *European Heart Journal, 34*(33), 2636–2648, 2648a–2648d.

Carli, M. F. D., Geva, T., & Davidoff, R. (2016). The future of cardiovascular imaging. *Circulation, 133*(25), 2640–2661.

Cetin, I., Petersen, S. E., Napel, S., Camara, O., Ballester, M.Á. G., & Lekadir, K. (April 8–11, 2019). A radiomics approach to analyze cardiac alterations in hypertension. In *2019 IEEE 16th international symposium on biomedical imaging (ISBI 2019)* (pp. 640–643).

Cheng, S., Fang, M., Cui, C., Chen, X., Yin, G., Prasad, S. K., et al. (2018). LGE-CMR-derived texture features reflect poor prognosis in hypertrophic cardiomyopathy patients with systolic dysfunction: Preliminary results. *European Radiology, 28*(11), 4615–4624.

Chen, C., Qin, C., Qiu, H., Tarroni, G., Duan, J., Bai, W., et al. (2020). Deep learning for cardiac image segmentation: A review. *Frontiers in Cardiovascular Medicine, 7*, 25.

Choi, B. W. (2006). Differentiation of acute myocardial infarction from chronic myocardial scar with MRI. *Korean Journal of Radiology, 7*(1), 1–3.

Collins, G. S., Reitsma, J. B., Altman, D. G., & Moons, K. G. M. (2015). Transparent reporting of a multivariable prediction model for individual prognosis or diagnosis (TRIPOD): The TRIPOD statement. *Annals of Internal Medicine, 162*(1), 55–63.

Davies, E. R. (2018). Chapter 7 – Texture analysis. In E. R. Davies (Ed.), *Computer vision* (5th ed., pp. 185–200). Academic Press.

Dettori, L., & Semler, L. (2007). A comparison of wavelet, ridgelet, and curvelet-based texture classification algorithms in computed tomography. *Computers in Biology and Medicine, 37*(4), 486–498.

Devi, S. G., & Sabrigiriraj, M.' (March 1–3, 2018). Feature selection, online feature selection techniques for big data classification: – A review. In *2018 international conference on current trends towards converging technologies (ICCTCT)* (pp. 1–9).

Donal, E., Lip, G. Y. H., Galderisi, M., Goette, A., Shah, D., Marwan, M., et al. (2016). EACVI/EHRA Expert Consensus Document on the role of multi-modality imaging for the evaluation of patients with atrial fibrillation. *European Heart Journal – Cardiovascular Imaging, 17*(4), 355–383.

Engan, K., Eftestol, T., Orn, S., Kvaloy, J. T., & Woie, L. (2010). Exploratory data analysis of image texture and statistical features on myocardium and infarction areas in cardiac magnetic resonance images. *Conference Proceedings in IEEE Engineering in Medicine and Biology Society, 2010*, 5728–5731.

Esposito, A., Palmisano, A., Antunes, S., Colantoni, C., Rancoita, P. M. V., Vignale, D., et al. (2018). Assessment of remote myocardium heterogeneity in patients with ventricular tachycardia using texture analysis of late iodine enhancement (LIE) cardiac computed tomography (cCT) images. *Molecular Imaging and Biology, 20*(5), 816–825.

Gabor, D. (1946). Theory of communication. Part 1: The analysis of information. *Journal of the Institution of Engineers: Electronics Part III: Radio and Communication Engineering, 93*(26), 429–441.

Galloway, M. M. (1975). Texture analysis using gray level run lengths. *Computer Graphics and Image Processing, 4*(2), 172–179.

Gerber, B. L., Belge, B., Legros, G. J., Lim, P., Poncelet, A., Pasquet, A., et al. (2006). Characterization of acute and chronic myocardial infarcts by multidetector computed tomography: Comparison with contrast-enhanced magnetic resonance. *Circulation, 113*(6), 823–833.

Ger, R. B., Zhou, S., Chi, P.-C. M., Lee, H. J., Layman, R. R., Jones, A. K., et al. (2018). Comprehensive investigation on controlling for CT imaging variabilities in radiomics studies. *Scientific Reports, 8*(1), 13047.

Gillies, R. J., Kinahan, P. E., & Hricak, H. (2016). Radiomics: Images are more than pictures, they are data. *Radiology, 278*(2), 563–577.

Goetti, R., Feuchtner, G., Stolzmann, P., Donati, O. F., Wieser, M., Plass, A., et al. (2011). Delayed enhancement imaging of myocardial viability: Low-dose high-pitch CT versus MRI. *European Radiology, 21*(10), 2091.

Goetti, R., Kozerke, S., Donati, O. F., Sürder, D., Stolzmann, P., Kaufmann, P. A., et al. (2011). Acute, subacute, and chronic myocardial infarction: Quantitative comparison of 2D and 3D late gadolinium enhancement MR imaging. *Radiology, 259*(3), 704–711.

Guo, Y., Wang, Y., Kong, D., & Shu, X. (2015). Automatic classification of intracardiac tumor and thrombi in echocardiography based on sparse representation. *IEEE Journal of Biomedical and Health Informatics, 19*(2), 601–611.

Haralick, R. M., Shanmugam, K., & Dinstein, I. (1973). Textural features for image classification. *IEEE Transactions on Systems, Man, and Cybernetics, 3*(6), 610–621.

Hinzpeter, R., Wagner, M. W., Wurnig, M. C., Seifert, B., Manka, R., & Alkadhi, H. (2017). Texture analysis of acute myocardial infarction with CT: First experience study. *PloS One, 12*(11), e0186876.

Isgum, I., Prokop, M., Niemeijer, M., Viergever, M. A., & van Ginneken, B. (2012). Automatic coronary calcium scoring in low-dose chest computed tomography. *IEEE Transactions on Medical Imaging, 31*(12), 2322–2334.

Kagiyama, N., Shrestha, S., Cho, J. S., Khalil, M., Singh, Y., Challa, A., et al. (2020). A low-cost texture-based pipeline for predicting myocardial tissue remodeling and fibrosis using cardiac ultrasound. *EBioMedicine, 54*, 102726.

Kay, F. U., Abbara, S., Joshi, P. H., Garg, S., Khera, A., & Peshock, R. M. (2020). Identification of high-risk left ventricular hypertrophy on calcium scoring cardiac computed tomography scans: Validation in the DHS. *Circulation: Cardiovascular Imaging, 13*(2), e009678.

Kickingereder, P., Neuberger, U., Bonekamp, D., Piechotta, P. L., Götz, M., Wick, A., et al. (2018). Radiomic subtyping improves disease stratification beyond key molecular, clinical, and standard imaging characteristics in patients with glioblastoma. *Neuro-Oncology, 20*(6), 848–857.

Kim, S. M., Chung, M. J., Lee, K. S., Choe, Y. H., Yi, C. A., & Choe, B. K. (2008). Coronary calcium screening using low-dose lung cancer screening: Effectiveness of MDCT with retrospective reconstruction. *American Journal of Roentgenology, 190*(4), 917–922.

von Knobelsdorff-Brenkenhoff, F., & Schulz-Menger, J. (2012). Cardiovascular magnetic resonance imaging in ischemic heart disease. *Journal of Magnetic Resonance Imaging, 36*(1), 20–38.

Kolossváry, M., Karády, J., Kikuchi, Y., Ivanov, A., Schlett, C. L., Lu, M. T., et al. (2019). Radiomics versus visual and histogram-based assessment to identify atheromatous lesions at coronary CT angiography: An ex vivo study. *Radiology, 293*(1), 89–96.

Kolossvary, M., Karady, J., Szilveszter, B., Kitslaar, P., Hoffmann, U., Merkely, B., et al. (2017). Radiomic features are superior to conventional quantitative computed tomographic metrics to identify coronary plaques with napkin-ring sign. *Circulation: Cardiovascular Imaging, 10*(12).

Kolossvary, M., Kellermayer, M., Merkely, B., & Maurovich-Horvat, P. (2018). Cardiac computed tomography radiomics: A comprehensive review on radiomic techniques. *Journal of Thoracic Imaging, 33*(1), 26–34.

Kolossvary, M., Park, J., Bang, J. I., Zhang, J., Lee, J. M., Paeng, J. C., et al. (2019). Identification of invasive and radionuclide imaging markers of coronary plaque vulnerability using radiomic analysis of coronary computed tomography angiography'. *European Heart Journal - Cardiovascular Imaging, 20*(11), 1250–1258.

Kolossvary, M., Szilveszter, B., Karady, J., Drobni, Z. D., Merkely, B., & Maurovich-Horvat, P. (2019). Effect of image reconstruction algorithms on volumetric and radiomic parameters of coronary plaques. *Journal of Cardiovascular Computed Tomography, 13*(6), 325–330.

Kotu, L. P., Engan, K., Borhani, R., Katsaggelos, A. K., Orn, S., Woie, L., et al. (2015). Cardiac magnetic resonance image-based classification of the risk of arrhythmias in post-myocardial infarction patients. *Artificial Intelligence in Medicine, 64*(3), 205–215.

Kotu, L. P., Engan, K., Eftestøl, T., Ørn, S., & Woie, L.' (August 30–September 3, 2011). Segmentation of scarred and non-scarred myocardium in LG enhanced CMR images using intensity-based textural analysis. In *2011 annual international conference of the IEEE engineering in medicine and biology society* (pp. 5698–5701).

Lambin, P., Leijenaar, R. T. H., Deist, T. M., Peerlings, J., de Jong, E. E. C., van Timmeren, J., et al. (2017). Radiomics: The bridge between medical imaging and personalized medicine. *Nature Reviews Clinical Oncology, 14*(12), 749–762.

Larroza, A., Lopez-Lereu, M. P., Monmeneu, J. V., Gavara, J., Chorro, F. J., Bodi, V., et al. (2018). Texture analysis of cardiac cine magnetic resonance imaging to detect nonviable segments in patients with chronic myocardial infarction. *Medical Physics, 45*(4), 1471–1480.

Larroza, A., Materka, A., López-Lereu, M. P., Monmeneu, J. V., Bodí, V., & Moratal, D. (2017). Differentiation between acute and chronic myocardial infarction by means of texture analysis of late gadolinium enhancement and cine cardiac magnetic resonance imaging. *European Journal of Radiology, 92*, 78–83.

Lee, J. M., Bang, J. I., Koo, B. K., Hwang, D., Park, J., Zhang, J., et al. (2017). Clinical relevance of (18)F-sodium fluoride positron-emission tomography in noninvasive identification of high-risk plaque in patients with coronary artery disease'. *Circulation: Cardiovascular Imaging, 10*(11).

Lopes, R., & Betrouni, N. (2009). Fractal and multifractal analysis: A review. *Medical Image Analysis, 13*(4), 634–649.

Lurz, P., Eitel, I., Adam, J., Steiner, J., Grothoff, M., Desch, S., et al. (2012). Diagnostic performance of CMR imaging compared with EMB in patients with suspected myocarditis. *JACC: Cardiovascular Imaging, 5*(5), 513–524.

Lurz, P., Luecke, C., Eitel, I., Föhrenbach, F., Frank, C., Grothoff, M., et al. (2016). Comprehensive cardiac magnetic resonance imaging in patients with suspected myocarditis: The MyoRacer-trial. *Journal of the American College of Cardiology, 67*(15), 1800–1811.

Mackin, D., Fave, X., Zhang, L., Fried, D., Yang, J., Taylor, B., et al. (2015). Measuring computed tomography scanner variability of radiomics features. *Investigative Radiology, 50*(11), 757–765.

Mandelbrot, B. (1967). How long is the coast of britain? Statistical self-similarity and fractional dimension. *Science, 156*(3775), 636–638.

Mannil, M., von Spiczak, J., Manka, R., & Alkadhi, H. (2018). Texture analysis and machine learning for detecting myocardial infarction in noncontrast low-dose computed tomography: Unveiling the invisible. *Investigative Radiology, 53*(6), 338–343.

Maurovich-Horvat, P., Ferencik, M., Voros, S., Merkely, B., & Hoffmann, U. (2014). Comprehensive plaque assessment by coronary CT angiography. *Nature Reviews Cardiology, 11*(7), 390–402.

Maurovich-Horvat, P., Hoffmann, U., Vorpahl, M., Nakano, M., Virmani, R., & Alkadhi, H. (2010). The napkin-ring sign: CT signature of high-risk coronary plaques? *JACC: Cardiovascular Imaging, 3*(4), 440–444.

Mewton, N., Liu, C. Y., Croisille, P., Bluemke, D., & Lima, J. A. (2011). Assessment of myocardial fibrosis with cardiovascular magnetic resonance. *Journal of the American College of Cardiology, 57*(8), 891–903.

Meyer, M., Ronald, J., Vernuccio, F., Nelson, R. C., Ramirez-Giraldo, J. C., Solomon, J., et al. (2019). Reproducibility of CT radiomic features within the same patient: Influence of radiation dose and CT reconstruction settings. *Radiology, 293*(3), 583–591.

Mojsilovic, A., Popovic, M. V., Neskovic, A. N., & Popovic, A. D. (1997). Wavelet image extension for analysis and classification of infarcted myocardial tissue. *IEEE Transactions on Biomedical Engineering, 44*(9), 856–866.

Nam, K., Suh, Y. J., Han, K., Park, S. J., Kim, Y. J., & Choi, B. W. (2019). Value of computed tomography radiomic features for differentiation of periprosthetic mass in patients with suspected prosthetic valve obstruction. *Circulation: Cardiovascular Imaging, 12*(11), e009496.

Neisius, U., El-Rewaidy, H., Nakamori, S., Rodriguez, J., Manning, W. J., & Nezafat, R. (2019). Radiomic analysis of myocardial native T1 imaging discriminates between hypertensive heart disease and hypertrophic cardiomyopathy. *JACC: Cardiovascular Imaging, 12*(10), 1946–1954.

Nesković, A. N., Mojsilović, A., Jovanović, T., Vasiljević, J., Popović, M., Marinković, J., et al. (1998). Myocardial tissue characterization after acute myocardial infarction with wavelet image decomposition. *Circulation, 98*(7), 634–641.

Nguyen, L. H., & Holmes, S. (2019). Ten quick tips for effective dimensionality reduction. *PLoS Computational Biology, 15*(6), e1006907.

Nishimura, R. A., Otto, C. M., Bonow, R. O., Carabello, B. A., Erwin, J. P., Fleisher, L. A., et al. (2017). 2017 AHA/ACC focused update of the 2014 AHA/ACC guideline for the management of patients with valvular heart disease: A report of the American College of Cardiology/American Heart Association Task Force on Clinical Practice Guidelines. *Circulation, 135*(25), e1159–e1195.

Oikonomou, E. K., & Antoniades, C. (2019). The role of adipose tissue in cardiovascular health and disease. *Nature Reviews Cardiology, 16*(2), 83–99.

Oikonomou, E. K., Marwan, M., Desai, M. Y., Mancio, J., Alashi, A., Hutt Centeno, E., et al. (2018). Non-invasive detection of coronary inflammation using computed tomography and prediction of residual cardiovascular risk (the CRISP CT study): A post-hoc analysis of prospective outcome data. *The Lancet, 392*(10151), 929–939.

Oikonomou, E. K., Siddique, M., & Antoniades, C. (2020). Artificial intelligence in medical imaging: A radiomic guide to precision phenotyping of cardiovascular disease. *Cardiovascular Research*. https://doi.org/10.1093/cvr/cvaa021

Oikonomou, E. K., Williams, M. C., Kotanidis, C. P., Desai, M. Y., Marwan, M., Antonopoulos, A. S., et al. (2019). A novel machine learning-derived radiotranscriptomic signature of perivascular fat improves cardiac risk prediction using coronary CT angiography. *European Heart Journal, 40*(43), 3529–3543.

Parmar, C., Grossmann, P., Bussink, J., Lambin, P., & Aerts, H. (2015). Machine learning methods for quantitative radiomic biomarkers. *Scientific Reports, 5*, 13087.

Pepi, M., Evangelista, A., Nihoyannopoulos, P., Flachskampf, F. A., Athanassopoulos, G., Colonna, P., et al. (2010). Recommendations for echocardiography use in the diagnosis and management of cardiac sources of embolism: European association of echocardiography (EAE) (a registered branch of the ESC). *European Journal of Echocardiography, 11*(6), 461–476.

Pinamonti, B., Picano, E., Ferdeghini, E. M., Lattanzi, F., Slavich, G., Landini, L., et al. (1989). Quantitative texture analysis in two-dimensional echocardiography: Application to the diagnosis of myocardial amyloidosis. *Journal of the American College of Cardiology, 14*(3), 666–671.

Raisi-Estabragh, Z., Izquierdo, C., Campello, V. M., Martin-Isla, C., Jaggi, A., Harvey, N. C., et al. (2020). Cardiac magnetic resonance radiomics: Basic principles and clinical perspectives. *European Heart Journal - Cardiovascular Imaging, 21*(4), 349–356.

Roffi, M., Patrono, C., Collet, J.-P., Mueller, C., Valgimigli, M., Andreotti, F., et al. (2016). 2015 ESC guidelines for the management of acute coronary syndromes in patients presenting without persistent ST-segment elevation: Task force for the management of acute coronary syndromes in patients presenting without persistent ST-segment elevation of the European society of cardiology (ESC)'. *European Heart Journal, 37*(3), 267–315.

Saeys, Y., Inza, I., & Larrañaga, P. (2007). A review of feature selection techniques in bioinformatics. *Bioinformatics, 23*(19), 2507–2517.

Saric, M., Armour, A. C., Arnaout, M. S., Chaudhry, F. A., Grimm, R. A., Kronzon, I., et al. (2016). Guidelines for the use of echocardiography in the evaluation of a cardiac source of embolism. *Journal of the American Society of Echocardiography, 29*(1), 1–42.

Schofield, R., Ganeshan, B., Fontana, M., Nasis, A., Castelletti, S., Rosmini, S., et al. (2019). Texture analysis of cardiovascular magnetic resonance cine images differentiates aetiologies of left ventricular hypertrophy. *Clinical Radiology, 74*(2), 140–149.

Shannon, C. E. (1948). A mathematical theory of communication. *Bell System Technical Journal, 27*(3), 379–423.

Shaw, L. J., Raggi, P., Schisterman, E., Berman, D. S., & Callister, T. Q. (2003). Prognostic value of cardiac risk factors and coronary artery calcium screening for all-cause mortality. *Radiology, 228*(3), 826–833.

Siontis, G. C. M., Tzoulaki, I., Castaldi, P. J., & Ioannidis, J. P. A. (2015). External validation of new risk prediction models is infrequent and reveals worse prognostic discrimination. *Journal of Clinical Epidemiology, 68*(1), 25–34.

Stary, H. C., Chandler, A. B., Dinsmore, R. E., Fuster, V., Glagov, S., Insull, W., Jr., et al. (1995). A definition of advanced types of atherosclerotic lesions and a histological classification of atherosclerosis. A report from the Committee on Vascular Lesions of the Council on Arteriosclerosis, American Heart Association. *Circulation, 92*(5), 1355–1374.

Steyerberg, E. W., Bleeker, S. E., Moll, H. A., Grobbee, D. E., & Moons, K. G. M. (2003). Internal and external validation of predictive models: A simulation study of bias and precision in small samples. *Journal of Clinical Epidemiology, 56*(5), 441–447.

Steyerberg, E. W., & Harrell, F. E., Jr. (2016). Prediction models need appropriate internal, internal-external, and external validation. *Journal of Clinical Epidemiology, 69*, 245–247.

Steyerberg, E. W., Harrell, F. E., Jr., Borsboom, G. J., Eijkemans, M. J., Vergouwe, Y., & Habbema, J. D. (2001). Internal validation of predictive models: Efficiency of some procedures for logistic regression analysis. *Journal of Clinical Epidemiology, 54*(8), 774–781.

Steyerberg, E. W., & Vergouwe, Y. (2014). Towards better clinical prediction models: Seven steps for development and an ABCD for validation. *European Heart Journal, 35*(29), 1925–1931.

Sudarshan, V. K., Acharya, U. R., Ng, E. Y., Tan, R. S., Chou, S. M., & Ghista, D. N. (2016a). Data mining framework for identification of myocardial infarction stages in ultrasound: A hybrid feature extraction paradigm (part 2). *Computers in Biology and Medicine, 71*, 241–251.

Sudarshan, V. K., Acharya, U. R., Ng, E. Y. K., Tan, R. S., Chou, S. M., & Ghista, D. N. (2016b). An integrated index for automated detection of infarcted myocardium from cross-sectional echocardiograms using texton-based features (part 1). *Computers in Biology and Medicine, 71*, 231–240.

Sudarshan, K. V., Ng, E. Y. K., Acharya, U. R., Chou, S. M., Tan, R. S., & Ghista, D. N. (2015). Computer-aided diagnosis of myocardial infarction using ultrasound images with dwt, GLCM and HOS methods: A comparative study. *Computers in Biology and Medicine, 62*, 86–93.

Sun, L., Li, Y., Zhang, Y. T., Shen, J. X., Xue, F. H., Cheng, H. D., et al. (2014). A computer-aided diagnostic algorithm improves the accuracy of transesophageal echocardiography for left atrial thrombi: A single-center prospective study. *Journal of Ultrasound in Medicine, 33*(1), 83–91.

Sun, C., & Wee, W. G. (1983). Neighboring gray level dependence matrix for texture classification. *Computer Vision, Graphics, and Image Processing, 23*(3), 341–352.

Tak, T., Visser, C., Rahimtoola, S. H., & Chandraratna, P. A. (1992). Detection of acute myocardial infarction with digital image processing of two-dimensional echocardiograms. *American Heart Journal, 124*(2), 289–293.

Tanis, W., Habets, J., van den Brink, R. B., Symersky, P., Budde, R. P., & Chamuleau, S. A. (2014). Differentiation of thrombus from pannus as the cause of acquired mechanical prosthetic heart valve obstruction by non-invasive imaging: A review of the literature. *European Heart Journal – Cardiovascular Imaging, 15*(2), 119–129.

Thibault, G., Fertil, B., Navarro, C., Pereira, S., Lévy, N., Sequeira, J., et al. (2009). *Texture indexes and gray level size zone matrix application to cell nuclei classification.*

Thomas, A. M. K., & Banerjee, A. K. (2013). *The history of radiology.* Oxford University Press.

Versteylen, M. O., Kietselaer, B. L., Dagnelie, P. C., Joosen, I. A., Dedic, A., Raaijmakers, R. H., et al. (2013). Additive value of semiautomated quantification of coronary artery disease using cardiac computed tomographic angiography to predict future acute coronary syndrome'. *Journal of the American College of Cardiology, 61*(22), 2296–2305.

Virmani, R., Burke, A. P., Kolodgie, F. D., & Farb, A. (2002). Vulnerable plaque: The pathology of unstable coronary lesions. *Journal of Interventional Cardiology, 15*(6), 439–446.

Weng, Z., Yao, J., Chan, R. H., He, J., Yang, X., Zhou, Y., et al. (2016). Prognostic value of LGE-CMR in HCM: A meta-analysis. *JACC: Cardiovascular Imaging, 9*(12), 1392–1402.

Xinli, W., Albregtsen, F., & Foyn, B. (1995). Texture analysis using gray level gap length matrix. *Theory and Applications of Image Analysis II*, 65–78.

Zhao, B., Tan, Y., Tsai, W. Y., Qi, J., Xie, C., Lu, L., et al. (2016). Reproducibility of radiomics for deciphering tumor phenotype with imaging. *Scientific Reports, 6*, 23428.

Zoghbi, W. A. (2012). President's page: Cardiovascular imaging: A look to the past, present and future. *Journal of the American College of Cardiology, 60*(22), 2331–2334.

CHAPTER

13 Automated interpretation of electrocardiographic tracings

Mohammad Kachuee, Majid Sarrafzadeh

Computer Science Department, University of California, Los Angeles, CA, United States

Chapter outline

Electrocardiography

Electrocardiogram (ECG or EKG) is a recording of the electrical activity of the heart measured using electrodes placed on the skin. Since its invention in the early 1900s, ECG has been widely used as a tool to diagnose various heart conditions, such as arrhythmias, ischemic heart disease, structural abnormalities, heart failure, and pacemaker dysfunction. A typical clinical ECG machine records from 12 leads based on 10 electrode contacts. Each lead provides a unique viewpoint to the electrical stimulation patterns across the heart (Dubin, 2000).

Fig. 13.1 shows a typical ECG waveform with important key-points and cardiac cycles annotated. The P wave represents atrial depolarization where both atria contract simultaneously. The QRS complex marks the beginning of ventricular depolarization, initiating ventricular contraction. The pause after the QRS complex, the ST segment, marks repolarization of the ventricles. Lastly, the T wave corresponds to the end of ventricular repolarization and the end of ventricular contraction (systole).

ECG recordings can be used to measure major electrophysiologic characteristics such as rate, rhythm, and axis. The heart rate is a measure of the time it takes for the heart to complete one cycle. The normal resting heart rate is within the range of 60–100 beats per minute. An irregular waveform pattern on the ECG signal can be indicative of different arrhythmias or pacemaker dysfunction. Usually, multiple leads are used to find the electrical axis of the heart. Any physical displacement or structural abnormality of the heart also results in the deviation of the electrical

Machine Learning in Cardiovascular Medicine. https://doi.org/10.1016/B978-0-12-820273-9.00013-0

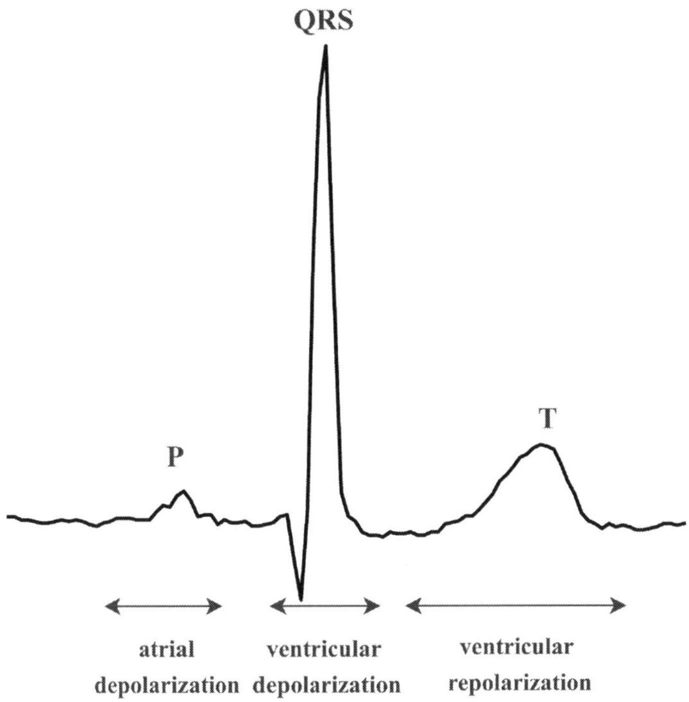

FIGURE 13.1

A typical electrocardiogram (ECG) waveform with annotation of inflection points as well as important segments of the cardiac cycle.

axis. Furthermore, the ECG can be used as a tool to diagnose important heart conditions such as hypertrophy in which myocardial tissue gets enlarged/thickened, causing disruption in ventricular systolic and diastolic function. Myocardial infarction is another important condition in which a part of myocardial tissue becomes ischemic and is replaced by akinetic scar tissue as a result of diminished blood flow, usually as a consequence of coronary artery disease. Further detail regarding ECG recordings and corresponding clinical cardiac conditions are outside the scope of this chapter, and are available in specialized textbooks (Dubin, 2000).

ECG signals are susceptible to different types of noise and artifacts from various physiological, instrumental, and environmental sources. Baseline wander, powerline interference, muscular activity, and electrode motion are considered as major noise and artifacts impacting real-world ECG recordings. Usually, a preprocessing stage is included in an automated analysis system to enhance the quality of signals before any analysis is performed. For this purpose, different filtering and denoising approaches have been suggested in the literature, spanning from very simple moving average or median filters to complex deep learning models trained to recover the clean signal from noisy inputs (Kameenoff, 2017):

1. **Baseline wander** is considered as very low-frequency artifact caused by breathing or slow body movements. A very simple high-pass filter can be used to filter signal components of less than 0.5 Hz. While it is straightforward to remove baseline wander, it is a very important step due to the often-large drifts it may cause to the signal baseline.
2. **Powerline interference** on ECG circuitry is another source of noise causing 50 or 60 Hz (and their harmonics) sinusoidal interference. Notch filters are usually designed to remove frequencies related to the powerline and its harmonics, and can be used to reduce the impact on signal quality. However, care must be taken as important signal components exist within the same frequency range, and excessive filtering may cause signal distortion especially to low-frequency waves such as P and T waves.
3. **Muscular activity** originating from the myocardium can cause significant interference with recorded electrical activity on the ECG tracing. From the frequency domain, it is very difficult to distinguish and remove this type of interference. One possible solution is to consider the periodic nature of ECG signals and to use longer signal windows since other muscular activities are not periodic.
4. **Electrode motion artifacts** are caused by skin impedance changes during physical activity. This artifact has frequency components within the range of 1−10 Hz strongly impacting major ECG signal components. Similar to muscular activity interferences, removing electrode motion artifacts requires more sophisticated filtering and analysis such as adaptive filters, estimating motion signals, wave averaging, and so forth.

Traditional approach to automated ECG analysis

Moving average (mean filtering) is frequently used to smooth ECG signals by removing high-frequency oscillations. Median filtering is often useful for removal of noisy samples captured by analog to digital converter (ADC) circuitry. While moving average and median filters are easy to implement and effective in practice, it should be noted that using these filters, especially in large window sizes, may result in oversmoothing or adverse nonlinear impacts for the case of median filters.

Finite Impulse Response (FIR) filters can be used to design different band-pass or band-rejection filters. However, designing a powerful FIR filter requires using large window lengths that may cause considerable latency and require extra memory to store a large signal segment during the process. Infinite Impulse Response (IIR) filters are usually much more compact and require very small filter windows; however, the phase response for IIR filters does not have a proper phase response as in FIR filters and may cause distortions to the original signals being processed. One solution to using the compactness of IIR filters without any adverse impact on the phase response is to use IIR filters in a bidirectional forward-backward fashion. A forward-backward filter simply applies an IIR filter on the signal window twice by inverting the signal window across the time axis after the first filtering and inverting again after the second filtering. One possible drawback of this approach is that bidirectional filters are not causal, which in practice is handled by using signal buffers causing a certain amount of latency.

Time-Frequency domain methods such as discrete wavelet decomposition (DCT) can be used to extract different time and frequency components and to apply different filtering or denoising functions. Additionally, DCT can be used as an effective compression method as the decomposition of natural signals usually tends to be sparse having many zero components and a few large nonzero components. For a more detailed discussion of these methods, we refer interested readers to M. Misiti et al. (1996).

Recently, deep learning methods have been suggested as powerful tools for ECG denoising, imputation, and generation. Denoising autoencoders implemented as either a convolutional encoder-decoder pair or as sequence-to-sequence architectures are natural choices to make denoising models. Generative adversarial networks (GANs) are suggested to generate synthesized data, applied to ECG inputs; one can use GANs for data augmentation or as a representation learning objective. Another use case would be to impute missing values which are in this case missing or noisy ECG samples based on partially observed data points (Rajan and Thiagarajan, 2018).

Fig. 13.2 shows the general pipeline used in traditional automated ECG analysis systems consisting of delineation, feature extraction, and predictive analysis. In the delineation step, ECG key-points are detected for a signal window at hand, usually P, Q, R, S, and T waves for typical ECG signals. Note that based on certain clinical conditions it may not be always possible to detect all the typical key-points for all leads, or the morphology of waves can be drastically different from normal healthy signals.

Afterward, based on the detected key-points, different features that are considered relevant to the final prediction task are extracted from the input window. It may include physiologically well-known features such as ST-segment length, the relative amplitude of key-points or any other expert-defined features such as frequency domain features, power spectrum, specific compression coefficient, and so forth (Richman & Moorman, 2000; Park, Lee, & Jeon, 2009; Alcaraz et al., 2010; Sutar, Kothari, & Keskar, 2012; Costa, Davis, & Goldberger, 2017). Fig. 13.3 shows

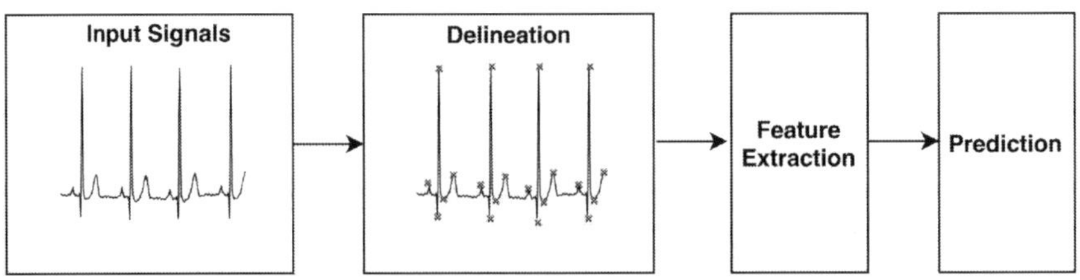

FIGURE 13.2

Traditional pipeline for ECG classification consisting of signal delineation, feature extraction, and prediction.

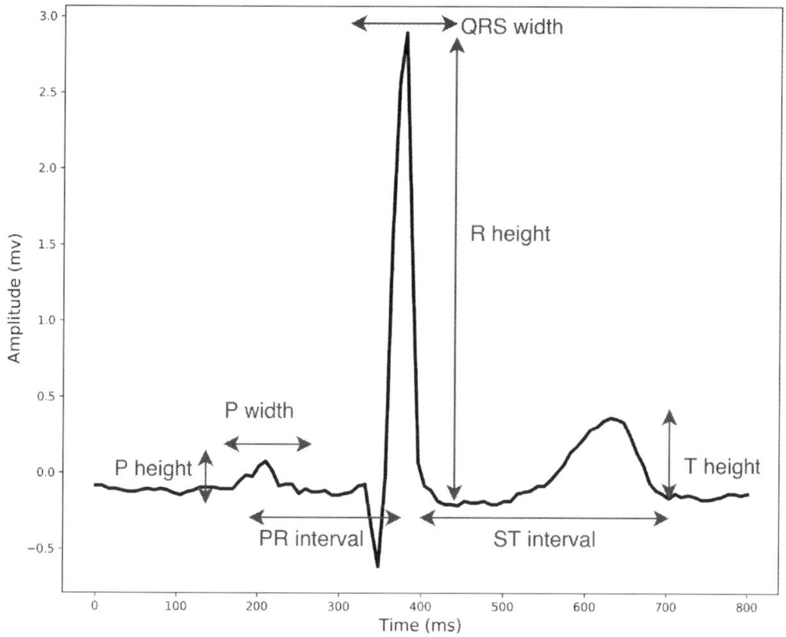

FIGURE 13.3

An example of different hand-engineered features extracted from an ECG waveform.

a visual example of such features extracted from an ECG signal. The common practice is to start with a set of relevant features and to expand the feature set while paying close attention to the final downstream task performance. To systematically enhance the efficiency of this process, feature selection and dimensionality reduction methods are usually employed in the literature (Ayar & Sabamoniri, 2018; Mar et al., 2011; Maršánová et al., 2017).

As the final processing step, extracted feature vectors are fed to a predictor model. In the literature, various model families have been used to learn the target feature relationship including logistic, support vector machine, decision tree, and random forest classifiers. As introducing different machine learning models is not the focus of this chapter, we refer interested readers to recently published review work (Bishop, 2006).

End-to-end automated ECG analysis

As discussed in the previous section, traditional automated ECG processing techniques require finding important signal key-points within a certain processing window. However, due to different signal morphologies, it is not always straightforward to detect these key-points. For instance, based on specific clinical conditions, certain points may not appear in ECG signals or may appear in different forms as they usually appear in a normal population. As an example, a heart axis inversion may cause R waves to appear as minima in certain ECG channels, or an elevated ST segment may impose waveform changes that are far from normal morphologies (Dubin, 2000). Moreover, the coexistence of different physiological conditions as well as noise and artifacts can further complicate this issue, as a superposition of different causal factors appears in the observed signal.

From another perspective, recent advances in pattern recognition suggest that even a completely accurate delineation followed by hand-engineered feature extraction would lead to nonoptimal solutions (Goodfellow, Bengio, & Courville, 2016). The rationale behind this statement is that these features are designed based on clinical studies suggested to help physicians to detect certain conditions or features selected based on heuristic and manual trial-and-error procedures. However, the recent availability of massive datasets and powerful machine learning models enables us to learn and exploit different signal patterns that are not necessarily detectable by human experts (Kachuee et al., 2016). More specifically, the recent advances in deep learning provide the ability to train large recognition networks in an end-to-end manner, taking raw signals as input and outputting the final classification result. In the next section, we present a brief introduction to neural network architectures that are most relevant to ECG analysis. For more in-depth analysis of network architecture, the reader is referred to Goodfellow, Bengio, and Courville (2016).

Convolutional Neural Networks (CNNs) are among the main drivers of advances in the field of deep learning. In the image recognition domain, where CNNs are becoming ubiquitous, CNNs exploit intrinsic properties of natural images such as scale, location, and rotation invariance of objects appearing in images (Goodfellow, Bengio, & Courville, 2016). For instance, a car may appear at different locations in an image with different viewpoints and distance to a camera. Alternatively, CNNs are applicable to timeseries such as voice recognition and vital signal analysis. In timeseries, CNN filters are applied along the time axis to exploit the coherence and invariance present along the time dimension.

For the case of 1-D convolution, we can formulate each convolution layer as:

$$s(t) = (x * \omega)(t) = \sum_{a=-\inf}^{a=+\inf} x(a)\omega(t-a),$$

where x and ω denote the input signal and the convolution kernel (i.e., filter weights), respectively. The output, $s(t)$, is the result of applying the kernel on the output signal. Usually, for a given signal frame, multiple kernels are applied at a certain convolution layer, each creating a different feature map providing different higher-level feature representations. The convolutional layer outputs also can be interpreted as the results of different pattern detectors applied to a certain input.

A typical CNN architecture consists of several convolution layers, each followed by nonlinearity, normalization, and pooling functions, as well as one or two classification layers. For instance, batch normalization (Ioffe and Szegedy, 2015) is frequently used to normalize the distribution of output activations to help the stability of the optimization and help the flow of gradients. Pooling operations such as max or mean pooling reduce the spatial size of representations and enhance the location invariance in CNNs. Furthermore, residual connections have been widely adopted to enhance the training in very deep networks of a 100 layers or more by providing skip connections to reduce vanishing gradients (He et al., 2016). Fig. 13.4 shows an example of a deep convolutional network with residual connections for ECG classification. The dotted area is repeated five times to produce a total of 13 weight layers.

Recurrent Neural Networks (RNNs) are models that read an input sequence one at a time while holding a state representation which is a function of inputs observed by the model from the initial state to the current input. RNNs can be implemented based on different input/output sequence timing. For instance, an RNN may

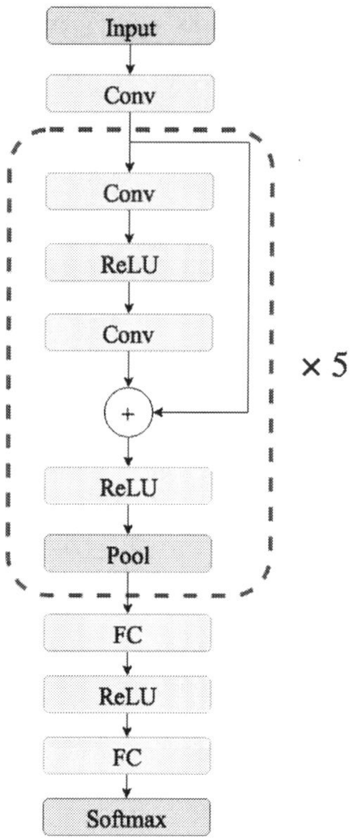

FIGURE 13.4

An example of a deep convolutional network with residual connections for ECG classification.

Graphics reproduced with permission Kachuee, M., Fazeli, S., & Sarrafzadeh, M. (2018). Ecg heartbeat classification: A deep transferable representation. In 2018 IEEE International Conference on Healthcare Informatics (ICHI), *pp. 443–444.*

take in all the input sequence and generate another output sequence one step at a time, or alternatively, it may generate outputs at each time-step while reading each input.

From the functional perspective, in general, each RNN hidden state is a function of the previous hidden state and the current input (Markov assumption). Similarly, each output is a function of the current hidden state and the current input:

$$h^t = f\left(h^{t-1}, x^t; \theta\right),$$

$$y^t = g\left(h^{t-1}, x^t; \theta\right),$$

where h^t is the hidden state at time-step t, and θ is a set of parameters defining the state transition (f) and output generation (g) functions.

The main drawback of RNNs stems from the limitations of gradient-based learning which effectively prevents RNNs from capturing representations from more than a dozen steps due to severe vanishing gradient issues. Note that to train an RNN, the computation graph of the recursive operations is unrolled, so the resulting networks are at least as deep as the number of time-steps. Long Short-Term Memory (LSTM) and Gated Recurrent Unit (GRU) have been proposed in order to circumvent the issue by using gating mechanisms to help holding the internal state representation (Goodfellow, Bengio, & Courville, 2016).

Architectures discussed earlier require a fix-sized input; however, ECG signals are usually recorded in varying lengths. Also, due to modeling and computational reasons applying learning models on very long signal sequences would be impractical. More specifically, it would cause massive memory usage and model complexity in CNNs or severe vanishing gradient in RNNs. In the literature, ECG records are commonly split into smaller fixed-size windows or segments. The size of processing windows usually spans from a single heartbeat to less than a minute based on the sampling rates, dataset sizes, and most importantly clinical conditions to be detected. It is well known that certain heart conditions may cause a consistent impact on ECG (e.g., myocardial infarction), while others only cause transient changes such as certain arrhythmia or pacemaker dysfunction. Also, each signal window can be aligned and preprocessed, or it can contain raw signals without any initial preparation.

Certain models only permit a limited input length. For instance, LSTMs and GRUs are only effective in processing up to 50 time-steps and rapidly decline in performance for lengthier sequences. On the other hand, the size and distribution of the training dataset at hand determine how complex the trained prediction models can be. If only a relatively small set of data is available for the training, using a smaller window size and a more compact model would result in better final performances. Another factor to be considered is the sampling frequency. ECG signals are usually sampled at rates as low as 100 Hz to as high as 4 KHz. It is often very helpful to resample the raw signal to a reasonable sampling rate that provides the necessary signal features while minimizing the number of samples in each window. Lastly, the window size directly impacts the amount of memory and compute required to make predictions, as well as the classification latencies. To summarize, the following factors should be considered in selecting the proper window size:

- Model architecture
- Dataset size
- Sampling frequency
- Cardiac conditions to be detected
- Computational, memory, and latency limitations

It is worth noting that CNN and RNN models are often combined such that an initial CNN network summarizes the usually lengthy input sequence to a set of more compact features while an RNN takes in these compact features as input to make the final predictions. This architecture can be particularly useful for detecting certain conditions that require considering a large signal window that simple CNNs or RNNs would usually fail at. Such a network is trained in end-to-end error signal backpropagating from the classification layers to unfolded RNN layers as well as CNN layers.

Evaluation of ECG classifiers

Evaluation of ECG analysis systems can be very challenging due to different factors such as limited quantity and quality of expert annotations, huge class imbalance, and coexisting health conditions to name a few. Another important aspect that is mostly unanswered in the current literature is considering the importance and severity of medical conditions in the evaluation of diagnostic systems. For instance, a system that is detecting several different diseases should be evaluated and compared based on the importance of weighted average of these disease types rather than simply using the average performance across all conditions (Goldstein et al., 2019).

In terms of the number of True Positive (TP), True Negative (TN), False Positive (FP), and False Negative (FN) predictions, typical evaluation measures including accuracy, precision, recall, and F1 score are defined as:

$$Accuracy = \frac{TP + TN}{TP + TN + FP + FN}$$

$$Precision = \frac{TP}{TP + FP}$$

$$Recall = \frac{TP}{TP + FN}$$

$$F1 = 2\,\frac{Precision \times Recall}{Precision + Recall}$$

Intuitively speaking, *Accuracy* measures the ratio of correct predictions to the total number of predictions. *Precision* measures the ratio of positive predicted cases which are detected correctly to the number of all predicted positive cases, showing what percentage of samples predicted as positive are actually positive samples. *Recall* is the ratio of correctly detected cases to the number of actual positive cases, basically measuring how reliable a system is in detecting a less frequent positive condition. *F1* score is reporting a harmonic mean of precision and recall values, which is quite helpful in comparing different systems using a single value rather than considering different measures at the same time (Powers, 2011).

Apart from these numeric measures, other more illustrative evaluation tools are often adopted in the literature. Confusion matrix is one of the most well-known evaluation measures that are frequently used in multiclass classification problems. In a confusion matrix, each row represents the number of actual samples in each class and each column represents the number of predicted samples for each class. Ideally, a perfect classifier would have zero elements for nondiagonal positions that are representing incorrectly classified samples. Furthermore, off-diagonal values can be used to investigate how different classes are being classified incorrectly with respect to other classes, i.e., confused or mixed-up with each other. Please refer to (Fawcett, 2006) for more information. Fig. 13.5 is an example of a confusion matrix used to evaluate a 5-class arrhythmia classifier.

Receiver Operating Characteristic (ROC) curve demonstrates the rate of true positives (TPR) versus the rate of false positives (FPR) to visually evaluate the operation of a binary classifier at different decision threshold settings. Here:

$$TPR = \frac{TP}{TP + FN}$$

$$FPR = \frac{FP}{FP + TN}$$

Intuitively, the ROC curve shows the trade-off between the ability to detect positive samples versus the rate of samples detected as positive by error. Often area under the ROC (AU-ROC) curve is used to measure how close a certain model is to the ideal case, where high recall (or, TPR) is achieved without inducing any false positives, in which the AU-ROC approaches to one (Fig. 13.6).

Alternatively, the precision versus recall curve has been used to illustrate the precision-recall trade-off and the area under this curve is used as a quantitative performance measure to compare classifiers. There is evidence suggesting that the precision-recall curve is more informative in binary classification tasks with a considerable class imbalance (Saito & Rehmsmeier, 2015).

Research directions

In the previous sections, we presented different approaches for automated analysis of ECG recordings. Many of these approaches rely on machine learning models that require training based on a dataset. Practically, the training dataset determines the complexity of models to be trained which translates to the complexity of patterns

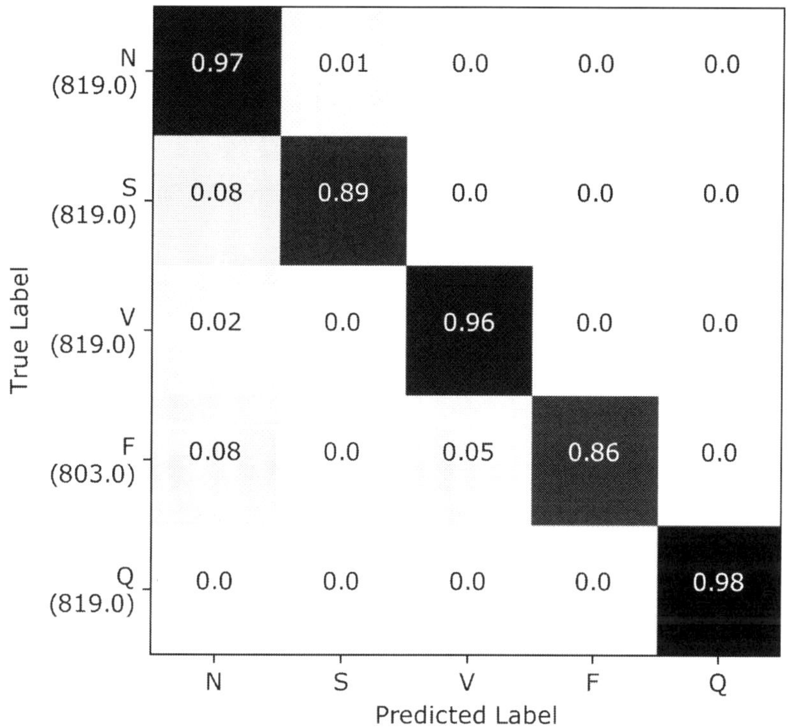

FIGURE 13.5

An example of a confusion matrix used to evaluate a 5-class arrhythmia classifier. Note that numbers reported inside the matrix are normalized and the actual numbers are provided in the y-axis ticks.

Graphics reproduced with permission Kachuee, M., Fazeli, S., & Sarrafzadeh, M. (2018). Ecg heartbeat classification: A deep transferable representation. In 2018 IEEE International Conference on Healthcare Informatics (ICHI), *pp. 443–444.*

to be detected and the final performance of the model. A common approach that is widely used in the deep learning literature is to augment a dataset with altered samples from the original dataset to effectively produce a larger training set. For instance, imaging data can be augmented by horizontal flipping, randomly cropping different parts of an image, changing pixel colors, etc. However, data augmentation for ECG signals is more challenging and most of the studies either use no data augmentation or use very basic augmentation such as adding a small Gaussian noise. A more comprehensive study on data augmentation for ECG signals can potentially unlock the use of more powerful and more complex architectures compared to the models being used today.

While the number of annotated recordings is usually very limited due to limited access and the costs associated with human experts, there are ample unlabeled recordings available. For instance, Physionet MIMIC III (Johnson et al., 2016) consists

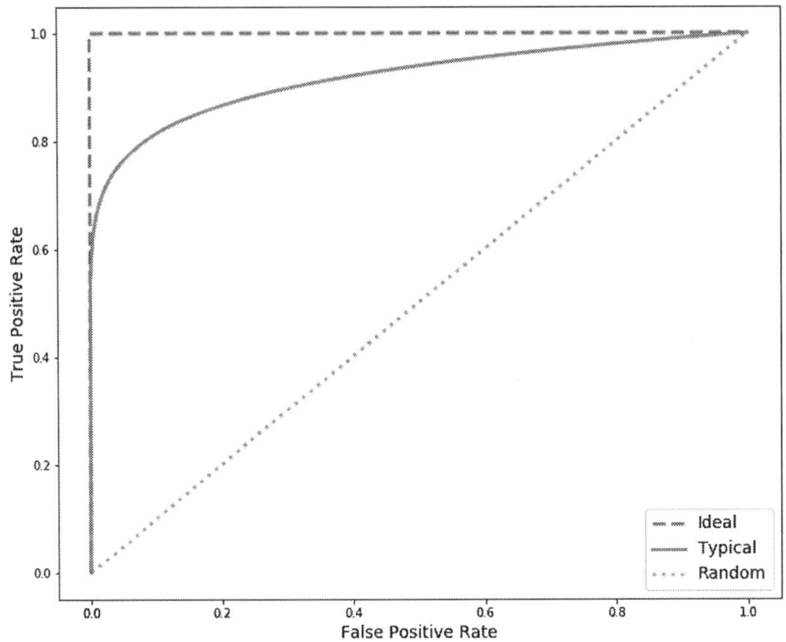

FIGURE 13.6

ROC curve can be used to evaluate classifiers at different decision thresholds. The figure shows a few curves including a perfect classifier, a random classifier, and a typical case.

of about 2.4 TB of recordings from bedside monitoring devices. This is a promising research avenue for the use of unsupervised, semi-supervised, as well as transfer learning applied to ECG analytics. A recent study showed the effectiveness of transferring ECG representations learned for a certain arrhythmia classification task to myocardial infarction detection (Kachuee, Fazeli, & Sarrafzadeh, 2018). As another example, Zhang et al. (2013) used wavelet features and a large unlabeled dataset to learn an unsupervised representation and suggested a semi-supervised ECG pattern recognition system that generalizes well using relatively small sample sizes.

Another avenue of research is using attention-based models for ECG analysis. Attention models are currently considered state-of-the-art for tasks that require detecting patterns within long intervals. Also, the attention mechanism can be used to reduce the computational load of processing signals. For example, Yao et al. (2020) suggested a network consisting of convolutional and recurrent layers that generate intermediate signal representations which are ultimately classified using an attention-based classifier, deciding which signal parts are relevant to the classification of each target category.

Interpretability is considered one of the main challenges for any automated learning-based health analytic system. There are many legal and ethical concerns associated with block-box models that only provide statistically appealing performances but do not provide any reasoning based on each analyzed instance. There have been many recent endeavors to address this by either using models that provide insights about their operations or modeling dynamics of a black-box model. However, the case of timeseries input domain is more challenging than usual tabular data as explanations for a signal window should present visually appealing and yet accurate interpretations. This requires extensive research in the interpretation of timeseries models. A recent work by Goodfellow et al. (2018) proposed utilizing activation patterns in a convolutional network to analyze the relevance of each signal segment for the final classifications. Alternatively, attention maps have been suggested by multiple papers to visualize important parts of the signal that are attended by a particular model (Hong et al., 2019; Kuvaev and Khudorozhkov, 2019). Despite the recent progress, understanding the reasoning behind each classification requires clearer and more concrete interpretations and continues to be a challenging research direction.

Conclusions

In this chapter, we presented an overview of the methods used for the automated interpretation of ECG records. We covered basic concepts about the ECG signal as well as different established signal processing, feature extraction, and recognition approaches. In the end, this chapter presents a few directions for future research that the authors believe to be promising avenues of extending the current literature.

References

Alcaraz, R., et al. (2010). Optimal parameters study for sample entropy-based atrial fibrillation organization analysis. *Computer Methods and Programs in Biomedicine, 99*(1), 124–132. Elsevier.

Ayar, M., & Sabamoniri, S. (2018). An ECG-based feature selection and heartbeat classification model using a hybrid heuristic algorithm. In *Informatics in medicine unlocked* (Vol. 13, pp. 167–175). Elsevier.

Bishop, C. M. (2006). *Pattern recognition and machine learning*. Springer.

Costa, M. D., Davis, R. B., & Goldberger, A. L. (2017). Heart rate fragmentation: A new approach to the analysis of cardiac interbeat interval dynamics. *Frontiers in Physiology, 8*, 255.

Dubin, D. (2000). *Rapid interpretation of EKG's: An interactive course*. Cover Publishing Company.

Fawcett, T. (2006). An introduction to ROC analysis. *Pattern Recognition Letters, 27*(8), 861–874. Elsevier.

Goldstein, O., et al. (2019). *Target-focused feature selection using a Bayesian approach*. arXiv preprint arXiv:1909.06772.

Goodfellow, I., Bengio, Y., & Courville, A. (2016). *Deep learning*. MIT Press.

Goodfellow, S. D., et al. (2018). Towards understanding ecg rhythm classification using convolutional neural networks and attention mappings. In *Machine learning for healthcare conference* (pp. 83–101).

He, K., et al. (2016). Deep residual learning for image recognition. In *Proceedings of the IEEE conference on computer vision and pattern recognition* (pp. 770–778).

Hong, S., et al. (2019). *MINA: Multilevel knowledge-guided attention for modeling electrocardiography signals*. arXiv preprint arXiv:1905.11333.

Ioffe, S., & Szegedy, C. (2015). *Batch normalization: Accelerating deep network training by reducing internal covariate shift*. arXiv preprint arXiv:1502.03167.

Johnson, A. E. W., et al. (2016). MIMIC-III, a freely accessible critical care database. *Scientific Data, 3*, 160035. Nature Publishing Group.

Kachuee, M., et al. (2016). Cuffless blood pressure estimation algorithms for continuous health-care monitoring. *IEEE Transactions on Biomedical Engineering, 64*(4), 859–869. IEEE.

Kachuee, M., Fazeli, S., & Sarrafzadeh, M. (2018). Ecg heartbeat classification: A deep transferable representation. In *2018 IEEE International Conference on Healthcare Informatics (ICHI)* (pp. 443–444).

Kameenoff, J. (2017). Signal processing techniques for removing noise from ECG signals. *Biomedical Engineering and Research, 1*(1), 1. JScholar Publishers.

Kuvaev, A., & Khudorozhkov, R. (2019). An attention-based CNN for ECG classification. In *Science and information conference* (pp. 671–677).

Mar, T., et al. (2011). Optimization of ECG classification by means of feature selection. *IEEE Transactions on Biomedical Engineering, 58*(8), 2168–2177. IEEE.

Maršánová, L., et al. (2017). ECG features and methods for automatic classification of ventricular premature and ischemic heartbeats: A comprehensive experimental study. *Scientific Reports* (Vol. 7,(1), 11239. Nature Publishing Group.

Misiti, M., et al. (1996). *Wavelet toolbox* (Vol. 15, p. 21). Natick, MA: The MathWorks Inc.

Park, J., Lee, S., & Jeon, M. (2009). Atrial fibrillation detection by heart rate variability in Poincare plot. *Biomedical Engineering Online, 8*(1), 38. BioMed Central.

Powers, D. M. (2011). *Evaluation: From precision, recall and F-measure to ROC, informedness, markedness and correlation*. Bioinfo Publications.

Rajan, D., & Thiagarajan, J. J. (2018). A generative modeling approach to limited channel ecg classification. In *2018 40th annual international conference of the IEEE engineering in medicine and biology society* (pp. 2571–2574). EMBC).

Richman, J. S., & Moorman, J. R. (2000). Physiological time-series analysis using approximate entropy and sample entropy. *American Journal of Physiology-Heart and Circulatory Physiology, 278*(6), H2039–H2049. American Physiological Society Bethesda, MD.

Saito, T., & Rehmsmeier, M. (2015). The precision-recall plot is more informative than the ROC plot when evaluating binary classifiers on imbalanced datasets. *PLoS One, 10*(3). https://doi.org/10.1371/journal.pone.0118432. Public Library of Science.

Sutar, R. G., Kothari, A. G., & Keskar, A. G. (2012). ECG feature extraction using LCAD. In *2012 international conference on communication systems and network technologies* (pp. 100–103).

Yao, Q., et al. (2020). Multi-class Arrhythmia detection from 12-lead varied-length ECG using attention-based time-incremental convolutional neural network. *Information Fusion, 53*, 174–182. Elsevier.

Zhang, H., et al. (2013). A 12-lead clinical ECG classification method based on semi-supervised discriminant analysis. In *2013 6th international conference on biomedical engineering and informatics* (pp. 177–181).

CHAPTER

Machine learning in cardiovascular genomics, proteomics, and drug discovery

14

Ming Wai Yeung[1], **Jan-Walter Benjamins**[1], **Pim van der Harst**[1,2,3], **Luis Eduardo Juarez-Orozco**[1,3]

[1]*University of Groningen, University Medical Center Groningen, Department of Cardiology, Groningen, The Netherlands;* [2]*University of Groningen, University Medical Centre Groningen, Department of Genetics, Groningen, The Netherlands;* [3]*Department of Cardiology, Division Heart and Lungs, University Medical Centre Utrecht, University of Utrecht, Utrecht, The Netherlands*

Chapter outline

Introduction

Modern cardiovascular medicine and research have seen the development of novel state-of-the-art techniques and analytics. This convergence is currently helping unravel the mechanistic determinants, pathways, and potential therapeutic targets in cardiovascular diseases.

On the one hand, three decades of refinements (including the completion of the Human Genome Project) in the study of genomics and its downstream influence into the study of proteomics have provided notions of the deterministic but also dynamic nature of this continuum (see Fig. 14.1) in cardiac health and disease. On the other hand, the recent advent of modern machine learning (ML)-based artificial intelligence has provided the opportunity to explore, integrate, and analyze large amounts of data in terms of their inherent complex interrelations and nonlinear dependencies.

Machine Learning in Cardiovascular Medicine. https://doi.org/10.1016/B978-0-12-820273-9.00014-2

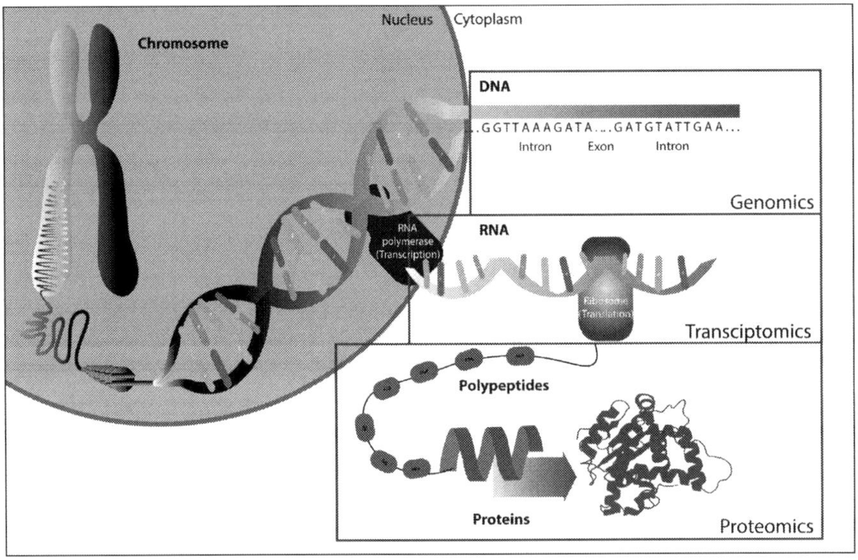

FIGURE 14.1 The continuum of -omics levels of analysis.

Machine learning methods can be applied on the studies of different -omics levels such as genome, transcriptome, and proteome, which are major fields in molecular biology. Combination of evidence from different -omics levels contribute to new insights of our understanding of cardiovascular diseases.

In essence, the very-high-throughput nature of -omics data at different analytical and mechanistic levels represents a suitable target for ML and is expected to boost our capacity to generate valuable insights in the study of cardiovascular disease. This chapter summarizes and interprets the current evidence on the application of ML in the genomic-proteomics horizon and in modern drug discovery initiatives.

Machine learning in genomics

ML involves use of computational algorithms for a specific task, which is usually achieved through extracting characteristics/patterns from data. One key feature of ML is that the performance of the learner or model (also called classifier when the objective is data classification, i.e., a binary output) is expected to improve with exposure to more data (experience). Thanks to the increased availability of genomic data in recent years, various ML algorithms are being adopted to help identify "patterns" in the genome and to predict functional consequences of specific genetic variations.

Molecular genetics

ML algorithms, using supervised and unsupervised learning methods (where the output label for the output is known or unknown, respectively), have yielded fruitful results in annotation of genomes ranging from identifying various genomic elements to predicting their functional consequences (Beer & Tavazoie, 2004; Bhasin & Raghava, 2004; Degroeve et al., 2005; Heintzman et al., 2007; Hertel & Stadler, 2006; Pertea, 2001; Vinayagam et al., 2004) (reviewed by (Libbrecht & Noble, 2015)). A genomic annotation model built from supervised learning is usually done by submitting a set of labeled data represented by features characterizing the patterns associated with these genomic elements to the algorithm. The model then learns to recognize those genomic components through the differences between those labeled as elements and those labeled as nonelements in the training data (Bhasin & Raghava, 2004; Degroeve et al., 2005; Hertel & Stadler, 2006; Pertea, 2001; Vinayagam et al., 2004). Conversely, unsupervised algorithms such as k-means clustering can be performed on regions with unknown functions (unlabeled data) along with regions with known functions to infer the function of target regions (Beer & Tavazoie, 2004; Heintzman et al., 2007).

The choice of the algorithm should be made based on the task to be solved but also the characteristics of the data at hand. Importantly, data transformation (i.e., generation of features to represent the data) is also critical for the end-performance of the models. For example, Pertea et al. developed a decision tree-based model called GeneSplicer to classify splice sites in *Arabidopsis thaliana* and in the human genome (Pertea, 2001). Decision trees are an ML model with a structure similar to a flowchart in which a decision is made based on a set of sequential rules. Pertea et al. tuned the parameters of the model namely the rules and their sequences using known true and false splice sites in 1323 genes from *Arabidopsis thaliana* and 1115 human genes. Performance assessed through fivefold cross-validation showed GeneSplicer could detect up to 97% of true splice sites at a cost of around 10% false positive rate. Using the same dataset Degroeve et al. (2005) built a support vector machine (SVM) classifier for the same task of splice site recognition. SVM is a popular modeling method in this area. It represents a supervised learning method that, given labeled data represented in a feature space, computes a hyperplane to best separate data members of different classes. The classifier developed by Degroeve et al. called SpliceMachine, took three types of features of a target genomic region, engineered to capture positional, compositional, and coding potential information, respectively, to predict the existence of a splice site. It was shown that SpliceMachine outperformed GeneSplicer, especially in predicting acceptor site in both species in fivefold cross-validation. The authors showed SpliceMachine also outperformed another SVM-based model on a same set of human splice sites. This demonstrated the importance of feature engineering in predictive ML models.

While elaborate feature engineering incorporating extensive domain knowledge can improve performance, it may be not always efficient/feasible due to time constraints or limited understanding of the underlying mechanistic structures of the subject at hand. In recent years, ML methods with capability to learn features from data such as Deep Learning are gaining popularity (reviewed by Zou et al., 2019). Deep learning is a subclass of ML methods which involves artificial neural networks. An artificial neural network consists of multiple layers of processing units (artificial neurons) with prespecified mathematical operations. The artificial neurons between layers are connected and the strength of these connections is adjusted during training. With sufficient amounts of data, artificial neural networks transform the data and learn to extract features relevant to the target task. However, it is worth mentioning that auto-extraction of features comes with the cost of steep data size (training set) requirements, which is often much higher than in other ML methods. In a recent study by Jaganathan et al. (2019), a 32-layer deep neural network called SpliceAI was developed to recognize splicing sites in human genome. The model was trained on a training set of 130,796 donor-acceptor pairs from 13,384 genes in contrast to aforementioned studies.

Besides functional annotation of a specific genomic region, ML can also be applied to predict the biological effect of individual genomic variations. Pathogenicity prediction is highly relevant in cardiovascular clinical research, especially in rare genetic diseases. Annotations of the variation in different contexts including evolutionary conservation and location on genomic level, regulatory information on transcription level, and functional change on translation/protein level have served as input features for training ML methods (Bao, Zhou, & Cui, 2005; Bromberg & Rost, 2007; Ioannidis et al., 2016; Kircher et al., 2014).

Thus far, benchmarking studies have been carried out to compare the performance of different in-silico prediction tools for variants involved in different classes of diseases like cancer (Hicks et al., 2011) and cystathionine beta synthase deficiency (Wei et al., 2010) (reviewed by Ohanian, Otway, & Fatkin, 2012), but no tools have achieved an absolute superior performance in all studies. Additionally, Ohanian et al. examined the performance of eight prediction tools with nine rare and nine common variants associated with cardiac disorders. For the nine rare variants, in spite of clinical and experimental evidence to support their role in cardiomyopathies and arrhythmias, prediction results were not unanimous in all but one (the Y315S variant in KCNQ1 in cases with congenital long QT syndrome). One reason for the discrepancies in predictions may be attributed to the stringency/differing criteria for defining pathogenicity. Notably, ML models were tuned using different training datasets. If a certain type of variant is overrepresented in the training data (a phenomenon known as *class imbalance*), the resulting models may carry bias toward such class. In connection with this issue, as all of these tools were trained on relatively rare variants with large effects, they may have limited utility for predicting weaker common variants. This is supported by the mostly benign prediction of three common variants involved in cardiac ion channelopathies—KCNH2 K897T, KCNE1 S38G, and SCN5A H558R, which have reported experimental evidence of pathogenicity as reviewed by Ohanian et al. (2012). Overall, the majority of ML prediction tools are more likely to be useful for prioritization of rare variants, and a consensus prediction may be considered to alleviate potential sampling bias.

Conversely, studies have also been carried out to build specific prediction models with the aim of achieving a better performance than generic predictors. Stead, Wood, and Westhead (2011) developed KvSNP, an ML classifier, to predict the probability of pathogenicity of a nonsynonymous single nucleotide polymorphism (SNP) for the voltage gated potassium channel gene family. Mutations in the voltage-gated potassium channel gene are the cause of various cardiac diseases such as long Q-T syndrome. Using five features characterizing the gene family such as the subfamily membership, a classifier was trained on around 1600 manually curated nonsynonymous SNPs for each of five different ML algorithms. A random forest classifier, a predictor based on an ensemble of decision trees was ultimately chosen as the final model and was compared to three generic predictors on an independent test set consisting of approximately 180 SNPs. While a specific predictor like KvSNP may give better discrimination on a specific class of variants, one should also keep in mind the balance of specificity and generalizability. In a recent study, Li et al. (2017) developed a model specific to one gene—KCNQ1 using artificial neural network.Taking features similar to KvSNP, the model called Q1VarPred was trained on 107 KCNQ1 variants in which almost 90% were pathogenic. While Q1VarPred was shown to have higher performance metrics than KvSNP and seven generic predictors assessed by cross-validation, an independent test set that represents the true distributions would offer a more appropriate assessment of the actual predictive values. Additionally, when training in small datasets, ML models are prone to overfitting (i.e., the model represents the training set very well but generalizes poorly to new data drawn from the same distribution). One way to avoid overfitting is to select a method with complexity which matches the underlying data structure. Interestingly, Li et al. (2017) conducted pathogenicity classification of KCNQ1 variant based on two features through simple regression models, and demonstrated comparable performance with Q1VarPred. This suggests that in some cases simpler models might already suffice for such task (Li et al., 2017).

Unsupervised ML may be an alternative if a reasonable labeled training set is not available. In a recent paper by Walsh et al. (2019)an unsupervised clustering algorithm was implemented to identify "hotspots" within protein coding regions for rare variants. By comparing the results in large cohorts of large hypertrophic cardiomyopathy (HCM) against those in population controls, they found a substantial difference in the prevalence of rare variants in those hotspots within four genes, namely MYH7, MYBPC3, TNNI3, and TNNT2, in cases as compared to controls, suggesting a high likelihood of pathogenicity for variants in these regions. Further, they applied the approach on variants of RYR2 from a cohort of catecholaminergic polymorphic ventricular tachycardia (CPVT), where the algorithm detected three case-enriched clusters in RYR2, all of which partially overlapped with previously recognized hotspots for CPVT. The study demonstrated an alternative way to utilize ML for extraction of local information to gain insight in variant prioritization and interpretation in rare genetic cardiovascular diseases. Fig. 14.2 depicts analytical targets in genomics for the implementation of ML.

Population and complex genetics

The role of genetic markers in cardiovascular disease risk is an active area of research in population genetics. Multifactorial or complex diseases are those that

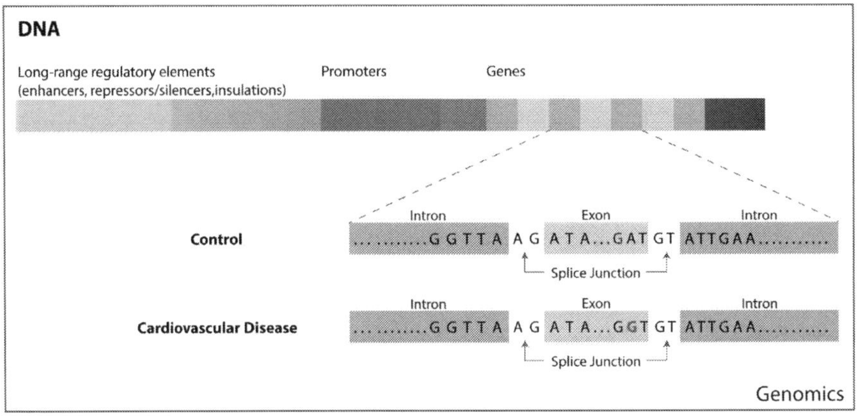

FIGURE 14.2 ML Targets in genomics.

ML models have been developed to identify functional elements such as long-range regulatory elements and splicing sites on the genome. They can be further customized for specific disease associated genes and/or diseases. For instance, prioritization of a splicing variant among a large number of variants to be associated with a cardiovascular trait by comparing the clustering results between case and control.

are influenced by both genetic and environmental factors in contrast to Mendelian disorders whose cause can be attributed to single mutations with large detrimental effects. Complex diseases are often linked to multiple genetic variants, which are usually weaker in effect as compared with those discovered in Mendelian diseases. Associations between complex disease and genetic variants are discovered by genome-wide association study (GWAS) which look at the difference in allele frequency of variants between cases and controls.

Many cardiovascular disorders such as coronary artery disease (CAD), atrial fibrillation (AF), hypertension (HTN), and stroke are complex diseases (Giri et al., 2019; Malik et al., 2018; Roselli et al., 2018; van der Harst and Verweij, 2018). As the individual effect of most genetic variants discovered in GWAS is often relatively small, the generation of genetic or polygenic risk scores (PRS) combining the effects of multiple variants have been studied. PRS is conventionally constructed as sum of risk alleles of all selected variants, each weighed by its effect size.

In recent years there has been interest in personal disease risk profiling using genetic information (Khera et al., 2018). While PRS has been shown to be strongly associated with disease, its discriminative power and clinical utility is at the moment discrete (Curtis, 2018; Fullerton & Nurnberger, 2019, p. 8). Nonetheless, given the complex genetic architectures of these diseases, the aim of replacing simple additive models by sophisticated ML analytical tools is an area of increasing interest. Celiac disease, for instance, is an autoimmune disease with a strong genetic component. Genetic test on human leukocyte antigens (HLA) in celiac disease offers high negative predictive value but low positive predictive value. To improve the predictive performance of such genetic testing, Abraham et al. (2014) constructed a PRS using L1-penalized SVM. L1 penalty enforced on model coefficients is a form of regularization that reduces the chance of overfitting when the number of features is much higher than the number of samples in the data. Abraham et al. reported that a PRS built using L1-penalized SVM model offered a greater improvement in performance than an additive PRS model in a large external multiethnic test set. In another analysis on a large population cohort, Paré et al. (2017) proposed a heuristic ML method to improve the discriminative performance of PRS. The proposed method, GraBLD, fine-tuned the weights of variants provided by external summary statistics with respect to the target cohort using gradient boosted regression trees; the weights were then corrected for linkage disequilibrium to produce the final PRS. They showed that GraBLD had superior performance over the conventional PRS method for all three traits: height, myocardial infarction, and diabetes mellitus. Results of these studies highlight the potential of improving the accuracy of genetic prediction with ML methods that incorporate additional information (Abraham et al., 2014; Paré et al., 2017).

Since complex diseases are influenced by both genetic and environmental factors, it is sensible to combine information from all aspects such as genetic risk profile, clinical risk factors and lifestyle to maximize the predictive power of risk models (Torkamani, Wineinger, & Topol, 2018). In fact, it has been shown that genetic and lifestyle factors contribute independently to various cardiovascular diseases namely CAD, AF, arterial HTN, and stroke (Khera et al., 2016; Said, Verweij, & van der Harst, 2018). Appropriate ML algorithms may be deployed to extract useful information from various data sources, such as "-omics" data, clinical characteristics, demographic profiles, and medical imaging data. Thereon, all data may be combined to build a general risk model. While such multimodal approaches are likely to require a large sample size to overcome the "curse of dimensionality," ML models on hospital-wide patient samples are gaining traction in recent years (Miotto et al., 2016; Wong et al., 2018). An exemplary study in the cardiovascular field is by Li et al. (2018) on Abdominal Aortic Aneurysm (AAA). They proposed a hierarchical ML framework to model AAA: nonsynonymous variants identified in whole genome sequencing data in 401 people with and without AAA were first annotated and assessed with existing pathogenicity prediction models using a bioinformatic tool named ANNOVAR (Yang & Wang, 2015). Then the mutation burden on a genic level was computed by summing up the average prediction scores of variants within each gene for each individual. Such mutation burden from each gene serves as one feature and hence each individual was represented by a vector with 17,443 mutation burden scores (i.e., number of genes considered in study). The mutation burden matrix including both cases and controls was then submitted to build an LASSO regression model (a regularized version of regression methods). The genetic model was complemented by a logistic regression model built on the electronic health records (EHR) containing lifestyle information and physiological measurements. Finally, a prediction was made by averaging the results of the genetic model and EHR model. Discriminative performance of the combined model, assessed by 10-fold cross-validation, was shown to be slightly higher than the individual models (area under receiver operating curve (AUC) = 0.80, while AUC EHR model alone = 0.775, AUC of the genetic model alone = 0.69). This illustrates the potential of a multimodular approach in ML.

Finally, ML can be used in genomic analysis beyond predictive purpose. Comparisons of diverse ML methods give new biological insights, such as the genetic architecture of complex diseases. Deep learning, for instance, is a method capable of abstracting nonlinear relationships that can be deployed to model gene-gene interactions (epistasis) (Günther, Wawro, & Bammann, 2009; Yeung, Maj, & Krawitz, 2019). Comparison of performance between linear and nonlinear methods offers insight in how certain genetic variants contribute to the disease (i.e., further understanding of the genetic architecture) (Romagnoni et al., 2019). Table 14.1 summarizes the machine learning methods in the genomic studies discussed previously. With further understanding both in capability and limitations on different ML methods, the use of novel analyses on genomic studies considering modern analytics is expected to keep increasing.

Table 14.1 ML methods used in genomic analysis

Tasks	ML methods
Genomic annotation	Supervised learning: Artificial neural network (Jaganathan et al., 2019) Decision tree/Random forest (Pertea, 2001; Rajagopal et al., 2013) Support vector machine (Degroeve et al., 2005; Hertel & Stadler, 2006; Vinayagam et al., 2004) Unsupervised learning: K-means clustering (Beer & Tavazoie, 2004; Heintzman et al., 2007)
Variant annotation ad prioritization: (Pathogenicity/Functional consequence prediction)	Supervised learning: Artificial neural network (Bromberg & Rost, 2007; Li et al., 2017) Decision tree/Random forest (Bao et al., 2005; Ioannidis et al., 2016; Stead et al., 2011) Support vector machine (Kircher et al., 2014) Unsupervised learning: Clustering analysis (Walsh et al., 2019)
Disease prediction	Supervised learning: Support vector machine (Abraham et al., 2014) Decision tree/Random forest (Paré et al., 2017; Wong et al., 2018) Lasso (L1-penalized regression) (Li et al., 2018) Unsupervised learning: Artificial neural network (Miotto et al., 2016)

Machine learning in proteomics

Following the line of modern system-levels approach to cardiovascular disease, emerging molecular profiling technologies such as proteomics could provide new insights into the pathophysiology of cardiovascular disease.

However, implementations of ML in proteomic studies have not yet witnessed an accelerated advance comparable to the exploratory initiatives described in genomics. It is possible that this discrete (although steadily increasing) development relates to the position of proteomics within the continuum of translational research and medicine (genomics—proteomics—metabolomics—clinical implementation). In this sense, either from a bottom-up or top-bottom perspective, proteomics lies further down the line as ML optimization of data-driven tasks permeates into the analytical arsenal at our disposal. Clearly however, high-throughput molecular (protein) profiling data represents fertile ground for deep profiling of such varying molecular entities in relation to cardiovascular disease.

Proteomics deals in multifaceted and dynamic profiles with complex signaling pathways and high levels of interconnectivity. In this sense, the potential of ML to explore nonlinear dependencies between large amounts of variables and to select, integrate, and interpret complex variable structures can ultimately refine classification and predictive performance, facilitating timely applicability in the evaluation of proteomic data. Fig. 14.3 depicts the continuum of applicability of ML in proteomics research. However, it is useful to contextualize that comprehensive data evaluation in this regard can benefit from a fair combination of statistical and ML approaches.

Very early, Almeida et al. (2005) opted for the utilization of an artificial neural network to approach the comprehensive modeling of residual variability in two-dimensional gel electrophoresis in order to estimate the confidence associated to protein expression quantification. In this case, ML was featured as a way to avoid bias given by spatial variability in the results obtained through such a technique, an early form of pattern recognition.

ML implementation in cardiovascular proteomics research has predominantly utilized urinary proteomes, with a few reports dedicated to blood sampling.

In CAD, for example, Delles et al. (2010) had already begun deploying ML algorithms (SVM) in a urinary proteomic biomarker validation analysis comprising 238 polypeptides significantly associated with CAD a decade ago. In their test performance, ML achieved an AUC = 0.87 [0.81, 0.92] in the identification of patients with CAD (at a base rate = 52%), a significant improvement from prior biomarker panels for the same purpose (Von Muhlen et al., 2009; Zimmerli et al., 2008). Fig. 14.4 depicts the analytical principle behind SVM.

Of note, this classifier (CAD238) has been later evaluated by Neisius et al. (2016) in the discrimination of patients with and without angiographic evidence of CAD in the setting of chronic coronary syndromes (formerly known as stable CAD (Knuuti et al., 2019)), where it was unable to achieve adequate differentiation of such invasively documented difference. Conversely, Htun et al. (2017) explored in a first multicohort study (Australia, Europe, and North America) the urinary proteome for an

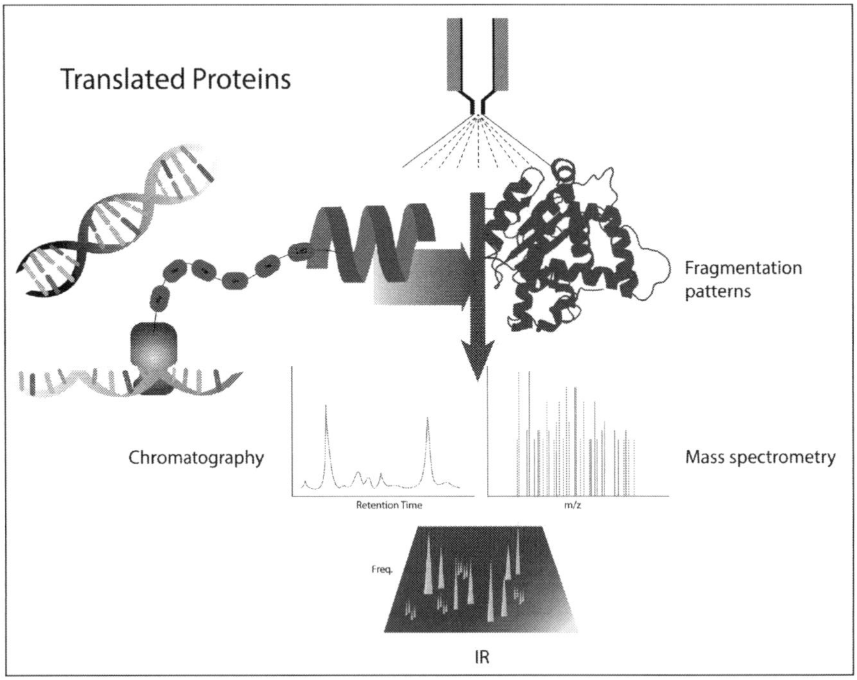

FIGURE 14.3 Targets in proteomics for ML analytics implementation (pattern and image recognition).

Typical steps involved in proteomic analysis include separation, detection, characterization, and quantification of proteins in which different methods such as chromatography, mass spectrometry, and Infrared spectroscopy (IR) are employed. ML can be integrated in any of the steps or clinical risk modeling based on the proteomic profiles.

acute coronary syndrome (ACS)-specific pattern in order to predict such events. The integrated model comprising 75 urinary peptides (named ACSP75) was only able to match the performance of the existing Framingham risk score (0.66 and 0.64, respectively). When the authors added the previously described CAD238 classifier and age, performance improved to an AUC = 0.75. The partial contribution of the classifiers and age was nevertheless not described, which leaves the possibility of some degree of overfitting. The ACSP75 classifier was a significant independent predictor in further survival analysis. Lately, McGilligan et al. (2019) investigated the utilization of a (quadratic) SVM in the integration of ST segment deviation with 20 plasma proteomic biomarkers in order to optimize the differentiation between patients with ACS or stable angina, between stable angina and controls, and between STEMI and NSTEMI with AUCs beyond the 0.90 threshold for good performance.

Another recent clinical study by Bom et al. (2019) utilized a deep stacking generalization framework of gradient boosting classifiers to generate a biomarker panel

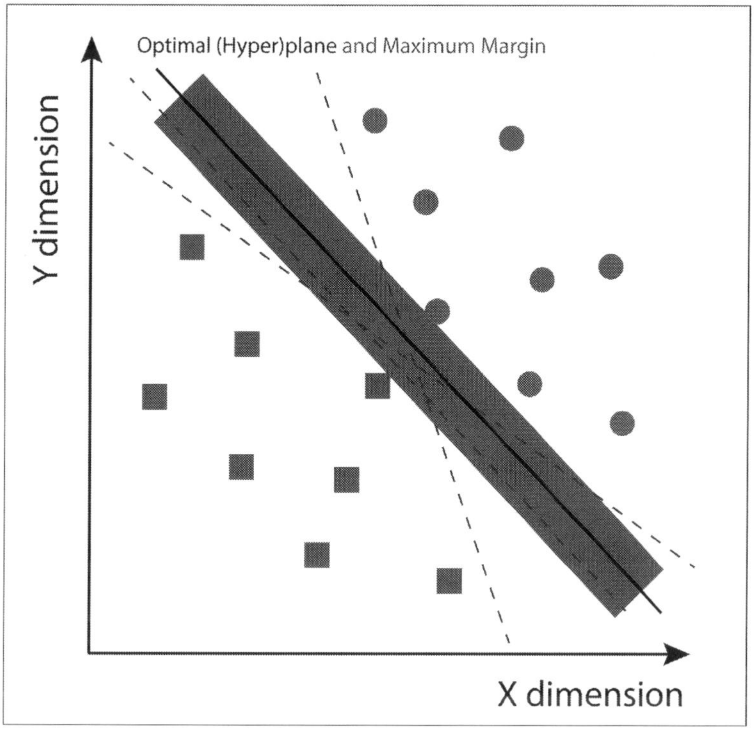

FIGURE 14.4 Support vector machine principle.

Data separation (classification) is optimized by finding the optimal separation between the classes (based on the maximum margin). Here a two-dimensional (variables) example shows a separation line, but as data dimensionality increases, so do the complexity of the separation hyperplane.

from plasma protein analysis to identify patients with high-risk atherosclerotic plaques or CAD from controls. Although performance was moderate (AUC = 0.79), ML surpassed the performance of common risk-descriptive clinical variables. Their results suggest the clinical potential for early detection of high-risk CAD and ACS through ML-analysis of proteomic signatures.

Urine proteomic analysis though ML has also been conducted in a "discovery" study by von zur Muhlen et al. (2016). Their report once again made use of SVM to explore, identify, and integrate urinary peptides in 49 patients with idiopathic deep vein thrombosis and/or pulmonary embolism. Interestingly, the highlighted peptides could then be validated in both human and murine tissue. Independent testing in a small independent sample of 47 subjects demonstrated high sensitivity and specificity (100% and 83%, respectively) of the SVM classifier, which was named DVT62. Notably, the current state-of-the-art ML standards have rapidly evolved to a degree where generalizability of these estimations may be reevaluated and novel analyses can be considered.

Thereon, integration of numerous urinary protein biomarkers obtained from capillary electrophoresis and mass spectrometry has also been performed to identify patients with asymptomatic left ventricular diastolic dysfunction in patients with arterial HTN. Kuznetsova et al. (2012) combined 85 features through SVM for this purpose, with AUC = 0.84 [0.70, 0.98] on the final test set. Although their study was also importantly limited by the sample size, these results support the potential benefit of ML in exploratory initiatives of high-dimensional dependencies that can optimize classification tasks based on proteomics data. Even more, this classifier was later validated in an independent Flemish patient sample (745 subjects) (Zhang et al., 2014), demonstrating good correlation with echocardiography-derived parameters of diastolic dysfunction including e' peak velocity and E/e' ratio, supporting its utility for early detection of diastolic dysfunction. Lately, the same group (Zhang, Nkuipou-Kenfack, & Staessen, 2019) has suggested a potential clinical research implementation roadmap through observational studies and randomized control trials of their multidimensional ML classifiers (HF1 and HF2). This supports and underlines the need for continuity within the novel insights that can be accessed through modern analytics such as ML for proteomic data.

Urinary proteomics has been also approached through ML for the identification of patients with heart failure (HF) with reduced ejection fraction. For instance, Rossing et al. (2016) studied samples from 62 patients to derive 103 peptides of interest to differentiate subjects with HF through a specific disease classifier. They later tested this SVM-based classifier (named HFrEF103) in 885 patients (552 controls, 127 with overt HF, and 176 with left ventricular diastolic dysfunction) demonstrating an AUC = 0.972 for the discrimination between HF patients and controls. Its performance was suboptimal when focusing on patients with diastolic dysfunction, which were not particularly selected in the derivation sample. This report underlines the relevance of the distribution of the input (training) data for optimizing specific distinctions and for achieving generalizability. Furthermore, comparative ML analysis has been also implemented in blood proteomics in chronic HF patients for the prediction of cardiovascular mortality by Lemesle et al. (2015) This group studied surface enhanced laser desorption ionization—time of flight—mass spectrometry data in 198 patients with HF (with a 50% of cardiovascular death at 3 years) through three ML algorithms: SVM, sparse partial least square discriminant analysis, and LASSO logistic regression. They built the classifiers from 42 mass spectrometry peaks and validated them in 344 patients. Interestingly, the three classifiers had very discrete although comparable performance AUC ~0.67 with directional advantage for SVM with a significant odds ratio for the stated endpoint (beyond 9) in the multivariate regression analyses (considering this was not survival analysis). They suggested that advances in computation biology may optimize risk stratification in HF patients although this has not yet been demonstrated.

Zhang et al. (2017) reported the development of another ML classifier (named heart failure predictor or HFP) from 277 subjects. Testing of the HFP classifier (based on 96 HF-specific peptides) in an independent 175 individuals showed only a modest AUC = 0.70 [0.56, 0.82] that was nonetheless superior to NT-proBNP (AUC = 0.57 [0.42, 0.72]); HFP was also shown with a marginal prognostic value for the progression from asymptomatic left ventricular dysfunction to overt HF (hazard ratio = 1.63 [1.04–2.55]). Although its utility is suggested, its performance may be considered in the best case discrete.

Aforementioned studies focusing on urinary proteomics have also underlined that coexistence of renal impairment may represent a major modifier of studied profiles (since excretion is by definition, hampered) and that results may not generalize to such subpopulations. Given the comorbid association of HF and chronic kidney disease, Farmakis et al. (2016) recently studied urinary peptides to constitute an ML-based biomarker feature pattern to identify patients with HF with reduced ejection fraction and CKD. The SVM classifier utilized 107 peptides (a majority of which were fragments of collagen type I and III) and achieved 84% sensitivity and 91% specificity in a 58-patient sample with only minor overlap (20%) of features with a prior CKD classifier, suggesting its use in further clinical trials, although this has not yet taken place.

In parallel, Emmens et al. (2018) have studied the proteomic "diversity" of high-density lipoprotein (HDL) as a predictor of clinical outcomes in HF. Although their study aligned with the notion of protein data exploration and integration for outcome classification (AUC for mortality = 0.82 [0.77, 0.87]), their focus was not on ML even though they utilized principal component analysis, a method within the theoretical ML family.

Quite recently, Captur et al. (2019) described an ML-integrated plasma biomarker panel for the identification of HCM. In their study 26 potential biomarkers were identified through plasma profiling as well as separate prior HCM myocardial tissue proteomic profiling experiments (Coats et al., 2018). Out of these, six circulating peptides were identified as statistically relevant and demonstrated an AUC of 0.89 [0.77, 0.93] and 0.87 [0.83, 1.00] in the training and testing sets, respectively. Interestingly, none of these panel components directly correlated with demographic, clinical, or echo-derived descriptors, while showing only moderate correlations to CMR parameters (e.g., mass and scar burden). On the other hand, this 6-peptide ML-integrated panel was associated with subclinical HCM, the development of nonsustained ventricular tachycardia and the estimated 5-year risk of sudden cardiac death. These results highlight an intrinsic advantage of the implementation of ML methods for the integration of weak univariate classifiers into stronger, significant, and potentially useful higher-dimensional discriminative tools for specific purposes.

Comparison of analytical approaches to characterize potential mechanistic profiles represents an interesting view in modern proteomics. In this sense, Ferreira et al. (2019) recently reported a clinically oriented investigation in the FIBRO-TARGETS (targeting cardiac fibrosis in HF treatment) consortium of hypertense, obese, and diabetic patients in order to explore plasma proteomic associations through various ML approaches (LASSO regression, AdaBoost, and complex network analysis). They studied 86 proteins, out of which 4 were (GDF-15, LEP, SORT-1, and FABP-2) related to HTN, 3 (CEACAM-8, LEP, and PRELP) to obesity, and 4 (GDF-15, REN, CXCL-1, and SCF) to diabetes. However, performance was rather discrete (AUC ~ 0.70) possibly due to the restricted sample and lack of fine-tuning. Yet, this report depicts the potential of advancing ML exploration in large available datasets with a clinical orientation.

An emerging aspect within the horizon of ML application in proteomic analyses is that of method standardization and quality control. Previously in this chapter, it has been explained how a comprehensive ML workflow comprises organized sequential steps that guarantee the robustness of the resulting models as well as their generalizability. One of such stages is adequate feature selection, which allows the user to reduce the dimensionality of the data, focus on consequent variables, and reduce computational needs to optimize training times. Very-highly correlated input variables may represent redundancy and hamper optimal workflow and modeling unnecessarily, while very weak classifiers (variables) may increase the risk of overfitting unduly. Since this balance in the input structure tuning is crucial, efforts have begun to discuss, experiment, and converge in general recommendations that guarantee seamless flow in ML workflows dedicated to high-throughput omics data (i.e., genomics and proteomics). For instance, Perez-Riverol et al. (2017) have discussed issues concerning feature selection methods with evaluation in several proteomics reference datasets.

Deep Learning has recently become the sharpest end of ML analytics due to its success in image and complex signal recognition. Its performance, once constrained by computational and storage capabilities, is now unbound and has begun to permeate into proteomic research. In 2017, Zhou et al. (2017) used Deep Learning in tandem mass spectrometry-based proteomics to predict theoretical peptide spectra. The approach (a bidirectional long short-term memory [LSTM] neural network) named pDeep was built to predict intensity distribution resulting from protein dissociation (peptide fragmentation) and was additionally able to learn physicochemical properties of amino acids in its intermediate layers and differentiate between very similar peptides, a clear advantage over traditional search methods. Notably, this approach demonstrated a better performance (>0.9 correlation coefficient with resulting spectra) than even prior ML-based (SVM, random forests, and shallow feed-forward neural networks) methods (e.g., PeptideART). Furthermore, it was trained in ~4,000,000 mass spectrometry spectra from several sources (e.g., ProteomeTools), contrasting the data size factor against non-Deep Learning ML studies. Thereon, an increasing amount of studies have further used Deep Learning methods to predict peptide sequences (DeepNovo (Tran et al., 2017) and DeepNovo-DIA(Tran et al., 2019)), fragmentation patterns for any peptide length (Tiwary et al., 2019), and properties such as retention time (DeepRT (Ma et al., 2018)) in mass spectrometry and liquid chromatography concerning data-dependent and data-independent acquisition sets with increasingly accurate results. Very recently, Zohora et al. (2019) explored the utilization of convolutional and recurrent neural networks (Deep Learning) in peptide feature detection and calculation of intensities from liquid chromatography with tandem mass spec (LC-MS) maps. Since modern LC-MS technologies generate very large amounts of high-speed and resolution data, they underlined how state-of-the-art Deep Learning analytics can advance protein identification and quantification. Their algorithm, named DeepIso, showed a >97% accuracy of peptide feature identification against the benchmark through the integration of a two-step (isoDetecting [convolutional and fully connected recurrent neural networks] and isoGrouping [convolutional and attention-gated recurrent neural networks]) workflow for raw LC-MS maps that delivers a feature table together with the mass-to-charge of its first isotope, charge, retention time range, and intensity. Parallel efforts in peptide identification and quantification of ready-to-use tools based in Deep Learning have been reported by Gessulat et al. (2019).

Furthermore, Guan, Moran, and Ma (2019) reported their Deep Learning approach through LSTM neural networks for improving the prediction of three key properties from peptide sequence characterization, namely: indexed retention times, survey scan charge state distributions, and sequence ion intensities of high-energy collisional dissociation spectra. Their study reported significant prediction improvements for the first and last, while charge state distribution prediction was comparable to prior ML-based methods.

Interestingly, novel maneuvers to economize data utilization and needs such as transfer learning have begun to be used in explorative studies for mass spectrometry analysis results (Zeng et al., 2019). Further results in the real utility of such implementations are expected, considering that proteomics has not been a research area necessarily constrained by data availability.

Furthermore, proteomics research also contemplates the analysis of a very large number of protein sequences for experimental functional characterization (e.g., using Gene Ontology), and ML-based approaches are expected to help in this multiclass and multilabel niche. Kulmanov, Khan, and Hoehndorf (2018) have reported on a Deep Learning approach (based on convolutional neural networks) named DeepGO for predicting protein function and cellular locations. Thereon, Zhang et al. (2019) have proposed their own approach named DeepFunc, which reports incremental gain in AUC (0.94) for protein function prediction through encoding information on domains, families, and motifs that are processed with topological information from protein-protein interactions and linkages. Nevertheless, all these potentialities and encouraging results are yet to be exploited in cardiovascular research initiatives.

In summary, the implementation of ML in the field of cardiovascular proteomics research has witnessed the progression between integration of high-dimensional classifiers (mostly through SVM) and modern deep learning approaches for high-throughput identification and function-prediction proteomic data. The suggested classifiers have been based on established software that could improve the identification of patients with cardiovascular disease (including CAD, DVT, HF, and cardiomyopathies), cardiovascular risk factors (dyslipidemia), or risk of adverse cardiovascular events, including death. However, such studies have been restricted in sample size and novel interest in analytical standardization and quality control for adequate ML model training and validation is only recently reemerging. On the other hand, Deep Learning-based approaches to proteomics are being explored as tools to optimize pattern recognition and functional identification/prediction, but only marginally in cardiovascular research. Considering the advancing technologies of molecular profiling, proteomics is eventually expected to consolidate as a standard component of our approach to cardiovascular disease characterization that could be routinely generated from patient samples processed and integrated into, for example, biomarker signatures that convey clinically relevant insights through ML analytics. Deep Learning is certainly influencing our analytical capacity in proteomic research (Tran, Zhang, & Li, 2018), and although its consolidation in cardiovascular research is yet to occur, it is clear that this knowledge area is steadily migrating from classic algorithm-centered to data-driven approaches.

Machine learning in drug discovery

Over the past century, drug discovery has evolved in multiple phases from the coincidental discovery of disease-treating substances—extracted from plants or found otherwise in biological organisms such as fungi—through a knowledge based, but manual approach to a fully automated industry. Following the initial extraction of active substances from plants, a first step in biochemical synthesis was taken to create the active molecular structures, isolated from their native organisms of origin. In the following decennia, wide ranges of chemical compounds were screened for activity in cells, without a prior chosen biological target, such as an amino-acid or protein. After discovery of a compound, a proper target playing a role in disease was then searched. This is what is known as phenotypic drug discovery or "classical" pharmacology. More recently, drug discovery has shifted toward a target-based approach, in which the disease modifying biological targets are first identified, followed by the search for chemical compounds with the required effect on the given targets. Based on available chemical property databases a preselection of chemical compounds can be made for inclusion in multiple screening rounds in which slight molecular modifications are made to optimize target modification and to minimize effect on closely related structures.

In de novo drug design, "in-silico" virtual prescreening is performed with computer models. Chemical properties of molecular substructures are thereby tested against the intended target, aiming for optimally targeting synthesized models. Most of these computer models are based on chemical compound descriptors, such as quantitative structure-activity relations (QSARs), giving an indication of how a synthesized molecule will alter biological targets.

Despite tremendous developments as seen in drug discovery over the past century, including the migration to a more target focused design process, the rate of positive treatment of disease is still relatively low. For example, Akyea et al. (2019) reported that 51% of patients receiving statin therapy do suboptimally respond, showing less than 40% reduction of low density lipoprotein cholesterol (LDL-C) within 24 months after start of the treatment. In a follow-up period up to more than 9 years, starting 24 months after initial prescription of statin, suboptimally responders were at a significantly higher risk (hazard ratio after correction for age and baseline LDL-C levels: 1.22, 95% CI from 1.19 to 1.25) of incident CVD events, such as coronary heart disease, stroke, TIA, peripheral vascular disease, or CVD-related death. Another example was shown by multiple studies (Verma et al., 2015; Yaseen, Farhan, & Abbas, 2017; Bobescu, Covaciu, Rus, Rogozea, et al., 2019; Pandey et al., 2019; Wójcik et al., 2019), investigating the responsiveness to antiplatelet drug clopidogrel after percutaneous coronary intervention (PCI), reporting 15%–18% of nonresponders, leaving the patients at higher risk of stent thrombosis and death after PCI. Studies from Yaseen et al. (2017) and Bobescu, Covaciu, Rus, Radoi, et al. (2019) show strong correlations between traditional cardiovascular risk factors and nonresponsiveness toward clopidogrel. In studies focusing on genetic variances, Pandey et al. (2019) and Wójcik et al. (2019) report a significant association between genetic variants in patients receiving clopidogrel and responsiveness toward the drug treatment. These studies indicate that individual genetic makeup and the dependent biological mechanisms are strong determiners in a patient's response to treatment.

Drug discovery also entails minimizing activity of synthesized substances in nonrelated biological targets, meaning to maximally reduce side-effects. In currently prescribed drugs, side-effects are numerous, and in some cases, serious adverse events (AEs) occur. An example is given by Schmidt et al. investigating side-effects in PCSK9 inhibitors. The investigators report that lipid lowering PCSK9 inhibitors are more effective than statin in lowering short- and mid-term LDL-C levels, as well in reducing long-term risk for CVD, showing no significant effect on mortality (Schmidt et al., 2017). However, an increase in AEs, such as cancer and type 2 diabetes mellitus, was seen in patients treated with PCSK9 inhibiting drugs.

Warfarin is an anticoagulation drug, known for its increased risk for bleeding and hemorrhage (Schulman et al., 2008; Sun et al., 2016). As described by D'Souza et al. (2017) and by Rozestraten et al. (2019), in effect, warfarin is teratogenic in the first trimester of pregnancies and associated with increased fetal loss throughout pregnancy, due to the fact that it crosses the placenta.

Additionally, Lutsey et al. (2019) show that patients using warfarin are at an increased risk of bone fractures compared to patients using other anticoagulants. This risk-increasing effect is higher in patients suffering from osteoporosis, meaning in this case, but not uniquely by itself, the adverse side-effect depends on the phenotypic characteristics of the patient.

Spironolactone is a diuretic that prevents dropping potassium-levels while facilitating excretion of water from the body. Beldhuis et al. (2019) reported that the use of spironolactone for the treatment of patients suffering from HF with preserved ejection fraction has been associated with hyperkalemia and worsening renal function, leading to increased drug discontinuation with suggested increased risk for declining renal function, especially in patients with chronic kidney disease. Reportedly, apart from AEs and overall worsened patient outcomes, drug related side-effects are responsible for a substantial rate of emergency department visits and hospital admissions, especially in older patients (Kongkaew et al., 2013; Zed et al., 2013). In a range of studies reviewed by Zed et al. (2013), 0.5%–3.3% of all emergency department visits and 0.16%–4.3% of all admissions were adverse drug event related, of which 20.3%–66.7% were deemed preventable. The side-effects account for an economic burden in healthcare, adding substantially to the total healthcare costs that cross US$100 billion per year in the United States.

Overall, the one-size-fits-all nature of most commonly used drugs lays the primary basis for reduced effects in parts of the addressed patient population, and for undesired, adverse side-effects reported frequently in literature.

Individual biological structures in humans differ due to individuals' genetic makeups and their dependent protein profiles. This not only affects their individual risk to develop cardiovascular diseases, but also their individual drug metabolisms, thereby determining the way they respond to their treatments, and what complications or side effects they face. Therefore, a more patient-tailored and better disease targeting approach is required in the development of new cardiovascular drugs to solve issues seen in current drug therapy. As described before, software models are currently used to perform a prescreening of components with potential disease target modifying chemical properties, with none or minimal activity toward non-targeted biological structures at the same time. Expert coded rules and traditionally simplistic regression or classification QSARs are used to predict chemical properties of newly synthesized substances in relation to their activity toward a drug target. Research into the application of expert coded rules by Segler et al. (2018), Segler, Preuss, and Waller (2018) show that molecules that are generated using these expert rules sometimes show unintended behavior.

ML and particularly deep learning techniques have the ability to "see" patterns in complex data structures, allowing humans to uncover information from complex data that would be otherwise hidden to the human eye. Thereby deep learning models may represent a powerful extension to traditional statistical models, such as linear and logistic regression models that perform well on one or a handful of independent variables, but gracefully fail to uncover dependencies in high-dimensional data.

Following the successes of advanced ML techniques in other fields such as image processing, the first examples of ML models have already been recently developed in the field of drug discovery. A first example that aims for increased drug target effectiveness, is shown by Onay and Onay (2019), who developed a drug discovery decision support tool, using the data already available in rulesets and chemical descriptor databases, and applied a combined set of tools and models to first select a set of chemical features that are most distinctive for candidate molecule selection, and then find an optimal set of candidate molecules for different drug targets. Here, Gaston algorithm (Nijssen & Kok, 2005) for graph mining was used to find common molecular fragments in drug datasets, which resulted in identifying approved as well as withdrawn cardiovascular, antiepileptic, and anti-Parkinson drugs.

Menden et al. (2013) described two approaches in the discovery of anticancer drugs. In an initial attempt, several single-drug neural networks were trained to predict the drug responsiveness on a type of cell line, with the cells' genomic background as inputs and the IC_{50} profile as output. Results of these genomics-based models were relatively poor. However, combining the genomic input data with SMILES-based chemical descriptors into a single multidrug model improved the model performance substantially. Random forest and neural nets were developed, resulting in comparable R^2 (0.72), RMSE (0.83), and Pearson correlation coefficients (R_p: 0.85) in both types of models. However, in an independent validation set, these models performed substantially worse (R^2: 0.64, RMSE: 0.97, R_p: 0.79), suggested to be induced by purposefully introducing substantial differences between training and validation cell line types.

Following more traditional feature selection strategies to infer QSAR models, ML techniques assist in the approach of feature learning on chemical compounds. It was reported by Ponzoni et al. (2017) that even though theoretically drug discovery could benefit from a feature learning approach or using artificial neural networks, neither method appeared to be superior in each scenario the researchers tested. Therefore, an ML-based hybrid solution was developed, using a combination of feature selection and feature learning with significantly better performance than achieved with both solutions independently of each other, elevating AUC values from 0.72 up to 0.865 for the best classifier. WEKA data mining tool (Hall et al., 2009) was used to detect the best applicable ML model per drug target, selecting between different strategies, e.g., Linear Regression, Random Committee, Decision Trees, Neural Networks, or Random Forest. Depending on the target, Random forests, Neural Networks, or Random Committee models demonstrated the best performance.

To overcome the poor predictive power of most rule-based systems and traditional QSARs, Segler et al. (2018) proposed a natural language processing inspired deep learning model. An LSTM, formerly used to generate sentences, e.g., to describe the contents of an image, was trained to generate molecular structures with desired treatment target/chemical behaviors. A model that takes SMILES strings as inputs and physicochemical properties as outputs was trained on a generic chemicals-dataset. Transfer learning was applied, by training a general model on a large generic dataset. For the purpose of this study, the pretrained model was then further trained on a smaller dataset with a specific target. The model was validated against an independent dataset, containing unseen molecules that were designed to treat malaria. 28% of the existing molecules in this validation set were generated by the model.

An example of research undertaken to diminish undesired side-effects in newly synthesized drugs can be found in the work of Cai et al. (2018), who developed a predictive model on the main cardiovascular outcomes (hypertension, heart block, arrhythmia, cardiac failure, and myocardial infarction) as a side-effect of newly developed drugs. The model is trained on four publicly available data sources (Comparative Toxicogenomics Database, SIDER, Offsides, and MetaADEDB) containing information about effects of drugs and other chemicals on biological structures. logistic regression, random forest, k-nearest neighbors, SVM, and neural network models were developed with relatively poor to moderate predictive performance (AUC: 0647–0.809), as well as an aggregate model, combining the different basal models, resulting in increased predictive performance (AUC:0.784–0.842). Validation of a combined classifier predicting adverse cardiovascular outcomes on 63 anticancer drugs yielded a success rate of 87%.

Wacker and Noskow (2018) developed ML-based QSAR models for the prediction of the cardiotoxicity of antiarrhythmic drugs. In this study an ML-platform was developed to target the problem of potentially lethal ventricular tachycardia, induced by HCN channel targeting antiarrhythmia. The developed model predicts the potassium channel, initiated by antiarrhythmic drugs. The proposed framework is a scalable solution with potential application in the development of other cardiovascular drugs, with the side note that sufficient data for any specific target is required to use the framework.

Most examples so far focus largely on better targeting biological structures that require modification in disease. However, except for Menden et al. (2013), no real step is made toward possible generation of personalized drugs, taking into account an individual's genetic makeup.

Rogers and Aikawa (2019) discussed combined work from genomics, proteomics, metabolomics, cell imaging a.o. in a review on cardiovascular calcification and its treatment, and proposed a more integral workflow, covering the origin of disease, subtypes of disease, and the treatment thereof. It was mentioned that mechanistic discovery of cardiovascular calcifications is still at a premature state, where other fields, such as cancer treatment, have presented multi-omics solutions, supporting mechanistic discovery in rare diseases and leveraging personalized patient treatment (Karczewski & Snyder, 2018). Results are promising in general, and applicable to most healthcare subdomains, for better drug target identification as well as improvement of identification of responders and nonresponders.

It should be noted that, even with more precisely targeted drugs, identification of a patient's genetic profile in relation to the targeted disease will remain crucial in administering drugs in a personalized way. Presently, the only part of a total solution seems to be to identify patient groups that are most likely to respond to a drug, and that are most unlikely to have any AEs in any of the non-disease-related, nontargeted biological structures, based on their genetic profiles.

Patel et al. (2014) reviewed the possible identification of statin-responders, based on GWAS analyses and candidate gene approaches on six different genes involved in lipid metabolism, clinical outcomes, and statin side-effects. Although various scientific reports suggest utility, conflicting results were also found between different studies, suggesting the necessity for further research on this topic. Another example from Claassens et al. (2019) showed a CYP2C19 genotype–guided strategy for the selection of oral P2Y12 inhibitor therapy in patients undergoing PCI.

Applying ML techniques may help to both improve patient selection and patient specific targeting of the disease, by development of personalized medications, improving overall patient outcomes, and preventing side-effects and adverse events

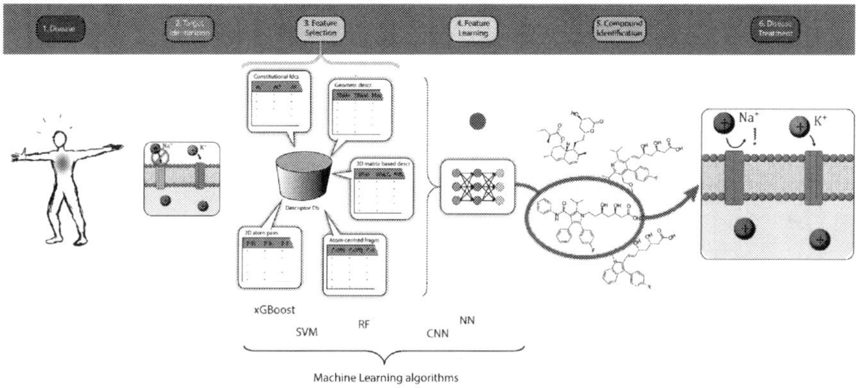

FIGURE 14.5 ML-Supported cardiovascular drug development horizon.

ML models have shown to be of increased predictive value compared to rule-based QSARs, by better inferring and matching the required biochemical properties of potentially useful chemical components in relation to the treated disease. Additionally, these models could assist to minimize adverse side-effects on nonrelated biological structures.

induced by pharmacological treatment. Fig. 14.5 summarizes the continuum in drug developments where ML implementation offers potential benefit.

Where the first steps have been taken to improve gene-based patient selection for certain drug treatment, and the first ML models have been used for biochemical feature learning as well as feature selection, huge improvements are to be expected, if ML models will be applied further on integral multiomics data, to gain knowledge of mechanisms underlying disease, and to improve identification of disease-related biological drug targets. Finally, combined improvement of disease-targeting medications and selection of responders should not only improve patient outcomes but also positively affect the economic burden of healthcare costs on society, by lowering unnecessary hospital admissions as well as readmissions.

Summary

Adaption of ML varies in the field of genomics, proteomics, and drug discovery in cardiovascular research. Drug Discovery. ML is widely used in the study of genomics in cardiovascular research for various purposes from variant annotation to disease risk prediction. It is less popular in proteomic analyses due to current limits in performance. ML has been used for QSAR analyses with a generally superior performance over simple rule-based and statistical methods, such as linear/logistic regression. While traditional ML methods namely SVM and random forest remain the most popular methods, emergence of deep learning in this field is expected with increasing data availability in the field. Efforts of developing better prediction/classification models with ML for disease status/risk have been seen in all three fields. An integral workflow/framework combining these data could be a promising approach to further enhance the predictive performance of ML models.

Acknowledgements

J.W.B and M.W.Y are supported by the Research Project CVON-AI (2018B017). The CVON-AI project is financed by the PPP Allowance made available by Top Sector Life Sciences & Health to the Nederlandse Hartstichting to stimulate public-private partnerships. This work reflects only the author's view, not that of the funders. Stichting LSH-TKI or Hartstichting or the Ministry of Economic Affairs is not responsible for any use that may be made of the information it contains.

References

Abraham, G., et al. (2014). Accurate and robust genomic prediction of celiac disease using statistical learning. *PLoS Genetics, 10*(2). https://doi.org/10.1371/journal.pgen.1004137

Akyea, R. K., et al. (2019). Sub-optimal cholesterol response to initiation of statins and future risk of cardiovascular disease. *Heart, 105*(13), 975–981. https://doi.org/10.1136/heartjnl-2018-314253

Almeida, J. S., et al. (2005). Normalization and analysis of residual variation in two-dimensional gel electrophoresis for quantitative differential proteomics. *Proteomics, 5*(5), 1242–1249. https://doi.org/10.1002/pmic.200401003

Bao, L., Zhou, M., & Cui, Y. (2005). nsSNPAnalyzer: Identifying disease-associated nonsynonymous single nucleotide polymorphisms. *Nucleic Acids Research, 33*, W480–W482. https://doi.org/10.1093/nar/gki372. Web Server.

Beer, M. A., & Tavazoie, S. (2004). Predicting gene expression from sequence. *Cell, 117*(2), 185–198. https://doi.org/10.1016/S0092-8674(04)00304-6

Beldhuis, I. E., et al. (2019). Efficacy and safety of spironolactone in patients with HFpEF and chronic kidney disease. *Journal of the American College of Cardiology: Heart Failure, 7*(1), 25–32. https://doi.org/10.1016/j.jchf.2018.10.017

Bhasin, M., & Raghava, G. P. S. (2004). ESLpred: SVM-based method for subcellular localization of eukaryotic proteins using dipeptide composition and PSI-BLAST. *Nucleic Acids Research, 32*, W414–W419. https://doi.org/10.1093/nar/gkh350. Web Server.

Bobescu, E., Covaciu, A., Rus, H., Radoi, M., et al. (2019). Correlation of cardiovascular risk factors and biomarkers with platelet reactivity in coronary artery disease. *American Journal of Therapeutics, 26*(5), e563–e569. https://doi.org/10.1097/MJT.0000000000000869

Bobescu, E., Covaciu, A., Rus, H., Rogozea, L. M., et al. (2019). Low response to clopidogrel in coronary artery disease. *American Journal of Therapeutics*, 1. https://doi.org/10.1097/MJT.0000000000001099

Bom, M. J., et al. (2019). Predictive value of targeted proteomics for coronary plaque morphology in patients with suspected coronary artery disease. *EBioMedicine, 39*, 109–117. https://doi.org/10.1016/j.ebiom.2018.12.033. The Authors.

Bromberg, Y., & Rost, B. (2007). Snap: Predict effect of non-synonymous polymorphisms on function. *Nucleic Acids Research, 35*(11), 3823–3835. https://doi.org/10.1093/nar/gkm238

Cai, C., et al. (2018). In silico pharmacoepidemiologic evaluation of drug-induced cardiovascular complications using combined classifiers. *Journal of Chemical Information and Modeling, 58*(5), 943–956. https://doi.org/10.1021/acs.jcim.7b00641

Captur, G., et al. (2019). Identification of a multiplex biomarker panel for hypertrophic cardiomyopathy using quantitative proteomics and machine learning. *Molecular and Cellular Proteomics*. https://doi.org/10.1074/mcp.ra119.001586. p. mcp.RA119.001586.

Claassens, D. M. F., et al. (2019). A genotype-guided strategy for oral P2Y 12 inhibitors in primary PCI. *New England Journal of Medicine, 381*(17), 1621–1631. https://doi.org/10.1056/NEJMoa1907096

Coats, C. J., et al. (2018). 'Proteomic analysis of the myocardium in hypertrophic obstructive cardiomyopathy', *circulation. Genomic and Precision Medicine, 11*(12), e001974. https://doi.org/10.1161/CIRCGEN.117.001974

Curtis, D. (2018). Polygenic risk score for schizophrenia is more strongly associated with ancestry than with schizophrenia. *Psychiatric Genetics, 28*(5), 85–89. https://doi.org/10.1097/YPG.0000000000000206

Degroeve, S., et al. (2005). Splicemachine: Predicting splice sites from high-dimensional local context representations. *Bioinformatics, 21*(8), 1332–1338. https://doi.org/10.1093/bioinformatics/bti166

Delles, C., et al. (2010). Urinary proteomic diagnosis of coronary artery disease: Identification and clinical validation in 623 individuals. *Journal of Hypertension, 28*(11), 2316–2322. https://doi.org/10.1097/HJH.0b013e32833d81b7

D'Souza, R., et al. (2017). Anticoagulation for pregnant women with mechanical heart valves: A systematic review and meta-analysis. *European Heart Journal, 38*(19), 1509–1516. https://doi.org/10.1093/eurheartj/ehx032

Emmens, J. E., et al. (2018). Proteomic diversity of high-density lipoprotein explains its association with clinical outcome in patients with heart failure. *European Journal of Heart Failure, 20*(2), 260–267. https://doi.org/10.1002/ejhf.1101

Farmakis, D., et al. (2016). Urine proteome analysis in heart failure with reduced ejection fraction complicated by chronic kidney disease: Feasibility, and clinical and pathogenetic correlates. *European Journal of Heart Failure, 18*(7), 822–829. https://doi.org/10.1002/ejhf.544

Ferreira, J. P., et al. (2019). Plasma protein biomarkers and their association with mutually exclusive cardiovascular phenotypes: The FIBRO-TARGETS case–control analyses. *Clinical Research in Cardiology*. https://doi.org/10.1007/s00392-019-01480-4. Springer Berlin Heidelberg, (0123456789).

Fullerton, J. M., & Nurnberger, J. I. (2019). Polygenic risk scores in psychiatry: Will they be useful for clinicians? *F1000 Research, 8*(F1000 Faculty Rev), 1293. https://doi.org/10.12688/f1000research.18491.1

Gessulat, S., et al. (2019). Prosit: Proteome-wide prediction of peptide tandem mass spectra by deep learning. *Nature Methods, 16*(6), 509–518. https://doi.org/10.1038/s41592-019-0426-7. Springer US.

Giri, A., et al. (2019). Trans-ethnic association study of blood pressure determinants in over 750,000 individuals. *Nature Genetics, 51*(1), 51–62. https://doi.org/10.1038/s41588-018-0303-9. Nature Publishing Group.

Guan, S., Moran, M. F., & Ma, B. (2019). Prediction of LC-MS/MS properties of peptides from sequence by deep learning. *Molecular and Cellular Proteomics : MCP, 18*(10), 2099–2107. https://doi.org/10.1074/mcp.TIR119.001412

Günther, F., Wawro, N., & Bammann, K. (2009). Neural networks for modeling gene-gene interactions in association studies. *BMC genetics. BioMed Central, 10*, 87. https://doi.org/10.1186/1471-2156-10-87

Hall, M., et al. (2009). The WEKA data mining software. *SIGKDD Explorations Newsletter, 11*(1), 10. https://doi.org/10.1145/1656274.1656278

van der Harst, P., & Verweij, N. (2018). Identification of 64 novel genetic loci provides an expanded view on the genetic architecture of coronary artery disease. *Circulation Research, 122*(3), 433–443. https://doi.org/10.1161/CIRCRESAHA.117.312086

Heintzman, N. D., et al. (2007). Distinct and predictive chromatin signatures of transcriptional promoters and enhancers in the human genome. *Nature Genetics*, 311–318. https://doi.org/10.1038/ng1966. Nature Publishing Group.

Hertel, J., & Stadler, P. F. (2006). Hairpins in a haystack: Recognizing microRNA precursors in comparative genomics data. *Bioinformatics, 22*(14), e197–e202. https://doi.org/10.1093/bioinformatics/btl257

Hicks, S., et al. (2011). Prediction of missense mutation functionality depends on both the algorithm and sequence alignment employed. *Human Mutation, 32*(6), 661–668. https://doi.org/10.1002/humu.21490

Htun, N. M., et al. (2017). Prediction of acute coronary syndromes by urinary proteome analysis. *PloS One, 12*(3), 1–18. https://doi.org/10.1371/journal.pone.0172036

Ioannidis, N. M., et al. (2016). Revel: An ensemble method for predicting the pathogenicity of rare missense variants. *The American Journal of Human Genetics, 99*(4), 877–885. https://doi.org/10.1016/j.ajhg.2016.08.016. Cell Press.

Jaganathan, K., et al. (2019). Predicting splicing from primary sequence with deep learning. *Cell, 176*(3), 535–548. https://doi.org/10.1016/J.CELL.2018.12.015. e24 (Cell Press).

Karczewski, K. J., & Snyder, M. P. (2018). Integrative omics for health and disease. *Nature Reviews Genetics, 19*(5), 299–310. https://doi.org/10.1038/nrg.2018.4

Khera, A. V., et al. (2016). 'Genetic risk, adherence to a healthy lifestyle, and coronary disease',. *The New England Journal of Medicine, 375*(24), 2349–2358. https://doi.org/10.1056/NEJMoa1605086. Massachusetts Medical Society.

Khera, A. V., et al. (2018). Genome-wide polygenic scores for common diseases identify individuals with risk equivalent to monogenic mutations. *Nature Genetics, 50*(9), 1219–1224. https://doi.org/10.1038/s41588-018-0183-z. Nature Publishing Group.

Kircher, M., et al. (2014). A general framework for estimating the relative pathogenicity of human genetic variants. *Nature Genetics, 46*(3), 310–315. https://doi.org/10.1038/ng.2892. Nature Publishing Group.

Knuuti, J., et al. (2019). 2019 ESC Guidelines for the diagnosis and management of chronic coronary syndromes. *European Heart Journal*, 1–71. https://doi.org/10.1093/eurheartj/ehz425

Kongkaew, C., et al. (2013). 'Risk factors for hospital admissions associated with adverse drug events. *Pharmacotherapy: The Journal of Human Pharmacology and Drug Therapy, 33*(8), 827–837. https://doi.org/10.1002/phar.1287

Kulmanov, M., Khan, M. A., & Hoehndorf, R. (2018). DeepGO: Predicting protein functions from sequence and interactions using a deep ontology-aware classifier. *Bioinformatics, 34*(4), 660–668. https://doi.org/10.1093/bioinformatics/btx624

Kuznetsova, T., et al. (2012). Urinary proteome analysis in hypertensive patients with left ventricular diastolic dysfunction. *European Heart Journal, 33*(18), 2342–2350. https://doi.org/10.1093/eurheartj/ehs185

Lemesle, G., et al. (2015). Multimarker proteomic profiling for the prediction of cardiovascular mortality in patients with chronic heart failure. *PLoS One, 10*(4), 1–12. https://doi.org/10.1371/journal.pone.0119265

Libbrecht, M. W., & Noble, W. S. (2015). Machine learning applications in genetics and genomics. *Nature Reviews Genetics, 16*(6), 321–332. https://doi.org/10.1038/nrg3920. Nature Publishing Group.

Li, B., et al. (2017). Predicting the functional Impact of KCNQ1 variants of unknown Significance. *Circulation: Cardiovascular Genetics, 10*(5). https://doi.org/10.1161/CIRCGENETICS.117.001754. Lippincott Williams and Wilkins.

Li, J., et al. (2018). Decoding the genomics of abdominal aortic Aneurysm. *Cell, 174*(6), 1361–1372. https://doi.org/10.1016/J.CELL.2018.07.021. e10 (Cell Press).

Lutsey, P. L., et al. (2019). Association of anticoagulant therapy with risk of fracture among patients with atrial fibrillation. *JAMA Internal Medicine*. https://doi.org/10.1001/jamainternmed.2019.5679

Ma, C., et al. (2018). Improved peptide retention time prediction in liquid chromatography through deep learning. *Analytical Chemistry, 90*(18), 10881–10888. https://doi.org/10.1021/acs.analchem.8b02386. American Chemical Society.

Malik, R., et al. (2018). Multiancestry genome-wide association study of 520,000 subjects identifies 32 loci associated with stroke and stroke subtypes. *Nature Genetics, 50*(4), 524–537. https://doi.org/10.1038/s41588-018-0058-3. Nature Publishing Group.

McGilligan, V., et al. (2019). An exploratory analysis investigating blood protein biomarkers to augment ECG diagnosis of ACS. *Journal of Electrocardiology, 57*(Supp.), S92–S97. https://doi.org/10.1016/j.jelectrocard.2019.09.002. Elsevier Inc.

Menden, M. P., et al. (2013). Machine learning prediction of cancer cell sensitivity to drugs based on genomic and chemical properties. *PLoS One, 8*(4), e61318. https://doi.org/10.1371/journal.pone.0061318. Edited by G. P. S. Raghava.

Miotto, R., et al. (2016). Deep patient: An unsupervised representation to predict the future of patients from the electronic health records. *Scientific Reports, 6*(1), 26094. https://doi.org/10.1038/srep26094. Nature Publishing Group.

von zur Mühlen, C., et al. (2016). Urine proteome analysis as a discovery tool in patients with deep vein thrombosis and pulmonary embolism. *Proteomics - Clinical Applications, 10*(5), 574–584. https://doi.org/10.1002/prca.201500105

Neisius, U., et al. (2016). Urine proteomics in the diagnosis of stable angina. *BMC Cardiovascular Disorders, 16*(1), 1–8. https://doi.org/10.1186/s12872-016-0246-y

Nijssen, S., & Kok, J. N. (2005). The Gaston tool for frequent subgraph mining. *Electronic Notes in Theoretical Computer Science, 127*(1), 77–87. https://doi.org/10.1016/j.entcs.2004.12.039

Ohanian, M., Otway, R., & Fatkin, D. (2012). Heuristic methods for finding pathogenic variants in gene coding sequences. *Journal of the American Heart Association, 1*(5), e002642. https://doi.org/10.1161/JAHA.112.002642

Onay, A., & Onay, M. (2019). A drug decision support system for developing a successful drug candidate using machine learning techniques. *Current Computer-Aided Drug Design, 15*. https://doi.org/10.2174/1573409915666190716143601

Pandey, C., et al. (2019). Aspirin & clopidogrel non-responsiveness & its association with genetic polymorphisms in patients with myocardial infarction. *Indian Journal of Medical Research, 150*(1), 50. https://doi.org/10.4103/ijmr.IJMR_782_17

Paré, G., Mao, S., & Deng, W. Q. (2017). A machine-learning heuristic to improve gene score prediction of polygenic traits. *Scientific Reports, 7*(1), 12665. https://doi.org/10.1038/s41598-017-13056-1. Nature Publishing Group.

Patel, J., et al. (2014). Genetics and personalized medicine—a role in statin therapy? *Current Atherosclerosis Reports, 16*(1), 384. https://doi.org/10.1007/s11883-013-0384-y

Perez-Riverol, Y., et al. (2017). Accurate and fast feature selection workflow for high-dimensional omics data. *PLoS One, 12*(12), 1–14. https://doi.org/10.1371/journal.pone.0189875

Pertea, M. (2001). Genesplicer: A new computational method for splice site prediction. *Nucleic Acids Research, 29*(5), 1185–1190. https://doi.org/10.1093/nar/29.5.1185

Ponzoni, I., et al. (2017). Hybridizing feature selection and feature learning approaches in QSAR modeling for drug discovery. *Scientific Reports, 7*(1), 2403. https://doi.org/10.1038/s41598-017-02114-3

Rajagopal, N., et al. (2013). 'RFECS: A random-forest based algorithm for enhancer identification from chromatin state. *PLoS Computational Biology, 9*(3), e1002968. https://doi.org/10.1371/journal.pcbi.1002968. Public Library of Science.

Rogers, M. A., & Aikawa, E. (2019). Cardiovascular calcification: Artificial intelligence and big data accelerate mechanistic discovery. *Nature Reviews Cardiology, 16*(5), 261–274. https://doi.org/10.1038/s41569-018-0123-8

Romagnoni, A., et al. (2019). Comparative performances of machine learning methods for classifying Crohn Disease patients using genome-wide genotyping data. *Scientific Reports, 9*(1). https://doi.org/10.1038/s41598-019-46649-z. Nature Publishing Group.

Roselli, C., et al. (2018). Multi-ethnic genome-wide association study for atrial fibrillation. *Nature Genetics, 50*(9), 1225–1233. https://doi.org/10.1038/s41588-018-0133-9. Nature Publishing Group.

Rossing, K., et al. (2016). Urinary proteomics pilot study for biomarker discovery and diagnosis in heart failure with reduced ejection fraction. *PLoS One, 11*(6), 1–15. https://doi.org/10.1371/journal.pone.0157167

Rozestraten, F. F. J., et al. (2019). A case report: Mechanical mitral valve thrombosis in pregnancy. *European Heart Journal - Case Reports, 3*(1). https://doi.org/10.1093/ehjcr/ytz024. Edited by T. Tan et al.

Said, M. A., Verweij, N., & van der Harst, P. (2018). Associations of combined genetic and lifestyle risks with incident cardiovascular disease and diabetes in the UK biobank study. *JAMA Cardiology, 3*(8), 693. https://doi.org/10.1001/jamacardio.2018.1717. American Medical Association.

Schmidt, A. F., et al. (2017). PCSK9 monoclonal antibodies for the primary and secondary prevention of cardiovascular disease. *Cochrane Database of Systematic Reviews.* https://doi.org/10.1002/14651858.CD011748.pub2

Schulman, S., et al. (2008). Hemorrhagic complications of anticoagulant and thrombolytic treatment. *Chest, 133*(6), 257S–298S. https://doi.org/10.1378/chest.08-0674

Segler, M. H. S., et al. (2018). Generating focused molecule libraries for drug discovery with recurrent neural networks. *ACS Central Science, 4*(1), 120–131. https://doi.org/10.1021/acscentsci.7b00512

Segler, M. H. S., Preuss, M., & Waller, M. P. (2018). Planning chemical syntheses with deep neural networks and symbolic AI. *Nature, 555*(7698), 604–610. https://doi.org/10.1038/nature25978

Stead, L. F., Wood, I. C., & Westhead, D. R. (2011). KvSNP: Accurately predicting the effect of genetic variants in voltage-gated potassium channels. *Bioinformatics, 27*(16), 2181–2186. https://doi.org/10.1093/bioinformatics/btr365

Sun, H., et al. (2016). 'Warfarin continuation vs interruption during procedures of cardiac rhythm devices: A meta-analysis of randomized controlled trials. *JPMA. The Journal of the Pakistan Medical Association, 66*(4), 458–465. Available at http://www.ncbi.nlm.nih.gov/pubmed/27122276.

Tiwary, S., et al. (2019). High-quality MS/MS spectrum prediction for data-dependent and data-independent acquisition data analysis. *Nature Methods, 16*(6), 519–525. https://doi.org/10.1038/s41592-019-0427-6. Springer US.

Torkamani, A., Wineinger, N. E., & Topol, E. J. (2018). The personal and clinical utility of polygenic risk scores. *Nature Reviews Genetics*, 581–590. https://doi.org/10.1038/s41576-018-0018-x. Nature Publishing Group.

Tran, N. H., et al. (2017). De novo peptide sequencing by deep learning. *Proceedings of the National Academy of Sciences of the United States of America, 114*(31), 8247–8252. https://doi.org/10.1073/pnas.1705691114

Tran, N. H., et al. (2019). Deep learning enables de novo peptide sequencing from data-independent-acquisition mass spectrometry. *Nature Methods, 16*(1), 63–66. https://doi.org/10.1038/s41592-018-0260-3. Springer US.

Tran, N. H., Zhang, X., & Li, M. (2018). Deep omics. *Proteomics, 18*(2), 1–5. https://doi.org/10.1002/pmic.201700319

Verma, S., et al. (2015). Should dual antiplatelet therapy be used in patients following coronary artery bypass surgery? A meta-analysis of randomized controlled trials. *BMC Surgery, 15*(1), 112. https://doi.org/10.1186/s12893-015-0096-z

Vinayagam, A., et al. (2004). Applying Support Vector Machines for gene ontology based gene function prediction. *BMC Bioinformatics, 5*. https://doi.org/10.1186/1471-2105-5-116

Von Muhlen, C. Z., et al. (2009). Evaluation of urine proteome pattern analysis for its potential to reflect coronary artery atherosclerosis in symptomatic patients. *Journal of Proteome Research, 8*(1), 335–345. https://doi.org/10.1021/pr800615t

Wacker, S., & Noskov, S. Y. (2018). Performance of machine learning algorithms for qualitative and quantitative prediction drug blockade of hERG1 channel. *Computational Toxicology (Amsterdam, Netherlands), 6*(1), 55–63. https://doi.org/10.1016/j.comtox.2017.05.001

Walsh, R., et al. (2019). Quantitative approaches to variant classification increase the yield and precision of genetic testing in mendelian diseases: The case of hypertrophic cardiomyopathy. *Genome Medicine, 11*(1), 5. https://doi.org/10.1186/s13073-019-0616-z

Wei, Q., et al. (2010). Testing computational prediction of missense mutation phenotypes: Functional characterization of 204 mutations of human cystathionine beta synthase. *Proteins: Structure, Function, and Bioinformatics*. https://doi.org/10.1002/prot.22722. NA-NA.

Wójcik, T., et al. (2019). Distribution of polymorphisms in the CYP2C19 and ABCB1 genes among patients with acute coronary syndrome in Lower Silesian population. *Advances in Clinical and Experimental Medicine, 28*(12). https://doi.org/10.17219/acem/110322

Wong, A., et al. (2018). Development and validation of an electronic health record-based machine learning model to estimate Delirium risk in newly hospitalized patients without known cognitive impairment. *JAMA network open, 1*(4), e181018. https://doi.org/10.1001/jamanetworkopen.2018.1018. NLM (Medline).

Yang, H., & Wang, K. (2015). Genomic variant annotation and prioritization with ANNOVAR and wANNOVAR. *Nature Protocols, 10*(10), 1556–1566. https://doi.org/10.1038/nprot.2015.105. Nature Publishing Group.

Yaseen, I. F., Farhan, H. A., & Abbas, H. M. (2017). Clopidogrel non-responsiveness in patients undergoing percutaneous coronary intervention using the VerifyNow test:

Frequency and predictors. *European Journal of Hospital Pharmacy*. https://doi.org/10.1136/ejhpharm-2017-001359

Yeung, M. W., Maj, C., & Krawitz, P. (2019). Capturing epistatic effect with deep learning – a comparison with polygenic risk score on phenotype prediction from SNP-based genetic data. In *30. Jahrestagung der Deutschen Gesellschaft für Humangenetik. Weimar: der Deutschen Gesellschaft für Humangenetik*.

Zed, P. J., et al. (2013). Medication-related emergency department visits and hospital admissions in pediatric patients: A qualitative systematic review. *The Journal of Pediatrics, 163*(2), 477–483. https://doi.org/10.1016/j.jpeds.2013.01.042

Zeng, W. F., et al. (2019). MS/MS spectrum prediction for modified peptides using pDeep2 trained by transfer learning. *Analytical Chemistry, 91*(15), 9724–9731. https://doi.org/10.1021/acs.analchem.9b01262. American Chemical Society.

Zhang, Z., et al. (2014). Left ventricular diastolic function in relation to the urinary proteome: A proof-of-concept study in a general population. *International Journal of Cardiology, 176*(1), 158–165. https://doi.org/10.1016/j.ijcard.2014.07.014. The Authors.

Zhang, Z. Y., et al. (2017). Novel urinary peptidomic classifier predicts incident heart failure. *Journal of the American Heart Association, 6*(8), 1–11. https://doi.org/10.1161/JAHA.116.005432

Zhang, F., et al. (2019). DeepFunc: A deep learning framework for accurate prediction of protein functions from protein sequences and interactions. *Proteomics, 19*(12), 1–7. https://doi.org/10.1002/pmic.201900019

Zhang, Z. Y., Nkuipou-Kenfack, E., & Staessen, J. A. (2019). Urinary peptidomic biomarker for personalized prevention and treatment of diastolic left ventricular dysfunction. *Proteomics - Clinical Applications, 13*(2), 1–9. https://doi.org/10.1002/prca.201800174

Zhou, X. X., et al. (2017). PDeep: Predicting MS/MS spectra of peptides with deep learning. *Analytical Chemistry, 89*(23), 12690–12697. https://doi.org/10.1021/acs.analchem.7b02566

Zimmerli, L. U., et al. (2008). Urinary proteomic biomarkers in coronary artery disease. *Molecular and Cellular Proteomics, 7*(2), 290–298. https://doi.org/10.1074/mcp.M700394-MCP200

Zohora, F. T., et al. (2019). DeepIso: A deep learning model for peptide feature detection from LC-MS map. *Scientific Reports, 9*(1), 17168. https://doi.org/10.1038/s41598-019-52954-4. Springer US.

Zou, J., et al. (2019). A primer on deep learning in genomics. *Nature Genetics, 51*(1), 12–18. https://doi.org/10.1038/s41588-018-0295-5. Springer US.

CHAPTER

Wearable devices and machine learning algorithms for cardiovascular health assessment

15

Abdallah Elshafeey, Omar Mhaimeed, Jamal Al Ani, Mohamed B. Elshazly
Department of Medicine and Medical Education, Weill Cornell Medicine-Qatar, Doha, Qatar

Chapter outline

Introduction

Smart wearables are smart, consumer-grade devices that can be worn on the body either as an accessory or part of clothing. They include clothing, ear-wear, watches, headbands, rings, sociometric badges, and wristbands. Smart wearables incorporate high processing power, advanced software, sophisticated sensors that can track activity, Global Positioning System (GPS), and biometric data. In the rapidly proliferating technological world of our modern day, the plethora of smart wearables in the market has made it rather difficult for one to keep up with this field. However, recent advancements in technology have begun to have implications on smart wearables

Machine Learning in Cardiovascular Medicine. https://doi.org/10.1016/B978-0-12-820273-9.00015-4

and their role on human health, and investing in the pursuit of the virtues of smart wearables may very well be a worthwhile endeavor.

Medical wearables date back to around 1286, when the first wearable eyeglasses were made by Salvino D'Armate (Desjardins, 2015). The rapid progression of technology in the last century has allowed for the advent of more complex and intricate wearable devices due to advances in computing. These devices include the pigeon camera which was used during the First World War, the calculator watch, and the Sony Walkman (Desjardins, 2015).

In the past few years, the smart wearables industry has been centered around the wrist, particularly in the form of smart watches, which have contributed to its massive growth (Richter, 2018). This most likely stemmed from the desire to be constantly connected, and to remain in touch with the virtual worlds that we have created for ourselves on various online platforms. In 2018, the top five smart wearable companies by shipment volume, market share, and year-over-year growth, shipped a total of 172.2 million devices, an increase from the 135 million shipments in 2017, and the market is only expected to grow (Perez, 2019). By the end of 2019, the industry is projected to have sold a total of 198.5 million units, and the market is expected to become a whopping $60-billion-dollar market with 279 million-unit sales by 2023 (Burns, 2019; Russey, 2019).

Another major driver for the smart wearables market is healthcare. Wearables are significantly improving in their ability to capture and record real-time user activity and biometric data, and have an appeal in the healthcare sector as they allow for daily monitoring of activity and a number of clinical parameters (Russey, 2019). For instance, with rates of obesity and diabetes rising, there is an increased need for the utility of fitness trackers and fitness related-parameters to gather data, which can easily be gathered through the use of smart wearables.

Smart wearables also allow for remote patient monitoring of biometric data by measuring an individual's pulse throughout the day across a number of various activities. Additionally, some can monitor the blood pressure and may even record a patient's heart rhythm through a single-lead electrocardiograph (ECG). Although this is not as accurate as the typical 12-lead ECG used in the clinical workplace, it can still provide invaluable information of a patient's heart rhythm (Burns, 2019).

A systematic review of 22 subjects conducted in June of 2019 by a team in Germany concluded that mobile health devices can aid in the early diagnosis and treatment of atrial fibrillation (AF), and that successful application of mobile health technologies could significantly contribute to reducing the cost of illness associated with AF (Giebel and Gissel, 2019). Furthermore, wearables can collect a vast amount of data that can be used to establish patterns using machine learning and artificial intelligence (Buttice, 2019). These patterns can then be integrated within advanced models that can be used to predict health conditions before they arise, so that the focus can shift toward their prevention (Burns, 2019). Different models can also be applied to the automatic detection and diagnosis of disease such as hyperkalemia, as well as the specific management of patients with preexisting conditions such as heart failure.

Smart wearables can also aid in the issue of medication adherence. In case of adults with chronic illnesses who are prescribed medications, 30%–50% of the medications are not taken as prescribed (Kini and Ho, 2018). In the United States, it has been estimated that medication nonadherence is associated with 125,000 deaths, 10% of hospitalizations, and $100 million in healthcare services per annum. Smart wearables can notify patients to take their medication and alert their physicians when they do not comply (Kini and Ho, 2018). Overall, smart wearables allow for capturing a more holistic snapshot of patients' wellbeing and provide an opportunity for individuals to take charge of their own health.

Types of data collected

Smart wearables have evolved over the years and have become capable of monitoring a multitude of organs, thereby providing real-time information about the various functions of the human body. These include devices that collect epidermal, ocular, intracochlear, dental, biochemical, and physiological parameters that can subsequently be processed and analyzed by complex algorithms and artificial intelligence (AI) software (Yetisen, Martinez-Hurtado, Ünal, Khademhosseini, & Butt, 2018). The devices used for these tasks are some of the most technologically advanced, and include enormous amounts of processing power, as well as other features that allow for the delivery of perfectly interpretable data within seconds.

There are many different types of sensors, each of which can collect unique data about the human body such as heart rate, blood pressure, activity levels and much more. Some examples of such sensors include:

- Accelerometers, altimeters, and GPS systems that can determine characteristics such as acceleration, speed, pace, and altitude. This information can then be combined with age, sex, weight, and height to give an indication of activity levels at any point in time. This is done through complex machine learning algorithms that can determine parameters such as the number of calories burned, number of steps taken, distance walked, and others.
- Heart rate sensors that can continuously measure heart rate at rest and during exercise. Most devices use a technology known as photoplethysmography (PPG), which consists of a light-emitting diode that emits green photons toward the skin. The light interacts with blood and gets absorbed at various degrees based on the blood flow, and the reflected light is then received by a photodiode and interpreted by the receiver. Again, machine learning algorithms that detect time from pulse peak to peak are able to determine heart rate (Allen, 2007).
- ECG sensors that can provide single lead ECG recordings whenever needed, or on a regular basis. Most sensors like the one used by the Apple Watch do this through utilization of a vector between the left and right arms when the patient places a finger on the watch.
- Sensors that can measure blood pressure when prompted. These include devices that utilize PPG technology. Some other watches (Omron Heart Guide) use an inflatable cuff around the wrist to measure blood pressure.

These are only a few of the multitude of sensors currently available. Others include oxygen saturation sensors, respiratory rate sensors, and many more (Table 15.1) (Appelboom et al., 2014).

Table 15.1 A list of the different types of sensors currently in use.

Clinical applications	Location	Type of sensor	Marker
Cardiopulmonary and vascular monitoring	Wrist	Ultrasound	Blood pressure
		Multivariable (AMON)	Blood pressure, blood oxygen saturation, body temp., heart rhythm
		Photoplethysmographic and electrocardiograph	Blood pressure and heart rhythm
	Finger (ring sensor)	Optical (heart rate) Radio-frequency identification (pulse and temp.)	Heart rate and temperature
	Arm or thigh	Microwave reflectometric cardiopulmonary	Heart rate variability as a method to evaluate stress
	Can be used with various equipments units	Optical absorption sensor ECG electrodes	Blood oxygen, heart rate and rhythm, fatigue levels, and exercise stress
	Phone adapter	Single-channel ECG	Heart rate and rhythm
	Seat belt of a car	Wiretype strain gauge	Heart rate and respiration rate
Glucose home monitoring	Arm	Multivariable	Blood glucose
	Subcutaneous	Glucose	Tissue glucose
	Eye	Glucose	Ocular glucose
Neurological function monitoring	Clothes	Inertial sensors and accelerometers	Walking distance, stride distance, and step count during stair ascent and descent
	Visual feedback-glasses Auditory feedback-headphones	Inertia	Walking speed and stride length
	Wrist or ankle	Accelerometer and motion	Seizure activity
Physical therapy and rehabilition	Ankle	Pedometers and accelerometer	Walking distance and step counts

Originally published by BioMed Central (BMC).

Machine learning (ML)

The collection of data is the first part in this process. Technology and progress have permitted the gathering of enormous amounts of data while simultaneously allowing for the safe storage. The challenge, however, lies with the effective utilization of the data and turning it into clinically meaningful parameters to consumers and patients. This is where the implementation of machine learning comes into play.

ML is the complex interplay between computer science and mathematics, more specifically statistics (Deo, 2015). It aims at using massive amounts of data to learn relationships and trends, which it can then apply to other datasets to interpret raw results. It does this through complex computational algorithms and statistical models. The method by which the computer learns the algorithms and models is subdivided into two major categories: supervised and unsupervised.

Supervised learning is the type of learning a computer does to which the outcome is already known and well established. They are all tasks that a trained person can do well, and the computer simply tries to replicate the same outcome reached by humans. These can range from simple to complex tasks, and mostly involve recognition and classification. Examples of these include classifying documents based on their content, or even just recognizing whether an image shows a cat or a dog (Deo, 2015). Once the algorithm classifies a piece of data into a subgroup, it can then make interpretations based on the preconceived general understanding of that particular subgroup. The main example of this in medicine is the recognition and classification of ECG patterns. Algorithms are able to recognize patterns of an ECG and place it in one of the many different subgroups of diagnoses it has previously learned (Deo, 2015). The main advantage to this type of learning is that it is able to replace the need for a physician to interpret every piece of data, thus improving the efficiency of clinical workflow. Patients and consumers can simply receive information about their health directly and instantly from their devices. This is certainly not a perfect replacement for human analysis as of yet, but it definitely has the potential to be perfected and be relied on to a greater extent in the near future.

On the other hand, unsupervised ML is much more open-ended and has limitless possibilities. In this type of learning, there is no specific outcome being measured. The computer simply aims to manipulate the data in limitless ways, in hopes of identifying patterns or specific groupings that can later be tested using a supervised algorithm in order to determine its clinical use, if any (Deo, 2015). The potential of the field of precision medicine is greatly augmented when it comes to this type of ML. Health data from a group of individuals suffering from what seems to be the same disease can be manipulated by algorithms, and the slightest patterns could suggest some deviations between patients in pathophysiology, thereby promoting different treatment strategies based on the specific pattern identified. This was successfully implemented in the realm of asthma, when ML-based algorithms identified a certain eosinophilic subtype contributing to the bronchial hyper-responsiveness. The focus was then shifted to developing a novel therapy to target that subtype of eosinophils (Deo, 2015) The main disadvantage to this type of learning is that the identified patterns could be clinically insignificant, which, in fact, most are. However, as the algorithms start to become smarter and expand their learned datasets, one can expect to start reaching more clinically significant patterns.

Uses of smart wearable data

Now that we have discussed the multitude of different types of data that can be collected by smart wearables, as well as the different ways in which they can be manipulated and interpreted, we can delve into how smart wearables can be useful for cardiovascular health. The uses of wearable data can be broadly split into three different groups: lifestyle management in the general population, screening for cardiovascular disease, and disease management in patients suffering from cardiovascular diseases.

Lifestyle management in the general population

Several modifiable causes of cardiovascular disease (CVD) have long been identified. These include hypertension, obesity, smoking, and diabetes mellitus among many others (Barnes, 2013). Physical activity is one of the pillars of CVD prevention, and several studies have attempted to elucidate that relationship. One of the most renowned is the *Lancet* study published in 2017, which looked at 140,000 individuals from 17 countries with varying income levels. The study looked at the effects of physical activity on mortality and CVD outcomes and concluded that increased physical activity was associated with lower mortality and better CVD outcomes in low, middle, and high-income countries (Lear et al., 2017). Therefore, a coherent link can be divulged between wearable devices and the potential for prevention of CVD. Smart wearables would be able to provide real-time information about levels of physical activity. Studies have long established benchmarks for physical activity that have been proven to decrease the incidence of cardiovascular disease and associated mortality. Smart wearables can help consumers meet these benchmarks on a daily basis. A recently published study attempted establishing a link between average number of daily steps and all-cause mortality in women (Lee et al., 2019). After a follow-up interval of over 4 years, the investigators found that women who averaged 4400 steps a day had significantly lower all-cause mortality rates. They also found that the peak for that effect was at 7500 steps, after which there was no significant reduction in mortality rates.

The field of smart wearables and machine learning is relatively young, and there has only been a small amount of research into it. However, there is one study that has looked at exercise data and attempted to link it to several diseases including diabetes and hypertension. This study is known as the Henry Ford Exercise Testing (FIT) project (Sakr et al., 2018). We will consider the hypertension part of the study, since it is the most closely related to CVD. The project used data collected using exercise treadmill stress testing at Henry Ford Health systems between 1991 and 2009 and had a complete 10-year follow-up. This study used six different supervised models of machine learning in an attempt to determine whether any of them would be able to predict the incidence of hypertension, and if so, which was most accurate. Fig. 15.1 shows the results of the study. The results are presented as Area Under the Curve (AUC), which reflects how accurately the model was able to predict the risk of hypertension. All models had around 60%–70% accuracy, but the Random Tree Forest (RTF) method reached 91% accuracy. This highlights the ability of ML to accurately predict the risk of CVD using functional data that can be collected from exercise testing.

One challenge that arises when considering wearable technology as a primary tool for lifestyle management in the general population is the notion that many individuals may not be motivated enough to actively keep track of their fitness level on a daily basis. Behavioral change in general is extremely difficult to instigate. The Framingham Heart Study could provide some insight on how this issue can be tackled (Patel et al., 2017). This study was conducted in an effort to learn whether gamification could increase physical activities among families. The randomized control trial involved 200 adults from 94 families. The idea was to develop a game system for the interventional arm of the study. This would involve a reward system based on the level of physical activity. The rewards were completely arbitrary, involved a competitive points system, and did not involve any real-life material gains. The interventional arm showed significantly increased levels of physical activity as compared to the control arm, both during the study and the follow-up period.

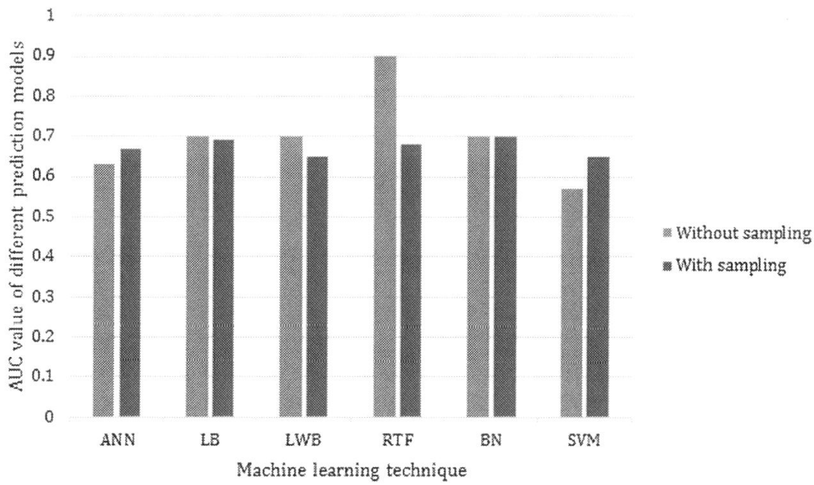

FIGURE 15.1

AUC (Area Under the Curve) for six different supervised machine learning models (FIT Study). The RTF (Random Tree Forest) method reached the highest accuracy at 91%. Abbreviations: *ANN*, Artificial Neural Network; *BN*, Bayesian Network classifier; *LB*, LogitBoost; *LWB*, Locally Weighted Naive Bayes; *RTF*, Random Tree Forest; *SVM*, Support Vector Machine.

This goes to show that individuals get increasingly motivated when a competitive component is included. This notion could be further exploited to encourage members of the population to actually use smart wearables, and in doing so, increase physical activity. In fact, this is already implemented by Apple in their Apple Watches. The activity app in the Apple Watch allows users to challenge friends or family, and whoever has the most steps by the end of the week earns a virtual trophy. Users can also work their way to different trophies by meeting their daily step goals.

Lifestyle management in the general population to prevent the incidence of CVD has been a constant challenge facing the healthcare system. However, as was demonstrated, wearable technology can play a vital role in how the healthcare community approaches such an important undertaking of disease prevention. Precision behavioral change may be difficult to accomplish; however, future endeavors must focus on identifying the most suitable, evidence-based, and personalized primordial prevention strategies to implement as early as possible in childhood, when behaviors are more pliable, and likely to last a lifetime.

Screening for CVD

With the prevalence of smart wearables across the general population, and the intrinsic feature of continuous data collection of those smart wearables, there is a wealth of data being acquired by these electronics that could prove instrumental at identifying certain diseases and pathologies. Continuous heart rate monitoring can allow these devices to screen their users for cardiac arrhythmias, and when asked to conduct an ECG, they can additionally screen for atrial fibrillation (AF) or long QT syndrome, for example. Asymptomatic AF presents with the same risks and complications as symptomatic AF, such as heart failure, kidney disease, stroke, coronary artery disease, and death.

Using their smartwatches, Apple launched the prospective Apple Heart Study in 2017, which aims to investigate the concordance of recorded pulse irregularities with simultaneous ECG recordings, and to assess the time it takes for the user to contact a healthcare provider within 3 months of pulse irregularity detection. The data collected by the Apple Watch, using a combination of LED (light-emitting diode) lights and light-sensitive photodiodes, measured the changes in blood flow to generate a photoplethysmogram that is subsequently used to estimate the pulse by recording intervals between photoplethysmography signal peaks (PSP) and producing a tachogram. The photoplethysmogram waveforms are then fed to an algorithm developed by Apple, the Irregular Pulse Notification Algorithm, to identify irregular pulses possibly indicative of AF. In the case of an irregularity that meets inclusion criteria, the algorithm will prospectively and opportunistically scan for photoplethysmography irregularities when there is minimal arm movement. Upon 4 serial conformations of an irregular photoplethysmography, the participant is informed of the irregular pulse. Out of a study population of 419,093 participants, an irregular pulse was detected in 0.52% of smartwatch wearers. About 34% of those with an irregular pulse were diagnosed with AF after wearing an ECG patch for up to

7 days. The results of this study showed a positive predictive value of 84% (participants who received a notification of an irregular pulse and also had an ECG showing AF) (Perez, Mahaffey, & Hedlin, 2019). Such screening studies would be even more effective with a lower false positive rate when conducted on moderate to high-risk populations that are more likely to have or develop AF, as opposed to a general screening.

Devices that screen for hypertension, such as the Omron HeartGuide, present numerous benefits, given the high rates of hypertension among the general population. The value of such devices lies in their function of conducting serial blood pressure measurements during the course of the day, and hence developing an extensive record of a user's blood pressure. This record can be used by a user's primary care physician with the assistance of ML algorithms to patients at risk for developing hypertension, and possibly predict cardiovascular events in patients with an established diagnosis of hypertension.

Disease management in CVD patients

Beyond the multitude of opportunities that data gathered by smart wearables have in lifestyle management within the general population, the data being collected also has the potential to be utilized in the management of patients with preexisting cardiovascular diseases. Examples of such opportunities that will be discussed in this section are monitoring for cardiac arrhythmias such as atrial fibrillation, detection of myocardial infarctions, and guiding postinfarction care, and monitoring as well as managing heart failure.

Management of atrial fibrillation

AF is one of the most common cardiac arrhythmias, particularly in older patients, and is characterized by an irregular heartbeat and loss of atrial contraction, consequently resulting in blood clots, stroke, and heart failure. There are several risk factors that predispose individuals to developing AF, such as increasing age, hypertension, obesity, and other underlying cardiovascular diseases. In the United States, it is predicted that the number of individuals expected to be diagnosed with this arrhythmia will reach up to 6 million by 2050, posing a significant pecuniary burden on the healthcare system (U.S. Census Bureau, 2018). Furthermore, AF is a long-term condition that requires close monitoring as well as hospital-based interventions such as direct current cardioversion when necessary.

There has been significant progress recently in the development of wearable cardiac rhythm recorders that can record rhythm strips using smartphone technology. After the initial success of AliveCor's single lead Kardia Mobile device in 2017, the company introduced the Kardia Band (KB), the first FDA-cleared Apple Watch accessory that patients can use to record 30 s rhythm strips. The KB operates synchronously with an application that uses machine learning ECG-recognition algorithms to provide an instant diagnosis of AF, informing the patient of the diagnosis, and providing them with the option of sending the results to their

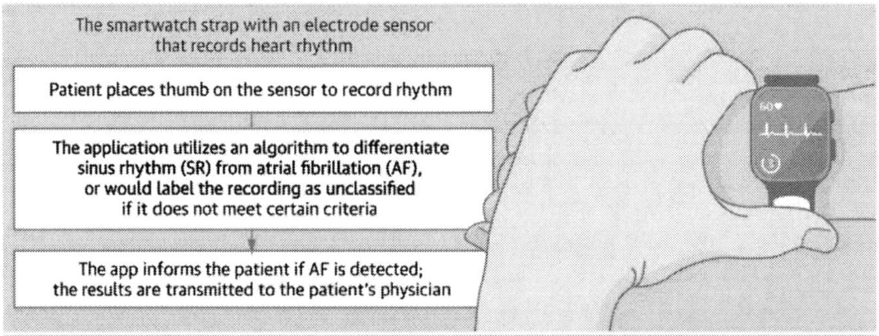

Method for interpreting the recording:	% of patients with interpretable results	Accuracy of AF diagnosis compared to 12 lead electrocardiogram
App algorithm only	66%	93% sensitivity; 84% specificity
Physician only	87%	99% sensitivity; 83% specificity
Recordings labeled as "unclassified" by the app algorithm when reviewed by physician	100%	100% sensitivity; 80% specificity

FIGURE 15.2

Illustration depicting the integration of the Kardia Band with the smartphone application, and the accuracy of AF (Atrial Fibrillation) diagnosis compared to 12-lead ECG (electrocardiogram) across different methods. Abbreviations: right (VnR) and left (VnL) limb leads.

physician (Fig. 15.2). Further, this technology has numerous applications in patients with previously established AF. A promising use to this technology in AF patients is confirming their AF status prior to a scheduled electrical cardioversion appointment, allowing them to make an informed decision on whether they should come to the hospital or not. On its own, the application's ML algorithm is able to accurately identify cases of AF up to 93% of the time, and electrophysiologists using the KB recording were able to accurately identify in up to 99% of AF cases. Not only does this help patients determine whether or not they should go to the hospital, but it can also help the hospital save valuable time and resources if a scheduled procedure such as direct current cardioversion is no longer necessary (Bumgarner et al., 2018).

Other beneficial applications of this technology is ECG monitoring in patients postcardioversion and ablation procedures that are common in AF, as there is an estimated recurrence rate of up to 50% within the first year following these treatments (Walker and Muhlestein, 2018). Hence, continuous monitoring allows physicians to gauge the success of the procedure and guide future management decision. In the iTransmit study, a small single-center trial comparing conventional transtelephonic monitoring to Kardia Mobile device for monitoring of arrhythmias found that the smart wearable had 100% sensitivity and 97% specificity in detecting AF and atrial flutter in postablation patients, and 92% of patients preferred smart wearable monitoring over conventional methods. Some patients' AF management may involve placing them on antiarrhythmic therapy that requires inpatient QT interval monitoring initially, specifically dofetilide, and it was demonstrated that the Kardia Mobile device can also be used to measure QT intervals accurately in comparison to the conventional ECG (Walker and Muhlestein, 2018).

Detection of ST-elevation myocardial infarction (STEMI)

Another significant application of smart wearable technology is the detection of ST-elevation myocardial infarction (STEMI), a possibly life-threatening cardiac event that requires urgent revascularization and is diagnosed largely using an ECG. In the United States, the occurrence of a coronary event in 2019 is expected to occur in about 1,055,000 individuals (Benjamin et al., 2019). About 47% of heart attacks occur outside of a hospital setting and only 27% of individuals are aware of all the major symptoms of a heart attack and know to call 911when someone is having one (Centers for Disease Control and Prevention, 2020).

Smart wearable technology, such as the Kardia Mobile, allows at-risk individuals to obtain an ECG earlier when presenting with STEMI, reducing their time for diagnosis and receiving critical intervention. The ST LEUIS pilot study used an adapted version of the Kardia Mobile device consisting of two wire leads, allowing the device to reconstruct a 12-lead ECG from serial lead measurements. The study showed excellent concordance of the reconstructed 12-lead ECGs with standard 12-lead ECGs for detecting an STEMI as well as non-ST elevation myocardial infarctions (NSTEMI), as seen in table 2 (Marvel, Wang and Martin, 2018). This has prompted an international multicenter study to validate the smartphone ECG application, and its ability to accurately identify the presence or absence of an STEMI in patients presenting with chest pain compared to the gold standard 12-lead ECG (Marvel et al., 2018). The potential applications of this technology is not limited to home use, but can also be implemented in emergency medical response teams (Walker and Muhlestein, 2018). Performing a remote ECG is only one part of the equation; however, it is also necessary to develop consumer-grade deep learning algorithms that can accurately classify acute coronary syndromes with high fidelity from ECG recordings.

Postmyocardial infarction management and recovery

A significant factor in recovery of any cardiovascular event is education and compli ance after hospital discharge. Unfortunately, most of this often occurs hastily during the discharge process, and considering the pervasiveness of technology in our lives, it seems intuitive to integrate technology in this process, and capitalize on the advantages artificial intelligence has to offer. The application Corrie Health, developed by a group at Johns Hopkins, is an example of how technology and smart wearables can be integrated into the hospital discharge process, and transform it into a patient-centered approach.

The target age group of Corrie Health are individuals between the ages of 40 and 85 years who are recovering from a myocardial infarction in the acute setting. The application accounts for this target cohort by working with Apple to create an optimal user interface. Some of the features of the Corrie Health application include informing users of their daily tasks, such as taking their prescribed medications and measuring their heart rate, blood pressure, and weight (Muhlestein et al., 2015). The application also asks users to log in their mood to track their mental health, records their daily steps, and provides reminders for various appointments, especially important for patients involved in a cardiac rehabilitation program. Cardiac rehabilitation has been demonstrated to have a significant impact in the recovery of acute MI patients, leading to decreased rehospitalization as well as reoccurrence of angina compared to control groups which were not engaging in rehabilitation. In a single-center study comparing patients who had undergone cardiac rehabilitation to those that have not after percutaneous coronary intervention for acute MI, only 1% of the rehabilitation group, compared to 12% in the control group, experienced recurrent anginalike pain, and only 2% compared to 19% were rehospitalized (Zhang, Cao, Jiang, & Tang, 2018).

Importantly, Corrie Health integrates with smart wearable devices and uses wearable data such as daily step or stairs count to provide personalized guidance to each patient. This information can be very valuable in structuring home-based cardiac rehabilitation programs that push patients to gradually increase their submaximal exercise capacity. Biometric wearable data, particularly heart rate, may even be used to optimize beta-blocker dosing in patients with chronic coronary artery disease, and correlate this data with their angina symptoms.

Detection of hyperkalemia in heart failure patients

Another innovative application of smart wearables is the detection of hyperkalemia from smartphone ECGs. Hyperkalemia is prevalent in multiple pathologies, including congestive heart failure, chronic kidney disease, and as an effect of several medications used to treat these conditions. Normally, the detection of hyperkalemia requires an invasive blood test, since only relatively severe hyperkalemia causes evident changes on an ECG. At mild to moderate levels, the PR interval is prolonged, and there is the development of peaked T waves (Mount, 2020). More severe hyperkalemia results in widened QRS complexes, and eventually the absence of a P wave as the QRS complex becomes a sine wave (Weiner, Linas; Wingo, 2010). (Fig. 15.3).

In a recent 2018 study, Mayo Clinic developed an algorithm utilizing artificial intelligence (AI) that was trained to detect hyperkalemia using ECG leads I and II from a training group of 484,310 patients admitted between 1994 and 2017 with a serum potassium collected within 12 h of their ECG. After an extensive learning phase, its performance was evaluated on an initial testing group of 224,690 patients. In the initial test group, the AUC was 0.87 (sensitivity 85%, specificity 72%). A secondary performance evaluation was conducted prospectively on 10 patients undergoing hemodialysis wearing the portable AliveCor smartphone ECG device. The sensitivity was 94% and specificity 74% across 5.4 h of ECG data in the hyperkalemic state and 44.1 h in a nonhyperkalemic state (Galloway et al., 2018). In another study, the same trained AI algorithm was used to detect hyperkalemia from the ECG of patients with chronic

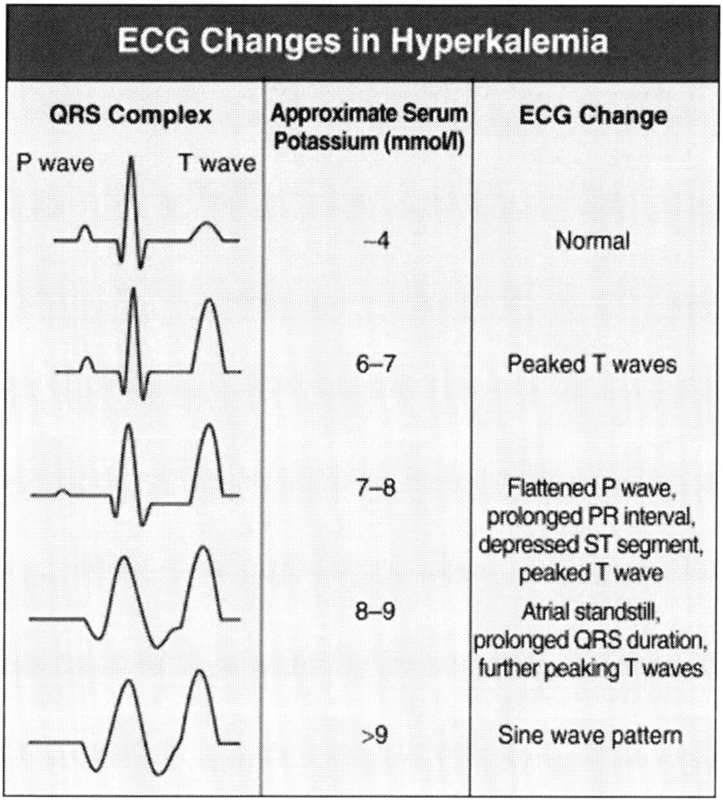

FIGURE 15.3

Electrocardiographic changes associated with progressive hyperkalemia. These include peaking of the T wave, flattening of the P wave, prolongation of the PR interval, ST-segment depression, and, eventually, progressing to a sine wave pattern.

kidney disease from Minnesota, Florida, and Arizona. The AUC using ECG leads I and II was 0.883 for Minnesota (sensitivity 90.2%, specificity 63.2%), 0.860 for Florida (sensitivity 91.3%, specificity 54.7%), and 0.853 for Arizona (sensitivity 88.9%, specificity 55.0%) (Galloway et al., 2019) Thus, the noninvasive detection of hyperkalemia using smartphone EEG has multifaceted applications across a variety of disease conditions, although several problems such as low specificity still exist, particularly if applied in screening of low-risk populations.

Monitoring chronic heart failure prognosis using physical activity

Similarly to how cardiac rehabilitation significantly improves recovery in post-MI patients, patients with chronic heart failure are seen to have low tolerance to physical activity, preventing them from engaging in daily tasks. Moreover, decreased physical capacity in heart failure patients portends a worse prognosis, with increased rehospitalization and mortality risk (Belardinelli, Capestro, Misiani, Scipione, & Georgiou,

2006). Unfortunately, despite encouraging patients to engage in low levels of physical activity, there is poor compliance among patients, and even those that are enrolled in supervised programs show poor adherence. However, walking has been described as an excellent exercise for heart failure patients as it meets the required moderately intense physical activity that helps improve outcomes. The 6-Minute Walk Test (6MWT) measures the distance a patient can walk within 6 min, and there is strong evidence demonstrating that the results of the test are responsive to clinical changes in patients participating in cardiac rehabilitation. Patients with decreased functional capacity (indicated by a distance less than 300 m for 6 min) have been shown to experience excessive mortality and morbidity (Vetrovsky et al., 2017).

These functional capacity tests, however, only occur in a hospital setting by scheduled appointments, and thus cannot pick up on health deteriorations that may occur between appointments. Smart wearables can close this gap by utilizing in-built accelerometer to calculate a patient's 6MWT score, and additionally, send the results to the patient's providers to monitor their progress. The benefit is that many smartphones currently have accelerometers built into their devices, making it even easier to collect and use the data necessary to assess a patient's functional capacity. Accelerometers were found to have high accuracy in detecting physical activity in heart failure patients, demonstrating their clinical applicability in performing out-of-clinic 6MWT that would allow for continuous patient monitoring and early detection of health deterioration. These results have instigated a multicenter trial to assess whether a smart wearable-based intervention improves functional capacity in patients with chronic heart failure with reduced (HFrEF) and preserved (HFpEF) ejection fractions (Vetrovsky et al., 2017).

Current challenges

As with any novel medical advancement, the principle of using smart wearables and artificial intelligence processed data to prevent, detect, and manage CVD is not without limitations. The first limitation that comes into play is the feasibility of collecting all this data for millions of people. The data for one person could exceed thousands of gigabytes, so how feasible would it be to collect and store all this data for an entire population? This would require lots of new data infrastructure that could accommodate these massive amounts of data. This is not something beyond reach but does need a lot of prospective planning before anything is implemented nationwide. This brings us to the second challenge we face which is confidentiality and privacy. With so much personal data being collected every second, patients deserve to know where that data goes and who has access to it. These are issues that companies need to be fully transparent about in order to maintain the everlasting trust between patients and the healthcare community.

Another important challenge will be deciding which types of data will be actually useful to collect and use in algorithms to predict CVD. Not all data is actionable and some of it may be redundant and the key to improving population health is generating clinically meaningful data. For example, is 24/7 monitoring of heart rate for the entirety of life meaningful? ML algorithms work more effectively when data is preoptimized. This can only be achieved through trial and error, as is the case with any development throughout human history. A lot more focus is shifting toward this novel field and more information will come to light in the coming years regarding which types of data are most useful and which types of algorithms work best. This also ties in with the next challenge, which is the need for extensive validation studies into the findings that preliminary studies report. These studies will come with time and will allow to further refine the methodology to achieve the most efficient data collection and analysis.

The final challenge is the economic aspect of an endeavor of such a scale. Studies and trials would need significant funding, since it is plausible to assume that individuals would require smart wearables in order to collect data, which is an expensive task. This will require a lot of reconsideration of national research funding allocation. It might also take time to convince health administrators and policy makers that this is a worthwhile national endeavor to pursue. To consider an example, the recently published Apple Heart Study recruited approximately 400,000 Apple Watch users over 2 years. Although this trial was the first of its kind and proved that such large-scale and rapid recruitment is possible, it had significant problems with retaining patients wherein 79% dropped out in the first few months. One has to be creative regarding improving such potentially revolutionary trials through integration of various clinical platforms and encouragement of patient involvement.

After that comes the huge challenge of applying what is discovered through research and trials to clinical practice and healthcare in general, especially disease prevention. Providing everyone in a population of 300 million with a smart wearable is not cheap. The research phase will be important in providing strong evidence as to whether or not a large-scale investment could potentially save healthcare providers' costs over the long term by lowering the incidence of CVD and lowering readmission rates for CVD patients.

Conclusion and future directions

The advent of smart wearables constitutes the latest trend in health technology. With improvements in computation and technology allowing for the incorporation of AI and ML, their clinical utility is only expected to rise. The design and implementation of sensors such as glucose, sleep, and breathing sensors would provide new and clinically significant health-related data. Parameters measured using different modalities can be used to expand already existing datasets in order to make ML algorithms more accurate in the detection of disease. These datasets can then be used in large studies in order to assess their effectiveness and their utility. Once the virtual infrastructure containing these parameters is built, physicians and healthcare professionals should be trained to retrieve and interpret the data from smart wearables. It is also imperative that cheaper, smart wearables are manufactured in order to be accessible by those of lower socioeconomic status.

Smart wearables undoubtedly add a new dimension to healthcare, and when utilized properly, have the potential to be extremely beneficial in a healthcare setting. They allow patients to take charge of their own health, be more engaged in the pursuit of good health, and build healthy habits. This can be achieved in numerous ways such as simply ensuring medication adherence and physical activity. With the extremely high rates of obesity and associated noncommunicable diseases such as CVD and cancer, physical activity would undoubtedly be a worthwhile endeavor, and help curtail the significant morbidity and mortality associated with such diseases. Their implementation in the clinical space would result in a paradigm shift, as the focus of disease can be aimed more toward prevention. Smart wearable technology is also ready for prime-time integration in the management of many cardiovascular diseases; however, we still need to improvise ways to streamline big data analysis, while making result visualization meaningful and seamless to both the physicians and patients.

References

Allen, J. (2007). Photoplethysmography and its application in clinical physiological measurement. *Physiological Measurement, 28*(3), R1−R39. https://doi.org/10.1088/0967-3334/28/3/r01.

Appelboom, G., Camacho, E., Abraham, M. E., et al. (2014). Smart wearable body sensors for patient self-assessment and monitoring. *Archives of Public Health, 72*(28). https://doi.org/10.1186/2049-3258-72-28.

Barnes, A. S. (2013). Emerging modifiable risk factors for cardiovascular disease in women: Obesity, physical activity, and sedentary behavior. *Texas Heart Institute Journal, 40*(3), 293−295.

Belardinelli, R., Capestro, F., Misiani, A., Scipione, P., & Georgiou, D. (2006). Moderate exercise training improves functional capacity, quality of life, and endothelium-dependent vasodilation in chronic heart failure patients with implantable cardioverter defibrillators and cardiac resynchronization therapy. *European Journal of Cardiovascular Prevention & Rehabilitation, 13*(5). https://doi.org/10.1097/01.hjr.0000230104.93771.7d.

Benjamin, E., Muntner, P., Alonso, A., Bittencourt, M., Callaway, C., Carson, A., et al. (2019). Heart disease and stroke statistics—2019 update: A report from the American heart association. *Circulation, 139*(10). https://doi.org/10.1161/cir.0000000000000659.

Bumgarner, J., Lambert, C., Hussein, A., Cantillon, D., Baranowski, B., Wolski, K., et al. (2018). Smartwatch algorithm for automated detection of atrial fibrillation. *Journal of the American College of Cardiology, 71*(21). https://doi.org/10.1016/j.jacc.2018.03.003.

Burns, M. (2019). Current and future role of wearables in healthcare. *Digitalist*. Retrieved from https://www.digitalistmag.com/customer-experience/2019/06/04/current-future-role-of-wearables-in-healthcare-06198759.

Buttice, C. (2019). How AI is enhancing wearables. *Techopedia*. Retrieved from https://www.techopedia.com/how-ai-is-enhancing-wearables/2/33627.

Centers for Disease Control and Prevention. (2020). *Heart disease facts & statistics*. Retrieved from https://www.cdc.gov/heartdisease/facts.htm.

Deo, R. C. (2015). Machine learning in medicine. *Circulation, 132*(20). https://doi.org/10.1161/CIRCULATIONAHA.115.001593.

Desjardins, J. (2015). The history of wearable technology. *Visual Capitalist*. Retrieved from https://www.visualcapitalist.com/the-history-of-wearable-technology/.

Galloway, C., Valys, A., Petterson, F., Gundotra, V., Treiman, D., Albert, D., et al. (2018). Non-invasive detection of hyperkalemia with a smartphone electrocardiogram and artificial intelligence. *Journal of the American College of Cardiology, 71*(11), A272. https://doi.org/10.1016/s0735-1097(18)30813-1.

Galloway, C., Valys, A., Shreibati, J., Treiman, D., Petterson, F., Gundotra, V., et al. (2019). Development and validation of a deep-learning model to screen for hyperkalemia from the electrocardiogram. *JAMA Cardiology, 4*(5). https://doi.org/10.1001/jamacardio.2019.0640.

Giebel, G., & Gissel, C. (2019). Accuracy of mHealth devices for atrial fibrillation screening: Systematic review. *JMIR Mhealth and Uhealth, 7*(6), e13641. https://doi.org/10.2196/13641.

Kini, V., & Ho, P. (2018). Interventions to improve medication adherence. *Journal of the American Medical Association, 320*(23), 2461. https://doi.org/10.1001/jama.2018.19271.

Lear, S., Hu, W., Rangarajan, S., Gasevic, D., Leong, D., Iqbal, R., et al. (2017). The effect of physical activity on mortality and cardiovascular disease in 130 000 people from 17 high-income, middle-income, and low-income countries: The PURE study. *The Lancet, 390*(10113). https://doi.org/10.1016/S0140-6736(17)31634-3.

Lee, I., Shiroma, E. J., Kamada, M., Bassett, D. R., Matthews, C. E., & Buring, J. E. (2019). Association of step volume and intensity with all-cause mortality in older women. *JAMA Internal Medicine, 179*(8). https://doi.org/10.1001/jamainternmed.2019.0899.

Marvel, F., Wang, J., & Martin, S. (2018). Digital health innovation: A toolkit to navigate from concept to clinical testing. *JMIR Cardio, 2*(1). https://doi.org/10.2196/cardio.7586.

Mount, D. (2020). *Causes and evaluation of hyperkalemia in adults*. UpToDate. Retrieved from https://www.uptodate.com/contents/causes-and-evaluation-of-hyperkalemia-in-adults?source=search_result&search=hyperkalemia&selectedTitle=2~150. (Accessed 24 November 2019).

Muhlestein, J., Le, V., Albert, D., Moreno, F., Anderson, J., Yanowitz, F., et al. (2015). Smartphone ECG for evaluation of STEMI: Results of the ST LEUIS pilot study. *Journal of Electrocardiology, 48*(2). https://doi.org/10.1016/j.jelectrocard.2014.11.005.

Patel, M., Benjamin, E., Volpp, K., Fox, C., Small, D., Massaro, J., et al. (2017). Effect of a game-based intervention designed to enhance social incentives to increase physical activity among families: The BE FIT randomized clinical trial. *JAMA Internal Medicine, 177*(11). https://doi.org/10.1001/jamainternmed.2017.3458.

Perez, M. V., Mahaffey, K. W., Hedlin, H., et al. (2019). Large-scale assessment of a smartwatch to identify atrial fibrillation. *The New England Journal of Medicine, 381*(20), 1909–1917. https://doi.org/10.1056/NEJMoa1901183.

Perez, S. (2019). IDC: Apple led wearables market in 2018, with 46.2M of the total 172.2M devices shipped. *TechCrunch*. Retrieved from https://techcrunch.com/2019/03/05/idc-apple-led-wearables-market-in-2018-with-46-2m-of-the-total-172-2m-devices-shipped/.

Richter, F. (2018). Infographic: The global wearables market is all about the wrist. *Statista*. Retrieved from https://www.statista.com/chart/3370/wearable-device-forecast/.

Russey, C. (2019). IDC forecasts double-digit growth in wearables with smartwatches and wireless headphones leading the way. *Wearable Technologies*. Retrieved from https://

www.wearable-technologies.com/2019/03/idc-forecasts-double-digit-growth-in-wearables-with-smartwatches-and-wireless-headphones-leading-the-way/.

Sakr, S., Elshawi, R., Ahmed, A., Qureshi, W. T., Brawner, C., Keteyian, S., et al. (2018). Using machine learning on cardiorespiratory fitness data for predicting hypertension: The Henry Ford Exercise Testing (FIT) Project. *PLoS One, 13*(4), e0195344. https://doi.org/10.1371/journal.pone.0195344.

U.S. Census Bureau. (2018). *The population 65 Years and older in the United States: 2016.* Washington D.C.: US Department of Commerce.

Vetrovsky, T., Siranec, M., Parenica, J., Griva, M., Stastny, J., Precek, J., et al. (2017). Effect of a 6-month pedometer-based walking intervention on functional capacity in patients with chronic heart failure with reduced (HFrEF) and with preserved (HFpEF) ejection fraction: Study protocol for two multicenter randomized controlled trials. *Journal of Translational Medicine, 15*(1). https://doi.org/10.1186/s12967-017-1257-x.

Walker, A., & Muhlestein, J. (2018). Smartphone electrocardiogram monitoring: Current perspectives. *Advanced Health Care Technologies, 4.* https://doi.org/10.2147/ahct.s138445.

Weiner, I., Linas, S., & Wingo, C. (2010). Disorders of potassium metabolism. *Comprehensive Clinical Nephrology.* https://doi.org/10.1016/b978-0-323-05876-6.00009-5.

Yetisen, A. K., Martinez-Hurtado, J. L., Ünal, B., Khademhosseini, A., & Butt, H. (2018). Wearables in medicine. *Advanced Materials, 30*(33), e1706910. https://doi.org/10.1002/adma.201706910.

Zhang, Y., Cao, H., Jiang, P., & Tang, H. (2018). Cardiac rehabilitation in acute myocardial infarction patients after percutaneous coronary intervention. *Medicine (Baltimore), 97*(8). https://doi.org/10.1097/md.0000000000009785.

CHAPTER

The future of artificial intelligence in healthcare

16

Nathan Radakovich[1], **Aziz Nazha**[2]

[1]*Cleveland Clinic Lerner College of Medicine, Cleveland, OH, United States;* [2]*Cleveland Clinic Taussig Cancer Center, Leukemia Department, Cleveland, OH, United States*

Chapter outline

Introduction and historical perspective

Intelligence is a difficult to encapsulate concept, one for which scientists and philosophers have proposed myriad definitions and metrics over the years. It is unsurprising, given the difficulty of reaching a consensus regarding our own intelligence, that

Machine Learning in Cardiovascular Medicine. https://doi.org/10.1016/B978-0-12-820273-9.00016-6

a firm definition of artificial intelligence (AI), a term first coined by the pioneering AI researcher John McCarthy in the 1950s (McCarthy, 1959), has been similarly elusive. McCarthy's initial writings referred to "intelligent machines," but did not detail what constituted intelligence and what was mere calculation. Alan Turing's contemporaneous paper "Computing Machinery and Intelligence" proposed that a machine could be considered intelligent if it could convince a human observer that it, too, was human (Turing, 1950). In a healthcare setting, a relatively narrow, task-oriented version of Turing's standard is typically used, defining AI as the computers' ability to perform tasks that would otherwise require uniquely human cognitive flexibility and creativity (Buch, Ahmed, & Maruthappu, 2018; Esteva et al., 2019; LeCun, Bengio, & Hinton, 2015; Miotto et al., 2018). We often apply an intuitive standard for "intelligence" when describing our own actions, and when we turn to computers to complete tasks, invoke similar standards both to describe our expectations and adjudicate the results. Thus unlike the terms "deep learning," "machine learning," and "neural networks," which carry specific, technical definitions, "artificial intelligence" is more flexible. It describes our wishes for how computers would mimic and complement—or surpass—our own intelligence.

Modern applications of AI fulfill previous definitions to various degrees. Autonomously driving cars, for instance, have debuted with some success, and have featured software both programmed a priori to understand and obey the rules of the road, as well as software whose driving abilities and decision-making is derived from observing hours of human driving. In these two different approaches to the same task, we see two data-processing and judgment-making faculties that we deem as signs of intelligence: the ability to learn the rules governing a task and then implement those rules; and the ability to learn from experience, thereby coming to an inductive understanding of a previously unknown task.

Turing's definition of AI required a computer to match human abilities, but in certain realms AI has begun not only to match human abilities, but to surpass them. Computational power, which has increased exponentially since the advent of microprocessors in the 1960s (Waldrop, 2016), now allows AI systems to accrue the equivalent of lifetimes' worth of domain-specific experience in mere days, thereby exceeding human performance in some well-circumscribed tasks. Regarding consistency and reliability, self-driving cars already hold some advantages over human drivers; the majority of reported accidents involving self-driving cars to date have occurred due to the error of human drivers, not their AI counterparts, although human faith in automated systems has itself been a precipitant of accidents (Hecht, 2018; Kokalitcheva, 2018). AI has also begun to take an edge in tasks that benefit from enormous quantities of sheer computing power. For instance, when trained on a database of 30 million unique scenarios in the board game Go, Google's Alpha Go software was able to consistently defeat world champion human players, a testament to the power of AI when applied to the right task (Silver et al., 2016).

With the diverse and rapid developments in AI, it is unsurprising that healthcare, which comprises one-fifth of the United States' economy and has data repositories measuring in the petabytes, has been a recent hotbed for AI research. The notion of computers as powerful tools—or rivals—for physicians is by no means recent, however. In a 1970 commentary, "Medicine and the Computer: The Promise and Problems of Change," in the *New England Journal of Medicine*, Dr. William Schwartz, an early advocate of AI in healthcare, envisioned not only the "housekeeping" role that computers would play in the form of electronic health records (EHRs), but the more direct role they would play in the intellectual work of medicine (Schwartz, 1970). Citing an "increasing shortage of physician manpower" and "expansion of medical knowledge [that widens] the gap between what a doctor should know and what he can retain and utilize," Schwartz envisioned computer systems that would participate in history taking, generating differential diagnoses, and selecting appropriate treatments, thereby "[freeing] the physician to concentrate on the tasks that are uniquely human" (Schwartz, 1970). A half-century later, physician scarcity and ever-expanding knowledge still pose challenges to the medical field. What has changed, however, are the tools at physicians' disposal, and the volume and sophistication of data that make up modern medical practice. This chapter provides a brief review of contemporary practices and tools in AI, a description of the state of emerging AI applications, and a discussion of the potential challenges and adjustments that will accompany AI's incorporation into clinical practice.

Supervised and unsupervised learning

AI tasks can be described as "supervised" or "unsupervised" according to the output desired by the user. In supervised learning tasks, an algorithm is given input data to make a prediction, which can be categorical (e.g., "does this biopsy contain malignant tissue?") or numerical (e.g., "what is the predicted survival of this patient, given their disease status and comorbidities?"). Source data are used to generate initial predictions by the algorithm. Predictions are then compared to actual outcomes, and the algorithm is iteratively refined based on the discrepancies between the two, yielding successively more accurate predictions (Deo, 2015; Saria, Butte, & Sheikh, 2018). In contrast, unsupervised learning tasks do not have a "correct" answer that they attempt to predict. Rather, they are given data without associated labels and are asked to identify patterns. For example, an algorithm that clusters a collection of similar DNA sequences may do without any mechanism for judging the correctness of its clustering, and while its output may reflect known patterns it may also identify novel ones (Buch et al., 2018; Deo, 2015). Alternatively, the output may be entirely spurious; in any case, the basis for adjudicating the algorithm's performance is external to the algorithm itself.

Neural networks: a brief overview

Neural networks are amongst the most widely used and successful tools in contemporary AI. Neural networks' namesake and inspiration come from biologic neurons, and in particular from the way that action potentials between neurons are used as the basic unit of communication in the nervous system. Neural networks process information by propagating data along a series of interconnected nodes, or "neurons," starting with some external input and generating an output after relaying the data through many layers of neurons (Esteva et al., 2019; LeCun et al., 2015). The output of a neural network may be as simple as a single neuron whose output, between zero and one, represents a single probability (e.g., determining whether or not a picture contains a particular object). It may also contain multiple such outputs, each corresponding to the likelihood of a different outcome (e.g., classifying a picture as one of many potential objects). More complex outputs are also possible, such as a series of vectors representing collections of words (e.g., in software for translating from one language to another), or an array of pixels (e.g., altering or generating images based on source data) (Fig. 16.1).

Because of their ability to create more abstract representations of data in hidden layers, neural networks are also capable of integrating several different types of data into a single model. Distinct types of data are fed into separate inputs, processed through several parallel hidden layers, and then connected by subsequent layers prior to yielding an output. This is of particular utility in medicine, where images, laboratory values, and free text reports all play a role in decision-making and patient care; rather than considering different data types in isolation, interactions between data can be captured.

The pace of change in AI research

The pace of innovation in healthcare-related AI is accelerating, and is on pace to continue doing so. A PubMed search of terms "deep learning," "machine learning,"

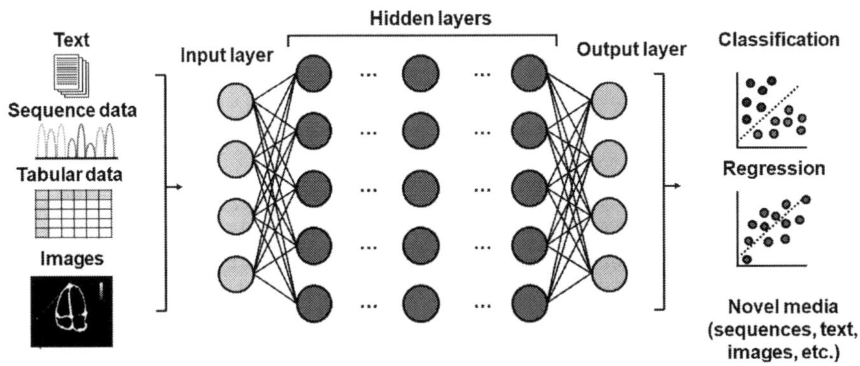

FIGURE 16.1

Schematic of neural networks. Data (pixels, words, etc.) are formatted and fed into neurons in an input layer. Subsequent hidden layers of neurons allow the network to capture nonlinearity and higher-level representations of source data. The network's output can be used for classification or regression tasks or to generate novel data.

"artificial intelligence," or "neural network" reveals a consistent increase in both the absolute volume of AI-related research, and in the proportion of the medical literature that it constitutes. In the year 2000, 1838 articles (0.38% of all articles published in 2000) met the aforementioned search criteria. In 2010, there were 5491 such articles (0.74% of all articles), and in 2018 there were 15,240 (5.20% of all articles) (Fig. 16.2).

These trends reflect the increasing recognition of AI's utility and promise in healthcare, as well as ongoing improvements in the hardware and software that facilitate increasingly sophisticated AI algorithms. They also reflect the democratization of the tools used to implement AI. Most industry-standard AI packages are freely available online, and cheap online computing power can be accessed for computationally demanding projects, eliminating the need for costly upfront capital investments. The convergence of these factors has lowered the threshold for successfully using AI, thereby allowing clinician experts to begin implementing AI across a wide array of domain-specific applications.

Computer vision

Computer vision is the use of AI to interpret and extract data from images. No medical application of AI has seen as much recent progress, or garnered as much popular attention, as medical image processing, especially in the fields of pathology and radiology. This progress has been facilitated by the advent of convolutional neural networks (CNNs), a subtype of neural network designed to process images in a

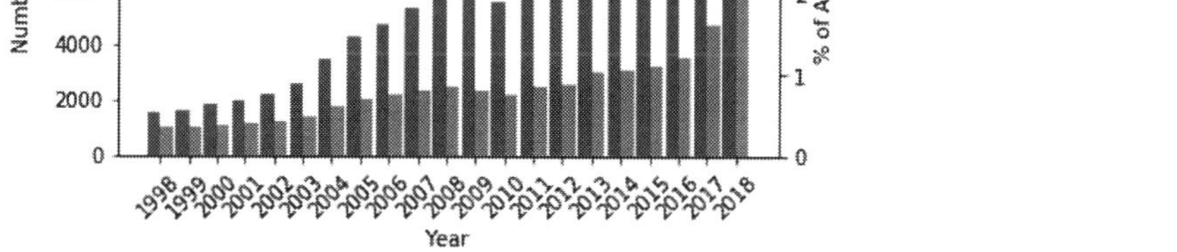

FIGURE 16.2 Healthcare-related AI publications' increase in absolute numbers and proportion.

In 2018, 15,240 articles were published related to AI in healthcare, or 5.20% of all articles in 2018. Publication numbers were generated via a Boolean PubMed search of the terms "deep learning," "machine learning," "artificial intelligence," or "neural network."

manner similar to human sight. CNNs begin with basic patterns and textures such as horizontal and vertical lines, then build successively more complex and abstract representations based on those initial steps, thereby learning a more generalizable version of the object in question (3,5). This allows CNNs to learn from training data, and to effectively apply what they have learned to previously unseen examples. The advent of transfer learning, which uses models that have already been trained to recognize basic shapes and contours, allows powerful models to be trained efficiently and effectively on novel tasks, facilitating the expansion of computer vision to increasing numbers of novel applications (Esteva et al., 2019).

AI in radiology

CNNs are approaching the point of clinical usefulness in several tasks in radiology, and in some cases have already passed that threshold. Routine screening mammography, detection of abnormalities on chest radiographs, bone age assessment, pulmonary nodule detection, and tuberculosis screening have all been successfully performed by CNNs, to name a few notable applications (Lakhani and Sundaram, 2017; Lee et al., 2017, 2019; Shen et al., 2019; Song et al., 2017). In cases where AI performs as well as human readers, some routine radiology tasks may become increasingly automated in the future, with human operators spending more time on complex cases, or attending to images where AI-based predictions fall below a certain threshold of certainty. Currently, radiologists' workloads constrain the amount of images that can be read in a health system, and reading fees add to medical costs. Automated systems designed to prompt human review for select images could improve throughput of image interpretation, allocate more time to challenging cases, and decrease costs (Saba et al., 2019).

Image segmentation, a tedious but important task used to identify areas of interest in images such as individual organs, tumors, fluid collections, or injured tissue, can also increasingly be performed by AI (Fig. 16.3) (Akkus et al., 2017; Chen et al., 2018; Dreizin et al., 2019; Trebeschi et al., 2017; Vigneault et al., 2018). Automating this process eliminates the painstaking work of manually delineating the borders of structures when scrolling through the many planes of a three-dimensional image. Not only does this decrease radiologists' workloads, it creates the potential for new applications in radiology such as whole-body volumetric tumor quantification, improved treatment planning in radiation oncology, and more accurate image-based body composition assessment (Bi et al., 2019; Dreizin et al., 2019; Ibragimov et al., 2017). Given these advances, the vast majority of practicing and trainee radiologists expect AI to be integrated into their practice within a decade (Waymel et al., 2019).

Beyond achieving remarkable performance on existing tasks in radiology, AI has demonstrated the potential to extract novel, clinically meaningful information from images. In lung cancer, AI has shown promise in using CT images to predict the invasiveness of tumors, assess tumors' response to chemotherapy, and distinguish between indolent and aggressive tumors (Li et al., 2017; Parmar et al., 2015; Tunali et al., 2017). Similarly, in gliomas, analysis of magnetic resonance imaging (MRI)

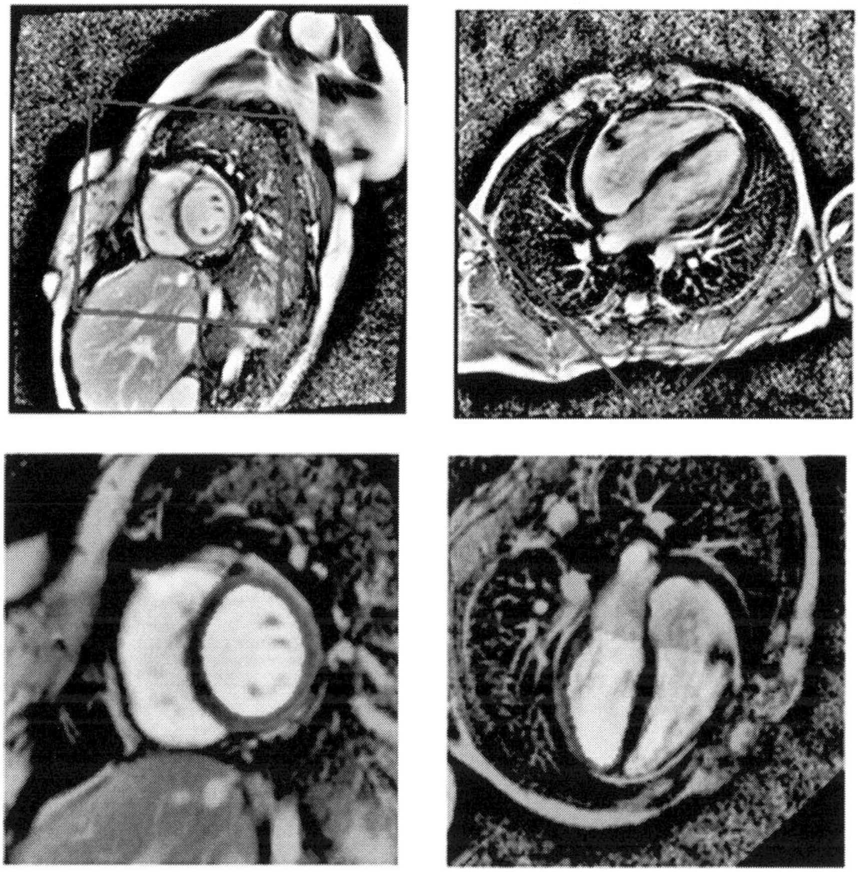

FIGURE 16.3 Automated cardiac segmentation by deep neural networks (24).

Myocardium, atria, and ventricles are distinguished from one another and highlighted by a neural network, facilitating automated measurement of cardiac dimensions.

alone can predict tumor grade, characterize mutations, and even provide an estimate of patient survival (Korfiatis et al., 2017; Zhang et al., 2017; Zhou et al., 2017). Insights from such radiomic studies provide added value to data that are already routinely collected, and may help in stratifying patients with regard to aggressiveness of therapy or in guiding the appropriate response to incidental findings (Bi et al., 2019).

Generative adversarial networks (GANs), a type of neural network designed to generate de novo data that resemble real data, have also found a foothold in radiology (Fig. 16.4). GANs have shown promise in several realms: reducing noise artifact in images, thereby making them more interpretable for human observers; synthesizing images across modalities, such as different MRI sequences; and creating synthetic instances of training data that can be used to improve the

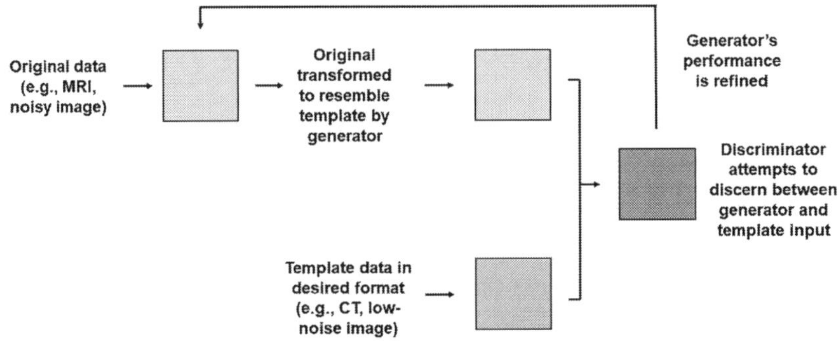

FIGURE 16.4 Style transfer with generative adversarial networks for cross-platform image generation.

Source data are processed by a generator, which attempts to transform the source data to resemble template data. A discriminator attempts to determine which data are original and which are transformed, allowing the generator's performance to be iteratively improved.

generalizability of image classifiers (Fig. 16.5) (Armanious et al.; Cai et al., 2019; Choi and Lee, 2018; Vigneault et al., 2018; Yi, Walia, & Babyn, 2019). These applications point toward GANs improving both the experience of human radiologists and the performance of AI models.

AI in pathology

Pathology has seen similar improvements as radiology in the use of AI for common tasks. While efforts to apply AI to tasks such as routine pap smears date to the early

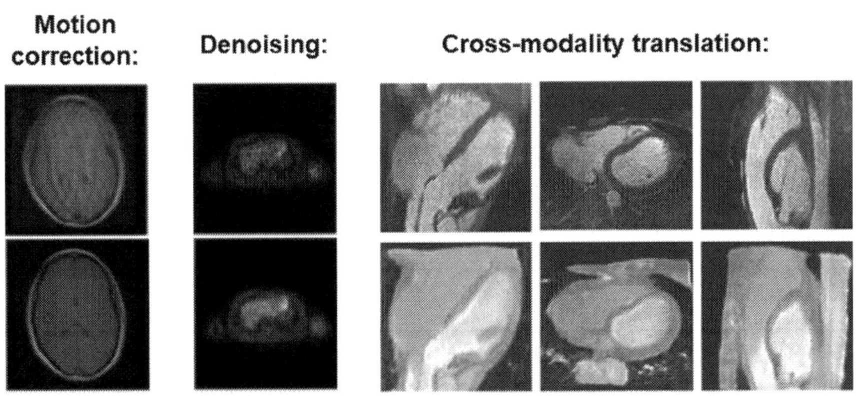

FIGURE 16.5 Applications of generative adversarial networks (GANs).

GANs can be trained to reduce motion artifacts and reduce noise in CT/PET images (36), or to alter images to resemble other modalities (MRI to CT translation is shown) (37).

1990s, contemporary work has yielded highly accurate and specific models capable of detecting abnormalities on routine smears (Boon and Kok, 1993; Bora et al., 2017; Zhang and Yao, 2017). AI has also moved into the more challenging realm of examining tissue slides rather than just cytology. Neural networks, in particular CNNs, have been able to approach, and in some cases match, human performance in detecting and classifying objects, including polyp classification, staging of liver fibrosis, detecting lymph node metastases in breast cancer, and subclassification of gastric tumors (Bejnordi et al., 2017; Komura and Ishikawa, 2019; Korbar et al., 2017; Sharma et al., 2017; Yu et al., 2018). AI also shows promise in precision medicine, for example, in predicting PD-1L status in tumors (Sha et al., 2019). Pathologists face the same limitations regarding workload, fatigue, and human error that radiologists do, and AI may similarly improve the throughput and consistency of pathology in the foreseeable future (Komura and Ishikawa, 2019).

Dermatology and ophthalmology

AI has also taken a foothold in clinical specialties, namely in ophthalmology, where AI is increasingly used to interpret fundoscopic images, and dermatology, where it is employed in classifying skin lesions. Importantly, recent work in these fields has also included implementation of AI into regular practice. In a 2018 trial, an AI-based diabetic retinopathy detection tool achieved 87.5% sensitivity and 90.7% specificity in a primary care setting, subsequently becoming the first AI-based system to gain U.S. Food and Drug Administration authorization (Abràmoff et al., 2018). Similar work using AI in community hospital settings has also demonstrated clinically useful performance as a retinopathy screening tool (He et al., 2019a). While these models fell short of previous findings in studies that used curated datasets (Abràmoff et al., 2013; Gargeya and Leng, 2017; Gulshan et al., 2016; Kermany et al., 2018; Poly et al.), they highlighted the ability to detect retinal disease across a variety of ages, races, and levels of glycemic control at a clinically meaningful level of performance. This ability to move from preselected data to the more imperfect setting of everyday practice represents an important step toward everyday clinical use of AI.

Prospective, implementation-oriented studies in ophthalmology demonstrate the potential for AI's generalizability and practical integration into clinical workflow, two factors that represent barriers to more widespread utilization of AI in medicine. Crucially, they also address an unmet healthcare need. Time, expense, and provider scarcity are barriers to success in any form of preventative care, and the patients affected by those barriers experience poorer health as a consequence (Hartnett et al., 2005; Jani et al., 2017). Effective point-of-care screening can help circumvent these barriers, and moreover may help allocate specialist time more effectively by assuming some of the clinical burden of primary prevention.

AI has made similar strides in dermatology, namely in classifying skin lesions. CNNs have continually improved over the last decade, with the latest generation of algorithms achieving parity with dermatologists in melanoma and multilabel lesion classification tasks (Cui et al., 2019; Esteva et al., 2017; Roffman et al., 2018). Indeed, in recent multicenter studies of dermatologist versus AI-based dermatoscopy interpretation, AI now outperforms human observers (Brinker et al., 2019; Tschandl et al., 2019). These advances may assist dermatologists in a manner similar to the work being performed in retinopathy screening, and given the quality and ubiquity of cell phone cameras, it may be even easier for patients to have lesions regularly evaluated. Smartphone-based screening apps have yet to achieve the sensitivity required to make them an acceptable screening option, in particular due to false negative rates that are too high for a screening test (Chuchu et al., 2018). However, with the appropriate improvements in model performance (perhaps achieved by uploading remotely captured images to secure servers with greater computational resources than a handheld device), smartphones could serve as a basis for triaging suspicious findings by uploading photos to be analyzed remotely.

Beyond single pictures of lesions, a logical next step in lesion imaging is longitudinal monitoring, since change over time is a consideration when evaluating potentially cancerous skin lesions. Dermatologists' sensitivity is only 80% for detecting melanomas when viewing a lesion at a single point in time, so even models that outperform humans will likely miss a significant number of melanomas, and interval monitoring may be required to improve screening (35). Prospective data are needed in order to create the datasets necessary to develop such a tool (Sondermann et al., 2019).

Time-series analysis

Understanding the chronology of patients' diseases is an integral part of the practice of medicine. Time-series analysis, which employs AI to process sequences of events, has seen significant strides in recent years. Time-series problems abound in medicine, from single variable tasks such as interpreting the pattern of voltages recorded in an electrocardiogram to complex multivariable predictions, such as determining whether an ICU patient is improving or deteriorating based on trends in vital signs, labs, written notes, and day-to-day physical exam findings.

Much of the recent research in time-series analysis has to do with event prediction. In the ICU setting this includes risk-stratifying patients by likelihood of mortality and promptly identifying sepsis (Nemati et al., 2018). Other work has focused on using EHR data for hospitalized patients at large, collecting vast amounts of multimodality data such as laboratory values, vital signs, and written notes in order to make predictions about common in-hospital events such as acute kidney injury and in-hospital mortality (Rajkomar et al., 2018; Tomašev et al., 2019). While these advances are intriguing, time-series models may not be as close to clinical implementation as other modalities. This is in part due to the challenge of integrating such a model into EHR, but also reflects the fact that to be adopted clinically, models need to provide information that will affect decision-making in a meaningful way.

Mobile device applications

Medical data are traditionally collected either in clinics or the inpatient setting, but for most patients this represents a small fraction of their lives. When we observe a tremor or assess a patient's gait in the clinic, it is with the assumption that we are seeing a representative sample of their daily lives, but this may not always be true, such as in patients with masked hypertension, or in Parkinson disease patients who experience fluctuations in symptoms throughout the day. With the advent of inexpensive and accurate portable monitoring devices, it is now feasible to directly gather information about patients' day-to-day activity outside of a medical setting, thereby greatly expanding the pool of potentially useful data about patients (Baig et al., 2017; Dinh-Le et al., 2019).

This wealth of data has been studied in many settings, such as monitoring for fall risk in the elderly (Hakim et al., 2017; Nait Aicha et al., 2018), arrhythmia detection (Sadrawi et al., 2017), and grading tremor severity in Parkinson disease patients (Jeon et al., 2017). Ambulatory monitoring provides a greater depth of actionable information for clinicians with minimal inconvenience to patients. Mobile devices also have the potential to facilitate immediate therapy adjustments. In Parkinson patients, for instance, AI can identify events such as falls or gait freezing in real time, and provide a basis for dynamically adjusting medication regimens rather than delaying adjustments until a follow-up appointment (Asakawa et al., 2019; Silva de Lima et al., 2017). Although concerns exist regarding information privacy and creating anxiety about medically insignificant findings, the pairing of AI with wearable devices has the potential to improve the sensitivity and personalization of medical care if applied properly.

AI in medical research

Classification of disease

While much popular attention is directed toward AI's applications in clinical practice, it also promises to alter how physicians research disease. For example, unsupervised clustering of asthma patients was able to group patients into distinct clinically meaningful cohorts that varied based on asthma severity, resource utilization, responsiveness to bronchodilators, and biologic variables such as eosinophilia, circulating inflammatory markers, and exhaled nitric oxide (Wu et al., 2014).

Not only do such clusters, which are generated without any a priori designation of outcomes or patient phenotypes, identify clinically relevant subpopulations of patients, they have clear diagnostic and therapeutic ramifications. Clustering approaches could be used in clinical practice to fine-tune aggressiveness of therapy, avoid ineffective treatments, and to identify those most likely to respond to novel targeted therapies. AI-based clustering may also inform future clinical trial design. In fields like oncology, where many new agents are designed with disease-specific biology in mind, AI may facilitate more targeted selection criteria. In cases where there is a strong biologic rationale for those selection criteria, this may allow for smaller trials to be conducted while still achieving statistical power, thereby reducing trial costs and sparing unnecessary interventions for those unlikely to benefit.

AI as a tool in data gathering

One of the challenges in clinical research that draws from EHRs is accurately sifting through disorganized data. Important information about patient outcomes often comes in the form of unstructured text in pathology reports or clinician notes, and such data need to be extracted manually in order to be incorporated into outcomes research. Obtaining information through manual review of EHR is time-consuming, and may place practical constraints on sample size. Conversely, methodologically convenient but less granular methods of chart review, such as querying International Classification of Diseases (ICD) codes without manual review of individual charts, facilitate larger sample sizes at the expense of detail. Natural language processing (NLP), another field where AI has yielded important strides, could be used in the future as a means of collecting data with the speed and throughput of computerized queries.

Traditional rule-based methods of natural language processing require explicit programming on how to extract information. Recurrent neural networks (RNNs), a type of network designed to take previous inputs into account, are capable of more flexible, context-sensitive performance, and can be applied to NLP; CNNs have also been used effectively in this setting (Esteva et al., 2019). AI has been used to process EHR with close to human accuracy in myriad settings, such as disease classification, determining disease severity, compiling multimodal data from reports on malignancy, classification of unstructured notes, making personalized assessment of risk factors, and automatically recognizing adverse drug reactions (Bacchi et al., 2019; Chen et al., 2019; Kaur et al., 2018; Leyh-Bannurah et al., 2018; Liang et al., 2019; Weng et al., 2017; Wi et al., 2017). These diverse applications highlight that while yet to be routinely employed in clinical practice, AI-based NLP has much to offer in both research and clinical practice.

Future challenges in implementation

Much of the focus on AI in healthcare has been on the development of AI models themselves: how to more accurately identify objects, classify sequences, or deal with the subtleties and inconsistencies of interpreting written language. However, as progress continues in each of these realms, thought also needs to be given to other factors that will impact the clinical utilization of AI.

Practical considerations

Broadly speaking, the enthusiasm for AI in healthcare is driven by two hopes: that AI will improve the quality of care that physicians deliver, and that it will do so in a way that improves patient and physician experiences. Thus, in arenas where AI achieves or surpasses human performance, the question turns from one of theoretical usefulness to one of practical implementation. Even if an AI-based model performs exceedingly well, it will not provide any clinical benefit unless it can reasonably be integrated into physicians' workflows. For many applications this likely means integration into existing EHR systems, which carries the attendant requirements of securely handling patient data, reliably standardizing the input data used by AI systems, and likely the ability to interface with other EHR-based AI programs (He et al., 2019b).

Interpretability

Any AI that plays a significant role in guiding patient care will have to include mechanisms for making its predictions explicable to clinicians and patients. Simple models such as logistic regression are easy to interpret because each variable is considered in isolation, making interpretation straightforward. However, simplistic models are rarely adequate, and medicine provides many examples of the adage that "for every complex problem there is an answer that is clear, simple, and wrong."

Medical data are never considered in isolation, and medical decision-making is almost never simple. For example, the same heart rate could have very different implications depending on patient comorbidities, other vital signs, and numerous other factors, all of which are implicitly or explicitly considered by human clinicians when delivering care and choosing a course of action. Moreover, clinicians are able to describe how they interpret complex sets of data and why they draw certain conclusions. If AI is to gain widespread acceptance by clinicians and patients, it will have to manage to be sophisticated while remaining similarly accountable.

Testing and overdiagnosis

The risk of overdiagnosis needs to be considered as AI-based diagnostic tools become increasingly sensitive. Overdiagnosis occurs when a test correctly identifies the presence of disease, but there is no benefit from intervening on the disease (Jenniskens et al., 2017). In such circumstances intervention may in fact be harmful, such as in aggressive treatment of otherwise indolent prostate cancer in older men or extensive workup of small, incidentally discovered thyroid nodules (Jenniskens et al., 2017; Jung, 2017). For instance, low-dose CT screening of long-term smokers is estimated to result in 1.38 cases of overdiagnosis for every lung cancer death averted (Patz et al., 2014). This is an acceptable trade-off given the upside of preventing a cancer death, but that trade-off is predicated on research performed using human radiologists, and there is no guarantee that AI-based systems would yield the same result (Bi et al., 2019). Similarly, AI advances in live-imaging endoscopy may provide high-sensitivity means for detecting polyps on screening colonoscopy, but given the propensity of existing models for false-positive findings, a firm understanding of their performance characteristics is needed before implementing them (Ahmad et al., 2019).

Outside of the scope of existing screening, overdiagnosis also needs to be considered as new sources of health information, such as wearable devices, become increasingly common. In some cases, such as atrial fibrillation or atrial high rate episodes in ambulatory patients, even asymptomatic events have unfavorable prognostic implications and require intervention (Chen-Scarabelli et al., 2015; Miyazawa et al., 2019). Conversely, there are some instances in which device-related findings are clinically insignificant or not actionable. Acting on such findings can lead to overuse of medical resources and inappropriate interventions (67). Care needs to be taken to ensure that clinical implementation of AI is geared toward applications that meaningfully guide clinician decisions and patient outcomes.

Barriers to implementation

Beyond practical barriers, the impact of AI on physicians themselves must be considered. It is often assumed that better technology necessarily improves the lives of its users, but studies of physician satisfaction with EHR and computerized order entry belie this assumption. Indeed, the use of computerized records and orders is an independent predictor of physician burnout, and if EHR-integrated AI increases the time spent interacting with EHR and detracts from physician-patient interactions, it could in turn become a contributor to the burnout problem (Shanafelt et al., 2016). Given that clinically practical AI would likely be implemented via EHR, care needs to be taken to ensure that such implementation does not become psychologically burdensome to clinicians.

Precautions about burnout aside, fluid and user-friendly EHR integration is also crucial to the successful uptake of AI. Numerous best practice alerts (BPAs) to warn clinicians about drug-drug interactions, adherence to guidelines, or health maintenance are already incorporated into EHR systems, with varying degrees of success and uptake, which in turn determines how much patients ultimately benefit from such tools. In cases where BPAs do not provide benefit, a major contributor is physician alert fatigue, which is driven in part by the sheer volume of BPAs encountered (with the HITEC Investigators et al., 2017). When dozens of alerts are triggered daily, physicians are less likely to attend to any of them, regardless of their utility. If AI's implementation becomes intrusive or disruptive to physicians' workflows, it risks being similarly ignored.

Despite posing potential problems, clinical AI may also offer solutions for physicians' struggles with burnout and burdensome EHR. The use of human scribes is known to ameliorate burnout and increase physician job satisfaction without detracting from patients' experiences, but their widespread use is limited by cost and scarcity (Gidwani et al., 2017). AI may represent a solution.

NLP in medicine currently functions as a dictation tool, and while dictation may be more efficient than typing, it does not substantially alter the current workflow for physicians; with or without dictation, a sizable amount of physician time is spent on documentation rather than actual patient care. This may change with the advent of increasingly sophisticated language-processing technologies. A system that can summarize physician-patient discussions and generate summary statements about patients problems would be able to capture the unique aspects of an individual visit, thereby freeing physicians from having to recapitulate an entire visit when writing or dictating a note (Coiera et al., 2018). Such a system would provide more time for direct patient-physician interaction while still maintaining robust documentation. Ideally, it would also have the ability to be continually trained by end users to more accurately reflect individual decision-making processes and efficiently summarize encounters in a provider-specific manner.

The challenge of accurate source data

AI models' performance will only be as good as their source data, and if flawed data are used, flawed predictions should be expected. For instance in cases such as pulmonary nodule classification where intraobserver variability is high, it is difficult to assign accurate labels to training data without a tissue diagnosis, limiting the ceiling of an algorithm's usefulness (Ciompi et al., 2017; Hamidian et al., 2017; Kang et al., 2017). Similar concerns exist for models derived from parsing EHR data. If inaccuracies are present in the medical records used to train a system, then it runs the risk of being trained to recapitulate the same mistakes as its source data.

Beyond the accuracy of source data, the same precautions about external validity need to be taken as with any other form of research. Many AI models achieve exquisite accuracy on internal datasets, but fail to recapitulate that accuracy when validated externally, meaning that stellar performance in one setting could be unacceptably, even dangerously, flawed in another (Zech et al., 2018). Federated learning, which allows multiple institutions to train an AI model without having to divulge private patient information to one another, could address validity concerns, either by helping to increase model generalizability or by allowing institution-specific fine-tuning of models (Miotto et al., 2018).

Beyond the prospect of inaccurate data, AI has the potential to unintentionally adopt human bias and prejudices. Photo recognition software, including some models developed for medical use, performs more accurately on light-skinned individuals than on dark-skinned ones due to underrepresentation in datasets; more troublingly, minority individuals are more likely to be viewed unfavorably by AI systems such as crime-prediction or loan-approval algorithms (Courtland, 2018; Hague, 2019; Zou and Schiebinger, 2018). Real-world instances of unintended bias have already occurred, for example, in a now-discontinued Amazon hiring algorithm that consistently favored male applicants over equally qualified female ones (Amazon scraps secretr). Human judgment contains numerous and often covert flaws; if we are going to use human experiences to train AI, then care needs to be taken that it does not internalize the uglier aspects of humanity.

Will AI make physicians obsolete?

The emergence of increasingly useful, clinically viable AI brings with it uncertainty about the future role of human physicians, including fears of being displaced or even replaced. This has been especially true in radiology, where much of the recent progress in medical AI has occurred. The majority of medical trainees anticipate that AI will reduce the workload required of radiologists, and in one study half of students contemplating careers in radiology cited this as a source of trepidation (Gong et al., 2019; Pinto dos Santos et al., 2019). However, both students with greater knowledge of AI and more clinical experience express less concern about being displaced by computers, suggesting that those with both domain-specific medical knowledge and a more nuanced understanding of AI tended to view AI as transformative and useful, but not a threat to the profession (Gong et al., 2019; Pinto dos Santos et al., 2019). Perhaps reflecting the same trend, a 2019 survey of practicing radiologists found that while 90% of respondents expected AI to be part of daily practice for radiologists and 26% expected that it would decrease the demand for radiologists, only 7% of all respondents thought it would have a negative impact on their personal practice, with 79% anticipating a positive impact (Waymel et al., 2019).

Practicing radiologists have expressed similar views in commentaries, acknowledging AI's incorporation into clinical practice as inevitable while simultaneously predicting that AI will support radiologists and augment their skills but not supplant them (Pesapane, Codari, & Sardanelli, 2018; Recht and Bryan, 2017; Waymel et al., 2019). In many ways, patients' views on AI in medicine are compatible with those of clinicians. Qualitative interviews with patients suggest that they welcome the incorporation of AI into clinical practice if it improves the quality of care and makes interpretable predictions, but still preferred that human physicians be involved in most aspects of their care; those preferring not to have AI involved in their care cite trust of AI systems as a major concern (Haan et al., 2019; Keel et al., 2018).

Meanwhile, complex decision-making remains the purview of human physicians, and AI systems capable of taking an active role in clinical decision-making have had decidedly mixed success. IBM's Watson, which attracted widespread attention for its ability to interpret semantically nuanced questions and compete on the TV quiz show *Jeopardy*, underwent several years of pilot testing as a decision support system in clinical oncology. However, after a period of use at select major cancer centers in the United States, Watson for Oncology (WFO) was discontinued, although it continues to be studied in several East Asian countries (Big Data Bust: Anderso; Ross and Swetlitz, 2017, p. 5).

Retrospective comparisons between WFO's recommendations and those of a multidisciplinary tumor board show that WFO's missteps occur more commonly in elderly patients, intermediate-staged disease, or in malignancies where there are significant regional or national variations in standards of care (Choi et al., 2019; Kim et al., 2019; Somashekhar et al., 2018; Zhou et al., 2019). Collectively, these shortcomings suggest that despite being able to incorporate many disparate data sources and guidelines into treatment recommendations, AI has considerable work to do in considering the nuances of medicine, such as how to best balance aggressiveness and palliation in unfit patients, interpret ambiguous results, or adapt to local practice patterns.

Nonetheless, the prospect of AI as a partner in clinical decision-making remains appealing, especially when considering the gaps in care provided by human clinicians. Medical error incurs billions of dollars in costs annually, and is estimated by some to be the third-leading cause of death in the United States, a sobering reminder of our own limitations (Makary and Daniel, 2016; Walsh et al., 2017). Implementing safeguards such as checklists, EHR alerts, and standardized handoffs reduces the likelihood individual error by creating a series of barriers to catastrophic errors. As it continues to mature, AI may have a similar role to play in preventing oversights in diagnosis or management.

AI in medical education

As AI becomes integrated into clinical practice, medical school and residency curricula will need to adapt in order to ensure that clinicians approach it in an informed manner. Since the reforms in medical education prompted by the Flexner report, one of the cornerstones of medical education has been literacy in research methods and statistics, as well as careful review of scientific literature (Cooke et al., 2006). Research continually alters the landscape of medicine throughout a physician's career, and prudent translation of research into clinical practice requires an understanding of how that research is conducted. While it is unrealistic—and unnecessary—for physicians to be fully fledged computer scientists, physicians do need to be savvy consumers of research that uses AI. This is especially true given the increasing ease of tackling AI projects, since the downside to the democratization of AI is that researchers may apply AI to their work without fully grasping potential pitfalls of the technology.

A knowledge gap likely already exists in the medical community. When surveyed about their understanding of AI, medical students' self-assessment of their level of knowledge outstrips their ability to actually answer questions about the basics of AI, suggesting that awareness and popular coverage of the topic alone does not confer a deeper understanding or the ability to critically appraise AI-based research (Gong et al., 2019). Perhaps reflecting this, most medical students favor the inclusion of AI into medical school curricula (Pinto dos Santos et al., 2019). Education has always had to adapt to match the pace of research, and the changes ushered in by AI are no exception.

Conclusion

The degree of computing power available in the modern era and the sophistication and diversity of artificial intelligence that has followed would have been unthinkable in prior decades. Modern AI has already made significant inroads in the fields of pathology and radiology, and is poised to become a routine part of clinical practice within the decade, with other branches of medicine likely to follow suit as the technology continues to mature. Beyond automating and improving on many existing tasks that physicians perform, AI also holds the promise of yielding new insights from the vast amount of medical data now available. As AI becomes more widespread, care needs to be taken to ensure that it is implemented thoughtfully, with an eye to unintended consequences, and to ensure that medical education promotes a basic working knowledge of AI for future physicians.

References

Abràmoff, M. D., et al. (2013). Automated analysis of retinal images for detection of referable diabetic retinopathy. *JAMA Ophthalmology, 131*(3), 351. https://doi.org/10.1001/jamaophthalmol.2013.1743.

Abràmoff, M. D., et al. (2018). Pivotal trial of an autonomous AI-based diagnostic system for detection of diabetic retinopathy in primary care offices. *NPJ Digital Medicine, 1*(1), 39. https://doi.org/10.1038/s41746-018-0040-6.

Ahmad, O. F., Soares, A. S., Mazomenos, E., Brandao, P., Vega, R., Seward, E., et al. (2019). Artificial intelligence and computer-aided diagnosis in colonoscopy: Current evidence and future directions. *Lancet Gastroenterol Hepatol, 4*, 71–80.

Akkus, Z., et al. (2017). Deep learning for brain MRI segmentation: State of the art and future directions. *Journal of Digital Imaging, 30*(4), 449–459. https://doi.org/10.1007/s10278-017-9983-4.

Amazon scraps secret AI recruiting tool that showed bias against women - Reuters (no date). Retrieved from: https://www.reuters.com/article/us-amazon-com-jobs-automation-insight-idUSKCN1MK08G. (Accessed 30 September 2019).

Armanious, K., et al. (no date). Retrospective correction of rigid and non-rigid MR motion artifacts using GANs. arXiv:1809.06276.

Asakawa, T., et al. (2019). Can the latest computerized technologies revolutionize conventional assessment tools and therapies for a neurological disease? The example of Parkinson's disease. *Neurologia Medico-Chirurgica, 59*(3), 69–78. https://doi.org/10.2176/nmc.ra.2018-0045.

Bacchi, S., et al. (2019). Deep learning natural language processing successfully predicts the cerebrovascular cause of transient ischemic attack-like presentations. *Stroke, 50*(3), 758–760. https://doi.org/10.1161/STROKEAHA.118.024124.

Baig, M. M., et al. (2017). A systematic review of wearable patient monitoring systems – current challenges and opportunities for clinical adoption. *Journal of Medical Systems, 41*(7), 115. https://doi.org/10.1007/s10916-017-0760-1.

Bejnordi, B. E., et al. (2017). Diagnostic assessment of deep learning algorithms for detection of lymph node metastases in women with breast cancer. *Journal of the American Medical Association, 318*(22), 2199–2210. https://doi.org/10.1001/jama.2017.14585.

Bi, W. L., et al. (2019). Artificial intelligence in cancer imaging: Clinical challenges and applications. *CA: A Cancer Journal for Clinicians*. https://doi.org/10.3322/caac.21552.caac.21552.

Big Data Bust: MD Anderson-Watson Project Dies (no date). Retrieved from: https://www.medscape.com/viewarticle/876070. (Accessed 30 September 2019).

Boon, M., & Kok, L. (1993). Neural network processing can provide means to catch errors that slip through human screening of pap smears. *Diagnostic Cytopathology, 9*(4), 411–416.

Bora, K., et al. (2017). Automated classification of Pap smear images to detect cervical dysplasia. *Computer Methods and Programs in Biomedicine, 138*, 31–47. https://doi.org/10.1016/j.cmpb.2016.10.001.

Brinker, T. J., et al. (2019). Comparing artificial intelligence algorithms to 157 German dermatologists: The melanoma classification benchmark. *European Journal of Cancer, 111*, 30–37. https://doi.org/10.1016/j.ejca.2018.12.016.

Buch, V. H., Ahmed, I., & Maruthappu, M. (2018). Artificial intelligence in medicine: Current trends and future possibilities. *British Journal of General Practice, 68*(668), 143–144. https://doi.org/10.3399/bjgp18X695213.

Cai, J., et al. (2019). Towards cross-modal organ translation and segmentation: A cycle- and shape-consistent generative adversarial network. *Medical Image Analysis, 52*, 174–184. https://doi.org/10.1016/j.media.2018.12.002.

Chen-Scarabelli, C., et al. (2015). Device-detected atrial fibrillation. *Journal of the American College of Cardiology, 65*(3), 281–294. https://doi.org/10.1016/j.jacc.2014.10.045.

Chen, H., et al. (2018). VoxResNet: Deep voxelwise residual networks for brain segmentation from 3D MR images. *NeuroImage, 170*, 446–455. https://doi.org/10.1016/j.neuroimage.2017.04.041.

Chen, L., et al. (2019). Extracting medications and associated adverse drug events using a natural language processing system combining knowledge base and deep learning. *Journal of the American Medical Informatics Association*. https://doi.org/10.1093/jamia/ocz141.

Choi, Y. I., et al. (2019). Concordance rate between clinicians and Watson for oncology among patients with advanced gastric cancer: Early, real-world experience in Korea. *Canadian Journal of Gastroenterology and Hepatology, 2019*, 1–6. https://doi.org/10.1155/2019/8072928.

Choi, H., & Lee, D. S. (2018). Generation of structural MR images from amyloid PET: Application to MR-less quantification. *Journal of Nuclear Medicine, 59*(7), 1111–1117. https://doi.org/10.2967/jnumed.117.199414.

Chuchu, N., et al. (2018). Smartphone applications for triaging adults with skin lesions that are suspicious for melanoma. *Cochrane Database of Systematic Reviews*. https://doi.org/10.1002/14651858.CD013192. Edited by Cochrane Skin Group.

Ciompi, F., et al. (2017). Towards automatic pulmonary nodule management in lung cancer screening with deep learning. *Scientific Reports, 7*(1), 46479. https://doi.org/10.1038/srep46479.

Coiera, E., et al. (2018). The digital scribe. *NPJ Digital Medicine, 1*(1), 58. https://doi.org/10.1038/s41746-018-0066-9.

Cooke, M., et al. (2006). American medical education 100 years after the Flexner report. *New England Journal of Medicine, 355*(13), 1339–1344. https://doi.org/10.1056/NEJMra055445. Edited by M. Cox and D. M. Irby.

Courtland, R. (2018). Bias detectives: The researchers striving to make algorithms fair. *Nature, 558*, 357–360. https://doi.org/10.1038/d41586-018-05469-3.

Cui, X., et al. (2019). Assessing the effectiveness of artificial intelligence methods for melanoma: A retrospective review. *Journal of the American Academy of Dermatology*. https://doi.org/10.1016/j.jaad.2019.06.042. S0190962219310412.

Deo, R. C. (2015). Machine learning in medicine. *Circulation, 132*(20), 1920–1930. https://doi.org/10.1161/CIRCULATIONAHA.115.001593.

Dinh-Le, C., et al. (2019). Wearable health technology and electronic health record integration: Scoping review and future directions. *JMIR mHealth and uHealth, 7*(9). https://doi.org/10.2196/12861.

Dreizin, D., et al. (2019). Performance of a deep learning algorithm for automated segmentation and quantification of traumatic pelvic hematomas on CT. *Journal of Digital Imaging*. https://doi.org/10.1007/s10278-019-00207-1.

Esteva, A., et al. (2017). Dermatologist-level classification of skin cancer with deep neural networks. *Nature, 542*(7639), 115–118. https://doi.org/10.1038/nature21056.

Esteva, A., et al. (2019). A guide to deep learning in healthcare. *Nature Medicine, 25*(1), 24–29. https://doi.org/10.1038/s41591-018-0316-z.

Gargeya, R., & Leng, T. (2017). Automated identification of diabetic retinopathy using deep learning. *Ophthalmology, 124*(7), 962–969. https://doi.org/10.1016/j.ophtha.2017.02.008.

Gidwani, R., et al. (2017). Impact of scribes on physician satisfaction, patient satisfaction, and charting efficiency: A randomized controlled trial. *The Annals of Family Medicine, 15*(5), 427–433. https://doi.org/10.1370/afm.2122.

Gong, B., et al. (2019). Influence of artificial intelligence on Canadian medical students' preference for radiology specialty: ANational survey study. *Academic Radiology, 26*(4), 566–577. https://doi.org/10.1016/j.acra.2018.10.007.

Gulshan, V., et al. (2016). Development and validation of a deep learning algorithm for detection of diabetic retinopathy in retinal fundus photographs. *Journal of the American Medical Association, 316*(22), 2402–2410. https://doi.org/10.1001/jama.2016.17216.

Haan, M., et al. (2019). A qualitative study to understand patient perspective on the use of artificial intelligence in radiology. *Journal of the American College of Radiology*. https://doi.org/10.1016/j.jacr.2018.12.043. S1546144019300109.

Hague, D. C. (2019). Benefits, pitfalls, and potential bias in health care AI. *North Carolina Medical Journal, 80*(4), 219–223. https://doi.org/10.18043/ncm.80.4.219.

Hakim, A., et al. (2017). Smartphone based data mining for fall detection: Analysis and design. *Procedia Computer Science, 105*, 46–51. https://doi.org/10.1016/j.procs.2017.01.188.

Hamidian, S., et al. (2017). 3D convolutional neural network for automatic detection of lung nodules in chest CT. In S. G. Armato, & N. A. Petrick (Eds.), *SPIE Medical Imaging, Orlando, Florida, United States* (p. 1013409). https://doi.org/10.1117/12.2255795.

Hartnett, M. E., et al. (2005). Perceived barriers to diabetic eye care: Qualitative study of patients and physicians. *Archives of Ophthalmology, 123*(3), 387–391. https://doi.org/10.1001/archopht.123.3.387.

Hecht, J. (2018). Managing expectations of artificial intelligence. *Nature, 563*, S141–S143. https://doi.org/10.1038/d41586-018-07504-9.

He, J., et al. (2019a). Artificial intelligence-based screening for diabetic retinopathy at community hospital. *Eye*. https://doi.org/10.1038/s41433-019-0562-4.

He, J., et al. (2019b). The practical implementation of artificial intelligence technologies in medicine. *Nature Medicine, 25*(1), 30–36. https://doi.org/10.1038/s41591-018-0307-0.

with the HITEC Investigators, et al. (2017). Effects of workload, work complexity, and repeated alerts on alert fatigue in a clinical decision support system. *BMC Medical Informatics and Decision Making, 17*(1), 36. https://doi.org/10.1186/s12911-017-0430-8.

Ibragimov, B., et al. (2017). Combining deep learning with anatomical analysis for segmentation of the portal vein for liver SBRT planning. *Physics in Medicine and Biology, 62*(23), 8943–8958. https://doi.org/10.1088/1361-6560/aa9262.

Jani, P. D., et al. (2017). Evaluation of diabetic retinal screening and factors for ophthalmology referral in a telemedicine network. *JAMA Ophthalmology, 135*(7), 706–714. https://doi.org/10.1001/jamaophthalmol.2017.1150.

Jenniskens, K., et al. (2017). Overdiagnosis across medical disciplines: A scoping review. *BMJ Open, 7*(12), e018448. https://doi.org/10.1136/bmjopen-2017-018448.

Jeon, H., et al. (2017). Automatic classification of tremor severity in Parkinson's disease using a wearable device. *Sensors, 17*(9), 2067. https://doi.org/10.3390/s17092067.

Jung, M. (2017). Breast, prostate, and thyroid cancer screening tests and overdiagnosis. *Current Problems in Cancer, 41*(1), 71–79. https://doi.org/10.1016/j.currproblcancer.2016.11.006.

Kang, G., et al. (2017). 3D multi-view convolutional neural networks for lung nodule classification. *PLoS One, 12*(11), e0188290. https://doi.org/10.1371/journal.pone.0188290. Y. Deng.

Kaur, H., et al. (2018). Automated chart review utilizing natural language processing algorithm for asthma predictive index. *BMC Pulmonary Medicine, 18*(1), 34. https://doi.org/10.1186/s12890-018-0593-9.

Keel, S., et al. (2018). Feasibility and patient acceptability of a novel artificial intelligence-based screening model for diabetic retinopathy at endocrinology outpatient services: A pilot study. *Scientific Reports, 8*(1), 4330. https://doi.org/10.1038/s41598-018-22612-2.

Kermany, D. S., et al. (2018). Identifying medical diagnoses and treatable diseases by image-based deep learning. *Cell, 172*(5), 1122–1131.e9. https://doi.org/10.1016/j.cell.2018.02.010.

Kim, E. J., et al. (2019). Early experience with Watson for oncology in Korean patients with colorectal cancer. *PLoS One, 14*(3), e0213640. https://doi.org/10.1371/journal.pone.0213640. Edited by P. Orzechowski.

Kokalitcheva, K. (2018). In *People cause most California autonomous vehicle accidents, Axios, 29 August.*

Komura, D., & Ishikawa, S. (2019). Machine learning approaches for pathologic diagnosis. *Virchows Archiv, 475*(2), 131–138. https://doi.org/10.1007/s00428-019-02594-w.

Korbar, B., et al. (2017). Deep learning for classification of colorectal polyps on whole-slide images. *Journal of Pathology Informatics, 8*. https://doi.org/10.4103/jpi.jpi_34_17.

Korfiatis, P., et al. (2017). Residual deep convolutional neural network predicts MGMT methylation status. *Journal of Digital Imaging, 30*(5), 622–628. https://doi.org/10.1007/s10278-017-0009-z.

Lakhani, P., & Sundaram, B. (2017). Deep learning at chest radiography: Automated classification of pulmonary tuberculosis by using convolutional neural networks. *Radiology, 284*(2), 574–582. https://doi.org/10.1148/radiol.2017162326.

LeCun, Y., Bengio, Y., & Hinton, G. (2015). Deep learning. *Nature, 521*(7553), 436–444. https://doi.org/10.1038/nature14539.

Lee, H., et al. (2017). Fully automated deep learning system for bone age assessment. *Journal of Digital Imaging, 30*(4), 427–441. https://doi.org/10.1007/s10278-017-9955-8.

Lee, S. M., et al. (2019). Deep learning applications in chest radiography and computed tomography: Current state of the art. *Journal of Thoracic Imaging, 34*(2), 75–85. https://doi.org/10.1097/RTI.0000000000000387 [Miscellaneous Article].

Leyh-Bannurah, S.-R., et al. (2018). Deep learning for natural language processing in urology: State-of-the-Art automated extraction of detailed pathologic prostate cancer data from narratively written electronic health records. *JCO Clinical Cancer Informatics, 2*, 1–9. https://doi.org/10.1200/CCI.18.00080.

Liang, H., et al. (2019). Evaluation and accurate diagnoses of pediatric diseases using artificial intelligence. *Nature Medicine, 25*(3), 433–438. https://doi.org/10.1038/s41591-018-0335-9.

Li, Q., et al. (2017). CT imaging features associated with recurrence in non-small cell lung cancer patients after stereotactic body radiotherapy. *Radiation Oncology, 12*(1), 158. https://doi.org/10.1186/s13014-017-0892-y.

Makary, M., & Daniel, M. (2016). Medical error-ther third leading cause of death the the US. *BMJ, 353*(2139).

McCarthy, J. (1959). *Artificial intelligence*. Research Laboratory of Electronics (RLE) at the Massachusetts Institute of Technology (MIT).

Miotto, R., et al. (2018). Deep learning for healthcare: Review, opportunities and challenges. *Briefings in Bioinformatics, 19*(6), 1236–1246. https://doi.org/10.1093/bib/bbx044.

Miyazawa, K., et al. (2019). Atrial high rate episodes in patients with cardiac implantable electronic devices: Implications for clinical outcomes. *Clinical Research in Cardiology, 108*(9), 1034–1041. https://doi.org/10.1007/s00392-019-01432-y.

Nait Aicha, A., et al. (2018). Deep learning to predict falls in older adults based on daily-life trunk accelerometry. *Sensors, 18*(5), 1654. https://doi.org/10.3390/s18051654.

Nemati, S., et al. (2018). An interpretable machine learning model for accurate prediction of sepsis in the ICU. *Critical Care Medicine, 46*(4), 547–553. https://doi.org/10.1097/CCM.0000000000002936.

Parmar, C., et al. (2015). Machine learning methods for quantitative radiomic biomarkers. *Scientific Reports, 5*(1), 13087. https://doi.org/10.1038/srep13087.

Patz, E. F., et al. (2014). Overdiagnosis in low-dose computed tomography screening for lung cancer. *JAMA Internal Medicine, 174*(2), 269. https://doi.org/10.1001/jamainternmed.2013.12738.

Pesapane, F., Codari, M., & Sardanelli, F. (2018). Artificial intelligence in medical imaging: Threat or opportunity? Radiologists again at the forefront of innovation in medicine. *European Radiology Experimental, 2*(1), 35. https://doi.org/10.1186/s41747-018-0061-6.

Pinto dos Santos, D., et al. (2019). Medical students' attitude towards artificial intelligence: A multicentre survey. *European Radiology, 29*(4), 1640–1646. https://doi.org/10.1007/s00330-018-5601-1.

Poly, T. N., Islam, M. M., Yang, H. C., Nguyen, P. A., Wu, C. C., & Li, Y. J. (no date). Artificial intelligence in diabetic retinopathy: Insights from a meta-analysis of deep learning. 17th world congress on medical and health informatics, MEDINFO 2019, pp. 1556–1557.

Rajkomar, A., et al. (2018). Scalable and accurate deep learning with electronic health records. *NPJ Digital Medicine, 1*(1), 18. https://doi.org/10.1038/s41746-018-0029-1.

Recht, M., & Bryan, R. N. (2017). Artificial intelligence: Threat or boon to radiologists? *Journal of the American College of Radiology, 14*(11), 1476–1480. https://doi.org/10.1016/j.jacr.2017.07.007.

Roffman, D., et al. (2018). Predicting non-melanoma skin cancer via a multi-parameterized artificial neural network. *Scientific Reports, 8*(1), 1701. https://doi.org/10.1038/s41598-018-19907-9.

Ross, C., & Swetlitz, I. (2017). *IBM pitch its Watson supercomputer as a revoltuion in cancer care. It's nowhere close*. Stat News, 5 September. Retrieved from: https://www.statnews.com/2017/09/05/watson-ibm-cancer/.

Saba, L., et al. (2019). The present and future of deep learning in radiology. *European Journal of Radiology, 114*, 14–24. https://doi.org/10.1016/j.ejrad.2019.02.038.

Sadrawi, M., et al. (2017). Arrhythmia evaluation in wearable ECG devices. *Sensors, 17*(11), 2445. https://doi.org/10.3390/s17112445.

Saria, S., Butte, A., & Sheikh, A. (2018). Better medicine through machine learning: What's real, and what's artificial? *PLoS Medicine, 15*(12), e1002721. https://doi.org/10.1371/journal.pmed.1002721.

Schwartz, W. (1970). Medicine and the computer: The promise and problems of change. *New England Journal of Medicine, 283*(23), 1257–1264.

Sha, L., et al. (2019). Multi-Field-of-View deep learning model predicts nonsmall cell lung cancer programmed death-ligand 1 status from whole-slide hematoxylin and eosin images. *Journal of Pathology Informatics, 10*. https://doi.org/10.4103/jpi.jpi_24_19.

Shanafelt, T. D., et al. (2016). Relationship between clerical burden and characteristics of the electronic environment with physician burnout and professional satisfaction. *Mayo Clinic Proceedings, 91*(7), 836–848. https://doi.org/10.1016/j.mayocp.2016.05.007.

Sharma, H., et al. (2017). Deep convolutional neural networks for automatic classification of gastric carcinoma using whole slide images in digital histopathology. *Computerized Medical Imaging and Graphics, 61*, 2–13. https://doi.org/10.1016/j.compmedimag.2017.06.001. Selected papers from the 13th European Congress on Digital Pathology.

Shen, L., et al. (2019). Deep learning to improve breast cancer detection on screening mammography. *Scientific Reports, 9*(1), 12495. https://doi.org/10.1038/s41598-019-48995-4.

Silva de Lima, A. L., et al. (2017). Freezing of gait and fall detection in Parkinson's disease using wearable sensors: A systematic review. *Journal of Neurology, 264*(8), 1642–1654. https://doi.org/10.1007/s00415-017-8424-0.

Silver, D., et al. (2016). Mastering the game of Go with deep neural networks and tree search. *Nature, 529*(7587), 484–489. https://doi.org/10.1038/nature16961.

Somashekhar, S. P., et al. (2018). Watson for oncology and breast cancer treatment recommendations: Agreement with an expert multidisciplinary tumor board. *Annals of Oncology, 29*(2), 418–423. https://doi.org/10.1093/annonc/mdx781.

Sondermann, W., et al. (2019). Prediction of melanoma evolution in melanocytic nevi via artificial intelligence: A call for prospective data. *European Journal of Cancer, 119*, 30–34. https://doi.org/10.1016/j.ejca.2019.07.009.

Song, Q., et al. (2017). Using deep learning for classification of lung nodules on computed tomography images. *Journal of Healthcare Engineering*, 1–7. https://doi.org/10.1155/2017/8314740.

Tomašev, N., et al. (2019). A clinically applicable approach to continuous prediction of future acute kidney injury. *Nature, 572*(7767), 116–119. https://doi.org/10.1038/s41586-019-1390-1.

Trebeschi, S., et al. (2017). Deep learning for fully-automated localization and segmentation of rectal cancer on multiparametric MR. *Scientific Reports, 7*(1), 5301. https://doi.org/10.1038/s41598-017-05728-9.

Tschandl, P., et al. (2019). Comparison of the accuracy of human readers versus machine-learning algorithms for pigmented skin lesion classification: An open, web-based, international, diagnostic study. *The Lancet Oncology, 20*(7), 938–947. https://doi.org/10.1016/S1470-2045(19)30333-X.

Tunali, I., et al. (2017). Radial gradient and radial deviation radiomic features from presurgical CT scans are associated with survival among lung adenocarcinoma patients. *Oncotarget, 8*(56). https://doi.org/10.18632/oncotarget.21629.

Turing, A. (1950). Computing Machinery and intelligence. *Mind, 59*(236), 433.

Vigneault, D., et al. (2018). Ω-Net (omega-net): Fully automatic, multi-view cardiac MR detection, orientation, and segmentation with deep neural networks. *Medical Image Analysis, 48*, 95–106.

Waldrop, M. M. (2016). The chips are down for Moore's law. *Nature, 530*(7589), 144–147. https://doi.org/10.1038/530144a.

Walsh, E. K., et al. (2017). Economic impact of medication error: A systematic review: Economic impact of medication error. *Pharmacoepidemiology and Drug Safety, 26*(5), 481–497. https://doi.org/10.1002/pds.4188.

Waymel, Q., et al. (2019). Impact of the rise of artificial intelligence in radiology: What do radiologists think? *Diagnostic and Interventional Imaging, 100*(6), 327–336. https://doi.org/10.1016/j.diii.2019.03.015.

Weng, W.-H., et al. (2017). Medical subdomain classification of clinical notes using a machine learning-based natural language processing approach. *BMC Medical Informatics and Decision Making, 17*(1), 155. https://doi.org/10.1186/s12911-017-0556-8.

Wi, C.-I., et al. (2017). Application of a natural language processing algorithm to asthma ascertainment. An automated chart review. *American Journal of Respiratory and Critical Care Medicine, 196*(4), 430–437. https://doi.org/10.1164/rccm.201610-2006OC.

Wu, W., et al. (2014). Unsupervised phenotyping of Severe Asthma Research Program participants using expanded lung data. *Journal of Allergy and Clinical Immunology, 133*(5), 1280–1288. https://doi.org/10.1016/j.jaci.2013.11.042.

Yi, X., Walia, E., & Babyn, P. (2019). Generative adversarial network in medical imaging: A review. *Medical Image Analysis, 58*, 101552. https://doi.org/10.1016/j.media.2019.101552.

Yu, Y., et al. (2018). Deep learning enables automated scoring of liver fibrosis stages. *Scientific Reports, 8*(1), 16016. https://doi.org/10.1038/s41598-018-34300-2.

Zech, J. R., et al. (2018). Variable generalization performance of a deep learning model to detect pneumonia in chest radiographs: A cross-sectional study. *PLoS Medicine, 15*(11), e1002683. https://doi.org/10.1371/journal.pmed.1002683.

Zhang, X., et al. (2017). Optimizing a machine learning based glioma grading system using multi-parametric MRI histogram and texture features. *Oncotarget, 8*(29). https://doi.org/10.18632/oncotarget.18001.

Zhang, L., & Yao, J. (2017). DeepPap: Deep convolutional networks for cervical cell classification. *IEEE Journal of Biomedical and Health Informatics, 21*(6), 1633–1643.

Zhou, H., et al. (2017). MRI features predict survival and molecular markers in diffuse lower-grade gliomas. *Neuro-Oncology, 19*(6), 862–870. https://doi.org/10.1093/neuonc/now256.

Zhou, N., et al. (2019). Concordance study between IBM Watson for oncology and clinical practice for patients with cancer in China. *The Oncologist, 24*(6), 812–819. https://doi.org/10.1634/theoncologist.2018-0255.

Zou, J., & Schiebinger, L. (2018). AI can be sexist and racist — it's time to make it fair. *Nature, 559*(7714), 324–326. https://doi.org/10.1038/d41586-018-05707-8.

Further reading

Komorowski, M., & Celi, L. A. (2017). Will artificial intelligence contribute to overuse in healthcare? *Critical Care Medicine, 45*(5), 912–913. https://doi.org/10.1097/CCM.0000000000002351.

CHAPTER

Ethical and legal challenges

17

Emily Tat[1], **Mark Rabbat**[2]

[1]*Department of Internal Medicine, Columbia University Medical Center, New York, NY, United States;* [2]*Medicine and Radiology, Division of Cardiology, Loyola University Chicago, Chicago, IL, United States*

Chapter outline

Introduction

Artificial intelligence (AI) has significant potential to transform the field of cardiovascular medicine. Major advancements are seen throughout all subsets of AI from machine learning algorithms and neural networks, natural language processing, and robotics to cognitive computing computer vision. There is promise for these advancements to improve the morbidity and mortality associated with cardiovascular disease, from better disease surveillance, improved diagnostics, novel treatments, and an era of truly personalized medicine (Topol, 2019b). Especially in today's rise of big data with a need to make sense of an enormous amount of information,

Machine Learning in Cardiovascular Medicine. https://doi.org/10.1016/B978-0-12-820273-9.00017-8

AI could seemingly not be developing at a better time. Yet, the impact of AI in current clinical practice is still limited. This is largely because its role in healthcare has not yet been fully defined. Unlike other realms of AI utilization outside of healthcare that are now well integrated into daily life, the use of sensitive health information and unique concerns of patient safety, privacy, and autonomy place medical ethics at the forefront of the overarching discussion. Currently, the medical community remains ill-informed of such ethical intricacies, with clear policy and ethical guidelines yet to be implemented.

The contents within this chapter will examine some of the most pressing ethical and legal challenges that healthcare providers must be informed of in this new era of medicine. One major theme that will be addressed is big data and the issues involving its integration. This includes threat to patient privacy and safety, the limits of inconsistent electronic health records, and the catch-22 of balancing reductionism and holism, that is, which and how many input variables are required to ensure accurate results. A second theme that will be discussed is the potential for biases within the algorithms. In particular, bias that can worsen health disparities and bias by way of an algorithm's intent to either improve patient care or for financial gain. The final theme of the chapter discusses the algorithm's accuracy and the consequences of decisions made from its results. Specifically, the dilemma of the archetypal "black box" algorithm, liability in the face of adverse events, and the question of what to do when humans and machines disagree will be elucidated.

Underlying each theme is the emphasis for transparency in the algorithm and a need for a code of ethics and clear guidelines to promote patient safety and minimize harm. Most importantly, there is a large fear of AI's potential to replace the human physician and disrupt the physician-patient relationship (Cabitza, Rasoini and Gensini, 2017). This notion needs to be dispelled. While AI will undoubtedly become an integral component to the practice of medicine in the near future, the best results will occur when AI and humans work together.

Big data

The current state of big data

From whole-genome sequencing to wearable biometrics, physicians are now tasked with making sense of an enormous amount of data in an increasingly limited amount of time. Enter AI systems, well-oiled algorithms that are designed to interpret and operationalize larger volumes of health information in a fraction of the time it would take for a human physician. Using a machine learning based model, a personalized effect of health behavior on blood pressure can be derived (Chiang and Dey, 2018). Heart failure with preserved ejection fraction is no longer one disease, but instead composed of various phenotypically distinct groups (Shah et al., 2015). The promise for more personalized medicine is evident. This is especially true for cardiologists who tend to have access to more quantitative data compared to other specialties.

Yet, there should be a balance in data quantity and quality to ensure reliable outcomes. The three principles of data collection borrowed by marketing expert Seth Godin are as follows (Brown, 2019; Godin, 2019):

- "Don't collect data unless it has a non-zero chance of changing your actions."
- "Before you seek to collect data, consider the costs of processing that data."
- "Acknowledge that data collected isn't always accurate and consider the costs of acting on data that isn't correct."

These principles translate well to the use of artificial intelligence in healthcare. In the face of big data, information is of low utility unless it can improve clinical outcomes. To collect data for the sake of collecting data skews the balance of enormous costs outweighing the benefit of clinical outcomes. The distinction for accurate data should also be emphasized. Incorrect results and predictions in the realm of healthcare can lead to severe consequences, and meticulous quality reporting must be protocolized into AI development to ensure reproducible results. However, professional associations, government agencies, and AI developers may be unaware of how to proceed in this new era of uncharted AI territory.

The effect of AI on healthcare expenditure

The logistics of how AI can be integrated into medicine require a discussion of cost. Healthcare costs in the United States currently surpass all other countries. The United States spends over $10,000 per capita or 18% of its Gross Domestic Product (GDP) on healthcare with costs projected to increase at an annual rate of 4.1% from 2017 to 2021 (Centers for Medicare and Medicaid Services, 2018). With rising healthcare costs compounded to the flux of enormous data pools, healthcare organizations are forced to shift toward more personalized, value-based models. Hospitals are estimated to produce on the order of a petabyte of data a year, with human health data generated at multiple petabytes a day (Weintraub, 2019). Massive data collection requires massive energy consumption with significant financial implications due to the need to not only store data but also to restructure traditional healthcare models with emerging technologies. While the initial integration of AI technology may be costly, once integrated, some have predicted that AI applications can create $150 billion in annual savings for the healthcare economy by 2026 (Forbes, 2019). In addition, advancements in data storage and retrievability have helped lower costs. The majority of health data has already transitioned to cloud storage under Amazon Web Services, Microsoft Azure, and Google Cloud, with eventual transition to end-devices as soon to be discussed (Weintraub, 2019).

Privacy and security

In the transference of such large amounts of data, questions of confidentiality and security arise. An efficient and accurate AI system cannot be built in solidarity confined to a single hospital network. This is reminiscent of current electronic medical records that are limited to the institution's hospital network, which greatly impact the efficiency of outside hospital transfers and the like. Given how extensively patient data would be shared across an AI system, existing practices of obtaining consent would be unrealistic. The alternative option also proves difficult. The privacy rule in the Health Insurance Portability and Accountability Act (HIPAA) of 1996 allows the use of deidentified patient-level data without consent (Centers for Disease Control and Prevention, 2018). It could then theoretically be possible to allow patient data that has already been collected to be used without explicit consent, provided that the data is deidentified. Yet, it has been proven possible to reidentify a patient using as few as three data points (Osoba and Wesler, 2017). Additionally, AI algorithms, as complex as they are, are still vulnerable to cyberattacks (Finlayson, Chung, Johane, & Beam, 2018).

An incident involving Google DeepMind and the Royal Free London NHS Foundation Trust occurred in which identifiable patient records were transferred across the entire Trust without explicit patient consent (Powles and Hodson, 2017). While there is not yet an established solution to this very real threat to patient privacy and security, new technologies are being developed to counteract these challenges.

Google introduced Federated Learning in 2017, a "distributed machine learning approach which trains machine learning models using decentralized data residing on end devices such as mobile phones" (Zhang, 2019). As opposed to a centralized training approach where sensitive data is sent to the cloud to be analyzed, federated learning brings the algorithm to the data. A central server houses a global model which communicates with individual hospital systems as outlined in Fig. 17.1

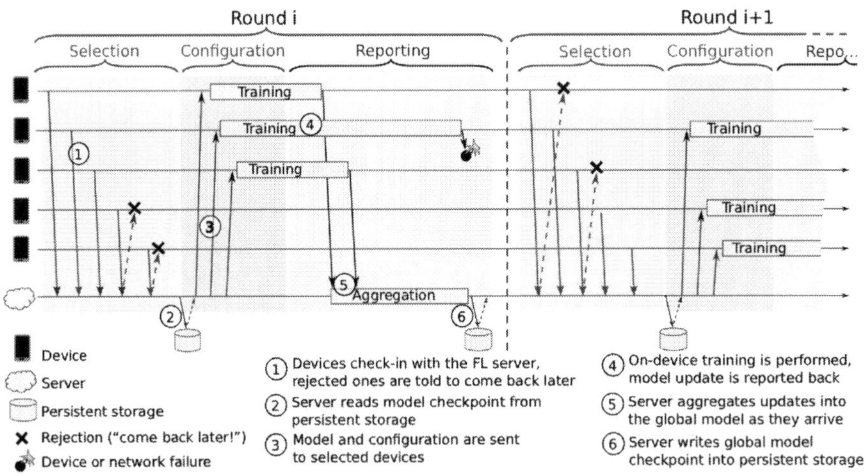

FIGURE 17.1 Federated Learning Protocol.

The communication protocol enables institutions (devices) to contribute to a single global machine learning model through successive computational rounds. Each round consists of three phases of data selection, configuration, and reporting which underlies constant communication between the model and the institution (Bonawitz et al., 2019).

(Bonawitz et al., 2019). This way, data does not leave the hospital and patient privacy is effectively preserved. Federated learning is a novel application with significant potential to safeguard patient privacy and security in the coming integration of AI. Of course, there are notable limitations to this new technology, including its dependency on communication bandwidth of participating hospital systems, storage and computational abilities of said systems, and variability and inconsistencies in the data being used across different hospitals (Kulkarni, 2019). Further, the complex nature of these expansive networks makes risk for data breach an even greater concern than ever before. For instance, it would be possible for a contributing client to intentionally contaminate the global model by contributing incorrect data (Kulkarni, 2019).

The issue with electronic health records

The issues with electronic health record (EHR) integration into AI systems should be discussed. Currently, EHRs are designed for billing, not clinical knowledge. Regulatory and billing requirements in the United States not only dampen EHR usability but also burden physicians with more administrative duties and contribute to physician burnout. In addition, EHRs are incomplete and ridden with errors. Data collection is highly variable, from vitals collected every 4–12 h, labs scheduled at inconsistent intervals, and chest X-rays and electrocardiograms (ECGs) ordered as needed to outdated medication and demographic data. Even worse, clinical notes are written in unstructured free-form text, awry with acronyms and misspelt words, and copy and pasting used in abundance to meet documentation requirements (Ghassemi, Naumann, Schulam, Beam, & Chen, 2019).

In order to effectively be used in AI systems, data derived from EHRs should ideally be streamlined and rid of inaccurate data, designed to be utilized as clinical knowledge, and kept up to date. If this can be accomplished, EHR data will serve as a rich source of information that can provide significant depth to our current medical knowledge. Given such paramount tasks, existing systems are integrating third-party AI vendors or installing in-house AI capabilities. Firms such as Epic, Cerner, Allscripts, and Athena have begun to add "natural language processing and machine learning capabilities for clinician decision support, integration with telehealth technologies, and automated imaging analysis" (Davenport, Hongsermeier and McCord, 2018).

Data sharing

Advancements in data sharing allow for the creation of data repositories on a global scale. International data harmonization initiatives such as the Observational Health Data Sciences and Informatics (OHDSI) program have shown success in aggregating individual datasets into large population-based analytics (2016). The development of these collaborative networks has created research studies of impressive sample sizes for use across multiple research centers, with hope to yield more accurate and generalizable results. Albeit a powerful tool, data harmonization is not without its limitations. The inconsistencies across institutional datasets have already been discussed. In addition, a feasibility study showed that only 15% of results from US-based clinical trials published in high-impact journals could be replicated through the use of real-world datasets (Bartlett, Dhruva, Shah, Ryan, & Ross, 2019). This may suggest that harmonization methods will unlikely replace the traditional clinical trial, but perhaps serve as an additional tool for conducting research. Further, multiple data harmonization repositories exist, and are not yet compatible with one another (Nordo et al., 2019).

Nonetheless, initial studies utilizing harmonization processes in healthcare are promising. The Biobank Standardisation and Harmonisation for Research Excellence in the European Union (BioSHaRE) project is an example of successful synchronization of data across eight databases that used both data harmonization and federated learning in evaluating healthy, obese individuals (Doiron et al., 2013). Federated learning, as previously discussed, is a unique privacy-preserving modality that also offers seamless cross-collaborative data exchange. This is illustrated in a study by Brisimi et al. (2018) in which federated learning achieved comparable accuracy to traditional centralized analytic models in predicting rehospitalization for patients with cardiovascular disease. Certainly, these two technologies will open the doors to unparalleled performance capabilities in cardiovascular research.

Data quality and quantity

Reductionist thinking in medicine was dispelled around 20 years ago when genome sequencing was thought to be the best method for predicting disease. It was assumed that the presence of a gene or genetic mutation was the key to finding disease. This reduction of a pathologic state to a molecular entity obviated environmental, developmental, and epigenetic factors that as we now know are large contributors to disease. The complexity of the natural world relies not only on these additional factors, but also the relationships between them. Holism, the antagonist to reductionism, is the concept that all variables matter. In the development of AI for cardiovascular application, it is easy to see how difficult it is to find the right balance of reductionism and holism—which variables matter? The accuracy of an AI algorithm to predict clinical outcomes is highly dependent on the data quality and quantity being inputted. A study using roughly 250 variables showed that three separate machine learning algorithms were not able to predict 30-day remission better than traditional logistic regression (Frizzell et al., 2017). Data with too much noise or too few of the important variables could potentially contribute to results such as this. AI models tend to overfit the dataset due to this difficulty to distinguish a true contributing factor from noise. An overfitted model shows high performance in the training data that is then not applicable to a new dataset. This is largely due to the model's tendency to capture noise that interferes with the ability to identify a true pattern in the data (Kagiyama, Shrestha, Farjo, & Sengupta, 2019). In addition, AI models have difficulty identifying and isolating confounders in their analysis. Results gleaned from

AI systems should be scrutinized and verified over trials of datasets to minimize overfitting. In light of this new age of enormous data quantity, a strong ethical framework and regulatory guidelines are necessary to ensure high quality datasets in order to yield reliable and reproducible results (Rockall, 2020).

The algorithm

Risk to worsen health disparities

Predictive AI models should be reflective of the entire population to create representative models of disease presentation and response to intervention or medical therapy (Ghassemi et al., 2018). One concern regarding the use of these technologies is the potential bias in the algorithm that may ultimately worsen the divide in health equity. Outside of medicine, AI systems have been shown to exaggerate racial bias in predictive policing, bail decisions, and credit scoring (Angwin, Larson, Mattu, & Kirchner, 2016; Verghese, Shah and Harrington, 2018). Current polygenic risk scores used to predict complex traits from genetic data perform far better in European-descent populations than other ancestries (Martin et al., 2019). A recent study by Obermeyer, Powers, Vogeli and Mullainathan (2019) found that an AI algorithm gave white patients higher risk scores and white patients were therefore more likely to be selected to receive extra care than black patients that were equally as sick. Similarly, a machine learning algorithm found that the number one predictor for prolonged hospital stay in Chicago was the patient's postal code, with low-income and predominantly African American neighborhoods correlated to longer hospital stays (Nordling, 2019).

The overarching problem is in the dataset training the algorithm. Not only is the data incomplete, with little representation from women and minority groups, but it is also well-ingrained with real-world biases. Historically, women, minority groups, and obese patients are underrepresented in research trials and as a result have worse health outcomes (Johnson, 2013; Phelan et al., 2015). The classic presentation of chest pain that radiates to the neck and left arm is applicable to the white middle-aged male but does not often present in such manner in other groups. If the algorithm were to screen for acute coronary syndrome in a Hispanic female, which often presents with different symptoms, there is a risk to not only miss the critical diagnosis but to further worsen these injustices in healthcare. Secondly, current electronic health records are ridden with bias with risk to perpetuate economic and social inequities. For instance, if patients with lower incomes "do worse after organ transplantation … machine learning algorithms may conclude such patients are less likely to benefit from further treatment—and recommend against it" (Khullar, 2019). Psychiatric clinical notes are narratives by diagnostic and therapeutic necessity, with a high degree of subjective information. When such unstructured data was analyzed using machine learning, insurance status was found to be the primary predictor for 30-day psychiatric readmission, irrespective of number of previous hospitalizations or severity of disease (Chen, Szolovits and Ghassemi, 2019). If an algorithm was designed to maximize efficiency or lower medical costs, it may discourage treatment or alter treatment regimens based on insurance policy.

AI algorithms should be validated on various minority and low-income groups to ensure accurate outcomes. Yet, to get access to this type of data is no easy task. To do so raises the concern of data protection and safety for vulnerable populations. Medical data is sensitive, and "information such as HIV status could be used to discriminate against certain populations" (Nordling, 2019). Further, the benefits of receiving advanced AI technology should not be limited to those that can pay, and instead be distributed equally to all. Use of community advocates in collaboration with AI programmers could help ensure that these inequalities areremoved. Governing structures that would supervise in the architecture of these predictive models and outcomes could include representatives of minority groups. Rigorous standards for validation should be in place to prevent misclassification of patients for when it really matters—at the patient's bedside where decisions are made. Exemplary of this is the recently introduced Algorithmic Accountability Act which "requires companies to study the algorithms they use, identify bias in these systems, and fix any discrimination or bias they find" (Ron Wyden United States Senator for Oregon, 2019). If these biases can be appropriately addressed, AI may help bridge health disparities that still remain from the days of traditional medical research.

The Algorithm's intent

An additional source of bias is in the intent of the algorithm. An algorithm designed to reduce costs will yield far different results than an algorithm designed to reduce mortality. Conflicts of interest are apparent in this coined approach to either generate profit as opposed to improving health outcomes in this very delicate relationship between the algorithm's architects, the purchasers of the machine, and the users of the machine. There is a clear need for transparency in the algorithm's design for all involved parties. At present, cardiovascular disease is the number one cause of death worldwide, with care especially lacking in lower-income countries (Rosengren et al., 2019). And, for a disease so widespread, the integration of AI has potential to create serious harm worldwide if strict measures to prevent bias are not implemented. If these barriers to bias can be alleviated, the use of AI has great potential in managing and treating cardiovascular diseases.

The results

The black box of AI

The "black box" of AI represents the lack of explanation regarding the machine's diagnostic and therapeutic conclusions. Whereas first-generation algorithms were unable to adapt beyond the strict parameters defined by the human programmer, advancements in deep learning allow for machines to create their own internal rules and patterns to provide predictions. Preliminary machine models used in ECG analysis, for instance, were once limited to discrete ST segment changes or degree of PR prolongation. Now, deep neural networks use ECGs to predict risk for developing arrhythmia and even risk of death within 1 year, as demonstrated in a recent study by Raghunath et al. (2019). The model was able to identify subtle changes to an ECG that would otherwise be missed and regarded as normal to the trained human eye.

Given the complexity of these algorithms, the exact programming and methodology of how such results arise and what these minute details within an ECG are are difficult to extract. Without "the knowledge within the machine," it would be inherently difficult for a clinician and the patient to carry forth with a therapy without an explanation as to *why* or *how* the particular therapy was chosen by the algorithm (Castelvecchi, 2016; Weinberger, 2017). Results should not be accepted to guide clinical decision-making unless validated by a clinician. The chain of logic in how the decision arose from the black box of convoluted neural networks must be deduced by the user of AI in their practice of medicine (Verghese et al., 2018). What the computer predicts, the physician should explain and ultimately decide with the patient, the best course of action. It is essential for the clinician to retain the prerogative of questioning the inputs and outputs of AI to preserve patient safety. Unfortunately, the more advanced the AI system, the harder it is to explain. For instance, probability-based Bayesian analyses are relatively more straightforward to understand, whereas deep learning networks are much more difficult to tease apart, as previously mentioned. The replacement of a prediction with an analysis that both clinicians and patients can understand would be a tremendous breakthrough toward optimizing their clinical use.

Autonomy and informed consent

While the AI designer understands the architecture of a deep learning system as it is being developed, what happens after the algorithm is developed is unpredictable. To receive a decision from this "black box" begs the question of both patient autonomy and liability in the case of an adverse event. In contrast to the physician describing the risks and benefits of each medical treatment option that the patient has, a single decision arises out of a black box without explanation. In this manner, the patient is left incompletely informed and almost psychologically obliged to follow suit an algorithm's command. This then begs the question of patient autonomy. As we know, autonomy is one of the four basic principles of medical ethics. It is the patient's right to thought, intention, and action when making decisions regarding their own health, and must be free of coercion or coaxing (Tauber, 2005). A decision made on behalf of the patient as the theoretically calculated "best course of action" by a machine risks diminishing the patient's autonomy.

The black box archetype combined with threat to patient autonomy complicates the informed consent process. In addition to the traditional aspects enclosed in an informed consent—purpose of treatment, potential risks and benefits, alternative treatments—the technical sophistication of AI has to be explained by the physician who fully understands the risks, possible biases, and error rates of the machine. In essence, the patient is asked to trust a new element of the unknown. Specifically, to trust a new technology that cannot explain its reasoning and lacks years of trusted use to support its use in making critical health decisions. AI users must have full transparency of the algorithm's capabilities and limits to close the gap of the black box phenomenon. In the progression of AI technologies, it is essential to regard patient care at the center of its development.

Medical malpractice and product liability

In the face of possible medical malpractice and product liability that results during the use of AI technology, who is then responsible? Currently, liability, as described by Yavar Bathaee "is built on legal doctrines that are focused on human conduct, which when applied to AI, may not function" (2017). In the face of a self-learning program, the designers of a said program are not able to foresee how the program will respond in real-world context. In this system of unforeseeable events, assigning responsibility and liability when errors occur becomes exceedingly complex, as liability cannot be so easily placed on the AI designer or the physician. Determining who is legally responsible with such technology is complicated by the "problem of many hands" (PricewaterhouseCooper, 2017). This problem refers to the challenge of having underlying moral responsibility distributed among multiple persons in a way that makes blame difficult to localize. This could allow individuals or organizations to "evade personal responsibility for wrongdoing" (Thompson, 1980).

Critical factors that the court or jury would consider in determining physician liability include: familiarity with the benefits and limitations of AI's use as well as the data used to train the algorithm, and if the algorithm's decision was substituted for their own clinical judgment. Extending beyond the physician using the technology is the hospital or network that could potentially be responsible for the integration of the AI software without proper investigation of its usability. Likely, rigorous screening and strict guidelines for safety and efficacy will become a part of their integration as AI further develops. Lastly, the AI company could be liable on multiple fronts, from design flaws, false advertising, negligent training, and lack of supervision (Bromfield and McMahon, 2019). However, courts have been reluctant to extend product liability to healthcare software, since previously software and the physician as decision-maker were distinct entities (Yu, Beam, & Kohane, 2018), compared to this new age of the "black-box" algorithm wherein lies a product with decision-making capacity (Price, 2017). Currently, numerous healthcare-based AI companies exist with willful attempts for rapid success among competitors. They must be scrutinized for usability and reliability to prevent premature integration. Especially since most algorithms remain proprietary, the ability to evaluate their quality and accuracy is significantly limited. Thus, AI systems should be standardized using the following principles (Francesca, 2016; Keskinbora, 2019):

- Transparency (operations visible to user)
- Credibility (outcomes are acceptable)
- Auditability (efficiency can be easily measured)
- Reliability (AI systems perform as intended)
- Recoverability (manual control can be assumed if required)

There is an imminent call for legal structure to better assign responsibility for medical errors. At present, irrespective of the remarkable potential shown thus far in AI technology, our current systems are limited to translational applicability due to the enigma of the "black box," questions of autonomy, safety, and liability.

When humans and machines disagree

An interesting dilemma arises when humans and machines disagree. Who can ultimately make better decisions? Human errors in judgment and decision-making can be broadly understood as either errors of bias or errors of noise (Kahneman, Rosenfield, Gandhi, & Blaser, 2016). Examples of bias include stereotyping (social bias) and overconfidence (cognitive bias). Errors of noise refer to decisions that are influenced by irrelevant entities such as time of day, mood, weather, or state of hunger. New technologies in deep thinking and machine learning are precisely designed to make decisions accurately, methodically calculating the best course of action without the sway of external factors. Yet research by Damasio's group shows this ability may not exactly translate well to appropriate real-world decision-making. In individuals that sustained brain damage (to the extent of an AI equivalent), it has been shown that while they do methodically consider all options in coming to a decision, their decisions were not necessarily better (Damasio, 2010). In fact, some of their decisions were easily recognized as inappropriate or unethical by healthy individuals. Perhaps, human intuition for empathy and moral responsibility are what delineate humans from the machine in the race of superiority. This inclination is highly specific to human nature and is a core aspect of what defines us. There is something to be said for the "art of medicine" and compassionate care that can only be provided by a human physician and cannot be substituted by a machine. The best potential that AI can deliver in medicine is to free physicians from the automated tasks of current healthcare demands while allowing more time to be spent with the patient.

With AI systems already surpassing human performance in several domains, there is great fear in their potential to replace the human physician. Ultimately, the best results occur when humans and machines work together as advocated for by AI proponent Eric Topol (2019a). In the interpretation of breast cancer screening mammograms, AI systems were found to be noninferior to the average radiologist, but not better than the most skilled radiologist. However, the combination of AI and radiologists in a hybrid model was significantly more accurate than either of the two separately (Wu et al., 2019). In the analysis of cardiac magnetic resonance imaging (MRI) scans for surgical planning, automated machine learning was found to be significantly faster and with comparable accuracy to human interpretation by trained cardiologists (Bhuva et al., 2019). Specifically, AI took about 4 seconds to interpret a cardiac MRI compared to 13 minutes which was the average time for a cardiologist in the study. Acceleration of data gleaned from AI systems may improve the speed of clinical decision-making, allocating more time to the patient's bedside. In fact, the investigators of this study calculated that using AI to interpret scans could potentially save 54 work-days annually, per hospital system. While this is certainly a benefit, the priority of the physician is to assure no harm comes to the patient. When human and machine are combined, the anxiety of false positive readings can be reduced. A cardiologist may be excellent in detecting what he or she expects to find but may miss what is not expected. This is the gap that AI systems can fill.

Conclusions

The role of AI is thus not to replace the cardiologist, but to augment the cardiologist's role as an advocate for patient care through time- and cost-saving benefits. Simple, repetitive tasks from electronic medical record documentation to the analysis of large quantities of data currently make up the majority of a physician's day to day. If physicians can be alleviated of these automated tasks with the integration of AI, more time can be spent enjoying the thrills of complex problem-solving with a return of meaning and purpose to the practice of medicine. With its ability to process massive datasets quickly, accurately, and inexpensively, AI holds potential to not replace, but to enhance the patient-physician relationship.

Of course, clear guidelines and a code of ethics must be placed to ensure there is no loss of patient integrity with maximal preservation of human rights. The U.S. Food and Drug Administration (FDA) has already issued a statement on a proposed regulatory framework that would allow "the FDA and manufacturers to evaluate and monitor a software product from its premarket development to postmarket performance" (2019). The FDA will also be holding its first public workshop in early 2020 to explore the risks and benefits of AI in radiology (Stempniak, 2019). In addition, a joint European and North American multisociety statement on ethics of AI in radiology was recently released (Geis et al., 2019). Furthermore, the White House proposed a set of 10 principles for governing the development of AI solutions in the United States: "public trust in AI, public participation, scientific integrity and information quality, risk assessment and management, benefits and costs, flexibility, fairness and nondiscrimination, disclosure and transparency, safety and security, and interagency coordination" (Walter, 2019).

While significant progress has been made, there are still many details to figure out as AI enters the realm of cardiovascular medicine. These questions of safety and efficacy—who is at fault in the event of medical error, its defined role in the physician's practice—have to be fully addressed before AI's integration into the practice of medicine. In the use of these new technologies, the most fundamental question that must be asked is—how will this benefit patient care? Ultimately, the core principles of medical ethics, including respect for persons, beneficence, nonmaleficence, and justice must be prioritized in the integration of AI technologies in the future of cardiovascular medicine.

References

Angwin, J., Larson, J., Mattu, S., & Kirchner, L. (2016). Machine bias. *ProPublica*. Retrieved from https://www.propublica.org/article/machine-bias-risk-assessments-in-criminal-sentencing.

Bartlett, V. L., Dhruva, S. S., Shah, N. D., Ryan, P., & Ross, J. S. (2019). Feasibility of using real-world data to replicate clinical trial evidence. *JAMA Network Open, 2*(10), e1912869.

Bathaee, Y. (2017). The artificial intelligence black box and the failure of intent and causation. *Harvard Journal of Law & Technology, 31*, 889.

Bhuva, N., Bai, W., Lau, C., Davies Rhodri, H., Ye, Y., Bulluck, H., et al. (2019). A multicenter, scan-rescan, human and machine learning CMR study to test generalizability and precision in imaging biomarker analysis. *Circulation: Cardiovascular Imaging, 12*(10), e009214.

Bonawitz, K., Eichner, H., Grieskamp, W., Huba, D., Ingerman, A., Ivanov, V., et al. (2019). Toward federated learning at scale: System design. *arXiv*. Retrieved from https://ui.adsabs.harvard.edu/abs/2019arXiv190201046B.

Brisimi, T. S., Chen, R., Mela, T., Olshevsky, A., Paschalidis, I. C., & Shi, W. (2018). Federated learning of predictive models from federated electronic health records. *International Journal of Medical Informatics, 112*, 59–67.

Bromfield, A., & McMahon, K. (2019). Artificial intelligence in healthcare: Preparing for a new legal liability landscape. *IronShore*. Retrieved from https://viewpoint.libertymutualgroup.com/wp-content/uploads/2019/09/LM-IS_Artificial-Intelligence_Healthcare.pdf.

Brown, E. (2019). *3 reasons data hoarding may not pay off*. The Enterprisers Project. Retrieved from https://enterprisersproject.com/article/2019/3/3-reasons-data-hoarding-may-not-pay. (Accessed 20 December 2019).

Cabitza, F., Rasoini, R., & Gensini, G. F. (2017). Unintended consequences of machine learning in medicine. *The Journal of the American Medical Association, 318*(6), 517–518.

Castelvecchi, D. (2016). Can we open the black box of AI? *Nature, 538*(7623), 20–23.

Centers for Disease Control and Prevention. (2018). *Health Insurance Portability and Accountability Act of 1996 (HIPAA)*. Retrieved from https://www.cdc.gov/phlp/publications/topic/hipaa.html. (Accessed 5 December 2019).

Centers for Medicare and Medicaid Services. (2018). *National health expenditure data*. Retrieved from https://www.cms.gov/Research-Statistics-Data-and-Systems/Statistics-Trends-and-Reports/NationalHealthExpendData. (Accessed 15 December 2019).

Chen, I. Y., Szolovits, P., & Ghassemi, M. (2019). Can AI help reduce disparities in general medical and mental health care? *AMA Journal of Ethics, 21*(2), E167–E179.

Chiang, P.-H., & Dey, S. (2018). Personalized effect of health behavior on blood pressure: Machine learning based prediction and recommendation. In *IEEE 20th international conference on e-health networking, applications and services*.

Damasio, A. (2010). *Self comes to mind: Constructing the conscious brain*. New York: Pantheon.

Davenport, T. H., Hongsermeier, T. M., & McCord, K. A. (2018). Using AI to improve electronic health records. *Harvard Business Review*. Retrieved from https://hbr.org/2018/12/using-ai-to-improve-electronic-health-records.

Doiron, D., Burton, P., Marcon, Y., Gaye, A., Wolffenbuttel, B. H. R., Perola, M., et al. (2013). Data harmonization and federated analysis of population-based studies: The BioSHaRE project. *Emerging Themes in Epidemiology, 10*(1), 12.

Finlayson, S. G., Chung, H. W., Johane, I. S., & Beam, A. L. (2018). Adversarial attacks against medical deep learning systems. *arXiv*. Retrieved from https://arxiv.org/abs/1804.05296.

Forbes. (2019). *AI and healthcare: A giant opportunity*. Retrieved from https://www.forbes.com/sites/insights-intelai/2019/02/11/ai-and-healthcare-a-giant-opportunity/#67534cea4c68. (Accessed 20 December 2019).

Francesca, R. (2016). Artificial intelligence: — potential benefits and ethical considerations — study. *Artificial Intelligence*, 2018.

Frizzell, J. D., Liang, L., Schulte, P. J., Yancy, C. W., Heidenreich, P. A., Hernandez, A. F., et al. (2017). Prediction of 30-day all-cause readmissions in patients hospitalized for heart failure: Comparison of machine learning and other statistical approaches. *JAMA Cardiology, 2*(2), 204—209.

Geis, J. R., Brady, A. P., Wu, C. C., Spencer, J., Ranschaert, E., Jaremko, J. L., et al. (2019). Ethics of artificial intelligence in radiology: Summary of the joint European and North American multisociety statement. *Radiology, 293*(2), 436—440.

Ghassemi, M., Naumann, T., Schulam, P., Beam, A. L., & Chen, I. Y. (2019). Practical guidance on artificial intelligence for health-care data. *The Lancet*, 157—159.

Ghassemi, M., Naumann, T., Schulam, P., Beam, A., Chen, I., & Ranganath, R. (2018). A review of challenges and opportunities in machine learning for health. *arXiv*. Retrieved from https://arxiv.org/abs/1806.00388.

Godin, S. (2019). *Data is expensive*. Seth's Blog. Retrieved from https://seths.blog/2019/02/the-am-pm-problem-the-curse-of-too-much-data/.

Johnson, K. S. (2013). Racial and ethnic disparities in palliative care. *Journal of Palliative Medicine, 16*(11), 1329—1334.

Kagiyama, N., Shrestha, S., Farjo, P. D., & Sengupta, P. P. (2019). Artificial intelligence: Practical primer for clinical research in cardiovascular disease. *Journal of the American Heart Association, 8*(17), e012788.

Kahneman, D., Rosenfield, A., Gandhi, L., & Blaser, T. (2016). Noise: How to overcome the high, hidden cost of inconsistent decision making. *Harvard Business Review*, 36—43.

Keskinbora, K. H. (2019). Medical ethics considerations on artificial intelligence. *Journal of Clinical Neuroscience, 64*, 277—282.

Khullar, D. (2019). A.I. Could worsen health disparities. *The New York Times*. Retrieved from https://www.nytimes.com/2019/01/31/opinion/ai-bias-healthcare.html.

Kulkarni, V. (2019). Federated learning for medical AI. *Towards Data Science*. Retrieved from https://towardsdatascience.com/federated-learning-for-medical-ai-65d0daeafe8b.

Martin, A. R., Kanai, M., Kamatani, Y., Okada, Y., Neale, B. M., & Daly, M. J. (2019). Clinical use of current polygenic risk scores may exacerbate health disparities. *Nature Genetics, 51*(4), 584–591.

Nordling, L. (2019). A fairer way forward for AI in health care. *Nature, 7775*, S103–s105.

Nordo, A. H., Levaux, H. P., Becnel, L. B., Galvez, J., Rao, P., Stem, K., et al. (2019). Use of EHRs data for clinical research: Historical progress and current applications. *Learning Health Systems, 3*(1), e10076.

Obermeyer, Z., Powers, B., Vogeli, C., & Mullainathan, S. (2019). Dissecting racial bias in an algorithm used to manage the health of populations. *Science, 366*(6464), 447–453.

Ohdsiorg. (2016). *Observational health data sciences and informatics*. Retrieved from https://www.ohdsi.org/. (Accessed 2 January 2020).

Osoba, O. A., & Wesler, W. I. (2017). *An intelligence in our image: The risks of bias and errors in artificial intelligence*. RAND Corporation. Retrieved from https://www.rand.org/pubs/research_reports/RR1744.html. (Accessed 20 December 2019).

Phelan, S. M., Burgess, D. J., Yeazel, M. W., Hellerstedt, W. L., Griffin, J. M., & van Ryn, M. (2015). Impact of weight bias and stigma on quality of care and outcomes for patients with obesity. *Obesity Reviews, 16*(4), 319–326.

Powles, J., & Hodson, H. (2017). Google deep mind and healthcare in an age of algorithms. *Health Technology, 7*(4), 351–367.

Price, W. N. (2017). Artificial intelligence in health care: Applications and legal implications. *The SciTech Lawyer, 17*(7), 10–13.

PricewaterhouseCoopers. (2017). *What doctor? Why AI and robotics will define new health: Data explorer.* Retrieved from https://www.pwc.com/gx/en/industries/healthcare/publications/ai-robotics-new-health/data-explorer.html#!/D/18/stackedbars?cut=Territory&Tecf=0. (Accessed 2 January 2020).

Raghunath, S., Cerna, A., Jing, L., vanMaanen, D., Stough, J., Hartzel, D., et al. (2019). Deep neural networks can predict 1-year mortality directly from ECG signal, even when clinically interpreted as normal. In *American heart association scientific sessions 2019.*

Rockall, A. (2020). From hype to hope to hard work: Developing responsible AI for radiology. *Clinical Radiology, 75*(1), 1–2.

Ron Wyden United States Senator for Oregon. (2019). *Wyden, Booker, Clarke introduce bill requiring companies to target bias in corporate algorithms*. Retrieved from https://www.wyden.senate.gov/news/press-releases/wyden-booker-clarke-introduce-bill-requiring-companies-to-target-bias-in-corporate-algorithms-. (Accessed 8 January 2020).

Rosengren, A., Smyth, A., Rangarajan, S., Ramasundarahettige, C., Bangdiwala, S. I., AlHabib, K. F., et al. (2019). Socioeconomic status and risk of cardiovascular disease in 20 low-income, middle-income, and high-income countries: The prospective urban rural epidemiologic (PURE) study. *Lancet Global Health, 7*(6), e748–e760.

Shah, S. J., Katz, D. H., Selvaraj, S., Burke, M. A., Yancy, C. W., Gheorghiade, M., et al. (2015). Phenomapping for novel classification of heart failure with preserved ejection fraction. *Circulation, 131*(3), 269–279.

Stempniak, M. (2019). FDA to probe public risks of AI's ever-growing role in radiology. *Radiology Business*. Retrieved from https://www.radiologybusiness.com/topics/artificial-intelligence/fda-probe-public-risks-ais-role-radiology?utm_source=newsletter&utm_medium=rb_monthly.

Tauber, A. (2005). *Patient Autonomy and the ethics of responsibility* (17th ed.). MIT Press.

Thompson, D. F. (1980). Moral responsibility of public officials: The problem of many hands. *American Political Science Review, 74*(4), 905–916.

Topol, E. (2019a). *Deep medicine: How artificial intelligence can make healthcare human again*. New York: Basic Books.

Topol, E. J. (2019b). High-performance medicine: The convergence of human and artificial intelligence. *Nature Medicine, 25*(1), 44–56.

U.S. Food and Drug Administration. (2019). *Artificial intelligence and machine learning in software as a medical device*. Retrieved from https://www.fda.gov/medical-devices/software-medical-device-samd/artificial-intelligence-and-machine-learning-software-medical-device. (Accessed 8 January 2020).

Verghese, A., Shah, N. H., & Harrington, R. A. (2018). What this computer needs is a physician: Humanism and artificial intelligence. *The Journal of the American Medical Association, 319*(1), 19–20.

Walter, M. (2019). *Wyden, Booker share concerns about the use of AI in healthcare*. AI in Healthcare. Retrieved from https://www.aiin.healthcare/topics/policy/wyden-booker-senators-about-use-ai-healthcare?utm_source=newsletter&utm_medium=rb_news. (Accessed 8 January 2020).

Weinberger, D. (2017). Our machines now have knowledge we'll never understand. *Wired*. Retrieved from https://www.wired.com/story/our-machines-now-have-knowledge-well-never-understand/.

Weintraub, W. S. (2019). Role of big data in cardiovascular research. *Journal of the American Heart Association, 8*(14), e012791.

Wu, N., Phang, J., Park, J., Shen, Y., Huang, Z., Zorin, M., et al. (2019). Deep neural networks improve radiologists' performance in breast cancer screening. *IEEE Transactions on Medical Imaging*, 1184–1194.

Yu, K. H., Beam, A. L., & Kohane, I. S. (2018). Artificial intelligence in healthcare. *Nature Biomedical Engineering, 2*(10), 719–731.

Zhang, M. (2019). Federated learning: The future of distributed machine learning. *Synced Review*. Retrieved from https://medium.com/syncedreview/federated-learning-the-future-of-distributed-machine-learning-eec95242d897.

Glossary

A

ACC/AHA American College of Cardiology/American Heart Association.

Accelerometer A device that measures proper acceleration.

Adjunctive therapy One or more secondary interventions used concurrently with a primary intervention to enhance treatment effectiveness.

Algorithm A process or set of rules to be followed for calculations or other problem-solving operations, especially by a computer.

Altimeter An instrument used to measure the altitude of an object above a fixed level.

Amyloidosis The buildup of amyloid proteins in the heart, kidneys, liver, or other organs.

Angiography A medical imaging technique used to visualize the inside, or lumen, of blood vessels and organs of the body, with a particular interest in the arteries, veins, and the heart chambers.

Annotations A note of explanation or comment added to a text or diagram.

Anomalies Something that deviates from what is standard, normal, or expected.

Armamentarium A collection of resources available for a certain purpose.

Arrhythmia Improper beating of the heart, whether irregular, too fast, or too slow.

Artificial intelligence The theory and development of computer systems that are able to perform tasks that normally require human intelligence, such as visual perception, speech recognition, decision-making, and translation between languages.

Artificial neural networks (ANN) Computing systems vaguely inspired by the biological neural networks that constitute human brains.

ASCVD Atherosclerotic cardiovascular disease.

ASCVD risk score A calculation of 10-year risk of having a cardiovascular event, such as a heart attack or stroke.

Atrial fibrillation An irregular, often rapid heart rate that commonly occurs as a result of disruption of organized atrial activation.

Attenuation The reduction of the amplitude of a signal, electric current, or other oscillation.

B

Bayesian learning A particular set of approaches to probabilistic machine learning.

Biomarker A measurable indicator of the severity or presence of some disease state.

Boltzmann machines	A type of stochastic recurrent neural network.
Boosted trees	A machine learning technique for regression and classification problems.
Bootstrapping	Any test or metric that uses random sampling with replacement, and falls under the broader class of resampling methods.

C

Calibration	A procedure in statistical classification performed to determine class membership probabilities that assess the uncertainty of a given new observation belonging to each of the already established classes.
Cardiac decompensation	A combination of symptoms and signs that indicate that the heart because of its abnormal condition no longer is able to maintain an efficient circulation.
Cardiomyopathy	An acquired or hereditary disease of the heart muscle that makes it harder for the heart to pump blood to the rest of the body.
Cardiovascular medicine	The branch of medicine that deals with the diagnosis and treatment of diseases and disorders of the heart and blood vessels.
Caveat	A warning or proviso of specific stipulations, conditions, or limitations.
Chromosome	A threadlike structure of nucleic acids and protein found in the nucleus of most living cells, carrying genetic information in the form of genes.
Cine images	Short movies that are able to show heart motion throughout the cardiac cycle.
Consensus label	Labeling training data multiple times and only using that data if all the labelers agree on the label of the data.
Creatinine	A breakdown product of creatine phosphate from muscle and protein metabolism. It is usually measured in clinical practice to determine kidney function.
Curating	(Curate) Select, organize, and look after the items in (a collection or exhibition).

D

Data augmentation	A strategy that enables practitioners to significantly increase the diversity of data available for training models, without actually collecting new data.
Decision trees	A flowchart-like structure in which each internal node represents a "test" on an attribute (e.g., whether a coin flip comes up heads or tails), each branch represents the outcome of the test, and each leaf node represents a class label.
Deep learning	A subset of machine learning where artificial neural networks, algorithms inspired by the human brain, learn from large amounts of data.
Denoising	The process of removing noise from a signal.
Dermoscopic	Refers to the examination of the skin using skin surface microscopy.
DICE coefficient	A statistic used to gauge the similarity of two samples.

Dimensionality	The minimum number of coordinates needed to specify a point.
Downsampling	The process of reducing the sampling rate of a signal, or the sample size of a cohort.

E

ECG	Electrocardiogram, a test that measures the electrical activity of the heart.
Echocardiography	The use of ultrasound waves to investigate the action of the heart.
Edema	A condition characterized by an excess of watery fluid collecting in the cavities or tissues of the body.
Ejection fraction	A measurement, expressed as a percentage, of how much blood the left ventricle pumps out with each contraction.
Electrophysiology	The study of the electrical properties of biological cells and tissues.
Embeddings	An embedding is a mapping of a discrete, categorical variable to a vector of continuous numbers.
En masse	In a group; altogether.
Endocarditis	An infection of the heart's inner lining, usually involving the heart valves.
Ensemble models	A process where multiple diverse models are created to predict an outcome, either by using many different modeling algorithms or using different training data sets.
Eosinophilia	A high level of disease-fighting white blood cells known as eosinophils in the blood.
Epigenomics	The study of the complete set of epigenetic modifications on the genetic material of a cell, known as the epigenome.

F

F1 score	A measure of a test's accuracy in statistical analysis.
Fourier transform	A function derived from a given function and representing it by a series of sinusoidal functions.
Fragility	The quality of being easily broken or damaged.
Framingham risk score	A gender-specific algorithm used to estimate the 10-year cardiovascular risk of an individual.

G

Gabor transform	A special case of the short-time Fourier transform. It is used to determine the sinusoidal frequency and phase content of local sections of a signal as it changes over time.
Gamification	The application of game-design elements and game principles in nongame contexts.
Gaussian kernel	A weighted linear combination of the kernel function computed between a data point and each of the support vectors.
Gaussian model	A stochastic process (a collection of random variables indexed by time or space), such that every finite collection of those random variables has a multivariate normal distribution.

Genomics The branch of molecular biology concerned with the structure, function, evolution, and mapping of genomes.

Gradient boosting A machine learning technique for regression and classification problems, which produces a prediction model in the form of an ensemble of weak prediction models.

H

Heart failure (HF) A condition in which the heart cannot pump enough blood to meet the body's needs.

Hemodynamics The dynamics of blood flow.

Hemoglobin A red protein responsible for transporting oxygen in the blood of vertebrates. Its molecule comprises four subunits, each containing an iron atom bound to a heme group.

Heterogeneous Diverse in character or content.

Heuristics Any approach to problem solving or self-discovery that employs a practical method that is not guaranteed to be optimal, perfect, or rational, but which is nevertheless enough for reaching an immediate, short-term goal.

Hidden Markov models A finite set of states, each of which is associated with a (generally multidimensional) probability distribution.

Hierarchical clustering An algorithm that groups similar objects into groups called clusters.

High-dimensional data The number of dimensions is staggeringly high, so high that calculations become extremely difficult.

Histopathology Refers to the microscopic examination of tissue to study the manifestations of a disease.

HITECH Health Information Technology for Economic and Clinical Health Act.

HITECH Act An act was created to motivate the implementation of electronic health records (EHR) and supporting technology in the United States.

Holism The idea that various systems should be viewed as wholes, not merely as a collection of parts.

Hyperkalemia A high level of the electrolyte potassium in the blood.

Hypertension Elevated blood pressure.

Hypertrophy The increase in the volume of an organ or tissue due to the enlargement of its component cells.

Hypoechoic A hypoechoic mass is a tissue in the body that is more dense or solid than usual.

I

Image texture A set of metrics calculated in image processing designed to quantify the perceived texture of an image. Image texture gives information about the spatial arrangement of color or intensities in an image or selected region of interest.

Incidence rate It is a measure of the frequency with which some event, such as a disease or accident, occurs over a specified time period.

Inference time Inference times are often expressed as "X + Y," in which X is the time taken in reasonably well-optimized GPU code and Y is the time taken in unoptimized CPU code.

Infrastructure The fundamental facilities and systems serving a country, city, or other area, including the services and facilities necessary for its economy to function.

Interpretability A relation between formal theories that expresses the possibility of interpreting or translating one into the other.

Intervention Action taken to improve a situation, especially a medical disorder.

Interventional cardiology A branch of cardiology that deals specifically with the catheter-based treatment of structural heart diseases.

Intravascular Situated or occurring within a vessel or vessels of an animal or plant, especially within a blood vessel or blood vascular system.

Iterative The repetition of a process to generate a sequence of outcomes.

J

Jaccard index A statistic used for gauging the similarity and diversity of sample sets.

Jensen-Shannon divergence A method of measuring the similarity between two probability distributions.

K

K-space An extension of the concept of Fourier space; well known in MR imaging.

Kullbach–Liebler divergence A measure of how one probability distribution is different from a second, reference probability distribution.

L

Lasso models A regression analysis method that performs both variable selection and regularization to enhance the prediction accuracy and interpretability of the statistical model it produces.

Late gadolinium enhancement A useful tool for scar detection, based on differences in the volume of distribution of gadolinium, an extracellular agent.

Latent Dirichlet allocation A generative statistical model that allows sets of observations to be explained by unobserved groups that explain why some parts of the data are similar.

Logistic regression A statistical model that in its basic form uses a logistic function to model a binary dependent variable, although many more complex extensions exist.

M

Machine learning The study of computer algorithms that improve automatically through experience. It is seen as a subset of artificial intelligence.

Mathew's correlation coefficient (MCC) It is used in machine learning as a measure of the quality of binary (two-class) classifications.

Matrices	A rectangular array of numbers, symbols, or expressions, arranged in rows and columns (singular- matrix).
Melanoma	A serious form of skin cancer that begins in cells known as melanocytes.
Meta-analysis	A quantitative, formal, epidemiological study design used to systematically assess the results of previous research to derive conclusions about that body of research.
Methylation	A form of alkylation, with a methyl group, rather than a larger carbon chain, replacing a hydrogen atom.
Modalities	A mode in which something exists or is experienced or expressed.
Mutually exclusive	Two or more events that cannot coincide.
Myocardial infarction	A blockage of blood flow to the heart muscle.

N

Naive Bayes	An algorithm that uses Bayes' theorem to classify objects.
Nash equilibrium	A proposed solution of a noncooperative game involving two or more players in which each player is assumed to know the equilibrium strategies of the other players, and no player has anything to gain by changing only their own strategy.
Neural networks	A set of algorithms, modeled loosely after the human brain, that are designed to recognize patterns.
NEWS score	National Early Warning Score determines the degree of illness of a patient and prompts critical care intervention

P

Paradigm	A typical example or pattern of something; a model.
Parameter	A numerical or other measurable factor forming one of a set that defines a system or sets the conditions of its operation.
Parasternal	Situated beside the sternum.
Parkinson's disease	A progressive nervous system disorder that affects movement.
Pathogenicity	Refers to the ability of an organism to cause disease (i.e., harm the host).
Perceptron	An algorithm for supervised learning of binary classifiers.
Perfusion	The passage of blood, a blood substitute, or other fluid through the blood vessels or other natural channels in an organ or tissue.
PET	Positron emission tomography is a nuclear medicine functional imaging technique that is used to observe metabolic processes in the body as an aid to the diagnosis of disease.
Phantom model	A body of material resembling a body or bodily part in mass, composition, and dimensions and used to measure a particular effect.
Phenomapping	To classify a group of patients with one disease into separate "phenotypes" within that category.
Phenotype	The set of observable characteristics of an individual resulting from the interaction of its genotype with the environment.

Piezoelectricity The electric charge that accumulates in certain solid materials in response to applied mechanical stress.

Pixel The smallest controllable element of a picture represented on a screen.

Polar maps Maps that provide a two-dimensional view of the entire three-dimensional left ventricular myocardium, and are used to show abnormal left ventricular myocardial perfusion when coronary artery disease is suspected.

Polymorphism The occurrence of different forms among the members of a population or colony, or in the life cycle of an individual organism.

Predictive analytics A variety of statistical techniques from data mining, predictive modeling, and machine learning, that analyze current and historical facts to make predictions about future or otherwise unknown events.

Probabilistic model Model incorporating random variables and probability distributions of an event or phenomenon.

Probability A numerical description of how likely an event is to occur or how likely it is that a proposition is true.

Prognosis Predicting the likely or expected development of a disease.

Prognostication The action of foretelling or prophesying future events.

Propensity An inclination or natural tendency to behave in a particular way.

Proteomics The large-scale study of proteins.

Pseudoaneurysm A collection of blood that forms between the two outer layers of an artery, the tunica media, and the tunica adventitia.

R

Radiomics A method that extracts a large number of quantitative features from medical images using data characterization algorithms.

Random forest An ensemble learning method for classification, regression and other tasks that operate by constructing a multitude of decision trees at training time and outputting the class that is the mode of the classes (classification) or mean prediction (regression) of the individual trees.

Recurrent neural networks (RNN) A class of artificial neural networks where connections between nodes form a directed graph along a temporal sequence.

Reductionism The practice of analyzing and describing a complex phenomenon in terms of phenomena that are held to represent a simpler or more fundamental level, especially when this is said to provide a sufficient explanation.

Regression problem A problem where the output variable is a real or continuous value, such as "salary" or "weight."

Retinopathy A disease of the retina that results in impairment or loss of vision.

Risk stratification The process of assigning a risk status to patients, then using this information to direct care and improve overall health outcomes.

Robustness The quality or condition of being strong and in good condition.

S

Sarcoidosis An inflammatory disease that affects multiple organs in the body, but mostly the lungs and lymph glands.

Schizophrenia A serious mental disorder that affects a person's ability to think, feel, and behave clearly.

Sensitivity Sensitivity (also called the true positive rate) measures the proportion of actual positives that are correctly identified as such (e.g., the percentage of sick people who are correctly identified as having the condition)

Single-photon emission computerized tomography (SPECT) Nuclear medicine tomographic imaging technique that uses gamma rays to image myocardial tissue perfusion at rest and stress conditions.

Sinogram A set of many projections under different angles organized in 2D.

Sociometry A qualitative method for measuring social relationships.

Specificity Specificity (also called the true negative rate) measures the proportion of actual negatives that are correctly identified as such (e.g., the percentage of healthy people who are correctly identified as not having the condition).

Stochastic A randomly determined process.

T

^{201}Thallium A radiopharmaceutical used in myocardial perfusion imaging.

Technetium-99m A metastable nuclear isomer used in a number of medical diagnostic imaging scans.

Thrombectomy The interventional procedure of removing a blood clot (thrombus) from a blood vessel.

Thrombosis The formation of a blood clot, known as a thrombus, within a blood vessel.

Topological properties Properties of a geometric object that are preserved under continuous deformations, such as stretching, twisting, crumpling, and bending, but not tearing or gluing.

U

U-Net A convolutional neural network that was developed for biomedical image segmentation.

V

Vectors An object that has both a magnitude and a direction.

W

Wasserstein GAN An extension to the generative adversarial network that both improves the stability when training the model and provides a loss function that correlates with the quality of generated images.

Index

'*Note*: Page numbers followed by "f" indicate figures and "t" indicates tables.'

Made in United States
Orlando, FL
31 March 2022

16350967R00248